Advances in Science, Technology & Innovation

IEREK Interdisciplinary Series for Sustainable Development

Advances in Science, Technology & Innovation (ASTI) is a series of peer-reviewed books based on important emerging research that redefines the current disciplinary boundaries in science, technology and innovation (STI) in order to develop integrated concepts for sustainable development. It not only discusses the progress made towards securing more resources, allocating smarter solutions, and rebalancing the relationship between nature and people, but also provides in-depth insights from comprehensive research that addresses the **17 sustainable development goals (SDGs)** as set out by the UN for 2030.

The series draws on the best research papers from various IEREK and other international conferences to promote the creation and development of viable solutions for a **sustainable future and a positive societal** transformation with the help of integrated and innovative science-based approaches. Including interdisciplinary contributions, it presents innovative approaches and highlights how they can best support both economic and sustainable development, through better use of data, more effective institutions, and global, local and individual action, for the welfare of all societies.

The series particularly features conceptual and empirical contributions from various interrelated fields of science, technology and innovation, with an emphasis on digital transformation, that focus on providing practical solutions to **ensure food, water and energy security to achieve the SDGs.** It also presents new case studies offering concrete examples of how to resolve sustainable urbanization and environmental issues in different regions of the world.

The series is intended for professionals in research and teaching, consultancies and industry, and government and international organizations. Published in collaboration with IEREK, the Springer ASTI series will acquaint readers with essential new studies in STI for sustainable development.

ASTI series has now been accepted for Scopus (September 2020). All content published in this series will start appearing on the Scopus site in early 2021.

More information about this series at http://www.springer.com/series/15883

Claudio Gambardella • Claudia Cennamo •
Maria Luisa Germanà • Mohd Fairuz Shahidan •
Hocine Bougdah
Editors

Advances in Utopian Studies and Sacred Architecture

A culmination of selected research papers from the International Conference on Utopian and Sacred Architecture Studies (USAS) which was held in association with the University of Campania "Luigi Vanvitelli", Aversa, Italy 2019.

Springer

Editors
Claudio Gambardella
Department of Architecture and Industrial Design
University of Campania "Luigi Vanvitelli"
Aversa, Italy

Claudia Cennamo
Department of Architecture and Industrial Design
University of Campania "Luigi Vanvitelli"
Aversa, Italy

Maria Luisa Germanà
Department of Architecture
University of Palermo
Palermo, Italy

Mohd Fairuz Shahidan
Department of Landscape Architecture
Universiti Putra Malaysia
Serdang, Malaysia

Hocine Bougdah
School of Architecture + Cities
University of Westminster
London, UK

Publishing Editor
Haitham Mohammed
International Experts for Research Enrichment
and Knowledge Exchange—IEREK
Publishing Editor at IEREK Press
Alexandria, Egypt

ISSN 2522-8714 ISSN 2522-8722 (electronic)
Advances in Science, Technology & Innovation
IEREK Interdisciplinary Series for Sustainable Development
ISBN 978-3-030-50764-0 ISBN 978-3-030-50765-7 (eBook)
https://doi.org/10.1007/978-3-030-50765-7

Cover illustration: "Il popolo di Dio" terracotta artwork by Ugo Marano. "UGO MARANO: handmaker felice" exhibition curated by Claudio Gambardella Ceramic Trilogy first edition organized by I LOVE POMPEI and Polo museale della Campania at Museo Duca di Martina, Naples_2016 Luciano Basagni photographer

This Springer imprint is published by the registered company Springer Nature Switzerland AG
The registered company address is: Gewerbestrasse 11, 6330 Cham, Switzerland

Series Editor's Preface 1

Every country and every community is in constant search for peace and happiness. A search to counter wars, poverty and famine. A constant dream that fills the mind of each and every one to have a world were all societal ills are resolved. For a long time, our ancestors have sought and exerted efforts in an attempt to bring utopian imaginaries and practices to life. Their endeavors were clearly reflected in the spiritual buildings such as churches, mosques, gardens and emergent urbanisms and alternative and vernacular designs. Following in their footsteps is what gave birth to sacred architecture and a new emerging age of religious and somatic schemes.

This book, on Advances in Utopian Studies and Sacred Architecture, contains selected research from the International Conference on Utopian and Sacred Architecture Studies which was held in association with the university of Campania "Luigi Vanvitelli" of Italy. Research from the conference has undergone a rigorous review process thanks to the book editors and scientific committee of reviewers. The book explores the interdisciplinary field of utopian studies and specifically the relationship between architecture and utopia as one that is long-standing and pervasive. It namely draws upon the transition and even intersections between contemporary and historical geographies of sacred architecture and utopia while showcasing the different and emerging approaches with traces from the past.

Through this book, readers will be exposed to a number of case studies around the world: from Middle Eastern countries to Yugoslavia, Peru, Naples and more. It stimulates discussions on religiousness, symbolism and spirituality and their influences on the development of a Utopian City. It covers an extensive variety of topics such history of sacred spaces, sacred dimensions, visualization of utopian worlds and sacred spaces while considering culture, spirituality and different ideologies and approaches for a more sustainable utopian future.

The book will benefit an array of readers from different backgrounds due to its interdisciplinary nature. It succeeds in bringing together fields such as sacred and historical studies, architecture and urban planning, landscape architecture and lighting design, environmental preservation and other environmental processes. It is highly recommended to academics, postgraduate students and interested readers.

Cairo, Egypt — Mourad Amer

Series Editor's Preface 2

For thousands of years, human beings have dreamt of perfect worlds, worlds free of conflict, hunger and unhappiness. But can these worlds ever exist in reality?

Our ancestors embodied these dreams in spiritual buildings, in hopes of turning those dreams into a reality, until the appearance of sacred architecture. As a result, many utopian communities and cities have now emerged.

This book on Utopian and Sacred Architecture Studies aims to explore the theories and practices stimulated over the last five centuries in a wide range of areas of thought, historiography, political sciences, social sciences, literary and art studies, social activism and in the conception of possible worlds. Thus, it addresses directly the controversy surrounding Utopia and its successors from its origin to our present and explores the impact on and by the utopian religious and cultural ideals.

The importance of this theme came about over the past century as utopian thought provoked political and social change in most of our cities. It is also turned into a source of inspiration for innovation in many sciences and technologies. For this reason, this book is dedicated to exploring More's intellectual legacy of utopian theory and practice across various academic disciplines. The discussions, theories, as well as case studies of wide range of issues are included as always Utopias have reflected a wide range of topics that continuously need to be argued in the contemporary age.

This book covers a range of subjects with a majority of topics that are influenced by city fabric and their sustainable development. In addition, it discusses the employment of the utopias as a tool and an economic and social resource for a city's wealth which should be maintained for an enhanced future and more livable cities.

Finally, the theme of this book is an approach to the study of formal and informal religions that segregate between them and their rejection as they are embodied, through utopias, in the form of sacred building and architectural representations and vocabulary.

Cairo, Egypt Mourad Amer

Scientific Committee

Acknowledgements

We would like to thank the authors of the research papers that were chosen to be added to this book. We would also like to thank the scientific committee of reviewers who helped us select these papers and the editors of this book. Lastly, special thanks go to the IEREK team for supporting the publication of the best research papers submitted to the conference.

Aversa, Italy	Claudio Gambardella
Selangor, Malaysia	Mohd Fairuz Shahidan
Palermo, Italy	Maria Luisa Germanà
Aversa, Italy	Claudia Cennamo
London, UK	Hocine Bougdah

Word from the Chairman of the Conference

It is really an honor and a privilege to have been the chairman of the Utopian and Sacred Architecture Studies conference organized by the Department of Architecture and Industrial Design of Università degli Studi della Campania "Luigi Vanvitelli" in partnership with IEREK—International Experts for Research Enrichment and Knowledge Exchange, an Egyptian institution very active in Italy and other countries.

The idea of a conference focused on the Sacred Space grew out during the International Conference on Improving Sustainability Concept in Developing Countries, another IEREK event held in Cairo in December 2015: at the time, frequent terrorist attacks in the world were a threat to the peace, feeding a climate of fear and spreading a feeling of suspicion among citizens. Through street demonstrations, people were reacting by expressing their will to fight the terrorism. Intellectuals, scholars and academics had to ethically choose other forms of response to this phenomenon, more congenial to their cultural responsibility.

So, through the studies on the Sacred Space, developed in different areas of architecture, from the historical to the more technical and design ones, the conference aimed to contribute to rediscover the great topic of the UTOPIA, in an age characterized by the crisis of ideals and the alleged clash of civilizations, but also by the deeply rooted aspiration to build a new world. The presence at the conference of Keynotes Speakers representatives of Judaism, Christianity and Islam, the most important monotheistic religions in the world, is a meaningful sign of this. However, their contributions are only the incipit of the event. So many authors from all over the world contributed with their specific expertise to recreate the Sacred Space as a kaleidoscopic place: a place which is part of wider human experience. Culture is creativity, and other possible worlds are born from creativity.

In the Hagia Sophia, in Istanbul, the Muslim religion did not erase the original Christian imprint. So, here today we can see a wonderful architectural synthesis, where the two religions' signs coexist, and thus resulting in unpredictable outcomes. However, this is not the only great example given us by history of architecture.

Today, there is another suggestive place in the world. It is "Doumia-Sakinah" ("silence" in Hebrew and Arabic), the space of silence and prayer: a simple dome, entirely white, symbol of fullness and unity beyond any barrier. This space, in which Muslims and Jewish faithful pray their own God, is in Wahat al-Salam—Neve Shalom: It is an Oasis of Peace founded in 1970, with the intention of creating a place where the people of this land would live together despite their national and religious differences, by the Dominican Father Bruno Hussar, who would conduct educational work for peace. It is located at the same distance from Jerusalem, Tel Aviv and the Palestinian city of Ramallah. Neve Shalom Wahat as-Salam, this supposed "utopian experiment," ends up playing the unpredictable role of "champion of realism," thanks to its ability to avoid the insidious pitfalls of religious fundamentalism and political extremism and to lucidly prefigure a situation of reasonable and secularized coexistence among people who identify with different and conflicting religious traditions, cultures, nationalities.

The venue of the scientific event has been the ancient San Lorenzo ad Septimum abbey, the Vanvitelli Department of Architecture and Industrial Design (DADI) seat in Aversa, a Norman town located in Campania region between Naples and Caserta. The department had intensively worked for many months to ensure full success of the Utopian and Sacred Architecture Studies conference (USAS). In agreement with IEREK, the conference has been structured in five main different topics and thirty-one subtopics. Thanks to the multidisciplinarity feature of the conference, one hundred and fifty-one abstracts has been submitted in the first step. The full final program has provided for seventy-one presentations and one hundred and twenty speakers from Italy and a lot of different countries of the world. Thanks to the hard work of young Ph.D. students, the event was enjoyed by all of the participants.

Now, I would like to take this opportunity to thank the Rector of Vanvitelli University, Prof. Giuseppe Paolisso, the Deputy Rector, Prof. Gianfranco Nicoletti, the Vice-Rector for Culture, Prof. Rosanna Cioffi, who were the Honor Committee members of the conference; I would also like to thank again the coordinator of the Ph.D. program in Architecture, Industrial Design and Cultural Heritage, Prof. Paolo Giordano and, in particular, the Director of the Architecture and Industrial Design Department and one of the Scientific Committee members, Prof. Luigi Maffei, with whom I shared the whole gestation of the event.

Moreover, I wish to thank, especially the IEREK's CEO, architect Mourad Amer, for his strong commitment to promoting international conferences in Italy and in world. He and his staff effectively supported the USAS conference. I also wish to thank each of the foreigner and Italian colleagues for their appreciated contributions and everyone who participated in the helping and organizing for this event, in particular the Scientific Committee, the "Platform" Magazine, MEDIA partner of the event, and the Organizing Committee. Finally, I want to express my gratitude to Prof. Ferdinando Trapani, the coordinator of the inter-university Italian agreement signed last year by the directors of some important departments, including ours. With this agreement, the Italian university will certainly bring an important contribution to the scientific growth of IEREK.

Aversa, Italy Claudio Gambardella

Contents

Utopian and Sacred Architecture

In this introductory part, we have the five keynote speeches that were presented in the International Conference on Utopian and Sacred Architecture Studies. They discuss ideas and visions of how Sacred Architecture is a way of achieving a utopian city or community that is free of conflict and hunger.

In the first keynote speech titled "Sacral Buildings as an Expression of Hope", Prof. Alfred Jacoby discusses the question of the possibility of a Utopian world to exist and if the Sacred Buildings play a factor in achieving such worlds.

In the second keynote speech titled "The Holy Place: History of Catholic Liturgical Architecture", Prof. Salvatore Frigerio takes us into a historical journey through the Catholic Architecture and the progress of the liturgical movement.

In the third keynote speech titled "A Small Jerusalem in Milan", Prof. Gabriele Muhyiddin Bottiglioni discusses the tangible and intangible aspects of sacred places, and specifically the Islamic arts and presents the stages of recognition of Al Wahid Mosque as a place of worship.

In the forth keynote speech titled "Designing the Intangible", Prof. Niccolò Casiddu reflects on the contribution of the design discipline that acts within the constitution of sacral objects, he also investigates the relation of this specific kind of architecture with the wider nature of design principles in general.

In the fifth keynote speech titled "Sacred Places: The Interaction Between Space, Time and Faith" Prof. Mona Helmy investigates what makes a place sacred and if that sense of sacredness is temporary or a permanent one. She examines different models of sacred places for a variety of tangible and intangible contexts to identify the elements that creates the sacred place.

Sacral Buildings as an Expression of Hope

Alfred Jacoby

The motto set out in this conference on Utopia and Sacred Building has put two questions to any of us participants:

1. **Can Utopian worlds with man free of conflict, hunger, and unhappiness, really exist?**
2. **Is the reference to Sacred Buildings a factor for achieving that?**

At first glance, a spontaneous answer to both questions comes to mind after two seconds of reflection: It is for both cases: "No".—In fact even: Imposssible.

Looking at the world today, it seems even naive to ask.

Utopias are always man-made. They always need to be described in relation to the distopia they aim to overcome. Allowing thought its own spatial dimension, I would call Utopias *Critical Thought Palaces*. As intellectual space, such a palace always offers a structured projection of a future reality in relation to its specifically the experienced past. There are large spaces, in which the mind can wonder from one room to the next, from the past to the present as well as to the future. These palaces are documents of a very basic anthropological characteristic, which we all share.

Hope: It is a basic principle of human existence. From our earliest days on, we begin a search of who we are. As one facet of our identity search, the notion of hope explains, why we are always embarking on something new, using our own self as point of departure.

Indeed, Sacral Buildings can easily be looked at this way: They are unique physically existent *Thought Palaces* which help to include the moment, remember the past and offer hope for the future. One of the great thinkers, who saw hope as a principal human driving force in life, was Ernst Bloch. Born in Ludwigshafen, Germany, in 1885, Bloch died in Tübingen in 1977.

As a materialist philosopher, he lived through an age, which the British Historian Eric Hobsbawm, called "*The Century of Extremes*". In his Main Oevre "*The Principle of Hope*", Ernst Bloch wrote: "*In a defining sense, hope is, what man actually strives for and wants.*

> The exact identity of daydreams, hopes and utopias at first lie in the dark. At the same time, they are the goldmine, where concrete utopias can be found. Each such daydream has this double ground as his "Heimat" or home.
>
> Hope is the yet to be found, the experienced non-experience contained in each experience we live through.

To establish some common ground between the Principle of Hope and Sacral Buildings, I would like to turn to Neil McGregor. He writes:

> Every known society shares a set of beliefs and assumptions - a faith, an ideology a religion. It goes far beyond the life of the individual and is an essential part of a shared identity.
>
> Such beliefs have a unique power to define but also to divide peoples. They are a driving force in the politics of many parts in the world. Sometimes the shared beliefs are secular, as in the case of nationalism, but throughout history they have in the widest sense been religious. (Living with the Gods, 2019)

Each faith, ideology or religion has over the ages produced its own architecture (Fig. 1).

For this, the investigative space to enter is Europe. A region, we can still call "*Our part of the world*". That is because, until today, it serves as our transformative common cultural space. In order to develop such a common culture requires two things:

- a lot of changes
- a lot of permanence.

Indeed: Man's drive towards happiness, the implementation of his hopes has taken very different turns over time in this geographical arena. Cultures and beliefs and, in consequence, the way to make Sacral Buildings have all radically changed.

A. Jacoby (✉)
Faculty of Architecture, Facility Management & Geo-Information, Anhalt University, Köthen, Germany
e-mail: alfred@jacoby-architects.de

C. Gambardella et al. (eds.), *Advances in Utopian Studies and Sacred Architecture*,
Advances in Science, Technology & Innovation,
https://doi.org/10.1007/978-3-030-50765-7_1

Fig. 1 The ancient world

The very first high culture, ancient Egypt, was a naturalistic culture:

They worshipped natural appearances and believed in a static and eternal world. Their pharaos were considered living Gods. Their pyramids were houses for a life in a realm of death, which Ernst Bloch called "Crystals of Death".

About 1500 years later, these ideas were completely reversed: In ancient Greece, sacred monuments were Temples for life—not death.

The Gods lived in them.

In his "*The Earth, the Temples and the Gods*", Vincent Scully demonstrates that the chosen sites for Greek Temples were carefully searched for holy places. Using the idea of a permanent holiness, Scully shows us a very important quality of Sacred Buildings:

> The way that such spaces may operate can differ, but much of the experience within them remains as it has always been.

That is true for the Pantheon as much as for the Hagia Sophia. It has to do with the holiness a place has acquired over a long-time span in the memory of man. A first distopia which changed Europe was the Fall of the Roman Empire in 410 AD. It not only destroyed its capital city, but also its culture and beliefs. In 410 AD, antiquity and its natural religions came to an end. For centuries afterwards, its religious building types were either taken down or stood empty with no sensible use (Fig. 2).

Now the time of the Christian Church began.

To transform such disorder into order again, its first theological philosopher, Augustinus, succeeded in establishing a first Christian Utopia with his *City of God.*

Written just after the Fall of Rome in 410 AD "*City of God*" was a way out of the barbaric reality on Earth.

Augustinus thought of a hierarchy of Saints, who converted Messianic afterlife into an attractive alternative to the miserable life on Earth. Beyond his book, a practical example of the Hierarchy of Saints can be seen in Kings College Chapel, Cambridge. The Utopia of a City of God on Earth, finally came true in the Middle Ages and its Great Cathedrals and Cities. In them, new theological thought and built form became one (Fig. 3).

I think even today, we have good reasons to stay on this path to express and deepen our own individual self-identity. Even if the prototype representing traditional ways of co-habitation has changed from the cleric to the citizen after the French Revolution.

In essence, it has led mediaeval men to become men of science. They finally arrived at the statement that *There is no God.*

Fig. 2 The Earth, the Temples and the Gods, Scully

Fig. 3 The City of God, Kings College Cambridge

With the French Revolution in 1789, the idea ancient of a King sent by God was dead. The Utopia of a new society of equality, brotherhood and liberty (Egalite Fraternite Liberte) set up a new time. The Citoyen, the civilised burgher, replaced the priest and his Utopian promise of Messianic afterlife. It is Modern Man that trusts in science, technology and rational thought, rejecting any form of metaphysics.

It was the adherence to science, that created Modern Man (Fig. 4).

This alternative type of a Utopian has indeed brought us far. We have won many battles against incurable illnesses, science and technology have produced items that have made life much more comfortable.

At the same time, and for good reason, modernity has developed very pessimistic utopias about itself.

As a counter position to Marlowe's Dr. Faustus, as a first Modern Dystopia Mary Shelley presents us Dr. *Frankenstein. The* Monster, as the central character in this novel, is a man-made creature. A Monster still longs to be loved.

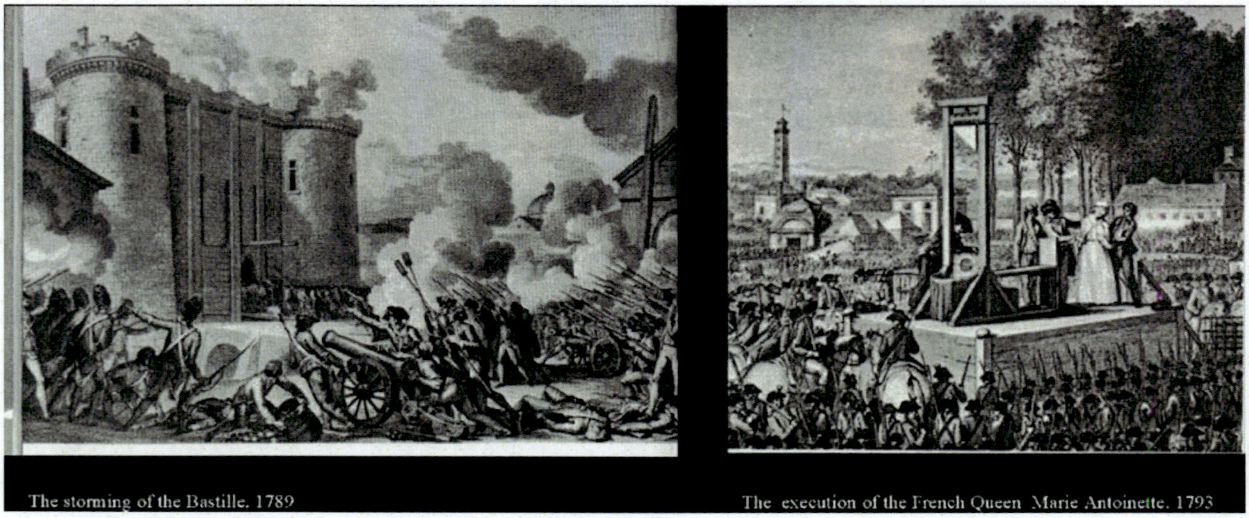

Fig. 4 The French Revolution

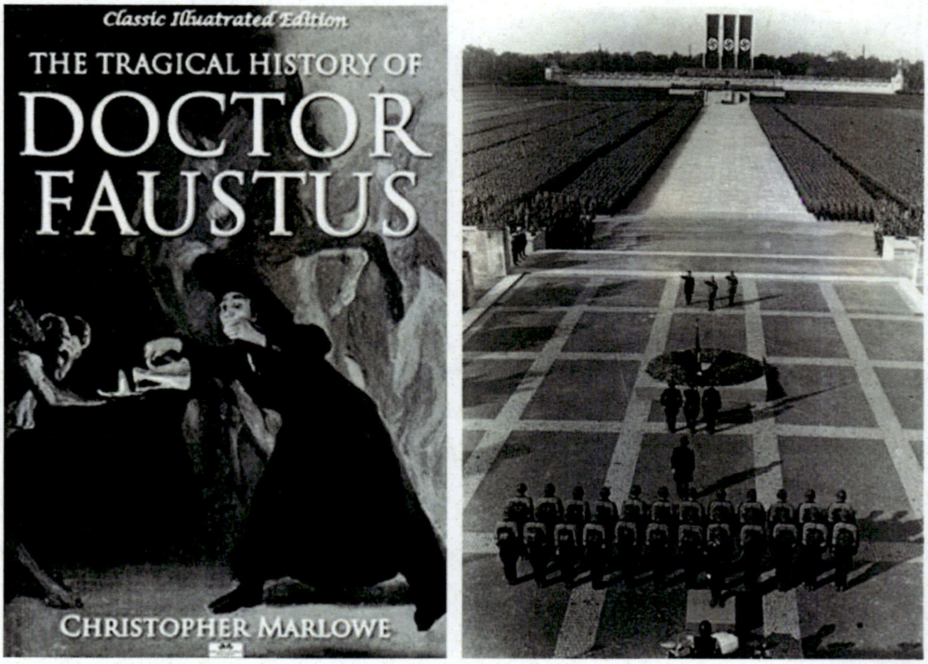

Fig. 5 Christopher Marlowe's Dr. Faustus 1604 and The Nuremberg Party Convention 1937

With her novel, Shelley for the first time puts the act of creation into our own hands. Something the Church considered an immortal sin.

Orwells "*1984*", Bradbury*'s* "*Fahrenheit 451*" *or* Aldous Huxley "*Brave New World*" follow in these pessimistic footsteps. They all tell the same story of brutal power and total control.

The first description of hell we can find in Dante's Inferno. Someone who shows us, how man can be seduced into such a hell was Christopher Marlowe. As we know from experience, even in our time, a German Doctor Faustus existed, who for 12 years transformed Earth into hell (Fig. 5).

So far, I have written about the power of Hope, the changes in traditions and belief, even to the point of a total rejection of any higher power in this world. That would in effect make Sacral Buildings unnecessary.

Yet, I tried to show, that in each of these periods, we as humans created our own valid *Thought Palaces*. As outcome of such new thought after World War II, Sacral Buildings again became real, tangible and materially built.

I argued that Holy Places have their own resilience. They keep that aura and a real permanence even when the built form on that location has been removed or the traditional belief once practiced within them has vanished. What happens in that moment is, that they become places of ***collective memory***.

> As such they are lieux – and that can mean places, sites or causes. They appear in three senses - material, symbolic and functional. In this way, certain places, objects or events can assume a special significance, related to a group's remembrance.
> (Les Lieux de la Memoire, Pierre Nora, 1991)

Such places carry a sense of what Ernst Bloch called the Primary Function of Architecture: To construct a Heimat—a permanent place to be at home. Both with oneself and within a built form.

Another architect, to speak about the idea of permanence of place, was Aldo Rossi. Alongside Vincent Scully and his permanence of Holy Places, Aldo Rossi in his "*Storia della Citta*", offered a secular point of reference for certain types of buildings, which he called *Monuments*.

Rossi demonstrated, that for him, form does not develop out of function. He shows, like Scully, that historically iconic monuments can adapt to various functions. What remains permanent, is their position as icons in the city.

For me, architecturally, all this was a starting point.

Bloch's *Utopian Horizon of Hope* has helped me, as a Jewish architect born after the Holocaust, to build holy spaces today.

I have pursued my work very much in resistance to Theodor Adorno's remark, that:

To write a poem after Auschwitz is barbaric. Meaning: hope after Auschwitz is futile.

Rossi's teaching was a starting point beyond Adorno's belief.

As a conclusion, I therefore include two of the answers I found, when confronted with a holy site, that had been destroyed and villified.

I mean the two Synagogues I built in Aachen (1996) and Chemnitz (2003).

Both buildings are monuments within a city. Even with some scepticism, I hope, that they will stand for a long time.

In the locations the stand, they carry with them the memory of their distopic counter-spaces:

The Old Synagogues that stood there before and were destroyed by a Faustian Regime of Hell. Both Synagogues are rebuilt on the actual place where the former Synagogue was burnt in 1938.

Both are concerned with repairing the groundplan of the city in which they stand (Figs. 6, 7, 8, 9, 10 and 11).

Fig. 6 Old Synagogue Aachen

Fig. 7 New Synagogue Aachen 1996

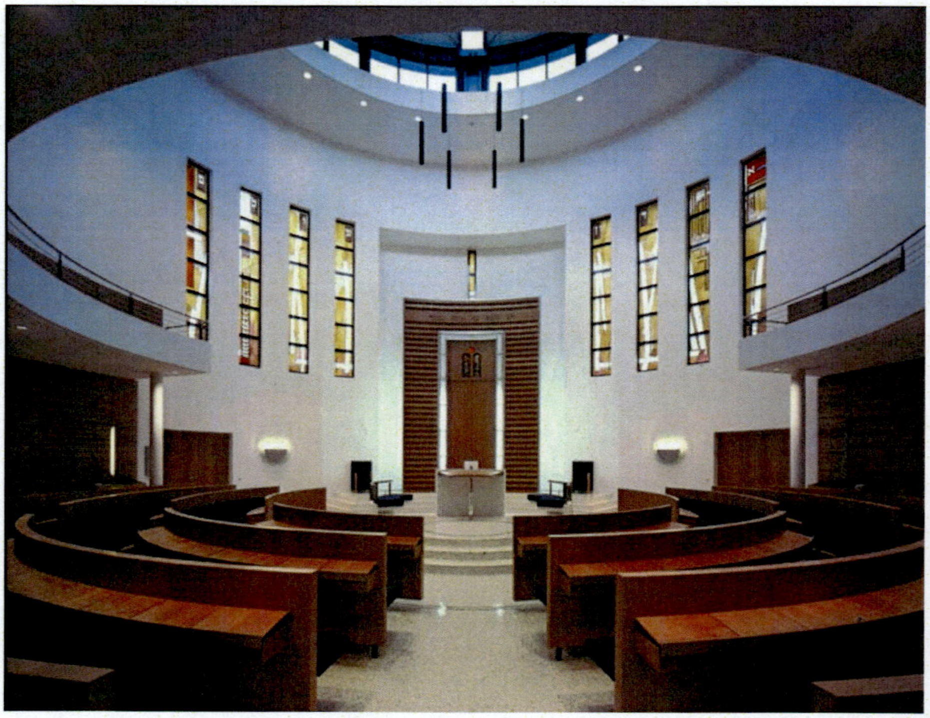

Fig. 8 New Synagogue Aachen 1996

Fig. 9 Old Synagogue in Chemnitz 2003

Fig. 10 New Synagogue in Chemnitz 2003

Fig. 11 New Synagogue in Chemnitz 2003

Beyond their urban quality, the main question I had to ask myself when designing these two buildings was:

What do you build for people, that have emerged from hell?

How do you create a space that is on one hand a place for remembrance but on the other a space of light, of hope and of a possibility of healing wounds that are very difficult to heal?

The Holy Place: History of Catholic Liturgical Architecture

Salvatore Frigerio

In the biblical Jewish-Christian complexity, the concept of "holy place" has also evolved, moving first of all from the same concept of "sacred". The religious man, to get in touch with the divine, selects from profane life gestures, people, spaces and times and loads them with a symbolic value, considering them privileged areas for the meeting with the divine. That gives birth to the bounds of the "sacred", which we can find in all the religions. We are dealing with something which is not God but has become an evocation of the divine, precisely the "sacred", different, separated from profane uses and an object of respect, veneration, self-examination in front of the totally "Other". Thus, "sacred spaces" are also born but certainly they run the risk of precisely delimiting the presence of God, imposing it under man's control, or even removing the profane space from God's presence.

Therefore, "the Tent of the Meeting", which accompanied the Israelites in the desert, met the needs of the closeness of God and His Transcendence. On this point, prophets were and are critical and watchful. God has accepted and accepts a temple but he conserves his freedom undiminished.

The great turning point of that tradition happened when, after the Christian faith was established, there began to spread the awareness that the real space for the presence of God among men, the Jewish Shekinà, is not the temple of Jerusalem any longer but the "body" of Christ (Jn 2, 21; 1, 14).

The last word of the Second Testament on the Temple is the amazing vision of the Apocalypse in which the "new Jerusalem, universal city without temple" is described. The new humanity fruit of the accomplishment of creative harmony of the Word (Gn 1, onwards) is in perfect communion with God, direct, transparent, without veils and mediation. The symbols that uncover and hide God at the same time have fallen away: we are before God.

S. Frigerio (✉)
Collegium Scriptorium Fontis Avellanae, Serra Sant'Abbondio, Italy
e-mail: salvatorefrigerio16@gmail.com

All this, though, is a long journey that is expressed, lived and documented by art and, particularly, by architecture. Even in Western art, the same principal styles, such as the Early Christian, the Romanesque, the Gothic, the Renaissance art, the Baroque and the Neoclassical, are represented by religious works and architecture. Each of them reflects a particular moment in the history of faith and in the way of living it. For this reason, we can say that architecture especially makes visible, in a plastic and spatial manner, the complex journey of the Christian Church that has been tightly linked to the historical events it has experienced and by which it has been conditioned. The "domus orationis" of the first Christian communities was the "domus familiae" that, inside its own spaces, celebrated the sacramental presence of the Risen Christ. After the edict of Constantine (113 A.D) that freed the act of Christian worship, the "domus orationis" was replaced, as the space for the public cult, by the Romanesque basilica (Fig. 1).

Initially, the community, freed from the persecution, met and celebrated its own rite in a nave, while in the others there buzzed the social life of the time. Little by little, the spaces of the basilica stopped being a civil public place, presided no longer by the emperor but by the local bishop. The architectural space of the basilica, wide and bright, was considered the sign of Christ's victory over death and persecutions, and, moreover, a sign of the eternal light of which the believers considered themselves heirs and witnesses. The mosaic art covering the walls made the assembled people aware of the celebration of a cult that involved the whole of history, created and saved.

The dimension of the luminous joy of Easter characterized the first centuries of the Christian communities until the Carolingian age (cenr. IX–XII) when the self-declaration by Charlemagne of being the representative on Earth of the divine justice was such to strike fear of the divine judgement, with the consequent constant invocation of forgiveness. Then the "domus orationis" became places emotionally more intimate, less bright, more adapted for the penitent's introspection, lower and "hut-shaped", even today noticeable

C. Gambardella et al. (eds.), *Advances in Utopian Studies and Sacred Architecture*,
Advances in Science, Technology & Innovation,
https://doi.org/10.1007/978-3-030-50765-7_2

Fig. 1 Basilica di Santa Sabina in Roma

to the visitor. Moreover, the sacredness of the priest class, also emphasized in the liturgical rite, was made evident by the stress on the separation between the presbytery, reserved for the clergy, and the nave, occupied by the people, architecturally rendered by large steps that divided the sacred from the profane, that is the clergy, who were celebrating, and the penitent people who were attending prostrated. Also the exposed stone construction made the environment emotionally more suited to that penitential rite. The "tales" of the Salvation history were replaced by the individual figure of saints who had to sustain the penitents.

It would be the monastic community of Cluny who would strongly re-propose the freeing cult of the community who raises the whole creation to God, celebrating while standing, as risen in Christ, with their hands lifted and opened up to receive the light of the Risen Christ. Then, the architecture would rise again, high and launched towards Heaven, creating that style which would be erroneously named "Gothic" instead of "Frank" by Giorgio Vasari of Arezzo (Fig. 2).

But the height of the perimeter walls would weigh on the Roman round arches, concentrating the load on the keystones that would not have held. The weight had to be distributed on the pillars or on the sustaining columns with pointed arches, that is precisely "Gothic". The outside walls were exposed to the violence of the wind, and to avoid their collapse were supported by external "fying buttresses", still evident or, later on (particularly in Italy) incorporated into the transept wall.

However, it remained the leading longitudinal/processional model of an architecture enabling the transmission of a "personal" spirituality which was increasingly less in favour of a community liturgy. The active involvement in the liturgy is conserved only inside monasteries that, for the elevated cultural level of the monks, still express the unified character of a community celebration.

The widely spread dilemma in the Western Church, between the theology of the cross and that of the glory, is solved in the humanistic period that acquires architecture in service to preaching and the glorious presence of the Church. Nicholas V, Julius II, Leo X and all the Popes of the Renaissance triumphalism underline their preference for grand buildings, for "the imperishable monuments, almost eternal testimony, and quasi divine" (Nicholas V). Therefore, the basilica model, up to the first quarter of the Cinquecento, is replaced by the central plant model, circular or with multiple symmetries, elaborated by the architectural research of the Quattrocento, as interpretation of the harmonic laws that regulate the universe. For the theorists of the "ideal city" of the Renaissance, the sacred building is meant to be the first organizer of the town, set in dominant positions with respect to the surrounding structures, emphasizing in these models of urban organization the hierarchy of values required by the religious authority.

After the crisis of the Lutheran Reformation, the Church deals with a new problem: that of establishing its own necessary presence to win over people's faith in the Catholic

Fig. 2 Cattedrale di Bruges

religion through the means of preaching. This would be obtained above all through the use of rhetoric and sentiment, introduced in all the used educational tools, primary among these architectures. As a result, the Baroque architecture renounces the mathematical–proportional researches on the symmetrical structures focusing instead on the development of new effects, through complex accumulated geometries, no longer used for their cosmological meanings but for the

Fig. 3 Arco incorporato nel transetto del Duomo di Orvieto

constant need to obtain emotional effects. In some cases—for example Borromini—it is also evident a high testimony of the deep conflict existing in the artist and in his contemporary world; nevertheless, as a rule, such a conflict is visible in the research of intended effects, but not for that reason necessarily experienced (Fig. 3).

Fig. 4 Particolare dalla Città ideale—Quadro della Galleria Nazionale di Urbino

The industrial revolution would be a complex period, full of widely contradictory attitudes and, at the same time, rich with new turmoil: industrialization, development of technology and natural sciences and, together, indifferentism, anticlericalism, liberalism, democracy, utopian socialism and marxist socialism, atheism, materialism. The attendance in the liturgical celebration was reduced to a "duty bound", passive presence that largely consisted in religious rites which diminished the mystery of salvation, in spite of the opposite testimony of great saints aware they were working in the above-mentioned complex context. Right among these difficulties, there developed a new interest for the liturgy, and towards the end of the nineteenth century there was a rebirth of theological studies. On the other hand, restrictive initiatives and measures of the Catholic hierarchy aimed at defending and strengthening the walls of the Catholic "ghetto", expecting to bring back modern society to Christianity. The Catholic hierarchy claimed to lead and limit the artistic research, showing in a plain or indirect way to have a

Fig. 5 Chiesa di S. Cristoforo in Fano

preference, in architecture, for the outdated Gothic, Baroque and also Neoclassical periods. The result of such an attitude can be seen in the careless achievements of the religious architecture of the period that caused the lack of commitment in the most qualified architects and craftsmen (Fig. 4).

Finally, with the progress of the liturgical movement and the publication of the encyclical "Mediator Dei" (20 November 1947), Pius XII declared that "the recent forms and images mustn't be rejected because of preconceived ideas, but it is absolutely necessary to give a clear expression to modern art, if it serves, with the proper reverence and honour for sacred buildings and holy rites".

The constitution of the sacred liturgy of the Second Vatican Council (SC, nm. 122–129) is the fundamental reference for an architecture that means to express in its shapes the character of community celebrations, thus underlining the rapport between the architectural space of the edifice and the liturgical rite. Therefore, I believe it is necessary to specify that the first real contribution of architecture can and must be that of removing the majority of technical and formal obstacles which intervene against a harmonious sequence of the rites, from the liturgical and paraliturgical celebrations to the forms of private and community piety. As a consequence, the typological adaptation of the religious architecture is possible when the meanings and the needs of the liturgical rite are analysed, in a tight relationship with the community, celebrating in the exercise of their own mysteries (Fig. 5).

Thus, the formal architect or craftsman can assign different values and meanings to the parts, through specific spatial relations—for example the use of materials connected with the shape, the dimension and the illumination—the patron is only required to identify the spiritual use rather than being concerned about architectural solutions.

Hence, the liturgical functionality, understood as a cluster of significant relationships between the human material elements and the divine ones that constitute the Edifice Church, comes from ecclesiology as theological doctrine of the Church. If the act of communication becomes a significant signal, this latter is therefore the result of a complex, architectural, artistic intuition of which the outcome can only be verified in the formation of every single piece of work with its own specific qualities.

References

Bouyer, L. (1964). *Il rito e l'uomo: sacralità naturale e liturgia*. Italy, Morcelliana: Brescia.

Bouyer, L. (1967). *Liturgy and architecture*. Notre Dame, USA: University of Notre Dame Press.

Varaldo, G. (1974). La chiesa casa del popolo di Dio (p. 15). *Quaderni di Rivista Liturgica*: Liturgia e architettura.

A Small Jerusalem in Milan

Imam Muhyiddin Gabriele Bottiglioni

Abstract

In sacred architecture, the spiritual centre constitutes the reflection of the archetypal centre, which in the traditions of Abrahamic monotheism is symbolically represented by the New Jerusalem. In this regard, we can say that the sacred place is of course tangible, but its raison d'être belongs to the invisible, and therefore, it is more properly a *u-topia*, a *non*-place, something that takes place elsewhere.

In sacred architecture, the spiritual centre constitutes the reflection of the archetypal centre, which in the traditions of Abrahamic monotheism is symbolically represented by the New Jerusalem. In this regard, we can say that the sacred place is of course tangible, but its raison d'être belongs to the invisible and therefore it is more properly a *u-topia*, a *non*-place, something that takes place elsewhere.

The nature of the reality that dominates Islamic art and the science of beauty that governs it comes directly from the Quran and the *aḥādīth,* the recorded words of the Prophet Muhammad. Being able to grasp this origin is undoubtedly a very difficult task because it is too subtle to be seen from the outside.

But what are these principles that have dominated all forms of Islamic art since its inception? The first and foremost is the *tawhid*, the doctrine of Unity. Each piece of authentic Islamic art must reflect the Divine Unity and this comes with a series of consequences. The first is that there must always be an integration of this concept in the artistic form. There is a centre, a centre that is expressed in architecture, in calligraphy, in miniatures and in carpet weaving—and all that is a reflection of the *tawhid*.

The doctrine of Unity acts on various levels in Islamic art: integration, absence of tension between artistic forms and elements, state of concentration, absence of dispersion by the observer of the work of art, etc.

In Islamic calligraphy, each letter has a precise numerical value but also a geometric one, and each verse offers an opportunity for learning, a deeper precision and exactness that only come with a divine science. The alphabet itself, for example, symbolises the birth of the universe: from the primordial vibration, from a dot expanding (which in Arabic corresponds to the number zero) we find *alif*, the first letter of the alphabet, a simple vertical line and the first letter of the name Allah. All Arabic writing is marked by that very alternation of horizontal lines, vertical lines and dots, which thereby trace the sign of the cross of which, according to statements from scholars, Christians keep the symbol while Muslims retain the pure doctrine. From these few examples, it is possible to infer that dedicating oneself to the art of Koranic calligraphy is like drawing the very essence of creation, allowing the symbols of that universal warp to emerge—the dot and the vertical and horizontal axes—whose respective geometric figures are the circle and the square.

And it is precisely in Muslim architecture that the essential nature of these two archetypal geometric shapes merges with the infinite richness of Koranic calligraphies, which constitute the only decorations on the walls in any mosque or place of worship.

We must stress the fact that geometric designs and mathematical models are not merely for decorative purposes, but are a means of remembering God, the centre that is omnipresent, and a proof of the well-known Koranic verse: "Wherever you turn, there is the Face of God" (2: 115). Islamic art is, in fact, the application of this verse, due to the role it plays of a faithful compass always pointing towards the centre, other than manifesting everywhere, through each of its productions—from the plate used on a daily basis to the great mosque—the "sign" of His Omnipresence.

I. M. G. Bottiglioni (✉)
COREIS, Milan, Italy
e-mail: gabriele@gbcreativestudio.com

C. Gambardella et al. (eds.), *Advances in Utopian Studies and Sacred Architecture*,
Advances in Science, Technology & Innovation,
https://doi.org/10.1007/978-3-030-50765-7_3

1 A Small Jerusalem on the Navigli Canals: Between Dialogue at Higher Level and Difficulties

Among the religious architectural utopias, we find a project proposed by Shaykh Abd al Wahid Pallavicini, who at the end of the 1980s made the current Al Wahid Mosque in Via Giuseppe Meda in Milan available for a project whose aim was to place three small places of worship for the three traditions of Abrahamic monotheism side by side: a synagogue, a church and a mosque (Fig. 1).

With no intention of commingling or confusion, they attempted to show the possible coexistence of the three distinct but united religions, with the opportunity for people to practice their faith without, however, and this should be emphasised, the risk of syncretism and mixing sacred rituals and places. The plan was for the various worshippers to find themselves in the peace of a place that shared their presence but nevertheless safeguarded their sacred specificity, to develop dialogue that was truly authentic because it respected each one of them.

The strong utopian value and the "signs of the time" meant that, although the initiative was presented to the religious authorities and to the press and given the slightly idealistic nickname "Little Jerusalem on the Navigli canals", because of it was located in Milan's Navigli district, nothing came of it, although its value remained strong.

It was a question of no longer speaking about principles in general, but about principles contained in a very specific faith which was actually being experienced and which would see itself being translated into a Western Islamic community, created in the meantime in those same spaces with the establishment of a small mosque. In this venue, the calling for dialogue among faiths was not lost, but now took form in practising concrete spiritual life.

Towards the end of the 1980s, the Centro Studi Metafisici was set up, a partnership aimed at soul-searching and whose aim was to intervene in the spiritual and religious debate of the contemporary world, characterised by a now pervasive secularism and by a spirituality that was often weak and confused. The starting point was therefore the conception of a "transcendent Unity" of all the traditions that have their roots in the same Primordial and immutable Tradition, the *Din al Qayyma*, the axial tradition of the Islamic religion but one that has its place in every orthodox denomination.

It was, therefore, a matter of bringing the focus back on the general metaphysical concepts that lie above every authentic spiritual manifestation and, at the same time, implementing these principles through a rediscovery of the meaning of the rituals and a practice of faith, whatever it may be, brought to life and made more aware of its sacred and transcendent content.

There is a clear utopian value in this project, also embodied in one of the possible applications of this "traditional thought".

If utopia indicates what does not have a location to take place in, the word is not suited to this project, since in the past there have been several places of close spiritual and ritual proximity among the faiths descending from Abraham, and from a spatial point of view, Jerusalem is still a clear example today. Perhaps the time was not ripe for the outward creation of a utopia like Jerusalem on the Navigli

Fig. 1 First visit of the Jewish community of Milan to the Al-Wahid mosque. Milan, 2006

canals, and the mosque has now occupied all of the venue's outdoor space. If we look at our interior life, however, the spiritual creation of this union between different religious forms constitutes the only inheritance and the special character of this place of pure and non-exclusive worship. So, if it is true that a Jerusalem on the Navigli canals, just like the City of the Sun by Tommaso Campanella, failed to materialise, it is equally true that on another level, on the level of Unity and authentic dialogue in the higher spheres, it did come about at the end.

What is lacking in these times is the deepest sharing of principles, a safe space for hearts, increasingly divided by ideological reasoning, by literalism or syncretism. This is the only real reason why this project did not see the light, despite the fact that the activities of the Italian COREIS (Islamic Religious Community) continue to carry on this work of testament, in an effort to purify the intellects and to achieve full convergence in the spirit.

Here, therefore, we find that we have moved on from a utopia, following another etymological path and replacing the negative prefix *ou* with the prefix *eu* meaning "good", to a eu-topia, i.e. a "good place". To reach this "good place", however, it was necessary to go through a series of very idealistic and not very real planning phases—which we could refer to as utopian if not sometimes dys-topian.

Just as, over time, the architectural projects of the mosque took on greater external essentiality, so the ecumenical perspective of the community that I represent, in recent years, has found its natural expression in the universal principles of the Islamic religion.

The last project for the construction of the Al Wahid Mosque, the work of Tuscan architect David Napolitano, a great expert when it comes to traditional symbolism and sacred geometry, dates back to 2006 (Fig. 2).

His project stood out thanks to its simple, essential nature, but also thanks to its ability to adapt the existing structure to liturgical requirements and to Islamic architectural canons, as can be seen from the use of the simple geometric shapes of the circle, the square and the octagon building the central colonnade, which provides the ritual orientation (*qibla*) of the structure. Moreover, a feature that stood out was the integration into the surrounding urban area through the use of materials.

Unlike a church or temple, the Islamic prayer hall does not have a centre to which worship is directed. In Islam, the concentric gathering of worshippers that characterises Christian communities becomes visible only at the time of the pilgrimage to Mecca, during the collective prayer around the Kaaba. In other places, when in prayer, believers turn towards this distant centre, outside the walls of the mosque. This is an essential feature of the spiritual attitude of Islam. In Muslims, the sense of the divine presence is based on the perception of limitlessness; they reject any objectification of the divine, but the one represented by unlimited space (Figs. 3 and 4).

Fig. 2 Redevelopment project for the Al-Wahid mosque of architect David Napolitano

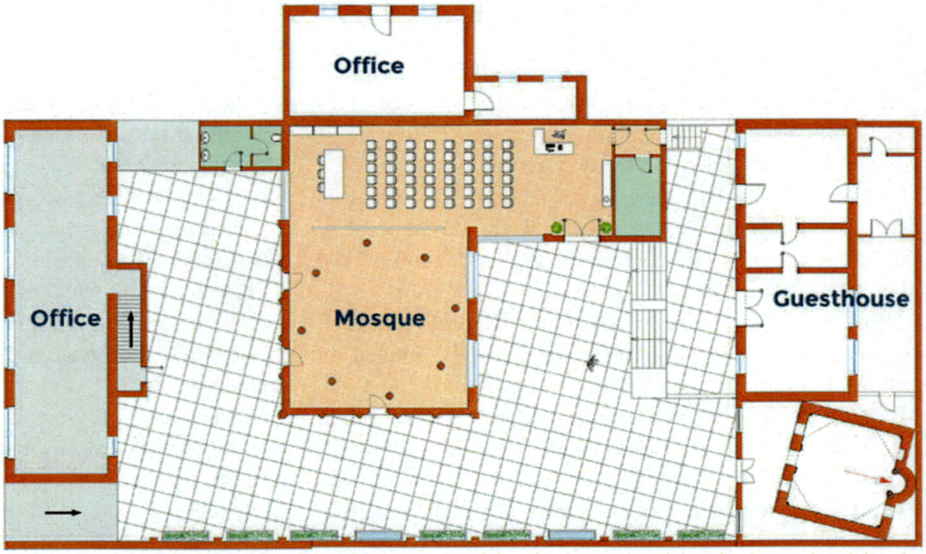

Fig. 3 Proposed spaces layout in the architect David Napolitano's project

Fig. 4 View of the Al-Wahid mosque complex

2 What the Al Wahid Mosque in Milan Looks like

At the operational level, in 2000, we had to purchase the existing building spaces, consisting of around 1000 square metres of land with three buildings standing on them, a mosque, guest quarters and offices, with relative courtyard area. It is a small mosque that does not aim to be the only one, nor to solve the problem of 100,000 Muslims in Milan, or nearly 400,000 in Lombardy, and the project does not involve any increase in volume but simply the renovation of the existing buildings, without any minarets being built. The buildings correspond to the typical Milan/Navigli-style architecture of the early twentieth century (Fig. 5).

The structure aims to blend in with the urban fabric of Via Meda, where it is located, which has private access and is well-connected to surface transport, near two large car parks and close to the centre. The gross floor area is approximately 220 m^2 for the place of worship, 300 m^2 for the guest quarters and classrooms and 110 m^2 for the offices. A significant aspect is that in the 220 m^2 intended for the mosque, there will also be a multi-purpose room to be used for conferences, interreligious or institutional meetings,

Fig. 5 View of the courtyard

seminars and round tables. The mosque has a single ritual space in which both men and women pray, without a women's gallery, while maintaining the distinction of the relative spaces (Fig. 6).

3 The Al Wahid Mosque, Stages of Legal Recognition as a Place of Worship

Since the beginning of the twenty-first century, the Al Wahid Mosque project has represented an international reference point for European Islam, thanks to its doctrinal transparency, the use of the Italian language in sermons and the fact that it is managed entirely by native Muslims, without exoticism, ideological contamination and political interference from foreign countries. To reach this point and achieve official recognition as a place of worship, it was necessary to follow a long legislative process. Italy is, in fact, perhaps the only country in Europe where Islam, despite over one and a half million worshippers, is still not officially recognised, unlike other religious confessions. Not being recognised by the Italian State means that there is no agreement governing relations between the State and the Islamic community. These relations are also of a practical nature, involving for example the construction of places of worship and how these should be included in the Territorial Government Plans (PGTs) (Fig. 7).

4 How the Italian COREIS Expressed Itself About Places of Worship in Milan

1. Favour and prioritise those Islamic entities that have signed the Charter of Values referred to in the Italian Ministerial Decree of April 23, 2007, present in the municipal territory and those that have been recognised for civic merit by the Municipality of Milan;
2. Checking the sustainability plans and traceability tools for original financial records funding the construction of places of worship.
3. Management of places of worship by an "Italian leadership", which is not dependent on the politics of foreign countries, and by a "religious leadership" devoid of ideological, radical or fundamentalist demands.
4. Focus on creating spaces and chances for promoting religious pluralism, also with conferences, seminars, workshops and round tables aimed at Milanese and Italian citizens;
5. Capacity for further social and cultural activities, also including in the project relations with the surrounding area, in particular with the institutions and city institutions, and dialogue with other religions;
6. Creating a (Register of Imams in Italy), as debated in the Council for Italian Islam, at the Ministry of the Interior, and approving ministers of worship through appropriate religious, civic and theological training; the approval of religious ministers is to be considered an ordinary means of connection and mutual recognition between public authorities and religious groups that wish to be integrated into the social fabric of the city;
7. Through the idea of a "mosque with glass walls", we promote accessibility to the comprehension of the content disclosed during religious activities, in particular of the *khutba* Friday sermon in Italian, or through the use of translations into Italian.

From this standpoint of recognition, compliance, or official legitimisation of places of worship, we could go beyond the trick of the **"mosque disguised as a cultural centre"**, aiming to transparency and legality. Therefore, in **full compliance with the building and urban planning legislation**, the spaces used for worship would also be recognised

Fig. 6 Believers praying in the mosque

Fig. 7 Inter-religious meeting at the Al-Wahid mosque

as intended for use as a mosque, but in these spaces, productive work of cultural and intellectual promotion would also take place, without passing off the cultural centre as a place of worship, due to a legislative gap or the lack of inclusion in the PGT of the reality of religious pluralism. This standpoint would also prevent pockets of illegality in which "Islamic centres" then find themselves located in run-down settings, such as garages, basements, old abandoned buildings, in which the neglect and the dignity of religious and spiritual needs are vilified, together with respect for the regulation in force. The last aspect we would like to underline, and on which the Italian COREIS has repeatedly expressed itself, is that we have never wanted to support the great "cathedral in the desert" mosque, where the danger of ghettoisation is insinuated, hoping instead for **more neighbourhood mosques**, integrated into the city in compliance with building and urban planning regulations, enjoying full dialogue and cultural and interactive open-mindedness. In this regard, it is not necessary for a mosque to have "Arabic-style" architectural features, not

Fig. 8 Al-Wahid Mosque. Main entrance portal

when the building itself, for its destined use, is an integral part of the Islamic tradition composed of nearly one and a half billion Muslims around the world. The architecture must be adequate for and compatible with the geographical, social, civil and cultural context in which the worshippers find themselves.

The financial resources for the construction of the "always under construction" Al Wahid Mosque, as it has been defined, will come *InshAllah*, when God wills, and we believe that, far from being a utopia, His project is actually coming to fruition through the prayers and invocations that our community practices in the same spaces. Those spaces where, even without high, showy minarets, the call to prayer, the *adhan*, resounds every day, the *muezzin*'s voice vibrating it in the heart of every worshipper: this is where the true niche of every prayer is kept and only here can the "always under construction" mosque be built before it can finally show itself to the world. A mosque that "lives on the inside", a mosque of believers still awaiting the second coming of Sayyidna Isa, our Lord Jesus, who said that the Temple of Jerusalem could have been destroyed and rebuilt in just three days (Fig. 8).

Designing the Intangible

Niccolò Casiddu

Abstract

It is here proposed a reflection about the contribution that the design discipline can act within the constitution of sacral objects, at the same time investigating the relation of this specific kind of artefact with the wider nature of design principles in general. By offering a standpoint about the importance of beauty and quality to be embedded since the very beginning in the design process, the statements will aid in the comprehension for the correct positioning of sacral design within the mass market and in respect to the constraints given by the credos. As well, other intangible aspects, such as trends and emotional values personally added, will be taken into account. Through proposing an overview over contemporary examples of what can be considered as part of sacral design and by understanding the reasoning of their birth, the purpose is to provide food for thoughts about the important matter of the relation between religion and design.

Keywords

Design • Artifacts • Design principles • Material culture • Aesthetics

1 Why Design and Religion?

Most of us, when talking about sacred space, are used to recall to mind great architectures, majestic mosques, and cathedrals. In short, the thought goes to the monuments of man for god, the immense.

Nevertheless, we often forget to consider that god also expresses himself in small things, just like man does (Fabietti 2014). However, this inaccuracy is historically supported by the position that the design discipline itself has covered since the early days of its existence, not equal in respect of those which have been considered the major arts, as painting, sculpture, architecture (Branzi et al. 2007).

Design considered to be culturally derived from the arts and crafts, indeed, for a long time has been named within the minors arts—as said—in an inaccurate manner: The concept of "minor," indeed, comes out of caring and taking into account as its own field of interest the small everyday things, supposed of no importance (Smock 2004; Branzi et al. 2007).

Researchers and schools have demonstrated, in not so recent times, how improper might this statement be: It happens that in one very little artifact, like a chair, a community can express its desires, beliefs, dictates, reaching to be as important as scholarly well-known institutions in the realm of the constitution of objects (Papanek 1984; Heskett 2005). We can consider as an example of what stated so far, over all the possible ones, the United Society of Believers in Christ's Second Appearing, more commonly known as the Shakers: originated as a religious closed community in the United Kingdom, and finally came to the United States after some time, within their book *The Millennial Laws*, they encoded and stigmatized in a very definitive and accurate way the consistency of objects, to be produced and performed within the community itself. Though even if being designed in their final forms for ethical reasons, their works succeeded so perfectly in merging functionality and aesthetical cleanness that brought to the interest of the world through the exhibition held in Philadelphia in 1876, it was inspiring and taken as a reference from Scandinavian design to the Austrian architect Adolf Loos.

2 Which Design for Religion?

As for what stated above is important to take the chance to understand why design is so difficult in finding a final definition of its positioning, and how can we reach a valuable

N. Casiddu (✉)
Department of Architecture and Design, University of Genoa, Genoa, Italy
e-mail: casiddu@unige.it

C. Gambardella et al. (eds.), *Advances in Utopian Studies and Sacred Architecture*,
Advances in Science, Technology & Innovation,
https://doi.org/10.1007/978-3-030-50765-7_4

statement, when we consider that this latter issue has been disputed over the last two hundred years—since the Industrial Revolution, according to most right to be regarded as the beginning of the time of the design—and why still this concept needs to determine some of the basic notions.

According to Francesco Trabucco, «in the current language "design" indicates both the craft of those who transfer aesthetic value and originality to a physical or virtual artifact and that artifact itself» (Trabucco 2015 p. 11).

This can easily be confirmed, as we are used today in our everyday life to indicate "things" as "design" if they look trendy or sleek, even if the exact meaning of the term designates more correctly other significances about the conception and realization of the designed object.

Bürdek, reports us a different definition of the term: «According to the Oxford Dictionary—the scholar says—the term "design" was used for the first time in 1588 and is so defined: a drawing or a scheme conceived by a person for something that needs to be produced; a first schematic representation for a work of art; an object of applied art, to be accurately followed in the realization of a work» (Bürdek 2008 p. 25).

If, in this latter sense, the concept of designing is evident and clear, is anyway pivotal to underline how much we are well ahead of the Industrial Revolution itself, so referring mostly to architecture, which brings us back to the introduction of this discourse (Smock 2004).

What has definitely changed the positioning of the architectural discipline in respect of the design one was the Ulm school. And these, the statements of the importance of the small items alongside to the big ones, let us return to more recent times, where they are assessed even in a cultural sense as traces of the contemporary world. Within the Ulm school, scholars collaborating with industries and factories determined the importance of quality and processes not only for bigger and "touchable" monuments but even for "minor" artifacts, witnesses of everyday little gestures and lives' silent companions.

We can recall, as a perfectly valuable emblem of this concept, the ten principles for good design developed by Dieter Rams, head of the Department of Product of Braun until the 90 s, already in close connection with the Ulm School: to him Jonathan Ive, designer of Apple, and Naoto Fukasawa, designer for Muji, have bestowed the importance of his teachings and achievements, to which they have referred firmly as for the exactness they always tried to embed in their design process.

Dieter Rams, to whom international museums also ascribe great importance, has always worked out projects that follow these ten principles, which are defined commonly "timeless commandments" but that he himself outlines as "implementable" and upgradable. This demonstrates the openness of the principles to the present time and their further validity for the future. So, according to Rams and referring to a largely intended product, the good design: is innovative; makes a product useful; is aesthetic; makes a product understandable; is unobtrusive; is honest; is long-lasting; is thorough down the last detail; is environmentally friendly; and least, but not last, involves as little design as possible.

On the basis of what has been said so far, how are we to think and, somehow, to evaluate—maybe, to excuse—the artifacts connected to religion? Question is not so obvious as so far, mostly we have considered functional aspects and theoretical discourses that, anyway, lead to an overall understanding of the discipline's principles, which themselves might appear very far when pertained to very specific or even useless objects as we commonly intend religious ones (Plate 2014). In particular, we will take into account as examples of this topic both the products specifically religious, both of products that fulfill a moment of union with the Sacred, whether it is special or everyday life one—for the three monotheistic religions.

Following this *fil rouge*, let's go back to the already mentioned Rams's ten principles and in detail, it is interesting to select among the others, three of them, seemly the farthest ones from our subject of interest, as for a further and deeper discussion. Rams says: "good design makes a product useful, it is aesthetic, and it is honest." What here stated appears to us completely far from the concept here reminded is because often the products related to the sacred are—at least in the dominant culture—not considered for their functionality, as they apparently do not encompass any and as well as they appear to concern, indeed, only for their aesthetic content, which, however, is often betrayed in the sense of honesty of the product itself (Rinallo et al. 2012). As a sign of this betrayal, we can remember to ourselves the holy rosaries of Pompeii made in China in poor plastic materials.

But, the design discipline always has something to teach and to deliver in its own meaning of existing. Death has always been a taboo, for instance, but still, products connected to the passing like urns and coffins are undoubtedly envisaged as designed objects. In this sense, one of the most important moments in the life of every one of us is, for example, the moment of the passing of a loved one, person or animal. While we are aiming to cope against the painful experience, we are forced to deal with the management of practical aspects; in the very moment, we would definitely prefer to avoid it as this concern could make us profoundly sadder and worried.

From the idea of turning into new life all "this," it comes *Bios urns*® studied either for human beings either for animals, with a system that makes feasible to follow the birth and the evolution of a chosen plant, thanks to an incubator which is connected through the mobile application to a personal smartphone. Biological conditions and the evolution of the growing process are made controllable and continue a bond through a different channel. The idea and the related project goes along with the wider vision of converting cemeteries into forests, and it will be precisely in Naples the first location where a part of the cemetery will be, in some way, transformed into woods, a living place for memory.

Another example is offered by EST SINE DIE, an Italian company offering specifically customization of their underground urns, studied in wood or honeycomb cardboard, environmentally friendly materials—as the aim of the firm is to strengthen the bond between men, animals, and nature—thought to be initial nourishment for the chosen plant, which seeds come along with the urn itself.

It shouldn't appear bizarre that animals are treated comparably to persons. The more and more, in very recent times, the culture of keeping memory and individual burial of domestic animals is becoming extensively widespread, as they are in all respects considered by many as part of the family. Thus, many companies have shown their interest in the possibilities related to rituals that can go over a "simple" cremation and took the chance to involve design for offering an opportunity to consider this moment not as an end but as the beginning of a new relationship. These two researches, evidently connected to the emotional concept of the painful experience that a severe caesura can bring, work on the possibility of generating new life from the ashes of the beloved animal (Rinallo et al. 2019).

Design can provide in the realm of so—considered religious object or accessories many and unexpected standpoints, able to be converted in products and to give touchable experience when "on-duty" in our religious moments (Sennet 2012). It is not new the idea of reaching every one interested in a religious rite, independently from distances and personal conditions. This is what happens for instance through an average radio when the Pope officiates a valid benediction *Urbi et orbi*. On this count, a prototype of a few years ago—dated back to 2006 and named *iRAuch* (obviously in the name connected to the Apple world)—developed at the Berlin University of the Arts. It is thought to make possible experiencing substantially in remote the blessing for those who are far away because of inability for physical or geographical reasons, sensing the benediction it while receiving through the presence of smoke, released in the exact moment by sensors. «*iRauch* allows absentees to be at the place of the happening» (Heller and Talarico 2009 p. 61) working on an extended concept of design for all, and setting out the conceptual basis for inclusion in a collective rite, devout, and participative even if from afar.

It is proper, for an additional and profounder understanding, to report Rams's thought: In one of the many interviews released about his ten principles, he takes the chance to add something more, explaining how "in the future, the un-spectacular things are the most important things, especially in the future." Perhaps, we can be doubtful about this latter concept when it matches the realm of design nowadays, as this appears confusing as an apt: It looks like, in fact, that the spectacle is, instead, part of a moment that mystifies the aesthetical contribution with the mere appearance. We are not far, sadly, from well-known speeches about the aesthetical validity of an object in the case of it being if simply decorated: But, still it happens and at the same time of the narrating territories and traditions to which it belongs to. It is a fact that designed religious objects can serve for their purposes as manifestation in substantial matter of the intangible one, alongside to the desirable connection with innovative technologies either for the improvement or the translation of the acknowledged heritage. This is the case of the Chalice and the Patenas made for the mass of Pope Benedict XVI, in Albissola ceramics, according to an ancient and consolidated tradition, but updated with 3D technology execution, directly from the drawing (Casiddu and Chimenz 2018).

Similarly, wish to consider how much significant can be the turning into incorporeal objects of the ones usually are perfectly physical, and which new means can give when this is related to sacral design. Let us think to the many apps for praying in all the three monotheistic religions: In their digital immateriality, they serve to help, and so operating is not less important task than to realize other services, featured by determined user—oriented design research. It is the case of the many smartphone apps available, as an example, for the Holy Qur'an: These apps make possible, in the feverish everyday life, to perform a duty as devout and honor our own beliefs.

But design also helps us to communicate about ourselves, and in our choices, in the expression of our taste, much of what and of who we are is said (Bodei 2009; Papanek 1984; Heskett 2005). So, Alessi, well-known Italian design brand, already working since the 80 s' with many international names has, more or less recently, found some interest in producing sacral objects. From this comes out the idea of creating a special collection, devoted to sacral objects and designed by bespoken international designers, uttering all the embedded tradition of the brand on one side and answering the call for a decorative high-aesthetic content on the other, carefully respectful of the original meaning of the symbol connected to the religious belief. After the mezuzah by Dror Benshitrit and the *Presepe* by Massimo Giacon, here is the *Cross* by Mario Trimarchi, in 2011: in metal, as simple as its symbol, and as contemporary as its appearance, perfectly suited to both the meaning of the brand, both to the object's religious significance (Plate 2015).

3 How to Design for Religion?

Still, for a better general comprehension of the concept of "design," it appears here considerably interesting to deem what Deyan Sudjic, current director of the Design Museum in London, says about the relationship between design and art—maybe, a reflection to which Alessi has brought us.

This tricky—might be said—connection is often the cause of misunderstanding of the aesthetical contribution of the first over the second. Sudjic says, «design has often been considered for what it is no, more than for what it is» (Sudjic 2014, p. 152) and continues explaining how design always has a principle of functionality, as its purpose and principle prime mover. Again, this concept leads us to estimate about appearance when this latter is related to religious items belonging directly or not to the credo and the Sacred (Branzi, Pederbelli, Biamonti 2007).

Continuing, however, is convenient to reason what Donald Norman, best known as the author of *The Coffee maker of the Masochist*, states in his further book, *Emotional design*. «Emotions—the scholar says—are inseparable by the cognitive process; they are a necessary part of it. Whatever we do, whatever we think, is imbued with emotions, largely unconscious» (Norman 2004 p. 4).

According to the psychologist, interested in understanding the aforementioned connection between aesthetics, functionality and users' needs, the issue is not new: What he has achieved in his long scholarly study is that our final disposition toward or against determined designs seemly responds to some kind of logic. Within this, emotional and visceral—related answers might eventually convey in the final assessment of an artifact and, therefore, influence it with unsaid values, connected with a variety of personal and social bounds. He reaches to assume with accuracy new finding in respect of his first earlier work and finally includes in its assessments the happiness coming from aesthetical content, because somehow it is proficient in satisfying an inescapable necessity for man. Norman tells, indeed, about a primal attraction, finally realized in the concept of beauty. «Attraction is a phenomenon of visceral level—the reaction is entirely due to the surface appearance of an object. Beauty comes from the reflective level. Beauty looks below the surface. Beauty derives from conscious reflection and experience. It is influenced by knowledge, education, and culture» (Norman 2004 p. 86).

It fits, in this regard, to quote how the Italian term of beauty, *bellezza*, is very close to the Sanskrit concept of "the place where God shines." So, bringing it back to a context belonging to the realm of the products and according as well to the words of Jay Chiat, already advisory for Apple and other well-known international brands, if the challenge is the design and the design of an object, which is linked to the sacred, perhaps it is correct to consider that "good enough" is not sufficient; in this case, the challenge is reaching and pursuing the highest excellence and is in this case more important than ever, no matter if the designer is religious or not. If the place where God shines is the purpose, then definitely "good enough is not enough."

References

Bodei, R. (2009). *La vita delle cose [The life of things]*. Roma-Bari: Editori Laterza.

Branzi, A., Pederbelli, M., & Biamonti, A. (2007). *Capire il design*. Firenze: Giunti.

Bürdek, B. E. (2008). *Design. Storia, Teoria e Pratica del Design del prodotto* [History, Theory and Practice of Product design]. Roma: Gangemi Editore.

Casiddu, N. & Chimenz, L. (2018). La variabile e la sua misura. Ragione, sostanza e tema per l'oggetto immaterialmente materiale [The variable and its measure. Reason, substance and subject for the immaterially material object]. In L. Chimenz, R. Fagnoni, M. B. Spadolini (Eds.), *Design su Misura. Atti dell'Assemblea annuale della Società italiana di Design* (pp. 115–130). Venezia: SID Società Italiana di Design.

Fabietti, U. (2014). *Materia sacra. Corpi, oggetti, immagini, feticci nella pratica religiosa* [Sacred matter. Bodies, objects, images, fetishes in the religious practice]. Milano: Raffaello Cortina Editore.

Heller, S., & Talarico, R. (2009). *Design School Confidential: extraordinary class projects from international design schools*. Beverly: Rockport Publisher.

Heskett, J. (2005). *Design: A very short introduction*. Oxford: Oxford University Press.

Norman, D. A. (2004). *Emotional design*. Milan: Apogeo.

Papanek, V. (1984). *Design for the real world: Human ecology and social change*. London: Thames & Hudson.

Plate, B. S. (2014). *Religion Is Alive and Well in Contemporary Art*. Retrieved March 10, 2017, from http://www.huffingtonpost.com/s-brent-plate/religion-is-alive-and-wel_b_5676134.html.

Plate, B. S. (2015). *Key terms in material religion*. London, New York: Bloomsbury Academic.

Rinallo, D., et al. (2012). When sacred objects go b®a(n)d. Fashion rosaries and the cointenporary linkage of religion and commerciality. In D. Rinallo et al. (Eds.), *Spirituality and consumption*. London: Routledge.

Rinallo, D., & Alemany Oliver, M. (2019). The marketing and consumption of spirituality and religion. *Journal of Management, Spirituality & Religion, 16*(1), 1–5. https://doi.org/10.1080/14766086.2019.1555885.

Sennet, R. (2012). *Together: The rituals, pleasures and politics of cooperation*. London: Penguin Books.

Smock, W. (2004). *The design ideal then and now. an illustrated guide to modern design*. Chicago: Academy Chicago Publishers.

Sudjic, D. (2014). *B is for Bauhaus. An A–Z of the modern world*. London: Particular Books an imprint of Penguin Books.

Trabucco, F. (2015). *Design*. Torino: Bollati Boringhieri.

Sacred Places: The Interaction Between Space, Time and Faith

Mona Helmy

Abstract

Sacred places are dedicated places and/or places that are tangibly constructed for a religious or holy purpose, or for a certain belief, such as mosques, churches or temples. Furthermore, sacred places could also be intangible places that are religiously or spiritually interpreted within various spaces, such as mountains, rivers or deserts. The presentation investigates what makes a place sacred, and if a sacred place is relative, or timeless; for example, if a place is considered sacred for all users, or only for a selected group of users. It also explores whether the sensation of the sacredness in places is temporary or permanent. Through examination of selected models of sacred places in a variety of tangible and intangible contexts, this presentation intends to identify the elements that shape the spirit of sacred places.

1 Sacred Places

Sacred places are places where individuals or groups are experiencing spiritual feelings or practicing certain activities. Bartle (2009) argues that '…sacred places represent socially or culturally a touchstone of sorts for a particular group of peoples. They might be places where unique activities happen on a regular basis, or where individuals congregate to explore their own meanings of spirituality.'

Sacred places are not limited to certain type of buildings or places, yet they could be sensed in spaces with no physical boundaries. Typically, religious buildings that are tangibly identified through their unique architecture, symbols, or landmarks, and are connected with indoor and outdoor plazas are acting as centers and hubs for a variety of activities. Other sacred places are intangibly and emotionally identified through the rituals that are practiced in them. In this context, Gojnik and Gojnik (2011) claim that a 'sacred place has the structure of existential space; it attains semantic character in its sacred symbols through spatialized meanings and is consequently also architectural space.'

Sacred places are places that provoke people's senses and minds through their trusted religions, beliefs, faith, values, traditions, memories or emotions. In all religions and beliefs, sacred places have different identifiable forms, moods or typologies. However, the definition of these places and why they are sacred remains shadowy, as they have connections between visible and/or invisible space attributes.

There are several definitions of sacred places. According to the *Cambridge English Dictionary*, 'sacred' is an adjective and it is 'considered to be holy and deserving respect, especially because of a connection with a god.'; As per *encyclopedia.com*, 'sacred space' is a place that people's ritualistic practices mark its sacredness and differentiate it from other defined spaces. *Encyclopedia Britannica* states that 'A place of worship became sacred and suitable by virtue of the holy's [sic] appearing at that place.'

Sacred places have a unique sense of place. They are often identifiable from their surrounding areas in several contexts and on different levels. Osbon (1991) debates that:

> A sacred space is any space that is set apart from the usual context of life. Sacred space has no function in the way of earning a living or a reputation. Practical use is not the dominant feature of anything in the space.

Lastly, sacred places are places where the articulation and use of architecture helps believers to enhance their spiritual experience for worship, prayer and rituals. In line with this, Candler (2010) states that 'Sacred places are where we meet something beyond ourselves. They are not always conventionally religious places.'

M. Helmy (✉)
Architecture Department, The British University in Egypt (BUE), Cairo, Egypt
e-mail: mona.helmy@bue.edu.eg

C. Gambardella et al. (eds.), *Advances in Utopian Studies and Sacred Architecture*,
Advances in Science, Technology & Innovation,
https://doi.org/10.1007/978-3-030-50765-7_5

2 Forms of Sacred Places

Sacred places are geographically defined and are religiously interpreted, i.e., landscape sites, mountains or rivers. They differ in their shapes, forms and contents. Obviously, religious buildings with their outdoor plazas, such as mosques, churches and temples, are considered holy and sacred places for their denominations. Furthermore, certain historical sites with specific memories or historical events are considered sacred for their communities. Moreover, sacred places could be natural sites, such as mountains, hills, rivers, deserts, caves or any other nature made sites. Figure 1 shows several types of forms of sacred places, which range from natural sites to special arrangements in some spaces, reaching the obvious religious buildings of different faiths.

3 What Makes a Place Sacred?

There are many aspects that mark the sacredness of any place. These aspects are often related to the location, history, physical characteristics or the construction of the place itself. Alternatively, these aspects could be related to activities happening in that place.

3.1 Location

There are places that are identified as sacred due to their geographical location, in which they usually are connected with certain memories or with occasional or continuous rituals practiced there. For example, the Mina Valley is a neighborhood of the holy city of Makkah, in the Makkah

Fig. 1 Sacred places: Important archetypal categories. *Source* Bartle, G. (2009). Sacred places: public spaces

Province in Saudi Arabia. Mina (the tent city) has a relatively small and limited area, in comparison to the number of pilgrims who are hosted there. During the three to four '*Tashreeq*'[1] days, which occur during the annual *Hajj* Pilgrimage, the number of Pilgrims hosted can reach up to four million pilgrims. Figure 2 shows the crowd of the pilgrims in Mina during the Hajj pilgrimage. In Islam, rituals practiced in Mina are considered to be the third stage of the Hajj Pilgrimage, which must be completed. The tent city is only inhabited for very short period every year, and it has no religious function for the remaining months of the year. Nonetheless, it is highly valued and considered as a sacred place for its location within the holy sites, both during Hajj in particular, and in Islam in general.

3.2 History/Memory

There are places that are identified as sacred, although no rituals are practiced there. They are considered sacred due to their history or the religious memories connected to them. Mount Sinai, or also called the Mountain of Moses, is an example of such a place. The mountain is located in the south-central Sinai, Egypt. Although the mountain has no identification as a certain sacred location, or place as shown in Fig. 3, the whole site is considered sacred in the traditions of Islam, Christianity and Judaism.

As per *Britannica.com*, Mount Sinai has witnessed historical religious events, such as being the principal site of divine revelation in Jewish history, where God is purported to have appeared to Moses and given him the Ten Commandments (Exodus 20; Deuteronomy 5). The mountain is also sacred in both the Christian and Islamic traditions; as it was interpreted as part of the route of the Israelite exodus from Egypt.

3.3 Building Type/Function

Religious buildings, such as mosques, churches, temples and other buildings dedicated to holy purposes including their indoor and outdoor plazas, are obviously considered sacred places. In this context, Estivill (2013) debates that a beautiful, functional and symbolically efficitive building could be a result of sacred architecture. The Vatican is an example to an unconditional sensation of sacredness to Christians. Figure 4 shows an event in St. Peter's Square in the Vatican City. The Church of Seed is located in the Luofu Mountain Scenery District—one of the seven famous Taoist mountains in China. The design of the church itself provokes a great sense of sacredness, due to its abstract design that uses strong symbolism of the shaped cross that illuminates the interior of the church using natural sunlight, as a metaphor for the cross as a source of light. This is seen in Fig. 5.

3.4 Practiced Rituals

Some places might not be categorized as sacred. However, human articulation of faith manifested in them turns them into a whole experience filled with a sacredness sensation. The Dervish traditional show at Hodjapasha, Istanbul, Turkey, is an example to this type of sacredness, whereby a spiritual ritual is performed. Originally, Hodjapasha is a fifteenth-century historical large-size Hammam (Turkish Bath), which was transformed into a cultural hub in Istanbul. Spectators from different cultures and various religions consider the ritual on display as a global spiritual manifestation (Fig. 6).

3.5 Objects Located in Place

The great mosque of Makkah is considered the most sacred place for all Muslims. The mosque itself has witnessed many religious and historical events. Above all, it has *al-Ka'bah al-Musharrafah*, the 'Kaaba', which is considered the most sacred spot on Earth for Muslims.

The cube-shaped structure is almost 15 meters high, and it is about 10 by 14 m at its base. It is oriented so that its corners roughly correspond to the main polar points of a compass. A holy stone from heaven, called the Black Stone, is housed in one of the Kaaba's corners. As part of the Umrah or Hajj for any Muslim, the Tawaf, a circling around the Kaaba in a counterclockwise direction for 7 laps, is an essential part of the pilgrimage. Figure 7 shows the physical Tawaf around the Kaaba performed by Muslims, whilst Fig. 8 is representing an illustration showing a metaphor on how the Kaaba is acting as a magnet that attracts Muslims.

3.6 Traditions

Human traditions are usually connected to historical backgrounds, such as following old habits, celebrating certain events that are anchored in culture, or recalling memories. Traditions that are performed in the form of rituals strongly affect the place that they are taking place in, which in turn

[1]*Tashreeq* days are the 11th, 12th and 13th of the Islamic month of *Dhul Hijjah*. All pilgrims should stay in Mina Valley during these day to remember Allah as an essential part of Muslim pilgrimage.

Fig. 2 Muslim pilgrims during Tashreeq days in Mina. *Source* Al Arabiya English. 3 Sept. 2017, english.alarabiya.net/en/News/gulf/2017/09/02/Muslim-pilgrims-enter-days-of-Tashreeq-as-Hajj-nears-its-end

Fig. 3 Mount Sinai, Sinai, Egypt. *Source* 'Mount Sinai.' Wikipedia, Wikimedia Foundation, 24 Feb. 2020, en.wikipedia

Fig. 4 A religious celebration in Vatican City. *Source* Chris Riotta New York @chrisriotta. 'Vatican Breaks Silence on Abuse of 1000 Children by More than 300 Priests in Pennsylvania.' The Independent, Independent Digital News and Media, 16 Aug. 2018, www.independent.co.uk/news/world/americas/pennsylvania-child-abuse-pope-francis-vatican-priests-catholic-church-latest-a8495371.html

Fig. 5 Sacred sense of place inside the Church of Seed. *Source* 'Gallery of Church of Seed /O Studio Architects—6.' ArchDaily, www.archdaily.com/253115/church-of-seed-o-studio-architects/5018ca6928ba0d5d5d000974-church-of-seed-o-studio-architects-image?next_project=no

are usually connected with the history of the tradition itself. The remembrance of Al Hussain's[2] birthday is an old celebratory tradition that happens annually in Cairo, Egypt. Often, people come from different regions of Egypt, and from different foreign countries, to Cairo to celebrate this event. The event takes place in the area around Al Hussain's mosque, which is the plaza and the surrounding streets of the mosque. The celebration itself has certain traditions and

[2]Al Hussain is the son of the son of Prophet Mohamed (PBUH).

Fig. 6 The Darvish Show at Hodjapasha, Istanbul, Turkey. *Source* http://www.istanbullife.org/Hodjapasha_SemaCeremony_045%20copyq.jpg

Fig. 7 Haram Area, Holy Kaʿbah, Makkah, KSA. *Source* https://cdn.pixabay.com/photo/2019/11/22/23/22/kaaba-4645916_960_720.jpg

Fig. 8 Embracing abstract of worship at Holy Kaʿbah. *Source* Riaz, K. (2018, August 16). Retrieved from https://almadinainstitute.org/blog/embracing-abstract-forms-of-worship/

rituals, which end with a parade in the areas surrounding the Mosque. During that time, the entire place is visually and emotionally reshaped with a high sense of sacredness. Figure 9 shows the parade that is performed as a closing to this religious tradition.

4 The Temporal Sensation of Sacredness

The sensation of a place's sacredness could be temporary and under certain conditions, such as the nature of the place itself, its original function, and the duration and nature of the rituals taking place in it. On the contrary, other places have a timeless sensation of sacredness according to their architectural type, recent function, and their connectivity with certain memories or past sacred events. In this context, it is argued that a sacred place comes into being when it is interpreted as a sacred place.

Burning Man is an annual event, which attracts tens of thousands of participants, who believe that *Burning Man* can produce a positive spiritual change in the world. The Burning Man believers spend 7 days annually to perform their rituals, forming a dense temporary urbanism in Black Rock City in the middle of Northern Nevada in the USA. The event promotes some principles, such as anti-consumerism, radical insulation, radical self-expression, communal effort and civic responsibility, among other principles that guide the behavior of the participants and their planned activities. The event ends by burning a large wooden sculpture representing the Burning Man. Figure 10 shows how an empty space in a middle of a desert land that has no unique land characteristics and has no connected memories or history is converted into a temporary planned urbanism spot. It shows how a neutral, natural site could be transformed into a temporary sacred place for a certain community.

5 Spatial Organization of Sacred Places

Spatial organization is the process of arranging different elements in a space. It has common methods, such as centralized, linear, radial, clustered, and grid organization. Spatial organization of sacred places has great impact on the way events or rituals are performed. From a different

Fig. 9 Parade of the remembrance of Al Husain's Birthday, Cairo, Egypt. Retrieved from https://www.youm7.com/story/2018/12/10/5

Fig. 10 Process of Burning Man temporary event, Black Rock City, Nevada, USA. Montaged photo. Retrieved from https://www.businessinsider.com/what-is-burning-man-theme-tickets-dates-outfits-2018-8#when-is-burning-man-2. https://www.flickr.com/photos/rafaelabreu/5765028376

perspective, the events/rituals performed in any place change its perception and may adapt its physical conditions, in which it may affect its original spatial organization. The two ways effect between the spatial organization of a sacred place and the rituals performed in it is considered an interactive dynamic process; aiming to interweave the event with the place to reinterpret them as sacred places. Table 1 shows examples of sacred places and their spatial organization.

6 Conclusion

The spirit of sacred places is a synthesis of tangible elements, such as the presence of holy sites, religious buildings, icons, and scenic landscape, as well as various intangible elements; past memories of a place, religious narratives, traditional rituals, and specific colors, odors, etc. These

Table 1 Spatial organization of sacred places' models

Sacred place	Spatial organization	Pattern	Effect
Mina Valley, Makkah Region, Saudi Arabia		Linear 	A primary path that connects several holy sites
St. Peters Square, Vatican City		Centralized with physical borders 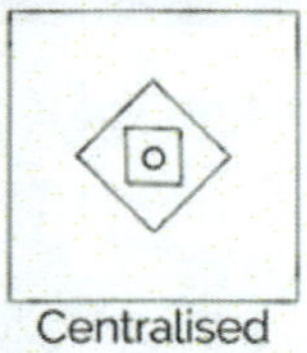	A maximum space enclosure that allows a variety of religious activities
Great Mosque of Makkah, Saudi Arabia	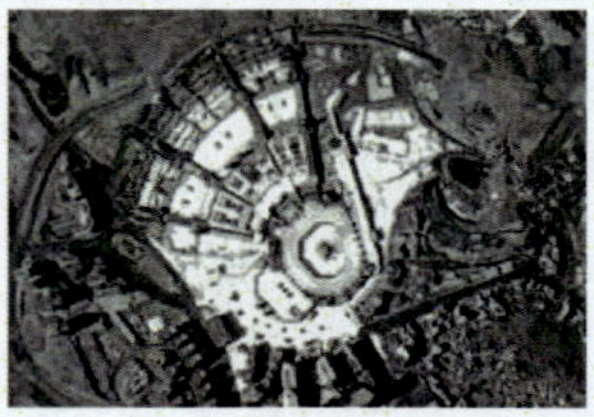	Radial ring 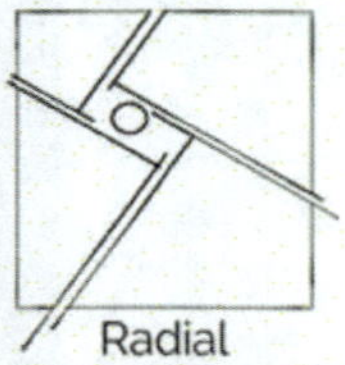	Loops or rings are surrounded by successively larger ones with a main holy object in the center
Al-Husain Plaza, Cairo Egypt		Clustered 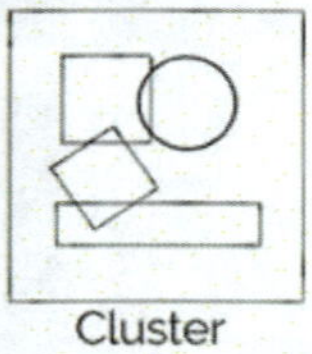	A combination of linear, clustered and central are guiding the rituals' route
Black Rock City, Nevada, USA		Centralized with no physical borders 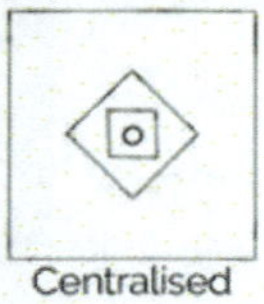	Rituals are forming a centralized place from an ultimate natural space

Satellite images source: Google Maps
Diagrams for patterns: By Franz D. K. Ching, Retrieved from www.firstinarchitecture.co.uk/space-planning-basics/

elements are consolidated in the physical structure of the place with its tangible and intangible elements, and the activities performed in the place that manifest its sacredness. Figure 11 shows a theoretical model to illustrate the relationship between the sacred places' elements that are performed, considering their temporal aspect.

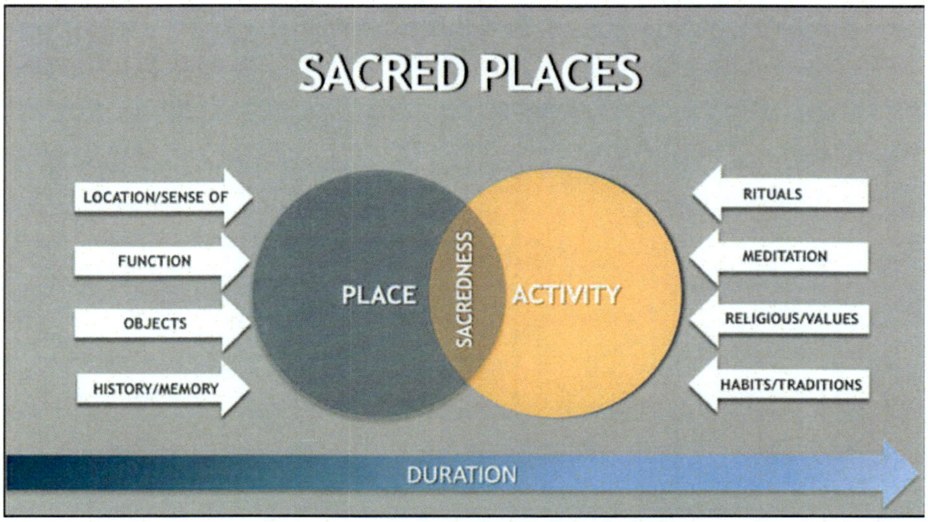

Fig. 11 A theoretical model of the shaping elements of sacred places

References

Bartle, G. (2009). *Sacred places: Public spaces*. Retrieved from https://www.mcgill.ca/urbandesign/files/urbandesign/SRP-FINAL-Bartle.pdf

Candler, G. S. (2010). *What makes a place sacred*. The Cathedral Times. Retrieved from https://www.cathedralatl.org/Sermons/what-makes-a-place-sacred/

Estivill, D. (2013). A reflection on contemporary sacred architecture. *Arte Cristiana Magazine, 878*.

Gojnik, Z. S., & Gojnik, I. (2011) *Meaning of place in sacred architecture*. In L. Kudumović, & A. Idrizbegović-Zgonić (Eds.), *4th H&mH Conference Importance of Place: Conference Proceedings*. Sarajevo, Bosna i Hercegovina (pp. 396–397). Retrieved from https://www.academia.edu/37481595/MEANING_OF_PLACE_IN_SACRED_ARCHITECTURE

Mount Sinai. (n.d.). Retrieved from https://www.britannica.com/place/Mount-Sinai-mountain-Egypt (Last accessed: 14.04.2020).

Osbon, D. K. (1991). *Reflections on the art of living. A Joseph Campbell Companion*. United States: Harper Perennial.

Sacred: Cambridge dictionary. (n.d.). Retrieved from https://dictionary.cambridge.org/dictionary/english/sacred (Last accessed: 14.04.2020).

Sacred place: Encyclopaedia Britannica. (n.d.). Retrieved from https://www.britannica.com/topic/worship/Times-and-places-of-worship#ref537651 (Last accessed: 14.04.2020).

Sacred Space: Encyclopaedia. (n.d.). Retrieved from https://www.encyclopedia.com/environment/encyclopedias-almanacs-transcripts-and-maps/sacred-space (Last accessed: 14.04.2020).

A Brief History of Historical Sacred Spaces

This notion of Sacred Space will take on a different meaning for readers in this book. This part specifically showcases examples from a variety of cultures and belief systems with of historic significance that are among some of the most studied works in the field of architecture. This part begins with research that that documents and virtually reconstructs the original spatial and artistic features of rare fourteenth-century religious buildings, focusing on, in particular, the frescoes. Methods of augmented reality are described, as a comprehension tool, which is then useful for a more active fruition.

Moreover, and due to new political conditions resulting in a national Italian unity, ancient religious monuments were seen in a different light. The restoration movement of the ancient churches that followed meant that the antiquity of construction was to be celebrated. The chapter titled "From Amalfi to Pompeii, Utopia of the Twentieth Century Sacred: The Architectural Epithelium for The House of God" deepens the understanding of the creation of sacred spaces in Campania region and the use of architectural languages that recall styles of the past through a synthetic exposition of three cases studies: Amalfi, Nola, and Pompeii.

The chapter "A Preliminary Structural Analysis of Typical Arches of Italian Gothic Churches" the behaviour of gothic arches under particular conditions is focus on. In their research, the authors have adopted a kinematic limit analysis approach to detect mechanisms and horizontal static multiplier for three different typologies of Gothic arches, namely drop, equilateral and lancet. The aim of the study is to evaluate the mechanical behaviour of these typologies, such as those of arches observed in typical Italian Gothic churches, by implementing an automatic kinematic approach. Outcomes obtained established a significant dependence on the main geometrical features of the investigated vaulted structures.

Finally, the part is wrapped up with the chapter "Changing the History of Architectural Heritage Palaeologan Renaissance—The Style that Never Was?" the Author provides a re-interpretation of the meaning and significance of the Serbian Medieval monastic heritage, followed by its cultural appropriation by the new Albanian authorities. It contributes further to deepening and severe inter-ethnic distrust existent between the Serbs and Albanians who share the same territory. The article juxtaposes modern attempts to re-write the meaning of religious heritage in Serbia against historical and political background in situ. The paper provides insightful conclusions regarding the relationship between cultural re-appropriation and scientific endeavours.

Re-construction and Virtual Fruition of a Fourteenth-Century Religious Building

Lia Maria Papa, Giuseppe Antuono, and Antonello Cerbone

Abstract

Old religious buildings in many urban centres often contain artistic treasures and perceptive enchantment. The latter trigger a journey of comprehension and involvement that users would be prompted to take if only the often precarious state of conservation and transformations in the architectural context did not prevent the broadening of the material environment, thereby precluding immersion in a spectacular, evocative space. The combination of several descriptive and communicative modes and a sensible use of the latest acquisition and digital elaboration technologies represent a fertile field of research and experimentation, especially as regards cultural heritage. In fact, they make it possible to generate images we could call "talking architectures"; these are images that visualise history and revive environments by creating temporally dynamic instruments that can critically manipulate data and develop tourist-immersive proposals. These tools include virtual and augmented reality; gesture-based interaction; localisation of multimedia devices and technologies; creation of itineraries combining real and virtual elements and turning fruition of a monument and all its decorative features into an active, engaging experience rather than a purely passive event. Within this methodological framework, the contribution illustrates the research conducted as part of the Workshop of Survey and Modelling (ReMLab) of the Department of Civil, Building and Environmental Engineering of the University of Naples. The study documents and virtually reconstructs the original spatial and artistic features of a rare fourteenth-century religious building—the Church of S. Maria dell'Incoronata—and what is left of its exquisite frescoes by the School of Giotto.

Keywords

Cultural heritage • Fruition • Enhancement • Immersive reality

L. M. Papa (✉) · G. Antuono · A. Cerbone
DICEA Università di Napoli Federico II, Naples, Italy
e-mail: lmpapa@unina.it

G. Antuono
e-mail: giuseppe.antuono@unina.it

A. Cerbone
e-mail: antonello.cerbone@unina.it

1 Introduction

Fresco is the most important pictorial technique used in wall paintings; after the thirteenth century, it was adopted extensively in Italy, especially in sacred buildings, to create extremely symbolic and spiritual, pseudo-perspective, polychrome images. Highly saturated colours without half-tones, with reference to the light and shadows generated on painted surfaces, bestow the expressive force of the "metaphysics of light" that characterised works by the School of Giotto in Angevin Naples where Giotto was active during the last years of his life and where he inspired numerous collaborators and followers. These works illustrate a faith that is logical construction, in which there is not necessarily a yearning to mystically identify the divinity, but dialectical clarity, inner control and adherence to everyday life always experienced as the expression of moral rigour.

Before perspective theories succeeded in producing absolutely rigorous and visually coherent images, several architectural representations with intuitive perspective connotations have been documented, for example in the Church of the Incoronata with its successful narrative of symbolic and allusive scenes. However, none of the frescoes (Figs. 1 and 2) that were present and conferred meaning and expressive importance to this space are still present on the wall surfaces of the Church. This superb architecture, built by Joanna I in the mid-thirteenth century presumably over the rooms of the Angevin court (Bernich 1904, p. 101), was later completed with the addition of a hospital. At present,

C. Gambardella et al. (eds.), *Advances in Utopian Studies and Sacred Architecture*,
Advances in Science, Technology & Innovation,
https://doi.org/10.1007/978-3-030-50765-7_6

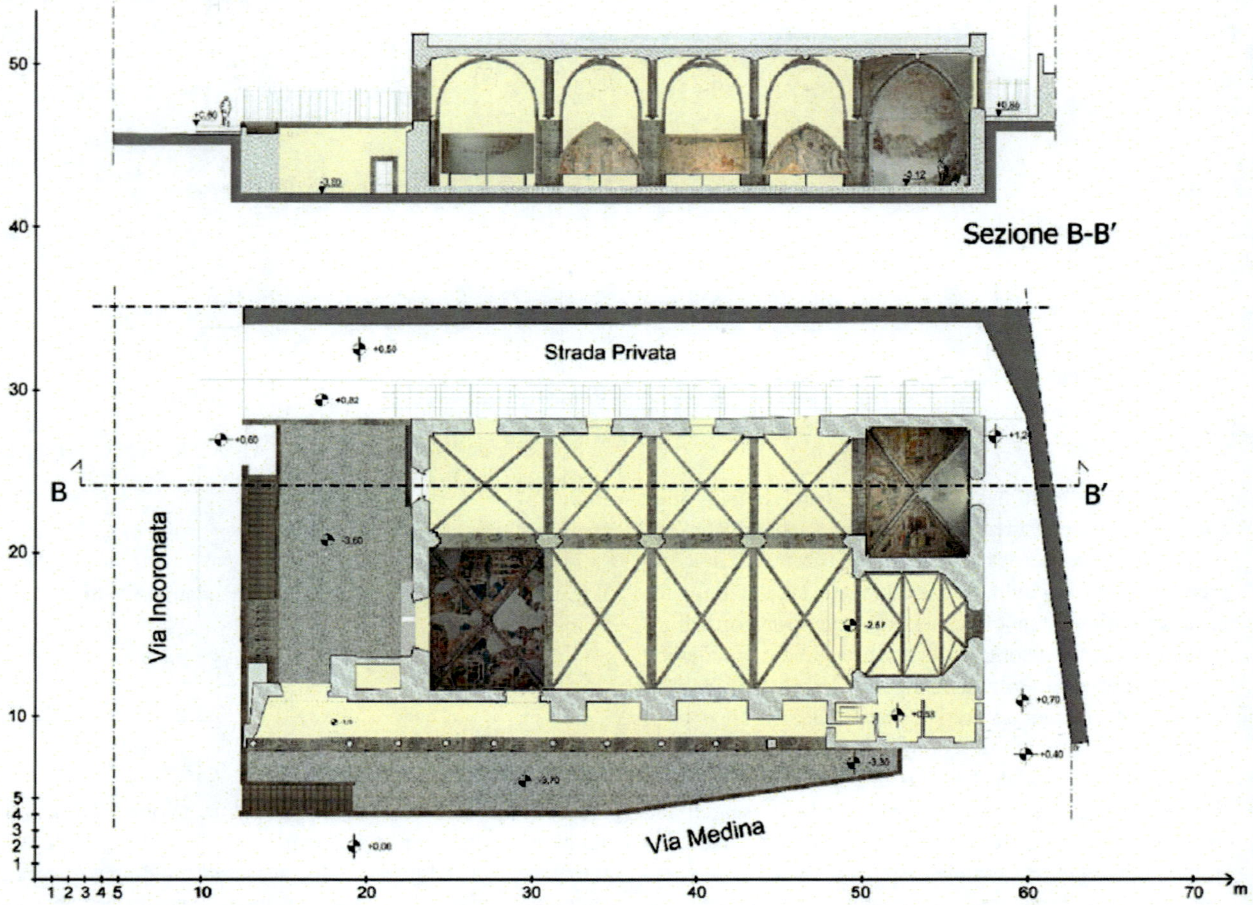

Fig. 1 Church of the Incoronata: plan, with the projection of the vaults, and sections. The drawings show the current location of the remains of the decorative panels removed during restoration

the Church is used only temporarily (Papa and Antuono 2016, pp. 1529–1534).

1.1 The Cycle of Frescoes Between Liturgy and Iconographic Tradition

The objective of the study was to make the site accessible in compliance with architectural and conservative constraints; its first target was to develop ameliorative design solutions and strategies.

The combination of technological innovation and cultural heritage is undoubtedly the *focus* that inspired the idea of achieving broader fruition of an asset currently lacking most of its chromatic-figurative components. The digital model produced by a critique and interpretation of the object in question, coupled with its figurative–sensorial components, establishes virtual contact with the chromatic and symbolic importance of the fourteenth-century sacred space of the Church. This process triggers communication with and knowledge of an asset which is now merely a closed container of the historical, architectural and artistic memory of the past.

Virtualisation and careful reinterpretation of archival and documentary material enabled the identification of most of the figurative–formal pieces and the order in which they were to be interpreted; it also led to their virtual fruition and full comprehension of the narrated episodes.

Obviously, any analysis of an artistic work must start with knowledge, in other words with an initial critical analysis of its history, as well as survey and measurement; these actions make it possible to draft the ensuing analytical documentation. Photogrammetric survey techniques were used to reconstruct the scenes and pictorial images; this enhanced comprehension of the complex relationship between real space, pictorial image, context and the artists' concept of space. It also facilitated a more complete iconographic interpretation since comprehension of the complex

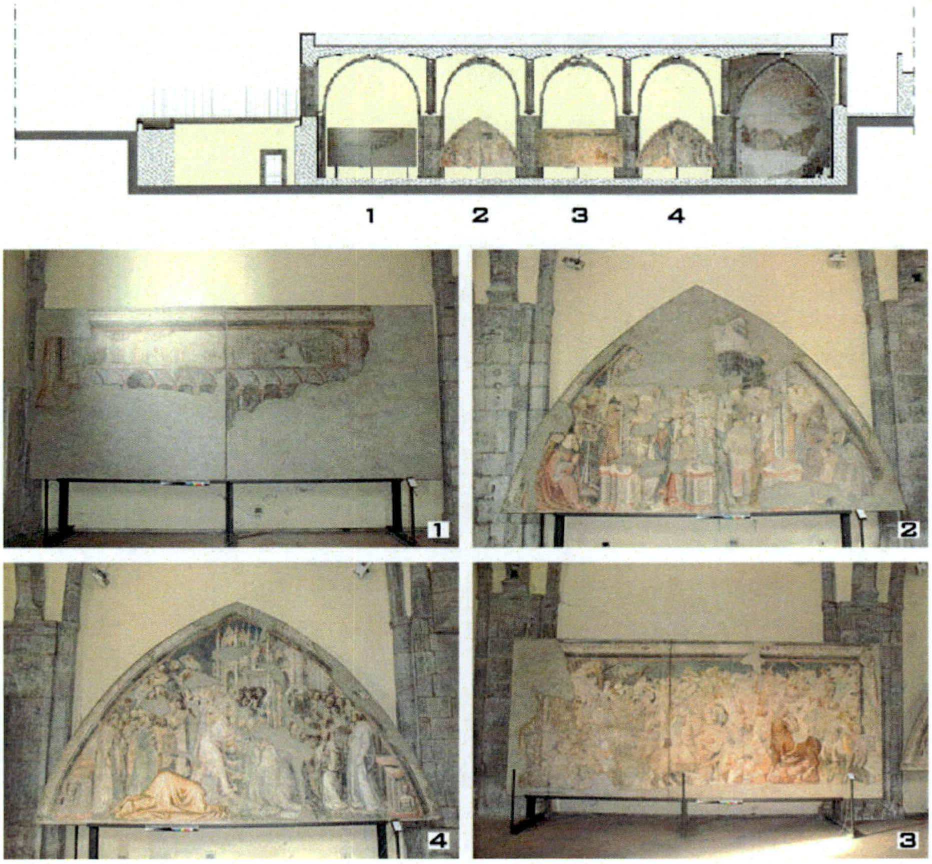

Fig. 2 Section showing the current location of the restored decorative panels

but extremely fragmented figurative scheme in the Church of the Incoronata does not facilitate contextualisation. In turn, what is still visible in the first span, next to the main entrance, was part of a much more extensive decorative scheme extending throughout the whole Church; this is evident in the traces of decoration in the adjacent vaults and was detected during the last restoration projects (Chierici 1930). Today, only the first span of the main nave (Fig. 3) and the last span of the side nave, the so-called Chapel of the Crucifix (Fig. 4), illustrates Christianity in fourteenth-century Naples.

Once past the entrance portal, the main theme of the frescoes is the "Triumph of the Seven Sacraments and Religion", the first known example of transposition of this theme in a monumental painting. Celebration of the "Sacraments" represents the most profound and ancient form of participation of the faithful in Christian rites: "Baptism" and "Confirmation", the sacraments of "initiation", are located in the panels above the main entrance where the biggest door is located: the Church entrance coincides with the beginning of the sacramental path. Baptism and Confirmation are followed by "Eucharist" and "Penance", the latter in the double form of auricular confession and corporeal penitence, and the two so-called free or voluntary sacraments, Marriage and Holy Orders, because there is no absolute need for these sacraments (Longère 1975, p. 275) and, unlike the others, are administrated like those of "initiation", only once in a lifetime. The cycle ends with the "Anointing of the Sick" and the "Triumph of the Church": this latter scene, with the Ecclesia supporting the chalice, represents the beginning and end of the cycle.

There are no forms of dramatic concitation in each narration, just a calm definition of methods, a conscious combination inputting into the effectiveness of the composition of the scene. Every panel of the groin vault depicts a moment of the Church dispenser of sacraments: in "Baptism", the newborn is carried by three godfathers while the father holds the candle to the right; in "Confirmation" the child is instead accompanied by his mother, and in her arms is anointed on his forehead by the Bishop; in the "Eucharist", the faithful participates in the sacrament together with the communicants, kneeling with folded hands, who receive the host from the priest; in "Penance", there is a representation of three flagellants; in the "Marriage" scene, the numerous individuals in the procession accompany the bride and groom, undoubtedly of high birth given their elegant

Fig. 3 Digital reconstruction of the first span of the main nave and its figurative images

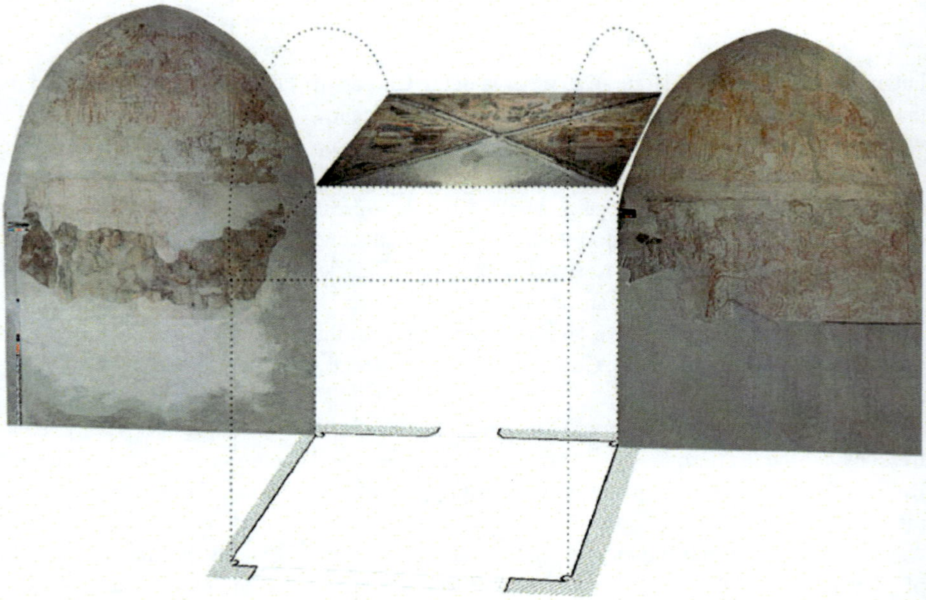

Fig. 4 Axonometric composition of the Chapel of the Crucifix with its figurative images. The sinopias became visible after removal of the frescoes, now positioned and exhibited along the side nave of the Church

clothes and precious ornaments, earrings and elaborate headdresses; in "Holy Orders", all the orders are present in front of the pope; in the "Anointing of the sick", the sick person, who appears rigid and vaguely unconscious, is held up by a relative while the priest, next to his assistant holding a lighted candle, anoints his senses and a group of women pray around him.

In the "Sacraments", the author Roberto da Oderisio, inspired by Giotto, but also influenced by the work of the School of Siena, inserts several iconographic solutions typical of Giotto's figurative repertoire: the choir of the singing monks in the "Holy Orders" and excerpts from other contexts; the 'Washing of the newborn" in the Baptism, normally present in scenes of the Nativity; the confirmant child caressing his mother, reminiscent of Jesus' gesture towards the Madonna in several fourteenth-century panels. The theme of the "Sacraments" is the climax of a narration that continues downwards, with symbolic rather than narrative traits, and illustrates episodes of "Biblical Stories" which are however fragmented and have no coherent narrative theme.

There is one particularly complex element in the unitary interpretation of the cycle of frescoes in the Chapel of the Crucifix, executed by an artist from the Marche region known as the Master of Stories of Saint Ladislaus. In the early fifteenth century, the artist painted the "Stories of the Virgin" in the panels of the vault, the scenes of the "life of Saint Ladislaus" on the walls and the "entrustment of the hospital Church of the Crown of Thorns" to the Carthusians by Queen Joanna I (the religious complex formerly acted also as a hospital). Although very little of the original decoration is still in place, the fact that the themes and level of interpretation are quite complex is more than noticeable: one involves the fourteenth-century phase, the other concerns the next phase, the early fifteenth century. Two distinct but perfectly coherent phases interconnected by numerous cross-references successfully conveying the building's double role. The themes involve the Crown of Thorns, source of salvation for men, in line with what was then the role of the Church as well as a symbol bestowing holiness on the temporal queenship of its foundress.

The main portal—with its hunting and grape-gathering scenes, the Eucharistic lamb on one side and the crown of thorns held up by angels on the architrave—is a powerful conceptual synthesis of the Christian message represented in the unitary design of the decorations inside, the starting point for a path of initiation and education towards Christianity.

1.2 Image Acquisition Methodology and the Geometric-Perspective Analysis of an Utopian Reality

A research and knowledge-gathering strategy to support the configurative analysis of the pictorial images was necessary in order to provide a more in-depth critical assessment of the figurative and symbolic images in the sacred area of the Church. An initial documentary and archival research phase was followed by acquisition of the metric and photographic data required to develop a photogrammetric model that could accurately establish the spatial features of the areas in question.

A digital *Reflex Canon EOS 600D* camera with a *EFS 18-55mm* lens and a *Reflex K&F Concept TM324* tripod were used in the survey campaign. The instruments were chosen and used to obtain high resolution images of the flat and vaulted surfaces, including a special focus on the survey of the colours; in fact a *Kodak Grey Scale* and a *Kodak Colour Control Patches* were inserted in each frame to mitigate the dominant colours. The survey, also as regards the instruments used, concentrated in particular on the acquisition distance in order to ensure a superimposition of at least 60% of the ensuing frames. Dedicated software (Agisoft Photoscan) made it possible to create a digital model complete with photographic textures subsequently used to obtain post-modelling elaborations in support of the geometric-perspective analysis of the frescoed surfaces (Figs. 5 and 6).

Let us now focus on the cycle of frescoes in the Chapel of the Crucifix (Minieri 1845). Note the very precise geometric partition of the side walls. Each pictorial composition is divided in two superimposed registers with scenes in which the human element prevails over the architectural element. This complicates identification of reference elements indicative of an intuitive perspective; instead the very few geometric elements and the connotation and position of the figures provide food for thought. For example, in the story "Saint Ladislaus going to Church for the coronation" (Fig. 7), the only architectural elements that can help establish the perspective construction are the bases in the foreground, on either side of the scene and another pair of bases behind them. Analysing these elements reveals the use of a multifocal approach; the lines orthogonal to the picture plane establish the lower part of the base in the foreground and converge in a central vanishing point on the vertical axis of symmetry of the painting; all the other lines defining the upper part of the base, or the one behind it, reveal a construction with several vanishing points in the upper left part of the fresco.

This kind of perspective does not correspond to codified rules, but does make it possible to govern optical depth and thereby preserve the recognisability of the painted space and figures present, each with their own roles and meanings. It was impossible to consider the geometry of the lower part of the bases because they are partially hidden by the fake frame around the fresco. The artist perhaps intended to compose, organise and give visibility to the figures in the scene as well as confer depth on the architectural–figurative space.

The author prefigures an intuitive-symbolic perspective, less rigid than an arrangement with a single vanishing point and capable of nuancing the foreshortening in the pictorial sections furthest from the focal point of the scene. Furthermore, the final effect is probably also related to the way in which the fresco is viewed and interpreted; the scene was

Fig. 5 Three-dimensional reconstruction of the current configuration (below) and original configuration (above) of the frescoes on the side walls of the Chapel of the Crucifix

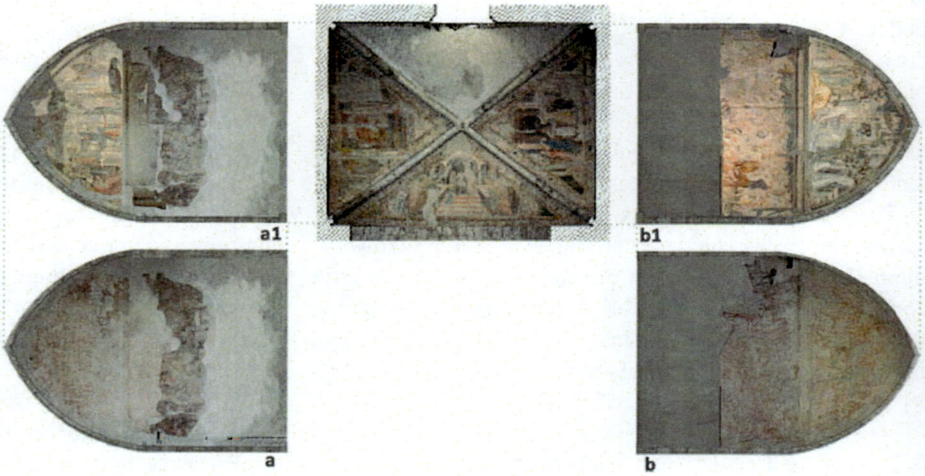

Fig. 6 Plan of the ceiling with rabattment of the side views of the Chapel of the Crucifix in the current configuration (**a** and **b**) and the virtual reconstruction (**a1** and **b1**)

originally painted on the side wall to the right of the entrance to the Chapel of the Crucifix: the author portrays most of the actors in a very forward position to the left of the scene, arranging them in a predetermined order on a virtual inclined plane. His intent was to ensure visibility when the scene was viewed from below and from the side.

Although the connotation and position of the figures convey simple messages, they reflect composition control techniques that make the episodes look as if they were designed as choreographies. The effect of depth and spatial dilation is also enhanced by the shrewd use of colours, brighter in the foreground and then gradually less brilliant.

Elements of geometric–compositional continuity are also present in the partition of "Saint Ladislaus venerating the crown of St. Stephen", part of "St. Ladislaus' battle against the Uzes" and in "Saint Ladislaus going to Church for the incoronation" that complete the pictorial cycle of the side walls of the Chapel of the Crucifix. While in the former, the scene is architecturally framed, and in the other two the perspective layout and narrative pathos are emphasised by the position and grouping of the figures as well as by the use of colour. Here again, the perspective construction appears to be linked to the perception of space vis-à-vis the observer's position and where he is located in the sacred space.

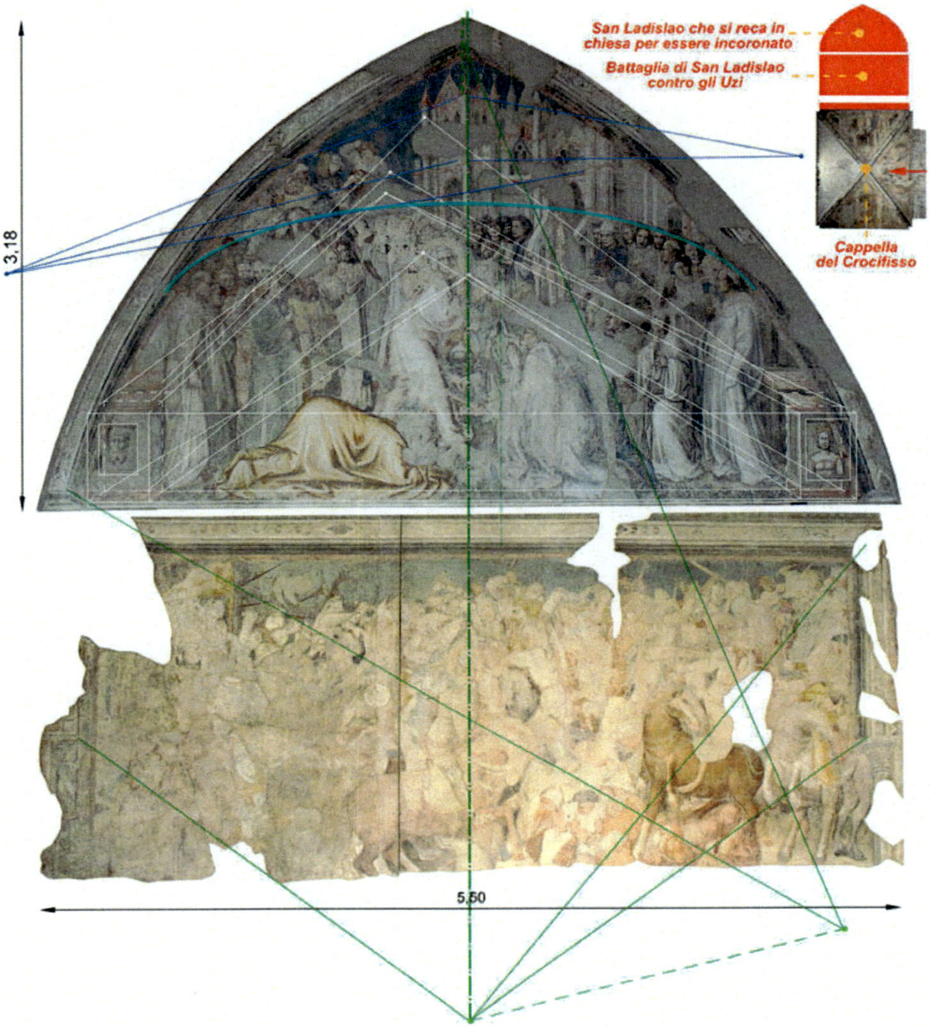

Fig. 7 Geometric-perspective analysis, in relation to the fruition mode, of the frescoes "Saint Ladislaus' battle against the Uzes" and "Saint Ladislaus going to Church for the coronation" in the Chapel of the Crucifix

This figurative choice is adopted, in particular, in the fresco on the left side wall depicting the old portico of the Church along Via Medina which can be seen from the main nave and seems to almost emerge from the plane of the fresco.

The relationship between the pseudo-perspective representation and the observer's position is particularly obvious even in the frescoes in the first span next to the entrance portal. Every story, considered as a instrument of religious persuasion, is embellished by an iconic–symbolic component and is positioned so as to ensure maximum visibility and easy interpretation.

Only the fake frames link the pictorial cells to the built space; thanks to their shadows, these frames appear three dimensional and establish visual and stylistic continuity with the Chapel, just like the stories painted by Giotto for the Basilica in Assisi or the frescoes in the Scrovegni Chapel in Padua (Fig. 8).

In some ways, this elegant expedient implemented between 1403 and 1414 by the anonymous artist from the Marche region links pictorial and real space; it appears to figuratively preceed Alberti's definition of a painting considered as a "window from which the historia is seen" (Grayson 1980, p. 7). The pictorial surface is no longer a limit but becomes a filter/threshold between reality and virtuality (Feiner 2002), required to establish the optical/perceptive relationships that were soon to be scientifically established.

This sacred space contains an ensemble of images that should be interpreted one after another like the pages in a book, in other words they are structured in a perfectly logical sequence: the use of an intuitive perspective sustains the goal of the narrative hidden behind the iconic images alluding to an ideal religious space characteristic of fourteenth-century culture (Galante 1872).

Fig. 8 Left "St. Francis in front of the Sultan (or Test of Fire)" the eleventh of the twenty-eight scenes in the cycle of frescoes in the Stories of St. Francis in the Upper Basilica in Assisi, attributed to Giotto; right, the compositions of the Triumphal Arch in the Scrovegni Chapel, with the following enlargements: "The Angel of the Annunciation", "Annunciation to Mary", "Judas' Betrayal" e "Visitation"

1.3 Augmented Reality, a Comprehension and Fruition Tool

Multimedia-interactive displays providing visual navigation of a digitally recreated simulated reality (Brusaporci 2006, pp. 16–21) satisfy the need to enhance and communicate precious historical–figurative assets linked to changes in the Church's interior, from its construction and alterations made over the years. Mobile systems and interactive applications create new perspectives and lead to extremely interesting applications in the field of simulations and virtual reconstructions (Empler 2015, pp. 60–69); they are also very helpful during restoration projects. If on the one hand, these visualisation instruments reveal details which would otherwise be difficult to see using two-dimensional representation, on the other there is still much to be done in studies and researches to reconstruct and visualise the reproduced scenes and chromatic data, especially as regards applications in which the original image has to be reconstructed (Ippoliti and Meschini 2010). Involvement in this experience and degree of interactivity are proportional to the quality of the sensorial information provided to users by the technological device in order to capture their attention and enhance the acquisition of knowledge in the illusory space in which they are immersed (Bisogni 2014).

The digital model is therefore integrated in the field of cultural heritage; by combining the virtual reconstruction of past environments, and enabling comparison between what exists and what existed, all users can interpret the changes wrought by time. It is still possible to be amazed by the recomposed cycle of frescoes in the intrados of the panels of the spans of the naves of the Church of the Incoronata. Only some of the figurative images removed for restoration have been replaced; however, the chromatic compositions of the ones *in loco* reflect the surrounding architectural forms and create overall environmental features that are often precious and unique, whether observed close-up or at a distance. Documenting and understanding these images can, albeit virtually, allow reinsertion in a reality devoid of its original figurations and become part of a unitary recomposition of the artwork as close as possible to the spirit of the place.

The exemplification implemented in the ReMLab advances in synch with the theme of "integration": integration of the omissions in documentary evidence, integration of communication, integration of our culture of accessibility (Papa et al. 2014). The feeling of immersion and presence in the recreated environment is even more meaningful when the asset can only be enjoyed episodically.

The repositioning of the recomposition of the cycle of frescoes in an application that can be used immersively is the last step in a process involving an initial phase of data collection and elaboration from an onsite survey and archival research. Together with the interpretation of the signs and traces still visible on the wall surfaces, this data enabled not only the reconstruction of the three-dimensional model of the Church, used for virtual fruition, but also simulation of an augmented reality application (Fig. 9), useful to enhance the user's sensorial perception thanks to data that in general is manipulated and electronically conveyed. The generated flow of video live images (taken by a smartphone or tablet) enabled modification of the vision of space and added virtual contents and animations in real time (Simonetta 2012).

The AR application uses computer vision techniques to determine the position and perspective of the user in the real world. This position can be absolute (a specific place in the world) or relative (when the position is deduced based on the

Fig. 9 Example of the augmented reality application, with a smartphone, to reinterpret the original configuration of the frescoes in the Chapel of the Crucifix

presence of a certain element). To activate computer vision, the sensor needs a camera that "sees" the real world and, based on the information it receives, can determine its position and direction compared to the overall scene; then the software analyses the images recorded by the camera and "positions" the virtual elements in the surroundings. When creating the augmented effects, called "AuRAs", it is important to establish the initial image that acts as a trigger as well as the multimedia contribution that will be overlaid on it; in this case the former are the sinopias actually present on the wall surfaces of the Chapel of the Crucifix while the latter are generated by the photogrammetric reconstruction of the restored frescoes currently positioned on special easels along the wall of the side nave (Fig. 2).

The new visualisation tools allow to recover the chromium, the spatiality and the symbolic value of the sacred fourteenth-century environment of the Church of the Incoronata, thanks to the virtual relocation of details utopically perceptible in reality that allow you to fully understand the episodes told. The implementation of such a guided route aims to lead the visitor through a utopian journey aimed at re-reading the frescoes, with an extensive multi-sensory approach between amusement and edutainment (Cervellini and Rossi 2011), in other words a source of education and comprehension of the historical and cultural value of the Church and its pictorial images.

1.4 Conclusions

Pursuant to the study of the multiple relationships between architectural and figurative elements, the virtual simulation experience is an original form of immersive communication enabling assessment of the most appropriate management tools required to communicate the material and immaterial components currently either partially compromised or lost. Integration between these descriptive and communicative modes contributes to the acquisition and dissemination of unusual data (Papa 2014).

The objectives of protection, conservation and requalification of historical–architectural heritage can be achieved by using modern digital technologies; the latter provide incredible options to visualise and "narrate" cultural assets based on the multiscalar organisation of knowledge and the construction of temporally dynamic tools that can manipulate data and structure tourist-immersive proposals. These technologies can also be increasingly used to provide disciplinary assessments and execute experiments regarding the effectiveness of communicative strategies, including from a remote position.

In this context, augmented reality, during the drafting of visitors' itineraries combining real and virtual elements, undoubtedly turns the fruition of a monument into an active and captivating experience rather than purely passive event.

The contribution was developed, in equal measure, by the authors L. M. Papa and G. Antuono, with the collaboration to the editing of the images of the Ing. Antonello Cerbone.

References

Bernich, E. (1904). La chiesa dell'Incoronata. *Napoli Nobilissima, 13.*

Bisogni, M. (2014). *Realta aumentata per la comunicazione di prodotto*. Milano: Tecniche nuove.

Brusaporci, S. (2006). *Sistemi informativi integrati per la tutela la conservazione e la valorizzazione del patrimonio architettonico e urbano*. Roma: Gangemi.

Cervellini, F., & Rossi, D. (2011). Comunicare emozionando. L'edutainment per la comunicazione intorno al patrimonio culturale. *Disegnarecon, 4*(8), 48–55.

Chierici G. (1930). Il restauro della chiesa dell'incoronata a Napoli. *Bollettino d'arte*, II, pp. 410–423.

Empler, T. (2015). *APP design con uso della realtà aumentata per la divulgazione dei Beni Culturali = APP design using augmented reality to disseminate Cultural Heritage*. Roma: Gangemi Editore.

Feiner, S. K. (2002). Augmented reality: A new way of seeing. *Scientific American, 286*(4), 48–55.

Galante, G. A. (1872). Guida sacra della città di Napoli. Stamperia del Fibreno.

Grayson, C. (1980). *De pictura / Leon Battista Alberti*. Bari: Laterza.

Ippoliti, E., & Meschini, A. (2010). Dal "modello 3D" alla "scena 3D". Prospettive e opportunità per la valorizzazione del patrimonio culturale architettonico e urbano. *Disegnarecon, 3*(6), 77–91.

Longère, J. (1975). *OEuvres oratoires de maîtres parisiens au XIIe siècle*. Paris: Études augustiniennes.

Minieri, R. C. (1845). *Saggio storico critico intorno alla chiesa dell'Incoronata di Napoli e suoi affreschi*. Napoli: Tipografia di Vincenzo Priggiobba.

Papa, L. M. (2014). Strumenti operativi e strategie descrittive. In L. M. Papa (Ed.), *Il disegno delle trasformazioni*. Napoli: Clean.

Papa, L. M., & Antuono, G. (2016). Il disegno tra conoscenza e fruizione. In S. Bertocci & M. Bini (Eds.), *Le Ragioni del Disegno*. Gangeni editore: Fienze.

Papa, L. M., D'Agostino, P., & Antuono, G. (2014). Urban spaces accessibility and improvement. An operative proposal. *Agribusiness Paesaggio & Ambiente, 17*(3), 242–247.

Simonetta, G. (Ed.) (2012). *Realtà aumentate. Esperienze, strategie e contenuti per l'Augmented Reality*. Milano: Apogeo Education.

Check for updates

From Amalfi to Pompeii: Utopia of the Twentieth Century Sacred. The Architectural Epithelium for the House of God

Saverio Carillo

Abstract

In the regional context of Campania from the second half of the Nineteenth century and for the following hundred years appears very conditioned by the use of architectural languages recalling the past. Despite the use of modern technologies for the building site, the final finishing layer of the architecture is inspired by the recovery of ancient ways and figural renditions of the architectural surfaces. Finishes in marble, entire wallsplastered with stucco imitating selected types of stones, soffits vaults decorated with frescoes, apses decorated by mosaics, paving made inspired by inlays with stones, roofs decorated with gilding, all these elements display an image of architecture conditioned by formal values directly derived from the past. The paper intends to examine some Neapolitan experiences to highlight technical skills and competences recovered from the more traditional technical culture of the Territory.

Keywords

Amalfi • Pompei • Nola • Bologna • Firenze • Napoli • Sacred architecture in nineteenth century

1 The Nineteenth Century: Religion, Homeland and Monuments

With the birth of the Italian state, following the struggles of independence, which allowed the creation of a unitary state in 1861, a new sensitivity in interpreting the ancient monuments of the nation also developed. In Italy, there is a strong movement of recovery and reworking of the ancient churches where the antiquity of construction was celebrated through restoration work.

The same restoration work was then supported by the widespread nationalist ideology aimed at claiming the important role of the new unitary State and with it of the individual local communities.

The attention that in the European context is dedicated to the historical revaluation of the Middle Ages, in Italy allows to focus attention on the spiritual value of religious architecture. Similar attention to the constructive capacities of the Middle Ages also appears for civic buildings such as Towers and Public Residences that define the cultural season that Italian historiography calls "Municipal Age".

In the regional context of Campania, some cities during the nineteenth century carried out an interesting job of recovering their civil or religious history with the explicit intention of claiming a modern role to be assigned to the single city within the framework of the new configuration that the territory assumed (Fig. 1).

2 The Facade of the Cathedral of Amalfi as a Monument of the Identity of Its Community

The case of Amalfi is very interesting precisely because of the great value that is attributed to the history of the city in relation to its ancient role as a maritime republic (Abbate et al. 2001). The rebirth of the Middle Ages for the coastal city becomes an occasion to remember its international role and its status as a historical city of considerable value that saw it compete, centuries ago, with cities like Genoa, Venice and Pisa (Fig. 2).

The ancient role for Amalfi of being one of the Four Maritime Republics, in the historical context of the Italian nineteenth century, was an interesting premise for a new future full of important expectations of prosperity. In this scenario the city had to identify, in a representative building,

S. Carillo (✉)
Università degli studi della Campania "Luigi Vanvitelli", Caserta, Italy
e-mail: saverio.carillo@unicampania.it

C. Gambardella et al. (eds.), *Advances in Utopian Studies and Sacred Architecture*,
Advances in Science, Technology & Innovation,
https://doi.org/10.1007/978-3-030-50765-7_7

Fig. 1 Amalfi, Sant'Andrea church, new facade under construction (1899)

the load of the history that allowed it to legitimize its role as an authoritative and prestigious community (Camera 1999).

The painful circumstance of the collapse of the portico of the cathedral in 1861 with the death of an elderly lady, offered the original reason to undertake the restoration of the frontispiece of the Cathedral. The particular environmental condition of the church, situated on a rocky outcrop with an altitude of about twenty meters above sea level, allowed to

Fig. 2 Amalfi, a panorama of the coast (1899)

define a new and more scenographic condition of elevation of the whole environment.

In the future capital of elite tourism there is an important series of conditions that allow for the creation of an urban landscape of great scenographic value, such as receiving the immediate success of popular consensus (Fiengo 1994). A new city shows itself to the men of the new century, the twentieth century, the century of modernity, despite the orographic condition of the site making it difficult to reach the city by land. The city also, with its choice to create a new façade for its most famous monument, identifies in its landscape—and in its landscape seen from the sea—the promotional tool of its future economy.

The cathedral of Amalfi with the construction of the new portico and the construction of the new facade designed by Errico Alvino chooses to consecrate in that monument, the symbolic image, and synthesis of itself (Fiengo 1991) (Fig. 3).

The stylistic elements of the medieval tradition are used to represent the greatness and historical value of this interesting architectural work. The Amalfi Middle Ages are not limited to recovering forms of the past, it invent new ones; is a language used in the composition of elements and forms of architecture that recall the relationships that the city has had with the East. Elements of Arab and Mediterranean culture are proposed in newly designed or restored buildings within the city and territory of the Amalfi Coast (Carillo 2018).

In the same years in which the restoration was carried out, according to stylistic languages of the cathedral of Amalfi, other important interventions were carried out in Italy over historic monuments of the nation. In nearby Naples, for example, the Cathedral of the city, by the same architect Alvino, is given a new façade that will complete the works to expand the new urban road that passes in front of the Duomo (Fiengo 1993).

Among the important national cases for the commitment of historical sacred architecture, we must mention the Cathedral of Florence, the cathedral of Milan and the Cathedral of San Petronio in Bologna (Zucchini 1933) (Fig. 4).

Along with this widespread interest in sacred architecture, a new sensibility also develops that reads, in these operations, a value of utopia for a renewed image of the city. Within European culture, a mode of attention to historical architecture is developed both in terms of identity and in the utopian perspective of the new planning for the future.

Fig. 3 Bologna, San Petronio, facade

New phenomena, even from a social point of view, are on the horizon. New expectations for the weaker social classes reveal new dynamics in relationships.

Sacred space seems to represent one of the architectural themes that best incorporates the contents and expectations of the new social utopias that are being defined.

Events considered exceptional, as the apparitions in the grotto of Lourdes, generate renewed interests for the spiritual datum, as well as historical research on ancient Christian foundations of the apostolic age, generates archaeological research and feeds the design vocabulary for the new figurative languages.

3 The Reconstruction of Nola Cathedral Following the Unification of Italy Is Exemplary

Built on an area characterized, since Roman times, by a specific cultic vocation, it was destroyed by arson in February 1861. The incident caused such a stir that, just four days later, the King's Lieutenancy for the southern provinces issued a decree for the reconstruction of the building (Carillo 1989).

Fig. 4 Bologna, San Petronio, project a new facade (Rubbiani, Collamarini)

The interest of the new unitary state for the Nolan church is difficult to understand both for its substantial extraneousness to the southern events and the hostility to the Catholic reality (Toscano 1998).

At first, the work was entrusted to the engineer Francesco Giordano, who, as has been widely documented (Carillo 1993), designed a three-nave basilica with six bays in each aisle, one more than the ancient monument.

The fire that had destroyed the Cathedral of Nola assumed, therefore, a clear political meaning, to be connected with the conflicts that had generated, in 1861, the new State Organization. The architectural consideration for which no surviving element had been recovered from the ancient building, destroyed by the fire, allows us to consider how, with the help of public funds, in a time of extreme crisis, the city intended only to acquire a new religious monument. In this way the city of Nola, the site where the philosopher Giordano Bruno was born, inaugurated a new path of life open to modernity (Jannelli 1998).

On the architect's death in 1878, the assignment passed to Oscar Capocci, a professor and very active patriot, who was preferred to the Neapolitan designer Gerardo Rega Angelini, proposed by the Supervisory Commission. His experience was quite limited: he decided to abandon the vault and cover the central nave with a trussed roof and a coffered ceiling, an idea already proposed by the Academy of Fine Arts of Florence in 1862, taking into account the seismic nature of the area (Fig. 5).

However, in the course of the works, disturbing injuries occurred in the columns of the side aisles, as shown by the correspondence of the same architect, so it was necessary to proceed with the consolidation of the works already carried out. The same architect was held responsible by public opinion for having neglected the completion of the work, because he was simultaneously engaged in other building sites.

Therefore, in 1884, he was sent, as a trusted technician of the Ministry of Grace and Justice, to the most prestigious Italian architect of the time, Pio Piacentini, who exonerated Capocci from accusations made against him.

The works, however, remained almost blocked due to lack of funds, until in 1891 a third phase began, with the

Fig. 5 Nola, Santa Maria Assunta, facade

appointment of Nicola Breglia as head of the construction site. The success of the Breglia project is due to several factors, including the keen interest of the municipal authority that it interpreted «l'operazione "cattedrale" quale occasione per trasformare la città e, soprattutto, lo spazio antistante l'edificio» (Carillo 1993: 368) (Fig. 6).

The result was a building typical of the Vesuvian environment of the nineteenth century, characterized by a direct rereading of the ancient, with the result of an understanding and acquisition of the formal characters and an attention to the historicized image of the same (Carillo 1998).

The cathedral of Breglia has an exquisitely neo-Renaissance taste, with a basilica plan with three naves, preceded by a portico. Each chapel is designed volumetrically as a square-based prism with an approximately double-height development, topped by a barrel vault, with the lacunar intrados. The transept connects the aisles, but also carries out a distributive task between the existing

Fig. 6 Nola, Santa Maria Assunta, high altar

structures—Immacolata Chapel and SS. Apostoli Church-and the new accessory rooms, the main sacristy and the staircase for the dome and the roof (Jubilate! 1909).

Although the Breglia project is conditioned by the surviving structures—in addition to those mentioned, the crypt of S. Felice and the bell tower located to the south—the architect's main concern is to define a unitary image of the architecture. The ancient crypt was very transformed (Campone 1998) however it is also true that the style chosen by Breglia allowed the conservation of important sculptures of the XV and late Renaissance which were present in the ancient basilica.

In order to distinguish the work done by the architects, the planimetric organization of the Cathedral must be ascribed to Francesco Giordano. The completion of the construction with the neo-Renaissance style finishes were carried out under the direction of the architect Nicola Breglia (Carillo 1993).

The decorative elements (moldings, cornices, friezes, squares) immediately highlight the recovery of the ancient and, in particular, of the Pompeian model, which Breglia knew well, since, during the years of its formation, between 1858 and 1860, it had studied and inspected, for the Pensioner of Architecture (today we would say Ph.D.), the Pompeian city.

Recovering the local decorative traditions, Breglia uses papier-mâché as a decorative element that, if necessary, takes the form of "noble" materials such as marble, bronze, wood, so that the monumental sculpture of the apse, by Salvatore Cepparulo, is made in papier-mâché.

Through the intervention of Breglia, the stylistic restoration, freeing or completing of the antique, renders, at the same time, the history docile to the needs of contemporary everyday life. So the papier-mâché is used not only for its flexibility and lightness (for example, for the coffered) or for the ability of local workers to work it or for its long local tradition, but also because it can be easily aged, so that the material becomes medium communicative of the ancient (Figs. 7, 8 and 9).

4 The Shrine of Pompeii: Utopia of Spiritual Rebirth of an Archaeological City

The shrine of the Virgin of the Rosary in Pompeii was born in the last decades of the nineteenth century on the initiative of an Apulian lawyer who, after years of practicing atheism, following a religious crisis, committed himself to creating a cult place dedicated to the Virgin Mary (Gambardella 2012). An important aim of the work that the founder of the sanctuary aims at is the social redemption of the poorest strata of the rural population of the Neapolitan territory

Fig. 7 Pompei, casts of bodies found in the ancient archaeological city

(Carillo 2001). Another aspect of the lawyer's commitment, concerns the opportunity to recover the children of prisoners through job training supported by participation in an educational activity at least at the elementary level.

Following his commitment the lawyer, Bartolo Longo, in the wake of other Italian Catholic thinkers such as the Veneto industrialist Alessandro Rossi, formed a nucleus of workers' houses in Pompeii almost reviving a social fabric that was not present on the territory (Iuliano and Federico 2000).

In the same years, the archaeologist Giuseppe Fiorelli, experimented in Pompeii with the technique of "casts"; pouring into the voids, that were found in the city buried by the eruption of Vesuvius, liquid chalk that, upon consolidating, restored the bodies of the inhabitants of the ancient city.

Fig. 8 Pompei, apse of sanctuary of the virgin of the rosary

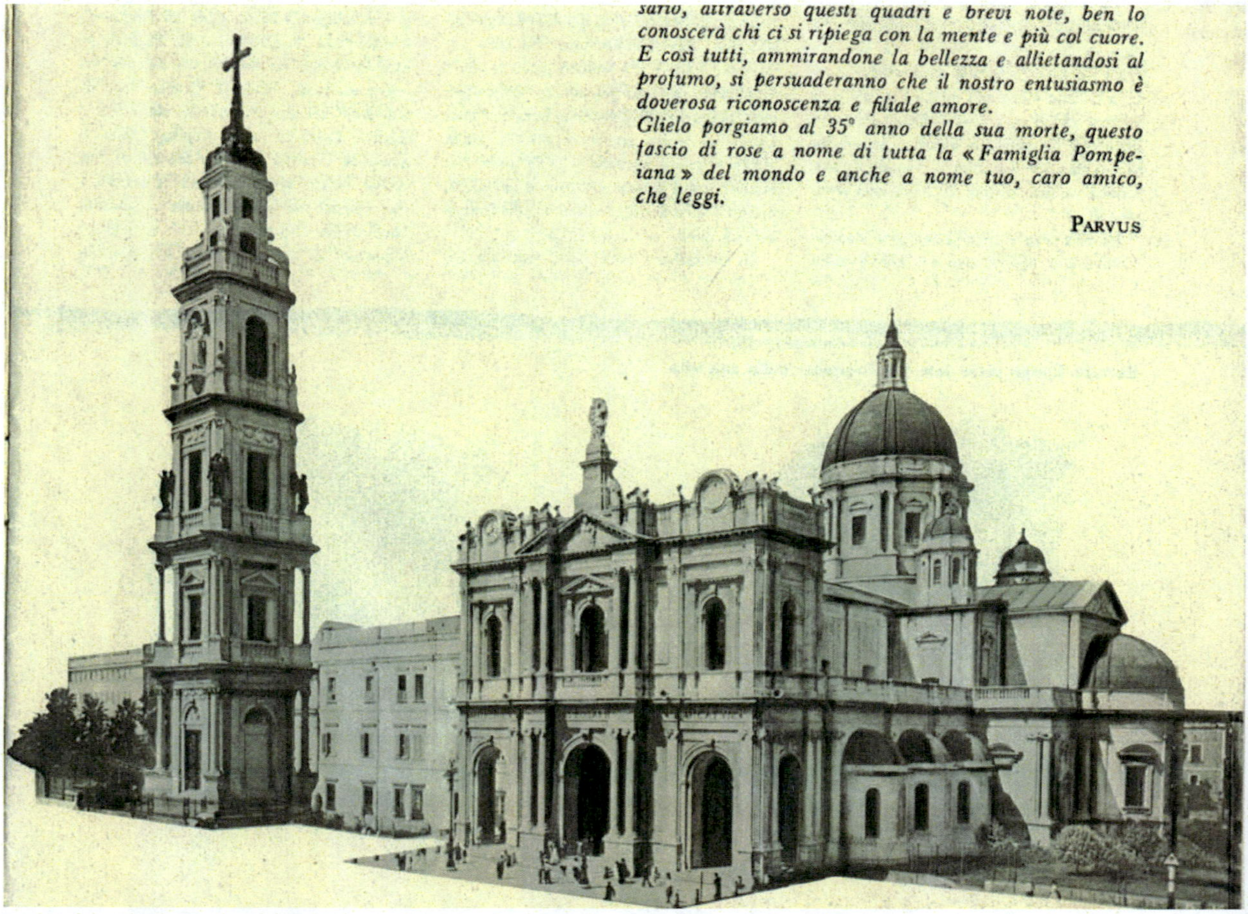

Fig. 9 Pompei, monumental facade and bell tower

Following this interesting experience, Longo commits to repopulate the territory that, inspired by the religious belief, will lead to revive the city that had been destroyed by the terrible natural event 1800 years prior. Longo's commitment, with the collaboration of the Neapolitan nobility, allowed the creation of an important religious center that philosopher Benedetto Croce called the Italian Lourdes (Fig. 10).

The expansion of the sanctuary of Pompeii is an exemplary story for the influential personalities involved and the reflection on the sacred art of the period (Carillo 2013).

The growing influx of pilgrims necessitated an intervention entrusted to Spirito Maria Chiapetta, a Lombard architect, who became a priest in adulthood (Carillo 1999). The style chosen was the baroque style, to confirm the adherence to the sacred Neapolitan tradition, in a place symbolizing paganism. At the same time, this taste satisfied the expectations of the Neapolitan nobility, who had supported the construction promoted by Bartolo Longo.

However, the designer's choice lies in a particular historical moment for architectural reflection, which, on the one hand, was controversial towards the Baroque figurative culture, and on the other, censored the stylistic completions in buildings (Carillo 2014). Thus, Croce defined the Baroque: «il barocco è una sorta di brutto artistico, e, come tale, non è niente di artistico, ma anzi, al contrario, qualcosa di diverso dall'arte, di cui ha mentito l'aspetto e il nome» (Carillo 2000: 168).

In solving the problems connected with enlargement, the intellectual dynamism of the Church of Pius XI stands out.

Chiapetta's project has preserved the nave of the small church, so as to recognize the value of "ancient", as can also be seen in the judgment of Gustavo Giovannoni,

Fig. 10 Pompei, new apse of sanctuary of the virgin of the rosary

Academician of San Luca and founder of the Higher School of Architecture in Rome: «Arduo The theme is indeed complex and complex, as we propose to dare worthy and wide form in a great sanctuary of our faith by grafting into a pre-existing construction such as the current church, which must be preserved in its essential lines as the nucleus of the new sacred building» (Carillo 2000: 166).

The architect, using modern materials, transforms the narrow church into an impressive three-nave basilica with three large transepts and a very functional presbytery area, even if it is weak in plan.

Thus, for example, since the previous dome constituted a strong element of environmental appeal, the designer, while replacing one with a changed position, re-proposed the landscape and urbanistic role, highlighting the profile of the sanctuary projected against that of Vesuvius (Carillo 2008).

Chiapetta manages to create a unitary organism, as can also be seen in the external appearance, where a basilical composition of clear Roman derivation emerges, as a probable homage to the Apostolic See, to which the complex was donated in 1906 by Bartolo Longo (Carillo, Sepe & Petillo 2010).

The "subversive" character of this experience is singular, whose designer, despite his "gothic" vocation, decides to adopt a neglected and criticized style, continuing the construction "in style", while respecting the unity of the whole (Fig. 11).

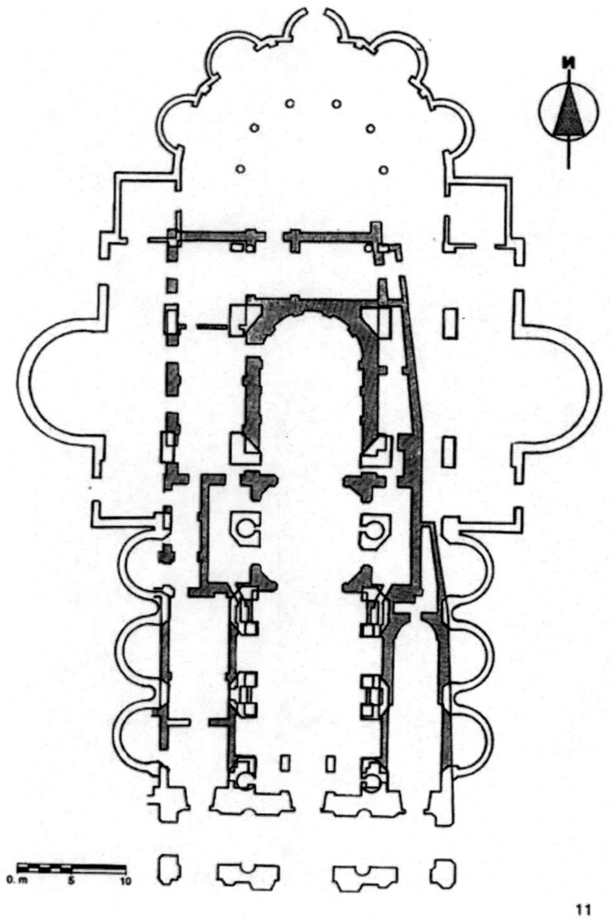

Fig. 11 Pompei, enlargement of the shrine (1933–39), plant

5 Conclusions

The synthetic exposition of the cases of historical sacred spaces subjected to transformation and restoration interventions during the nineteenth century, shows the utopian aspects, conditioned by religious values, which the Italian social context has realized after the National Unity.

These new buildings that recovered the ancient memories of the civilizations of Italian history gave an opportunity to show the ferment of novelties through which the new generation of Italians confronted the future and modernity of the new times.

References

Abbate, G., Carillo, S., & D'Aprile, M. (Eds.). (2001). *Luci ed ombre della Costa di Amalfi. I Beni Culturali ieri ed oggi* (Exhibition held in, Amalfi date 2000). Amalfi: Centro di Cultura e Storia Amalfitana.

Camera, M. (1999). *Memorie storico-diplomatiche dell'antica città e ducato di Amalfi*, Vol. I-II, Salerno 1876–1881, anastatic reprint, Centro di Cultura e Storia Amalfitana, Amalfi.

Campone, M. C. (1998). La commissione conservatrice casertana e i reperti paleocristiani dal duomo di Nola. In G. Jannelli, & S. Carillo (Ed.), Brevi cenni degli scrittori e uomini illustri della città di Nola e di quelli che si occuparono della medesima città. Nola.

Carillo, S. (1989). *La città attorno alla cattedrale. Il restauro del duomo di Nola e la sua influenza sull'assetto urbano*. Nola.

Carillo, S. (1993). La ricostruzione del duomo di Nola: 1861–1909, In G. Fiengo (Ed.), *Tutela e restauro dei monumenti in Campania 1860–1900* (pp. 355–379). Napoli.

Carillo, S. (1998). Progetti e trasformazioni urbanistiche di metà Ottocento nell'ambito della città e diocesi di Nola. Lettura dell'esperienza della "città cristiana" di Pompei. In T. R. Toscano (Ed.), *Nola e il suo territorio dal secolo XVII al secolo XIX. Momenti di storia culturale e artistica* (pp. 175–229). Castellammare di Stabia.

Carillo, S. (1999). Gustavo Giovannoni, Spirito Maria Chiapetta e l'ampliamento del pontificio Santuario di Pompei. *Arte cristiana*, 217–231.

Carillo, S. (2000). L'ampliamento del Santuario (1933–1939). In M. Iuliano & S. Federico (Eds.), *Bartolo Longo 'urbanista' a Valle di Pompei (1876–1926)* (pp. 163-172). Napoli.

Carillo, S. (2001). Il Duomo di Nola. Un'occasione metodologica di riflessione per la storia e il restauro dell'architettura tra Ottocento e Novecento. In T. R. Toscano (Ed.), *Nola fuori di Nola* (pp. 381–500). Castellammare di Stabia.

Carillo, S. (2008). L'ampliamento con struttura in c.a. del Santuario di Pompei (1933–1939): una tecnologia moderna a servizio dell'identità storica di un luogo sacro, (with V. Sepe, P. Petillo). In R. Ientile (Ed.). *Architetture in cemento armato. Orientamenti per la conservazione, Proceedings Study day on "Il ciclo di vita delle architetture in cemento armato: l'approccio ingegneristico e le ragioni della conservazione"* (pp. 152-163), Politecnico di Torino, 16 maggio 2007, Franco Angeli editore, Milano. (ISBN: 13: 9788846498861).

Carillo, S. (2013). Il restauro come genesi del progetto. in S. Carillo, V. Sepe, P. Petillo (Ed.), Un laboratorio sperimentale della modernità. L'ampliamento del Pontificio Santuario di Pompei. In C. Gambardella (Ed.), *Atlante di Pompei*, La Scuola di Pitagora editrice, Napoli 2012 (pp. 213–225) ISBN: 978-88-6542-171-0).

Carillo, S. (2014). Pompei, la città del turismo religioso. In *VisibileInvisibile. Percepire la città tra descrizioni e omissioni* (Vol. 6, No. VI. *Città immaginate: sguardi sulla città contemporanea*, pp. 1746–1755). Scrimm Edizioni (ISBN: 978-88-98547-15-9).

Carillo, S. (2018). Islamic influences in the nineteenth century restorations. The Amalfi coast. In C. Gambardella, & D. Listokin (Eds.), *Development and preservation in large cities: an international perspective* (V 2016–VI 2017 ed) (pp. 80–89). Napoli: La Scuola di Pitagora Editrice (ISBN: 978-88-6542-636-4).

Carillo, S., Sepe, V., & Petillo, P. (2010). The huge enlargement with r. c. structure of the Shrine of Pompeii in Italy (1933–1939): a technological, architectonic and cultural challenge. In *34th IABSE Symposium, Large Structures and Infrastructures for Environmentally Constrained and Urbanised Areas,* Venice, 22–24 September 2010. IABSE, Zurich (Switzerland) (pp. 802–810) (ISBN: 978-3-85748-122-2).

Fiengo, G. (1991). *Il Duomo di Amalfi: restauro ottocentesco della facciata*. Amalfi: Centro di cultura e storia amalfitana.

Fiengo, G. (Ed.). (1993). *Tutela e restauro dei monumenti in Campania, 1860–1900*. Napoli: Electa.

Fiengo, G. (1994). La divinazione della facciata medioevale del Duomo di Amalfi. In G. Fiengo, A. Bellini & S. Della Torre (Eds.). *La parabola del restauro stilistico nella rilettura di sette casi emblematici* (pp. 245–311), Politecnico di Milano - Facoltà di Architettura - Quaderni del Dipartimento di Conservazione e Storia

dell'Architettura, 7, Guerini studio, Edizioni Angelo Guerini e Associati, Milano (ISBN: 88-7802-492-9).
Gambardella, C. (2012). *Atlante di Pompei*. Napoli: La Scuola di Pitagora editrice.
Iuliano, M., & Federico, S. (Eds.). (2000). *Bartolo Longo "urbanista" a Valle di Pompei: 1876–1926*. Napoli: Edizioni Scientifiche italiane.
Jannelli, G. (1998). *Brevi cenni degli scrittori e uomini illustri della città di Nola e di quelli che si occuparono della medesima città*. In S. Carillo (Ed.), Nola.
Jubilate! (1909). Numero unico – Ricordo della riapertura del Duomo di Nola. Nola.
Toscano, T. R. (Ed.). (1998). *Nola e il suo territorio dal secolo XVII al secolo XIX. Momenti di storia culturale e artistica*. Castellammare di Stabia.
Zucchini, G. (1933). *Un disegno inedito di Alfonso Rubbiani e di Edoardo Collamarini per la facciata della Basilica di S*. Bologna: Petronio di Bologna. Stabilimenti Poligrafici Riuniti.

A Preliminary Structural Analysis of Typical Arches of Italian Gothic Churches

Mattia Zizi, Daniela Cacace, Valentina Corlito, Jafar Rouhi, and Gianfranco De Matteis

Abstract

The preservation of the built heritage is a challenge for contemporary societies. Arches and vaults represent a widespread structural system; since they were conceived exclusively to withstand gravitational loads, they are exposed with seismic damage and therefore the structural safety of the entire buildings could be compromised. To preserve these structures and to properly design retrofitting interventions, a careful assessment of their capacity is required both under gravitational and seismic loading. However, the complexity of the construction systems and their different arrangements make it difficult to fully understand the mechanical behaviour of such important structural elements. The aim of this study is to evaluate the mechanical behaviour of three different types of Gothic arches (namely drop, equilateral and lancet) observed in typical Italian Gothic churches by means of a parametric study based on limit analyses with a kinematic approach. Therefore, an automatic procedure has been implemented to calculate the horizontal load multiplier sufficient to activate a four-hinge collapse mechanism in each arch subjected to both its own weight and horizontal static equivalent forces arising from a mass proportional distribution of acceleration. As expected, since in the adopted procedure the mechanical characteristics of the base materials have not been accounted for, the obtained outcomes have returned a significant dependence on the main geometrical features of the investigated vaulted structures. Furthermore, the preliminary parametric study carried out in this article could be easily extended by means of the developed automatic procedure, which is a very powerful tool also to analyze specific practical cases.

Keywords

Gothic vaults • Gothic arch • Built heritage • Seismic assessment • Kinematic analysis

1 Introduction

The surveys conducted by the research group in the aftermath of recent Italian seismic events and the analyses carried out on homogeneous samples of damaged masonry churches based on a macro-element approach (De Matteis et al. 2019a, b; De Matteis and Zizi 2019) have pointed out that high percentage of activation of damage mechanisms are related to arches and vaults. The high vulnerability of these architectural elements strictly depends on their static configuration and the natural equilibrium established between the contrasting masonry voussoirs, as well as on the huge increase of global stresses under seismic actions. Due to the concentration of shear stresses (sliding between voussoirs) and/or support displacements, the activation of several damage mechanisms could be provoked, so that severe crack patterns, detachments and, in worst cases, total collapses, may occur.

Accurate structural analysis of masonry vaulted structures is decisive to properly design retrofitting interventions. However, being linked to different typological and structural parameters, it is not an easy task to perform. The structural analyses exclusively based on theoretical and numerical models are often not generalizable unless supported by experimental tests (De Matteis and Mazzolani 2010), which can provide an effective contribution to the calibration and interpretation of the numerical and analytical results. However, although the combined use of analytical methods, numerical modelling and experimental tests may represent the most suitable approach, the definition of reliable simplified methods for structural assessment may be more

M. Zizi · D. Cacace · V. Corlito · J. Rouhi · G. De Matteis (✉)
Department of Architecture and Industrial Design, University of Campania "L. Vanvitelli", Caserta, Italy
e-mail: gianfranco.dematteis@unicampania.it

C. Gambardella et al. (eds.), *Advances in Utopian Studies and Sacred Architecture*,
Advances in Science, Technology & Innovation,
https://doi.org/10.1007/978-3-030-50765-7_8

appropriate, being less computational demanding, costly and time-consuming.

In past and current literature, many studies have been carried out about the seismic issues of masonry arches. Nevertheless, most of them have tended to focus on semi-circular arches (Giordano et al. 2002; Alexakis and Makris 2015), while very few examples can be found about pointed arches (Aita et al. 2004; Huerta 2009; Romano and Ochsendorf 2010). Starting from an overview of the structural importance of Gothic constructions, the aim of this study is to estimate the seismic behaviour of three different typologies of Gothic arches (namely drop, equilateral or lancet) observed in typical Italian Gothic churches by means of a parametric study based on a limit analysis approach. At this aim, the authors developed an automatic procedure, which provides in the output the minimum horizontal load multiplier of an arch (semi-circular or pointed) subjected to its self-weight and uniform horizontal acceleration.

2 The Gothic Architecture

2.1 Structural Features

Gothic buildings are recognized as considerable feats of structural engineering, representing the most interesting issues related to masonry constructions (Heyman 1995). The so-called "Gothic style" was one of the highest examples of the ingenuity of the ancient builders, expressed by those constructions that contained experience, practical skills, social principles and stories of peoples: abbeys and cathedrals. In this era, everything was regulated by the church, considered as a community and a divine institution, and everything took place within the consecrated buildings. Time was marked by the opening of new construction sites, so much that *"in three centuries, from 1050 to 1350, several million tons of stone were quarried in France for the building of eighty cathedrals, five hundred large churches, and some tens of thousands of parish churches"* (Gimpel trans. 1982).

Between the twelfth and fourteenth centuries, cathedrals and abbeys were the driving force of the economic, social and cultural development of the cities, as well as the fulcrum of urban aggregation. Starting from the Ile-de-France, a slow and gradual transformation of compositional elements, constructive procedures and forms developed in Europe; they were influenced by specific geographical, economic, political, religious and social factors, but followed a unitary approach to cultural and technological methods. All the primary functional elements of the Gothic, which until then were not necessarily unknown, interact according to a new building logic, defining a specific construction system. The pointed arch and the cross vault—with ribs or ogival ribs that make up the skeleton—as well as the support systems of both the vaults and the roofing system, divide their function in three aesthetically and statically differentiated parts: the flying buttress, the slender pier and the counterfort (Tagliaventi 2009). The thrusts of the cross vaults, in fact, are conveyed to the large external counterforts, setting a new formal and technical relationship between load-bearing and carried structural elements. While the Roman cross vault was very heavy, needing wall masses of counter thrust in the corners as well as support at the springing, the cross ogival vault was lighter, transferring the lines of thrust on dimensionally contained areas. This allowed the Gothic master builders to launch the vaulted structure to a greater height than the Roman one and, being less the thrust on the piers with equal spacing, to open large windows in the side walls (Cacace and De Matteis 2018).

2.2 The Pointed Arch: Geometrical Considerations and Typologies

Viollet-le-Duc (1854–1868) defined "elasticity" as the essential quality of the Gothic arch, being the skeleton of a constructive system where "everything is equilibrium and, consequently, movement". His study of Gothic architecture started from the use of the pointed arch, directly linked to the genesis of the new ribbed ogival vaults that spread in this period, whose evolution stemmed from geometric and structural requirements. The quadripartite cross vault, built on a square plan, was already well-known in Ancient Romans time, even if it was configured in a massive and very pushing masonry structure, which required large masonry masses to counter thrust. In the Middle Ages, with the adoption of construction techniques based on cutting stones, it began a problem to build such vaults on a rectangular plan. In fact, at the intersection of the two constituting barrel vaults, two ellipses were created along the diagonals; it was much less practical to split an ellipse in masonry ashlars, as well as to build centerings systems that could follow such form (Gaetani et al. 2015). As a solution, the Gothic builders replaced the side semi-circular arches with pointed ones. In this way, the diagonals were semi-circular. The cross ogival vault, more slender and lighter, became the most used in Gothic constructions since the thirteenth century. Its use was expanded to cover wider spaces, reaching more daring forms and making possible much greater variations of proportion.

The use of the pointed arch instead of the semi-circular one also derives from its geometric nature, based on the equilateral triangle, which appears to be the constructive module of many Gothic cathedrals. In fact, from a geometrical point of view, the pointed arch is a two-centred arch, determined by the intersection of two half-circumferences with radius r, whose centres are placed at the springing level of the arch, at a certain distance. It was much easier for the Gothic master builders to draw a pointed arch—knowing the two centres on the springing line—rather than tracing the outline of a semi-circular arch, whose centre is on its central axis, which is priorly indeterminate (Bechmann 1989). Depending on the location of these centres, the following types of Gothic arches could be identified:

- *lancet* pointed arch, if the rise and the distance between the centres (i.e. the radius r) are greater than the span (S), meaning that $r > S$ (Fig. 1a);
- *equilateral* pointed arch, if the rise is equal to the span and the centres coincide with the points at the impost of the arch, so that $r = S$ (Fig. 1b);
- *drop* pointed arch, if the rise of the arch and the distance between the centres is smaller than the span and so $r < S$ (Fig. 1c).

2.3 Main Elements of Italian Gothic Churches

Compared to other European countries such as France, Germany and England, the Gothic style had a shorter life in Italy, where Romanesque architecture continued to spread until the appearance of a Cistercian Gothic, from the reconstruction of many villages and urban aggregates destroyed during the struggles between the Papacy and the Emperors. The first Cistercian Abbeys were built in Chiaravalle Milanese 1135–1221, Fossanova 1173–1210, Casamari 1203–1207 and San Galgano 1207–1288 (Tagliaventi 2009). They were generally characterized by a Latin cross plan, the absence of decorations and superfluous elements, the presence of only one level in elevation, and the use of cross vaults, which most likely separated the upper wooden roof from the interior of the church, protecting it from the fires (Como 2016).

The Italian Gothic has distinctive features, differentiated from the northern Gothic style. The Italian Gothic churches built between the twelfth and the fourteenth centuries—with completions up to the fifteenth century—have very few spires, almost never a triforium and almost always do not have flying buttresses (Enlart 1894). The perimeter walls are often massive and lacking large windows, requiring not much pronounced counterforts. The side aisles are divided into rectangular bays with the largest side equal to that of the square span of the central nave. The bell tower and the baptistery are frequently isolated and independent from the church. However, if in Western Gothic cathedrals the walls are devoid of pictorial ornamentation, there is no doubt that there are also wonderful examples of wall paintings, as in the Basilica of San Francesco in Assisi, 1228–1253.

Between Milan, Bologna, Venice, Florence and Verona there are very different styles and among the few common elements is the pointed arch. The Milan Cathedral (whose construction began in 1386), for which numerous foreign masters were called to solve static-dimensional problems related to its construction, is the most remarkable example of the persistence of the stylistic features of northern Europe. San Petronio in Bologna, whose construction began relatively late (1390), also had a project that was closer in size than any other Italian cathedral to the northern European models, even though the plan was never completed.

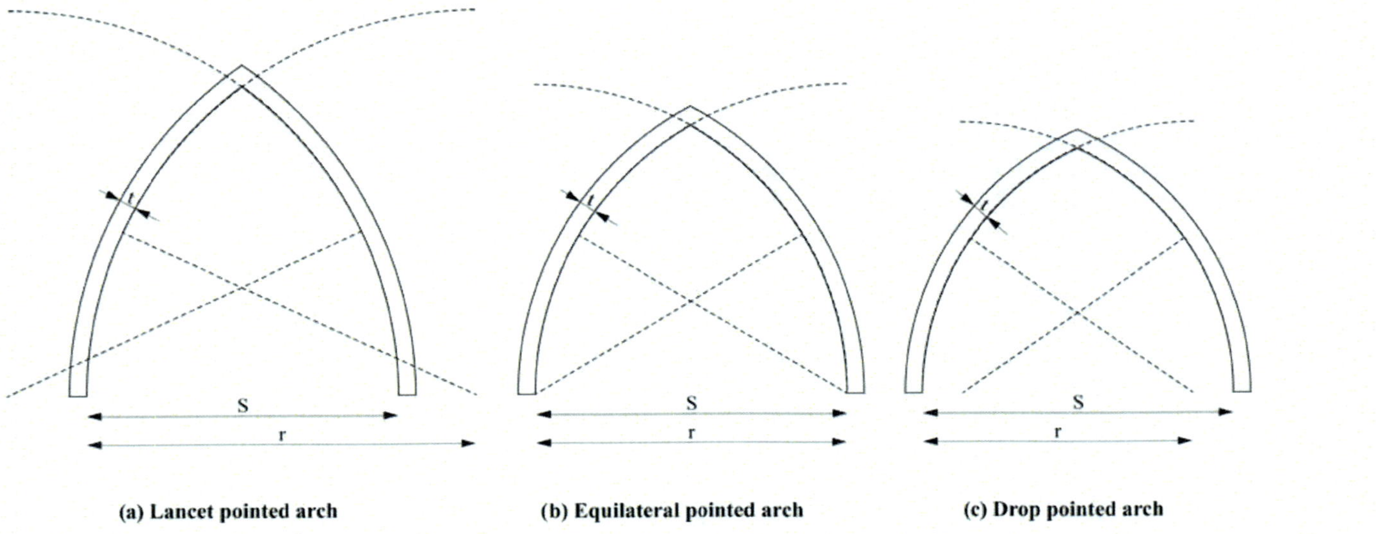

Fig. 1 Different types of Gothic arches based on geometrical radius

From these last two magnificent examples of Italian Gothic, geometrical characteristics were taken to set the parameters of the present study, as described in the following.

3 Structural Assessment of Masonry Arches

3.1 General

The origins of the widespread use of the pointed arch seem to be related to the replacement of the semi-circular arch due to construction problems. In the study of arches and vaults, geometric empirical rules have prevailed over scientific theories for a long time. From Antiquity to the Middle Ages and beyond, several attempts have been developed to evaluate the structural behaviour of vaulted masonry structures. The various theories of the masonry arch are exposed here according to the approach outlined by other references (Heyman 1972; Benvenuto 1991; Di Pasquale 1996) which dealt with the subject in detail.

3.2 Historical Notes

One of the earliest paper trails on vaulted structures is reported in Vitruvius's treatises. In *De Architectura* (trans. 1790), through the description of arches and vaults in residential buildings, he demonstrated to know the pushing effect of the arch on the abutments, emphasizing the importance of a correct dimensioning of the piers. The most widespread rule of thumb among medieval builders—only later theorized by Father Dérand (Rondelet 1832)—shows that, even in the absence of a true static theory, the ancient builders had sensed the need to contain the thrust of arches and vaults inside the thickness of the piers, to the point that these are greater for semi-circular arches than for pointed arches. For a long time after the Middle Ages, the geometric dimension was the most followed criterion by master builders. Examining the old treatises, it emerges that they were nothing more than a collection of constructive indications and rules, without considering a real theory, understood as an attempt not to find single solutions to single problems, but to assert general principles based on scientific facts.

The first attempt towards a *theory* of architecture was offered by Alberti who, in the study of arches and vaults, first provided a typological classification. In *De re aedificatoria* Alberti (trans. 1546) distinguishes the three types of arches: semi-circular, segmental (consisting of less than one-half of a circle) and pointed arch, asserting that the semi-circular one is the most stable of all. Almost at the same time, Da Vinci (1487) provided the first principles for the mechanical interpretation of the behaviour of an arch. Considering a model consisting of two linear arches separated by a certain thickness, he illustrates a condition of stability of an arch, so that an arch does not break if the chord of the external arch does not touch the inner arch. With this intuition Leonardo showed that he knew the need to have the thrust line contained entirely in the arch section (Fig. 2).

3.3 First Static Theories

The first mechanical approach to the arches, considering the masonry as rigid blocks with their own weight, comes from De la Hire (1695, 1712). He introduced the wedge theory for the voussoirs to interpret the masonry arch as the result of infinitely resistant rigid blocks (the voussoirs), placed on each other and kept in equilibrium by the mutual action of self-weight and reciprocal forces with adjacent voussoirs. De la Hire dealt with the double problem of the equilibrium of a vault (or an arch) independent from the piers (i) followed by the determination of the width of the piers themselves, as a function of the thrusts coming from vaults (ii).

To address the problem (i) De la Hire based his model on Hooke's catenary anticipating the concept of the polygon of forces. He considered a semi-circular arch with equal and frictionless voussoirs, determining the weight of each voussoir and the two forces transmitted by the adjacent blocks and applied in the centre of gravity, perpendicular to the joints (Fig. 3a). The thickness of the remaining voussoirs, by prefixing the one in the key and the two adjacent, is determined by imposing the equilibrium. As for the springer voussoir, which should thus have infinite weight, the contradictory explanation of friction is used.

For problem (ii) De la Hire proposes the first example of failure mechanism, dividing the arch into three rigid blocks, with a critical 45° section, and imposing equilibrium. According to the model, the central part slides downwards, exerting a wedge action on the remaining parts of the arch and pushing them outwards, causing its overturning, without

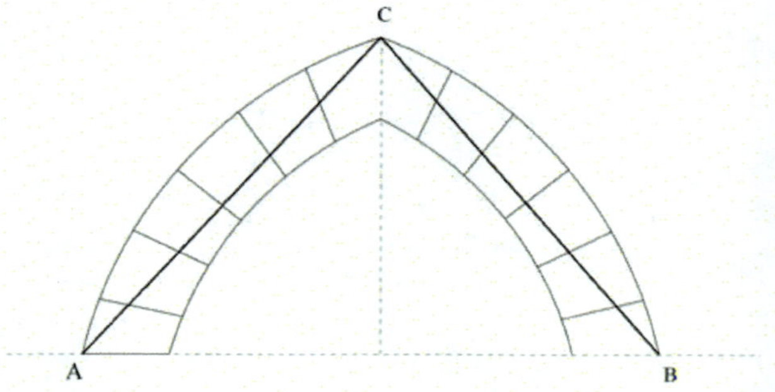

Fig. 2 Schematic representation of the masonry pointed arches' safety rule, according to Leonardo Da Vinci, after Benvenuto (1991)

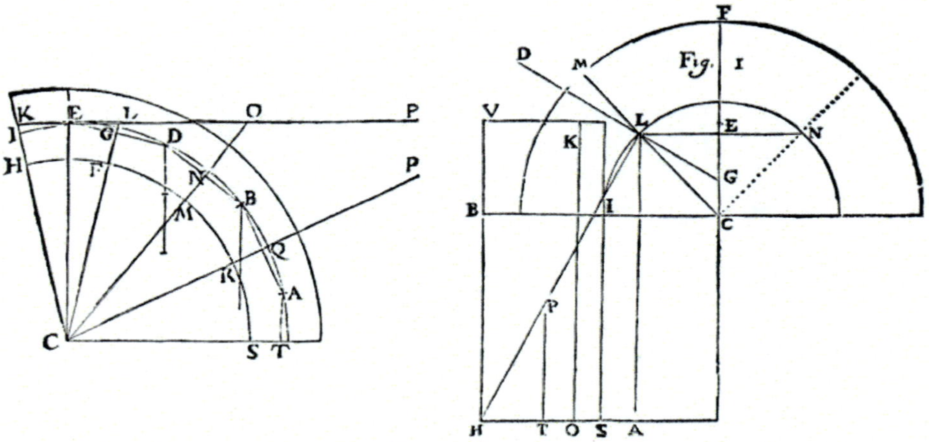

(a) (b)

Fig. 3 De la Hire (1695, 1712): masonry arch model, according to the wedge theory **a** and **b**

sliding. Thus, considering the half-arch, the abutment tends to overturn around a hinge placed at the external point of the base, for whose balance De la Hire uses a lever mechanism, where the stabilizing moment is given by the weight force of the piers, applied in the centre of gravity (Fig. 3b).

De Bélidor (1729) took up the theory of De la Hire, but with different hypotheses on the point of application of the overturning force and on the type of lever. Maintaining the position of the critical section at 45°, the thrust of the central wedge is applied in correspondence with the axis of the arch and not of the intrados.

The hypothesis of lack of friction caused the mechanism to be considered more dangerous than it actually is, conducing to a conservative assessment. In this context, Couplet (1732) addressed the problem to find the minimum thickness of a semi-circular arch subjected only to its own weight. The solution is obtained by hypothesizing for the first time a four-blocks collapse mechanism, with five hinges placed at the extrados (keystone and springer) and at the intrados, at a 45° section from the horizontal axis (Fig. 4).

Coulomb (1773) made several studies on the thrust of the arches taking into account the strength of the material and assessing the stability only with equilibrium considerations through maximum and minimum principles. He considers a portion of the arch between the keystone and a generic critic joint and, through the analysis of four collapse mechanisms, states that the failure of the arch could occur only with the formation of hinges between voussoirs. Coulomb identifies the problem as undetermined, showing that in a certain admissible range all solutions are equally acceptable, and that friction is sufficient to prevent sliding failure (Fig. 5).

Mascheroni (1785) proposed a failure mechanism of the arch that foresees a four-blocks rupture and the formation of hinges at the intrados (keystone) and at the extrados (haunches), overcoming the previous theories of the 45° critic joint.

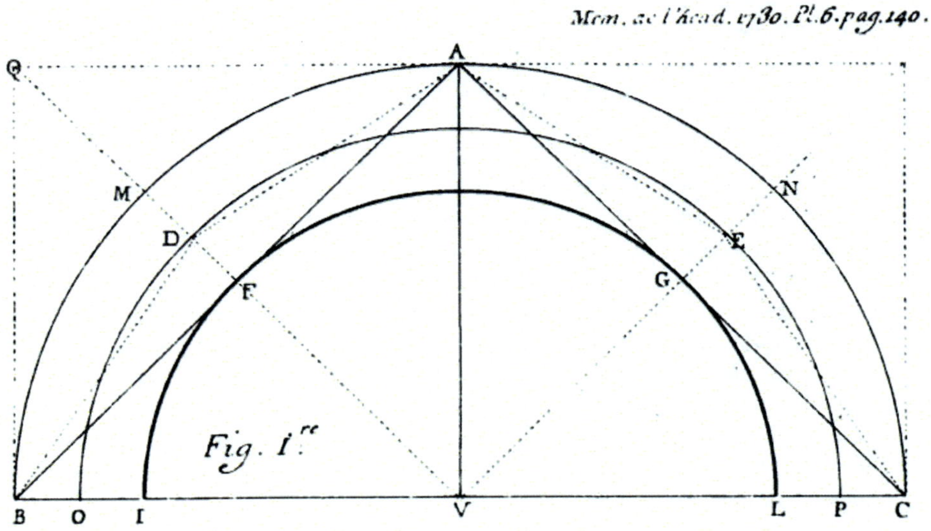

Fig. 4 Couplet (1732): four-blocks collapse mechanisms for masonry arches

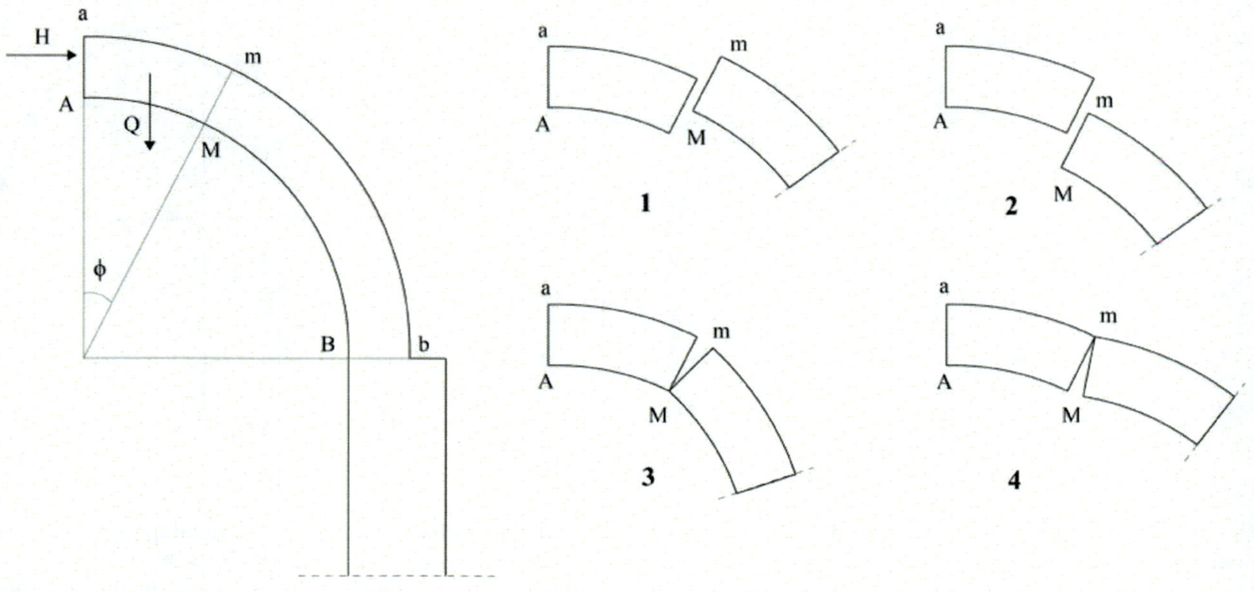

Fig. 5 Schematic representation of Coulomb's theory

3.4 From Modern Theories to Current Approaches

A true scientific formulation about the structural behaviour of masonry arches and vaults took place from the end of the eighteenth and during the nineteenth century when specific fields of the investigation were defined. Among these: (i) methods related to the definition of the so-called line of thrust; (ii) methods related to the application of the elastic theory; (iii) methods related to the plastic analysis.

The definition of the line of thrust in masonry arches constituted one of the most significant issues in the assessment of the stability of the vaulted structures. As the path of the resultants of all compressive forces through the masonry arch, the line of thrust must lie entirely inside the structure, to assure a prevalent compressive state in the section. Since the equilibrium solution is not unique, several historical researches were focused on determining the true position of the thrust line.

A different approach, based on the application of the elastic theory by Navier (1826), considered the masonry arches as elastic curved bars. Further elastic-linear and equilibrium equations useful to solve the hyperstatic problem were identified (Huerta 2008). The unsolvable problems in the conceptual rigid blocks are addressed: the effective capacity to withstand certain stress states and the actual path of the thrust line within the arch. Navier suggested considering the stability assessment proposed by Coulomb, hypothesizing that, in order to only have compression forces, the thrust line had to remain in the middle third of the resistant section. In this way, there was the limit condition of a triangular distribution of the compressive forces over the transversal cross-section, with a null value at the point where the sliding starts in case of rupture. Starting from the studies of Navier, Mèry (1840) showed that the static problem of an arch could be solved by forcing the polygon of forces to pass through two marked points: the lower middle third in the springer section and the upper-middle third in the key section, with a horizontal line of action (for symmetrically loaded and constrained arch). In this way, being known the external loads, it was possible to obtain the path of the pressure curve which, according to Mèry's theory, could avoid tensile stresses and ensures safety to the arch if included in the middle third of the arch (Fig. 6).

In 1867, Winkler argued that the real thrust line, based on the theory of elasticity, is the one that least deviates from the axis of the arch. However, the application of the theory of elasticity to masonry structures is not justified in a field of rupture calculation, for which the deformability is disregarded, and the voussoirs are considered as rigid.

The problem has been connected by Heyman (1982) to a formulation coherent with the methods of the rupture calculation using a theorem concerning the safety in the arch stability. In the early twentieth century, the new trend of the limit analysis theorem started to change the practical method for structural assessment of masonry vaulted structures. According to the plastic analysis, a complete understanding of the structural behaviour of masonry arches and vaults can be achieved only with the study of their possible collapse mechanisms. Heyman is recognized as a pioneer of the application of plastic limit analysis to masonry structures. Based on his theories, all methods adopted to describe the

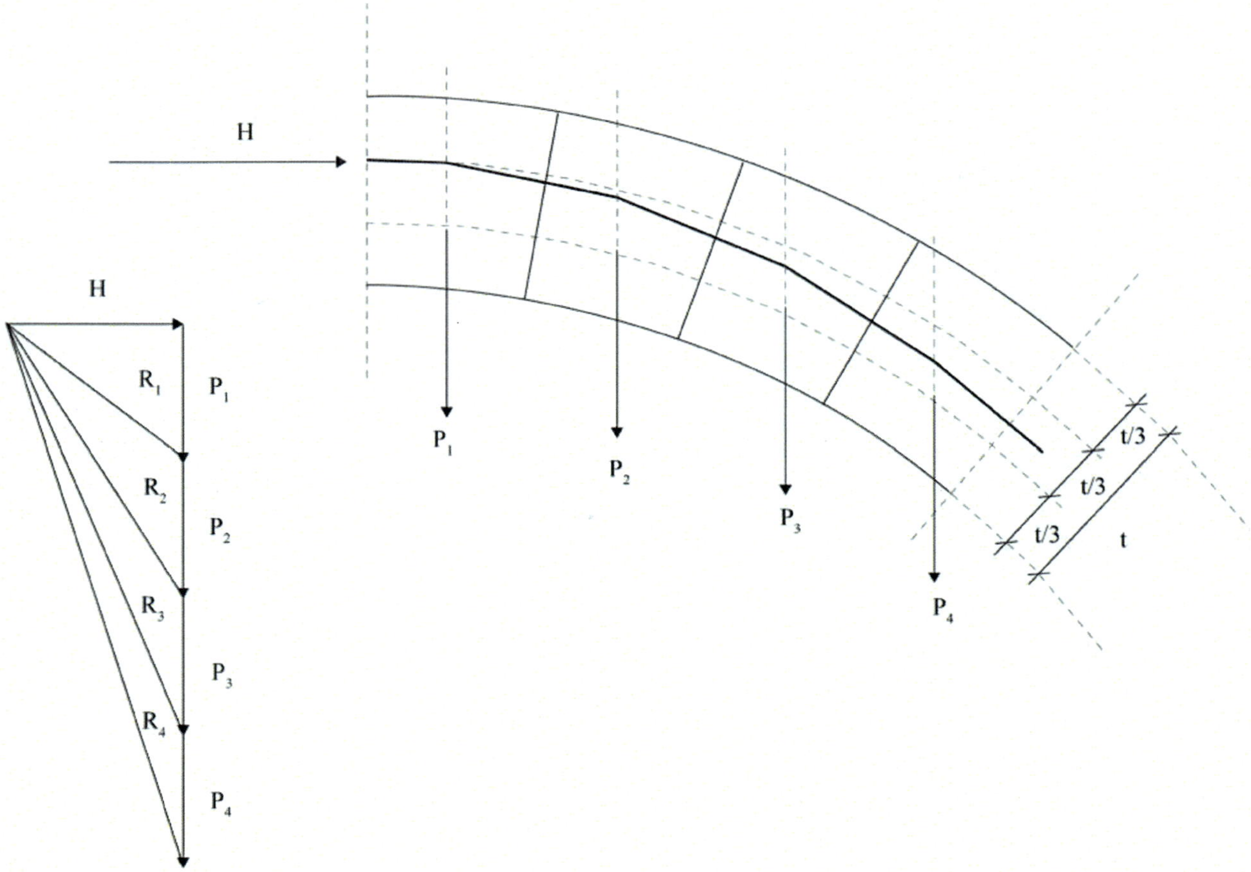

Fig. 6 Thrust line calculation, according to Mèry's theory

mechanical behaviour of masonry structures follow these assumptions: (i) masonry has an infinite compressive strength; (ii) masonry has zero tensile strength; (iii) sliding failure cannot occur. These simplified hypotheses were assumed within the Safe Theorem and the Kinematic Theorem, which refer to the concept of equilibrium and occurrence of failure mechanisms, respectively. According to the Safe Theorem, an arch is stable if the thrust line can be drawn inside the boundaries of the arch itself; on the contrary, when the forces are no more contained within the thickness of the arch, the section partialises and a phase of high deformations starts. The application of the Kinematic Theorem led to the so-called Limit Analysis for the study of masonry arches and vaults. In literature, some significant examples of the application of Limit Analysis can be identified. Among others Brandonisio et al. (2017), applied this approach to circular buttressed arches by means of a non-linear automatic procedure, aimed to identify the failure mechanism and the position of the hinges that minimizes the horizontal load multiplier.

Nowadays, although the Heyman's theory is accepted, the development of specific software in the field of structural engineering generates new horizons in the assessment of the structural behaviour of masonry curved structures (namely arches and vaults). The selection of an appropriate numerical model is an issue that greatly influences the reliability of the obtained results since it depends on the accuracy and availability of the data. The numerical models based on DEM (Discrete Element Method) and FEM (Finite Element Method) analyses are frequently used. Even if many examples of simulation methods for approaching the problem of arches and vaults (Pulatsu et al 2018; Rossi et al. 2015; Kamiński 2010) can be found in literature, these computational models suffer from some limitations. For instance, to describe the phenomena up to the scale of single bricks by using the FEM methods, a detailed micro-modelling (Lourenço 2002) is necessary, with enormous time effort increasing due to computational and modelling issues. On the other hand, being mainly based on the study of interactions between different blocks, also in case of large displacements, DE Models could be considered more suitable, even if they are preferred in the field of civil engineering. In any case, the use of these computational methodologies is still critical as they require high computational efforts and expert engineering judgment in the interpretation of numerical results.

In the wide context so far described, this research work aims to provide a contribution about the knowledge of collapse mechanisms occurring in the main three typologies of pointed arch (drop, lancet and equilateral), under the action of both vertical and horizontal loads, by investigating the influence of the main geometrical characteristics on their seismic behaviour.

4 The Developed Analysis

4.1 Geometric Parameters Setting

A preliminary analysis of the behaviour of three pointed arches, retrieved considering three different values of thickness-to-span (*t/S*) ratio, namely 1/18, 1/15 and 1/10, has been performed. In particular, the limit analyses with a kinematic approach have been carried out in order to evaluate the horizontal load multiplier sufficient to activate a four-hinge collapse mechanism of pointed arches (drop, lancet and equilateral), which were subjected to a uniform horizontal acceleration and self-weight. Therefore, nine arch typologies have been considered, according to the following assumptions:

- constant span (*S*) of the arches: $S = 19$ m, as a real length value observed in the examples from Milan Cathedral (equilateral arch) and San Petronio in Bologna (drop and lancet arches);
- three different values of radius-to-span (*r/S*) ratio: 0.87 (drop), 1.00 (equilateral), 1.25 (lancet), as real values taken from Milan Cathedral (equilateral arch) and San Petronio in Bologna (drop and lancet arches);
- three different values of thickness-to-span (*t/S*) ratio: 1/10 (ancient Roman bridges' mean thickness observed), 1/15 (Leon Battista Alberti's minimum thickness for masonry arches), 1/18 (Heyman's minimum thickness set for semi-circular arches).

4.2 The Adopted Limit Analysis Approach

Under Heyman's assumptions, a masonry arch is considered as an assemblage of non-deformable parts and the collapse is due to the formation of non-dissipative hinges, which induce a displacement distribution into the structure (i.e. kinematic mechanism). It is worth noting that, in the case of an arch subjected to uniform horizontal acceleration, the formation of four hinges placed at the intrados and the extrados alternatively is a necessary and sufficient condition to produce an admissible kinematic mechanism (Fig. 7a, b).

The displacements due to external loads produce two different works (i.e. vertical and horizontal), whose sum, according to the Principle of virtual works, must be equal to zero. In particular, once the positions of the non-dissipative hinges are chosen, it is possible to determine the horizontal multiplier λ sufficient to produce the collapse, by means of (1):

$$\sum_{i=1}^{3} M_i g v_i + \lambda \sum_{i=1}^{3} M_i g u_i = 0 \tag{1}$$

where M_i is the mass of each rigid part, g is the gravity acceleration and v_i and u_i are the vertical and horizontal virtual displacements, respectively, evaluated in the centroid of each rigid part.

The horizontal and vertical virtual displacements are defined according to the rigid block rotations, which are evaluated by imposing an infinitesimal rotation to the block between the first two hinges and by determining the kinematic chain (Fig. 8).

Nevertheless, considering that the position of the hinges is not known *a priori*, an iterative procedure is necessary, in order to evaluate the minimum horizontal multiplier, which corresponds to the real one. Such a procedure includes many attempts and, according to each admissible position of hinges, for each rigid part the rotation—and both the consequent vertical and horizontal displacements—the position of their centroid, and their masses should be evaluated. Therefore, to avoid time efforts, the authors developed a VBA (Visual Basic for Applications) automatic procedure, which takes into account all the aspects above mentioned. This specific tool calculates the horizontal multiplier sufficient to provoke the collapse mechanism activation for each hinge admissible position, providing in the output the minimum one, as well as the positions of the corresponding four non-dissipative hinges (α_A, α_B, α_C, α_D). Moreover, considering the presence of the gravity acceleration (g) in both the addends of (1), it is not necessary to define a specific weight of the masonry and therefore it is sufficient to introduce in input the span (*S*) of the arch, the thickness (*t*) and the radius (*r*).

4.3 Results and Discussion

The obtained results are summarized in Table 1 where, for each analyzed case, the position of the four non-dissipative hinges with the corresponding horizontal multiplier is provided. As shown in Fig. 7a the position of the hinges is defined through an angle between the negative horizontal

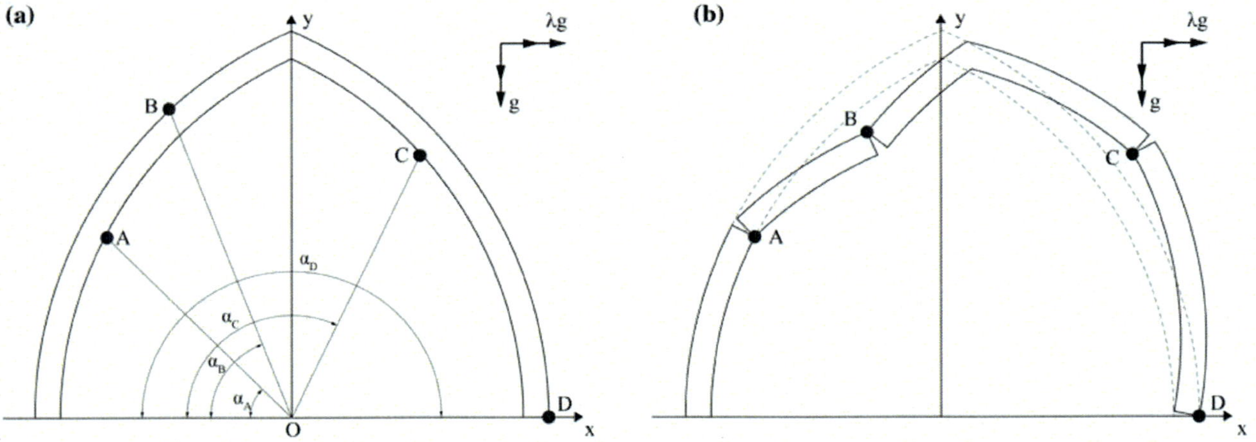

Fig. 7 **a** Possible position of non-dissipative hinges. **b** Collapse mechanism of a pointed arch due to horizontal uniform acceleration

Fig. 8 Kinematic chains and determination of virtual displacements

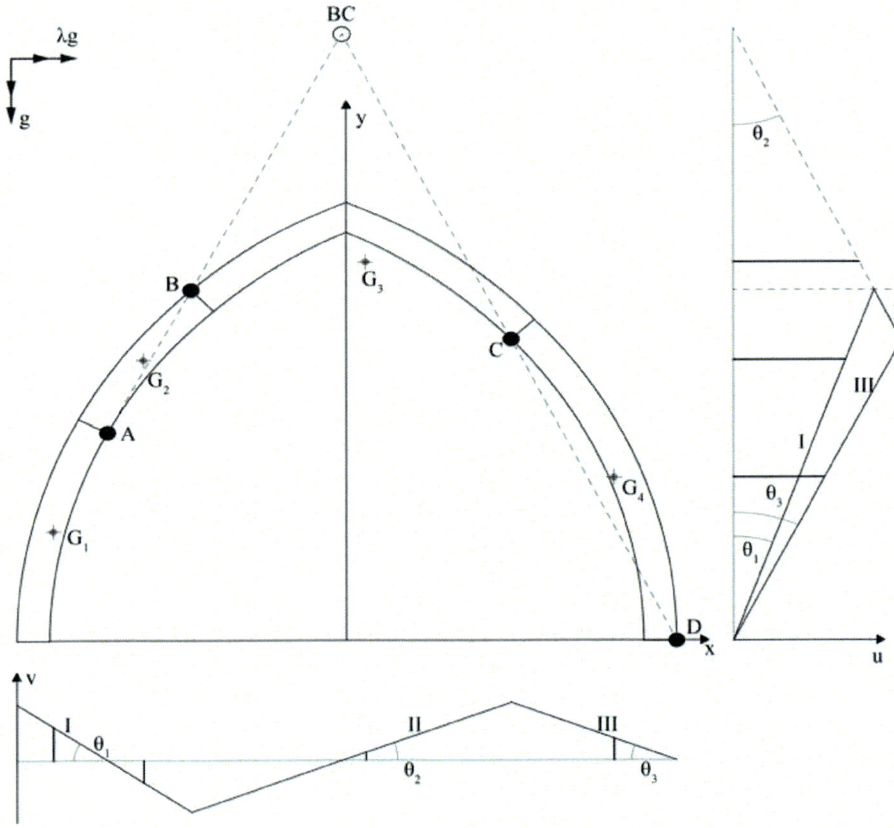

axis (at the impost of the arch) and the location of the hinge itself. In Fig. 9a, b, the results in terms of λ-*r/S* and λ-*t/S* curves are shown.

As expectable by applying the limit analysis approach, the obtained outcomes reveal that the minimum horizontal multiplier strictly depends on the geometric characteristics of the arch. In particular, for the considered cases, values of λ in the range of 0.03–0.2 *g* have been obtained. Moreover, it is possible to assert that, for arches having the same thickness, the seismic capacity of a pointed arch decreases

Table 1 Results of the parametric analysis

Arch type	r (m)	S (m)	t (m)	t/S	r/S	α_A (°)	α_B (°)	α_C (°)	α_D (°)	$\lambda(g)$
Drop	16.59	19	1.05	1/18	0.87	31	77	136	180	0.06
	16.59	19	1.27	1/15	0.87	27	76	132	180	0.10
	16.59	19	1.90	1/10	0.87	16	73	123	180	0.20
Equilateral	19.00	19	1.05	1/18	1.00	32	77	90	180	0.06
	19.00	19	1.27	1/15	1.00	27	76	129	180	0.10
	19.00	19	1.90	1/10	1.00	17	73	120	180	0.19
Lancet	23.75	19	1.05	1/18	1.25	36	79	90	180	0.03
	23.75	19	1.27	1/15	1.25	32	78	90	180	0.07
	23.75	19	1.90	1/10	1.25	20	76	90	180	0.15

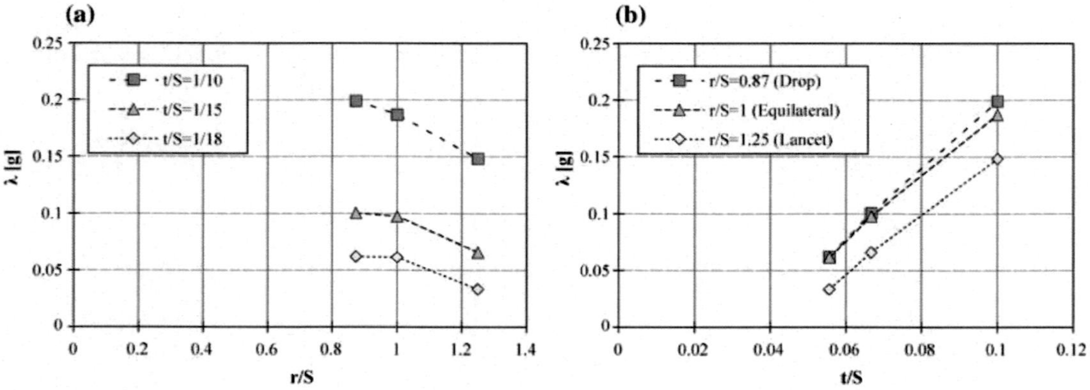

Fig. 9 **a** Obtained results in terms of horizontal multiplier-radius/Span ratio. **b** Obtained results in terms of horizontal multiplier-thickness/Span ratio

according to its slenderness (r/S). In fact, lancet arches showed a worse response against horizontal forces than equilateral and drop ones. On the other hand, the thickness obviously plays an important role in the response of the arch: for each analyzed typology of arch, the multiplier λ shows a similar direct proportion to the t/S ratio.

As regards the Italian Gothic style, some additional considerations have been carried out. According to the seismic hazard map provided in the current code, the Italian territory is divided into four different seismic hazard levels, depending on the expected Peak Ground Acceleration (PGA) level (expressed with a 10% exceedance probability in 50 years for rigid soils and ordinary buildings, i.e. return period $T_R = 475$ years):

- low seismic hazard (expected PGA $\leq 0.05\ g$);
- medium-low seismic hazard ($0.05\ g <$ PGA $\leq 0.15\ g$);
- medium-high seismic hazard ($0.15\ g <$ PGA $\leq 0.25g$);
- high seismic hazard ($0.25\ g <$ PGA $\leq 0.30\ g$).

Aiming at evaluating the appropriate geometrical characteristics necessary to guarantee a seismic safety for each seismic hazard level mentioned, further analyses have been carried out, also considering circular arches ($r/S = 0.5$).

The obtained results are depicted in Fig. 10: for low thickness values, the seismic behaviour of masonry arches is strongly influenced by the effects of vertical loads due to the

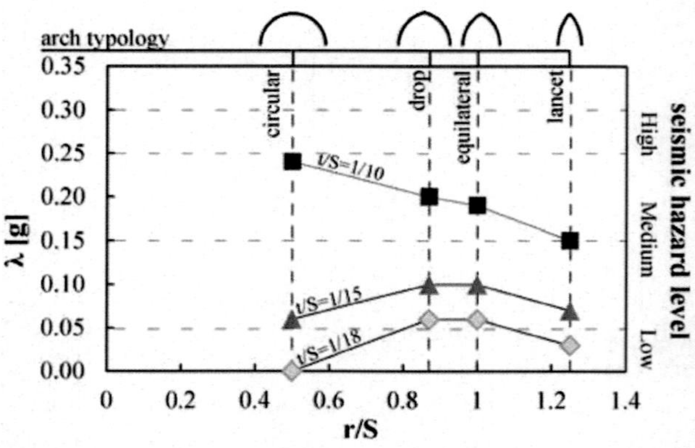

Fig. 10 Obtained results in terms of horizontal multiplier-radius/Span ratio related to the four Italian seismic hazard levels

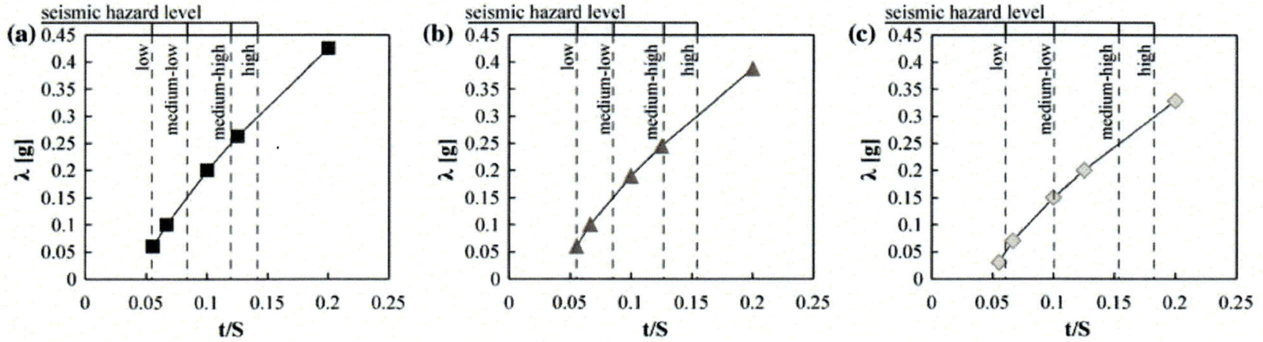

Fig. 11 Minimum thickness-to-span ratio according to Italian seismic hazard levels for considered drop (**a**), equilateral (**b**) and lancet (**c**) pointed arches

Table 2 Minimum thickness/Span ratio according to Italian seismic hazard levels for considered arches

Arch type	r (m)	S (m)	Minimum t/S			
			Low	Medium-low	Medium-high	High
Drop	16.59	19	0.055	0.083	0.12	0.14
Equilateral	19.00	19	0.055	0.085	0.13	0.15
Lancet	23.75	19	0.061	0.1	0.15	0.18

self-weight. For instance, if the minimum t/S ratio is adopted, the circular arches reveal a seismic capacity equal to zero. Moreover, the graph reporting the distinction of Italian seismic hazard levels highlights that, for example, the downer limit of high hazard level is never reached for $t/S = 1/10$.

The study has been further developed by attempting to evaluate, with the same procedure, the minimum t/S ratio of each considered pointed arch in the Italian seismic hazard levels. At this aim, the field of analyses has been extended by introducing $t/S = 1/8$ and $t/S = 1/5$. The obtained outcomes are reported in Fig. 11 and Table 2, where the minimum t/S ratio compatible for each seismic level is reported.

As it can be observed, only high values of t/S ratio can guarantee structural safety against expected seismic loads in the most hazardous Italian zone.

Although this approach may seem to be conservative, since it neglects the external vertical loads and the presence of the pillars, it should be noted that even the examined accelerations contain limitations, not considering the amplification effects due to the height of the structures, to the soil classes and the use coefficient of buildings. Therefore, the results obtained could be considered reliable in a perspective of a preliminary evaluation of the seismic safety of vaulted structures.

Referring again to the Italian Gothic style, it could be affirmed that although the ancient Italian builders did not have an accurate knowledge of seismic engineering, they had intuited the need to design structures less slender than those of other European countries (e.g. France, Germany or Spain), where expected PGA levels are less worrying. The Gothic style, spread from France to the south of the continent with significant stylistic contaminations, is affected by a low presence of examples in the Italian Peninsula. This aspect certainly concerns the greater seismic risk that characterizes the Italian territory, due to the high number of disastrous earthquakes that occurred in the past, as shown in the European Seismic Hazard Map (Fig. 12).

This hypothesis, which should not be considered of secondary importance and should be further investigated, can be corroborated by referring to the characteristics of the Gothic style in other European countries. The Catalan Gothic architecture, for example, is characterized by very large naves, with vaults and arches that reach daring dimensions (the largest Gothic vault in the world is located in the Gothic cathedral of Girona). However, it is known that the weight of the vaults is in the usual Catalan way transported by internal buttresses (diaphragm arches), which are not recognizable in other parts of the Iberian Peninsula, where the seismic risk is lower with respect to the Italian territory.

Fig. 12 European seismic hazard map—ESHM13 (Giardini et al. 2014)

5 Conclusions

Although affected by a strong regionalism and several contaminations, many examples of Gothic churches can be found in the Italian territory. Among all the elements that distinguish such architectural style, the pointed arch is the most common one. Since the Italian territory is affected by a high seismic hazard, this study focuses on a preliminary assessment of the influence of the main geometrical characteristics on the seismic behaviour of Gothic arches, by means of parametric analyses. In particular, on the basis of real measurements and empirical geometrical rules defined in the past, the behaviour of nine different pointed arches, obtained by varying the main geometric characteristics (i.e. slenderness and thickness) and subjected to self-weight and uniform horizontal acceleration, has been investigated.

At this aim, a VBA procedure has been implemented, which automatically evaluates the minimum horizontal multiplier and the corresponding position of the four non-dissipative hinges. The obtained outcomes have revailed that the main geometrical characteristics of a pointed arch (i.e. slenderness and thickness) strongly influence its seismic response. In particular, it has been noted that the seismic response of a pointed arch improves according to its thickness, while more slender arches exhibit worse behaviours than the less slender ones.

Some additional analyses have been performed to compare the slenderness of the Italian Gothic construction with the seismic risk of the considered area. Based on the results obtained, it is worth noting that large *r/S* values and low *t/S* values are not compatible with the worst seismic risk conditions. There is evidence to suggest the hypothesis of a minor presence of Gothic style in the seismic zones, an aspect that must be further investigated.

The implemented procedure could be profitably adopted for performing parametric analyses aimed at assessing the seismic capacity of masonry arches. This preliminary study will be extended to investigate other arches types as well as the effect of additional influential factors, such as the effect of the back-fill presence or of other external loads, the embrace angle and the boundary conditions.

References

Aita, D., Barsotti, R., Bennati, S., & Foce, F. (2004), The statics of pointed masonry arches between 'limit' and 'elastic' analysis. In P. Roca, & C. Molins (Eds.), *Arch bridges IV. Advances in assessment, structural design and construction* (pp. 353–262). Barcelona, Spain: CIMNE.

Alberti, L. B. (1546). Dieci Libri De L'Architettvra Di Leon Battista De Gli Alberti Fiorentino. *La commodità, l'utilità, la neceßità, e la dignità di tale opera* (trans: Lauro, P.). Vinegia: Appresso Vincenzo Vaugris.

Alexakis, H., & Makris, N. (2015). Limit equilibrium analysis of masonry arches. *Archive of Applied Mechanics, 85*(9–10), 1363–1381. https://doi.org/10.1007/s00419-014-0963-6.

Bechmann, R. (1989). *Le radici delle cattedrali*. Milano: Arnoldo Mondadori Editore.

Benvenuto, E. (1991). *An introduction to the history of structural mechanics—Part II: vaulted structures and elastic*. Berlin: Springer.

Brandonisio, G., Mele, E., & De Luca, A. (2017). Limit analysis of masonry circular buttressed arches under horizontal loads. In: *Meccanica*.

Cacace, D. & De Matteis, G. (2018). Typological classification of vaulted structures in masonry churches in WORLD HERITAGE and KNOWLEDGE, Representation, Restoration, Redesign and Resilience. In *Le Vie dei Mercanti, XVI International Forum*. Capri: Napoli.

Como, M. (2016). *Statics of historic masonry constructions* (2nd ed.). Springer.

Coulomb, C. A. (1773). Essai sur une application de maximis & minimis a quelques problemes de statique relatifs a l'architecture. In *Mémoires de l'Académie Royale des Sciences*. Paris.

Couplet, P. (1732). Seconde partie de l'examen de la poussée des Voûtes. In *Mémoires de l'Académie Royale des Sciences*. Paris.

Da Vinci, L. (1487–1505). *Forster Codex II*. London: Victoria and Albert Museum.

De Bélidor, B. F. (1729). *La Science des ingénieurs dans la conduite des travaux de fortification et d'architecture*. Paris: Jombert.

De La Hire, P. (1695). *Traité de Mécanique*. Paris.

De La Hire, P. (1712). Sur la construction des voûtes dans les édifices. In *Mémoires de l'Académie Royale des Sciences de Paris*.

De Matteis, G., & Mazzolani, F. M. (2010). The fossanova church: Seismic vulnerability assessment by numeric and physical testing. *International Journal of Architectural Heritage, 4*(3), 222–245. https://doi.org/10.1080/15583050903078903.

De Matteis, G., & Zizi, M. (2019). Seismic damage prediction of masonry churches by a PGA-based approach. *International Journal of Architectural Heritage*. https://doi.org/10.1080/15583058.2019.1597215.

De Matteis, G., Brando, G., Corlito, V., Criber, E., & Guadagnuolo, M. (2019a). Seismic vulnerability assessment of churches at regional scale after the 2009 L'Aquila earthquake. *International Journal of Masonry Research and Innovation, 4*(1–2), 174–196.

De Matteis, G., Cacace, D., & Rouhi, J. (2019b). Masonry vaults: Architectural evolution, structural behaviour and collapse mechanisms. In *Structural Engineers World Congress, 7th SEWC 2019*. Turkey: Instanbul.

Di Pasquale, S. (1996). *L'arte del costruire. Tra conoscenza e scienza*. Venezia: Marsilio.

Enlart, C. (1894). *Origines franqaises de l'architecture gothique en Italie*, Paris.

Gaetani, A., Monti, G., Lourenço, P. B., & Marcari, G. (2015). Masonry cross vaults: an overview of the historical developments. In *AID Monuments. Materials Techniques Restoration for Architectural Heritage Reusing, (July 2016)* (pp. 589–601).

Giardini, D., Woessner, J., & Danciu, L. (2014). Mapping Europe's seismic hazard. *EOS, 95*(29), 261–262.

Gimpel, J. (1982). *Les Bâtisseurs de cathédrales*. Paris. (trans: Italian Costruttori di Cattedrali). Milano. Original work published in 1980.

Giordano, A., Mele, E., & De Luca, A. (2002). Modelling of historical masonry structures: Comparison of different approaches through a case study. *Engineering Structures, 24*(8) 1057–1069. Modelling-of-historical-masonry-structures-comparison-of-different-approaches-through-a-case-study.pdf.

Heyman, J. (1972). *Coulomb's memoir on statics: An essay in the history of civil engineering*. Cambridge: Cambridge University Press.

Heyman, J. (1982). *The masonry arch*. Chichester: Ellis Horwood Ltd.

Heyman, J. (1995). *The stone skeleton: Structural engineering of masonry architecture*. Cambridge: Cambridge University Press.

Huerta, S. (2008), The analysis of masonry architecture: A historical approach. In: *Architectural Science Review*.

Huerta, S. (2009). The debate about the structural behaviour of gothic vaults. In *Proceedings of the Third International Congress on Construction History*. Germany: Cottbus.

Kamiński, T. (2010). Tests to collapse of masonry arch bridges simulated by means of FEM. In *IABMAS*.

Lourenço, P. B. (2002). Computations on historic masonry structures. *Programming of Structural Engineering Materials, 4*(2002), 301–319.

Mascheroni, L. (1785). *Nuove ricerche sull'equilibrio delle volte*. Bergamo.

Mèry, E. (1840). *Sur l'equilibre des voutes en berceu*. Paris, France: Annales de Ponts et Chaussess.

Navier, C. L. (1826). *Résumé des leçons données à l'École des Ponts et Chaussées sur l'application de la mécanique à l'établissement des constructions et des machines*. Paris: F. Didot père et fils.

Pulatsu, B., Erdogmus, E., & Lourenço, P. (2018). Simulation of masonry arch bridges using 3D discrete element modelling. In: *RILEM Bookseries 2018*, (Vol. 18, pp. 871–880). Netherlands: Springer.

Romano, A., & Ochsendorf, J. A. (2010). The mechanics of gothic masonry arches. *International Journal of Architectural Heritage, 4* (1), 59–82.

Rondelet, J. B. (1832). Trattato teorico e pratico dell'arte di edificare. Prima traduzione italiana sulla sesta edizione originale con note e giunte importantissime (trans: Soresina, B.). Mantova: A spese della Società editrice, coi tipi di L. Caranenti (Original work published in 1802).

Rossi, M., Calderini, C., Milani, G., & Lagomarsino, S. (2015). Numerical and experimental analysis of an in-scale masonry cross-vault prototype up to failure. In *International Conference of Computational Methods in Sciences and Engineering (ICCMSE 2015)*.

Tagliaventi, I. (2009). *La cattedrale gotica*. Alinea: Spirito e struttura della più grande opera d'arte della città occidentale.

Viollet-le-Duc, E. (1854–1868). Dictionnaire raisonné de l'architecture francaise du XI' au XVI' siècle. Paris, B. Bance–A. Morel.

Vitruvius Pollio, M. (1790). L'Architettura di M. Vitruvio Pollione, tradotta e comentata dal. Marchese B. Galiani. *Edizione seconda ricorretta* (trans: Galiani, B.). Napoli: Fratelli Terres.

Changing the History of Architectural Heritage Palaeologian Renaissance—The Style that Never Was?

Dragana Lazarević

Abstract

The Wars of Yugoslav Succession reintroduced the practice of cultural degradation of the defeated party—in this case the Serbs—not only by stripping them of any notion of humanity, but also of their cultural and religious heritage on the territory affected by military operations. By analysing the four UNESCO monuments in Kosovo and Metohija, this paper examines how the phenomenon of reinterpretation of meaning and significance of the Serbian Mediaeval monastic heritage, followed by its cultural appropriation by the new Albanian authorities, contributes further to deepening the already severe inter-ethnic distrust between the Serbs and Albanians who share the same territory. The paper juxtaposes modern attempts to rewrite the meaning of religious heritage in Serbia against historical and political background in situ. It argues that the politically enforced reinterpretation, so generously helped by the state and academic institutions from outside of the Balkans, aiming at cultural alienation of its Serbian roots does not provide permanent protection of historical religious heritage, so clearly evidenced by its inclusion on the List of World Heritage in Danger by UNESCO. The paper concludes that the attempts of the cultural re-appropriation have little to do with scientific endeavours, but with the political aims of the victorious parties.

Keywords

Architectural religious heritage • Architectural interpretation • National narratives • Cultural appropriation • Historical revisionism • Inter-ethnic conflicts • Utopian reading

D. Lazarević (✉)
Cardiff University, SHARE, Cardiff, UK
e-mail: lazarevic.dragana@gmail.com

1 Introduction

When the self-proclaimed independent Republic of Kosovo, a disputed southern province of the Republic of Serbia failed to win the necessary number of votes at the General Conference of UNESCO on 9 November 2015, the reactions of the two parties involved were diametrically opposed; the Albanian authorities in Kosovo and Metohija and in Albania argued that the outcome of the vote was "a direct result of the racist campaign against Kosovo, which was continuously maintained by Serbia" (AlJazeera Balkans 2015).[1] Serbian side, on the other hand, for the first time in more than twenty-five years since the beginning of Yugoslav disintegration, had a sense of victory "in a very unfavourable political situation in which the whole state found itself since the early 1990s" (Jevtić 2015).[2] Apart from being met with disbelief by both Serbian and Albanian delegations—because neither side expected such an outcome—it also stunned major Western powers which took active roles in Yugoslav disintegration by overwhelmingly supporting the Albanian cause.[3] The key-arguments put forward in Kosovo's application to be admitted to UNESCO (and other international institutions and agencies for which Kosovo's Albanian authorities applied since 2008 following the unilateral proclamation of its independence) were that the state

[1]http://balkans.aljazeera.net/vijesti/kosovo-nije-primljeno-u-unesco—The then-Kosovo Prime Minister, Isa Mustafa, in an interview given to the Al Jazeera Balkans immediately after the vote took place on 9 November 2015—Accessed on 1 March 2019.

[2]http://www.rts.rs/page/stories/sr/story/9/politika/2098204/kosovo-nije-primljeno-u-unesko-spiskovi-kako-su-glasale-drzave-clanice.html—The then-Serbian Prime Minister, Ivica Dačić in an interview to the Serbian national broadcaster on the same day—Accessed on 1 March 2019.

[3]How divisive the problem of the monuments in Kosovo and Metohija were within UNESCO voting process was best evidenced in the result of the vote, which the Albanian side lost with only three votes of difference.

C. Gambardella et al. (eds.), *Advances in Utopian Studies and Sacred Architecture*,
Advances in Science, Technology & Innovation,
https://doi.org/10.1007/978-3-030-50765-7_9

was recognized by the majority of the Western democracies and that the state of Serbia no longer holds the right to the territory of its southern province because of the alleged war crimes committed against the local Albanian population. Following the NATO led military campaign in 1999, the Albanian insurgents took over the province of 1.8 million people and are currently in the process of creating an independent state, free of its Serbian population and heritage (UNHCR 2019).[4]

However, it appears that the UNESCO experts took into consideration arguments different from those of the Albanian side, despite the relentless efforts put by the Albanian authorities in Priština and their Western supporters into proving Albanian rights and ability to take care of the heritage of Kosovo and Metohija. The main concerns expressed by the majority of the voting states in UNESCO in November 2015 were not about the Albanian ability, but their attitude towards the heritage of Kosovo and Metohija (Лазић 2019).[5] Those concerns predated the 2015 vote for almost a decade. Still vivid recordings and photographs of the organized destruction of the mediaeval Serbian Christian heritage and attacks on their communities in Kosovo and Metohija on 17 March 2004 put in serious doubt any Albanian claims to the stewardship of the World Heritage Sites on the territory of the troubled province (Figs. 1, 2 and 3). The organized violent attacks on the Christian edifices throughout Kosovo and Metohija which began on that day were conducted by the local Albanian population of the predominantly Muslim faith, with the KFOR peace-keeping missions standing by without interfering.

These attacks represented a continuation of similar violence against the heritage that was taking place for decades in Kosovo and Metohija and which had culminated during the 1990s. They significantly intensified after the NATO troops (now renamed into KFOR) occupied the province in June 1999 following the military intervention alongside the Albanian insurgents (Fig. 4).

Fig. 1 Serbian Orthodox Seminary in Prizren. Founded in 1871, near the fourteenth century Church of Our Lady of Ljeviš, UNESCO site, also destroyed in March 2004

Fig. 2 Church of the Holy Saviour, fourteenth century, Prizren. Destroyed on 17 March 2004

2 Why Kosovo Matters?

It can be argued that in its most rudimentary form, the conflict between the Serbs and Albanians in Kosovo and Metohija represents one aspect of "clash of civilizations" along the "civilizational fault-lines, which are particularly prevalent between Muslims and non-Muslims" as argued by Samuel Huntingdon in the early 1990s (Huntington 1996).[6] However, the support which the Albanian Muslims received from the West during the Yugoslav wars was difficult to incorporate into domestic politics of the main Western states, already struggling with growing unease towards their own

[4]Since June 1999, a large number of Serbs and other non-Albanians left province under coercion by the Albanian authorities. The UNHCR cited 252,523 persons of concern for the territory of Serbia, including Kosovo at the end of 2017, of which IDPs were counting to 217,453, predominantly from Kosovo and Metohija. See official web-page of the UNHCR: http://popstats.unhcr.org/en/overview#_ga=2.267661582.1431272768.1552401925-809441953.1552401925—Accessed on 1 March 2019.

[5]Former Serbian ambassador at UNESCO HQ, Darko Tanasković, in an interview „Насиље над српским светињама неће стати" to the Magazine „Православље", 19 February 2019.

[6]Huntington (1996), pp. 207–238.

Fig. 3 Destruction of the Monastery of St Archangel in March 2004, Prizren, fourteenth century

Fig. 4 Church of St. John The Baptist, thirteenth century, where, according to legend, Prince Lazar and his army received the last Holy Communion on the eve of the Kosovo Battle in 1389. It was destroyed during both World Wars, rebuilt, damaged again during the Albanian insurgencies in 1981, partly destroyed during the NATO attack in 1999 and finally demolished and turned into a scrap heap and public toilet (photograph 5 to the right) in March 2001

Muslim populations (Huntington 1996).[7] Deep divisions that existed and still exist among non-Muslim and Muslim population in the major Western states were thought to be easier to overcome if the political elites in the West actively present to their Muslim citizens that they respect Islam and its culture. The Yugoslav conflict became a convenient showcase, as it gave an opportunity to the Western leaders to depict themselves as protectors of Islam and Muslims, especially as the conflict itself was taking place in the part of Europe which was never considered part of the European civilization.

This was not a new approach to the Balkan area, because the position held by the major Western powers was that of an enduring "evolutionary belief in the superiority of orderly European civilization over barbarity, archaic predispositions, backwardness, squabbles, uncomforting and unpredictable behaviour—tribalism" (Todorova 1997).[8] The notion of tribalism, especially within European framework, as Maria Todorova eloquently argued, relegated the Balkans to a lower civilizational category, which is in itself "intrinsically passive, incompatible and imitative in nature". This perception enabled the maintenance of imperial principles towards the Balkans and "released the civilized world from any responsibility or empathy that it might otherwise bestowed on more reasonable people" (Todorova 1997).[9] Similarly, Balkan tribalism became "a convenient substitute for the emotional discharge that *orientalism* provided, exempting the West from charges of racism, colonialism, *eurocentrism* and Christian intolerance towards Islam" (Todorova 1997).[10] Therefore, the position of "tolerance towards Islam in Europe" by the Western political elites during the 1990s wars in Yugoslavia was a conscious political decision which, unleashing aggressive and unscrupulous propaganda in the media, required both social and academic justification. It is not surprising, thus, that the outcomes of such political decisions resulted in the revisionist interpretations of heritage in situ, which had little or nothing in common with its artistic meaning and historic significance. Subsequently, when the physical destruction of the Serbian Christian Mediaeval heritage in Kosovo and Metohija took place in March 2004 which the NATO troops stationed there failed to prevent, the defenders of the politics behind the Albanian political construct in Kosovo and Metohija argued that this was an understandable act of revenge and that the only side which needs to be "punished" should be Serbian (Petiffer and Vickers 2009).[11] The Albanian Muslims, encouraged by the combination of deliberate choice of political inaction by the major Western powers, understood that there will be no consequences for their conduct against the Serbs, no matter how destructive or inexplicable they may seem to the ordinary citizens of the Western states which supported them.

The disparity between political rhetoric of the Western elites and media reports that could no longer obscure and/or hide highly volatile situation in Kosovo and Metohija evidenced by the destruction of centuries-old heritage, prompted the key-policy makers in the West to launch a new explanatory campaign in order to defend their choice of sides in the conflict. Again, aimed primarily at their own citizens, but also to their new pet-nations in the Balkans, the political leadership of the West further exacerbated the problem by endorsing those academic arguments which relied on dubious evidence and selective use of historical sources.[12] This was not a novel approach to the Balkan political framework by the West; some initial theoretical work was already put forward before the military action took place in 1999, as part of the preparation for the interventionism.[13]

2.1 Revisionist Approach to Historical Narratives

In order to justify the intended military action against Serbia—singled out as a unique rogue state and nation, solely responsible for the disintegration of Yugoslavia, key-Western states engaged a significant number of academics and journalists to write, publish and promote new versions of local history, often relying on incomplete or deliberately distorted evidence to support their argument. The British scholarship produced several books which became influential worldwide primarily owing to the widespread usage of the English language rather than genuine academic efforts. Already during the Bosnian campaign in 1994, Noel Malcolm, a British historian working as a journalist for *The Spectator* and the *Daily Telegraph* wrote a short history of Bosnia which departed from the standard history textbooks that existed until then. The book *Bosnia, A Short History* was widely publicized as "*a brilliant work of history which set the terrible war in the Balkans in its full*

[7]Ibid., p. 209.
[8]Todorova (1997), p. 185.
[9]Ibid., p. 185.
[10]Ibid., p. 188.
[11]Petiffer and Vickers (2009), Preface.

[12]The term "pet-nation" was first introduced by Rebecca West in her travelogue *Black Lamb and Grey Falcon: A Journey through Yugoslavia*, published in 1941 in London. The term is self-explanatory and denotes the small nation favoured by one of the big powers.
[13]It began with pervasively biased international journalist reports from Yugoslavia when the wars broke out in 1991 and continued with "academic" presentation of the situation. In 1994, Noel Malcolm (b.1956), a British historian working as a journalist, wrote a short history of Bosnia which set the course of revisionism of Serbian history. See further discussion.

historical context" (Malcolm 1994).[14] Acknowledging that he never visited Sarajevo libraries (Malcolm 1994).[15] Malcolm wrote an eloquent and well-referenced account, based predominantly on a distorted use of secondary sources, written mostly by the late nineteenth and early twentieth century Austro-Hungarian and twentieth century Croat émigré historians close to the pro-Nazi NDH regime of the Second World War, openly hostile towards the Serbs (Ković 2011).[16] The *Bosnia, A Short History* set the course for the future revisionism of the Balkan history in the English Language literature and the British scholarship.[17] In 1998, a year before the open war between NATO and the remainder of Yugoslavia escalated, Malcolm published a second book *Kosovo, A Short History*, in which he formulated the basic postulate according to which the "pagan mediaeval Serbs" only occupied Kosovo inhabited by the Albanians in the late twelfth century (Malcolm 1998).[18] Again, using the same research methods which neglected the majority of the primary sources and emphasized the scholarship from the period of Austro-Hungarian expansion through the Balkans in the later nineteenth and early twentieth centuries, Malcolm created the revisionist narratives since followed by the majority of Western studies of the region.

After the wars, the newly established states on the ruins of Yugoslavia and the change of the political system were accompanied by the revision of national grand narratives within the new polities. The revised narratives, based on Malcolm's work, brought into existence a series of new, mutually opposing interpretations of the surviving religious architectural heritage, with sole aim to legitimize modern ethnic claims to territories within current borders. This forced conversion of the meaning of historic architecture, established the utopian realm in minds of its creators with little supporting evidence found in situ, either scientifically or socially. Although common politically, such interpretations prove to have long-lasting negative consequences in the conflict-prone Balkans. Both Malcolm's books asserted the argument of mediaeval Serbian expansionism over the non-Serbian, predominantly Muslim territories, and as a by-product brought into prominence the claim that the Serbs, since the beginning of their independence in the early nineteenth century, deliberately targeted not only the Muslims, but the Islamic heritage in all territories they attempted to conquer since the earliest days of their national history (Zulfikarpašić 1998; Muller 2009).[19] Drawing heavily on Malcolm's arguments, and openly supporting Albanian cause, two other British authors, James Petiffer and Miranda Vickers, asserted in the first year of Kosovo's self-proclaimed independence that "*Serbian destruction*" of the "*ancient monuments, mosques and houses in order to commit the culturcide of the Albanian ethnicity*" had to be "*rightfully punished by the International Community*" (Petiffer and Vickers 2009).[20] This manner of academic approach was further promoted by founding graduate and post-graduate programmes at the key-British universities which would educate new Albanian scholarly elite ready to disseminate new version of national narrative and its position in regards to the Serbian history and state.[21] Needless to say, the Albanian scholars in both Albania and Kosovo and Metohija welcomed Malcolm's initial postulates as the basis of their national narratives.

However, this way of interpretation, which freely uses anachronistic methods to satisfy current political relations unwittingly posed two sets of mutually exclusive arguments:

(1) Malcolm's assertion that the Serbs attempted to commit "cleansing" of the non-Serbian population from those territories prior to the Ottoman conquest (Malcolm 1998)[22]:
  - allowed political justifications for targeting Serbian history and heritage without fear for consequences in the short term, and
  - enabled its usage as substitute for everything negative in Europe (and even the world) by comparing it

[14] Malcolm (1994), cover page.

[15] Malcolm (1994), Acknowledgements.

[16] Ković (2011), p. 403.

[17] For a non-specialist, the book reads like a genuine work of scholarship. One problem was a total absence of references to any primary source and some mistranslations of the Serbian/Croatian words. Needless to say, Malcolm's books were highly praised by the Muslim scholars of Bosnia and Herzegovina and the Albanian scholars in both Albania and in those neighbouring states where the Albanian minority currently asserts territorial claims (Montenegro, North Macedonia and Greece). Subsequently, the Bosnian Muslim scholars of Bosnia and Herzegovina and the Albanian scholars in both Albania and Kosovo and Metohija adopted Malcolm's initial postulates as the basis of their newly established national narratives.

[18] Malcolm (1998), p. 26.

[19] Zulfikarpašić (1998), pp. 23–24. "In accordance with the principle *cuius regio eius religio* Serbia had been cleansed of Muslims, and even of those Serbs who had converted to Islam and who lived around Užice and Valjevo. In negotiations between Turkey and Serbia they had been declared Turks and forced to move, and so they had resettled in Bosnia. There are still hundreds of families in Tuzla, Šamac, Sarajevo and Foča who are descendants of these immigrants from Užice — Serbian speaking Muslims. This was all a repeat of what had happened a few centuries before in Slavonia and Lika." Zulfikarpašić greatly respected Malcolm's work. Also, Müller (2009), p. 70. Both Zulfikrapašić and Müller drew heavily on Malcolm, who even claimed that the "spread of Orthodoxy was favoured by the Ottoman authorities" and that the Serbian Orthodox monasteries in Bosnia were built as a result of Serbian expansionism. Malcolm (1994), p. 71.

[20] Petiffer and Vickers (2009), Preface.

[21] The university activities were supported by vivid publishing activities. During and after the conflict, IB Tauris Publishing of London, now part of the influential Bloomsbury Publishing was particularly active in advertising revisionist historical narratives.

[22] Malcolm 1998, p. 54.

with other, often dissimilar events of European history in the long run.[23]

(2) The second set of arguments dealing with revisionism of the Serbian national history and heritage is already resulting in disruptive discourse in those parts of the world where similar political tensions exist, as it sets a precedent for manipulating the facts of some future conflicts.[24]

2.2 The Problematic Evidence

The greatest obstacle facing the revisionist approach to the historical interpretation was total absence of evidence which would be able to support new grand narratives. In Kosovo and Metohija, there are no primary sources which refer to the presence of the Albanians in the territory prior to the late nineteenth century.[25] Similarly, there is no archaeological evidence which can be clearly connected to the presence of the Albanian ethnicity in the same territory prior to the early eighteenth century—and even then, it displays Islamic religious affiliation. On the other hand, the evidence of Serbian presence in Kosovo and Metohija predating the appearance of Islam in the disputed territories is numerous, both in written sources, as well as archaeologically. This inconsistency between the newly formed claims of the Albanian presence in Kosovo and Metohija prior to "the Serbian occupation in the late twelfth century" and material remains which indicate the opposite has been argued in anachronistic terms in the past three decades. One typical example is that of Naser Ferri, a professor of archaeology at the University of Priština, who quite seriously argued that: "*Their* (Serbian) *history and their acts have shown that Serbs were and still are one of the most destructive people in the world, who, during their migration from the Carpathians and occupation of the Balkans, destroyed everything where they stepped. One of the few things they accepted without destroying was Orthodox Christianity. After they accepted it, a long period began in which they embezzled sacred Dardanian*[26] *objects and adopted them for the needs of the Serb Church. Even today, the traces of this assimilation are evident in ancient churches throughout Kosovo…There is information that even the Patriarchy of Peć, the Dečani Monastery, Gračanica, St. Marko Monastery, Church of Korish in Lipljan and many other famous 'Serbian' churches were constructed precisely on the foundation of earlier autochthonous churches…*" (Herscher 2010)[27] The claims of "Serbian destruction" and "imposition of Serbian architecture" of the previous architectural edifices that used to belong to the "autochthonous", expressed in this manner, consciously generated the sense of being nationally victimized among the Muslim Albanians in Kosovo and Metohija. The vague term of "autochthonous churches" aiming to obfuscate the existence of the Serbs prior to the twelfth century, further served the later assertion that "disappearance" of autochthonous churches enabled quick conversion of the Albanians to Islam in the centuries after the Ottoman conquest.[28] However, this method of presenting the arguments can be disputed without difficulties by posing two mutually related questions:

(1) If there was indeed a national programme or intent to "embezzle the Dardanian objects and adopt them for the Serbian church" in the twelfth–fourteenth century period—why are there no edifices of similar architectural styles or, at least, the remains of them further south of Kosovo and Metohija, in Albania proper, where the majority of the Albanians live? (Malcolm 1998)[29]

[23]See works Margaret McMillan and Christopher Clark related to the outbreak of the First World War, for example.

[24]In Europe, the political question of Crimea raises tensions between Russia, Ukraine and the EU.

[25]In 1854 German diplomat Georg von Hahn (1811–1869) published in Jena three volumes of Albanian Studies (*Albanesische studien*) on Albanian history, language and culture, focusing on the Indo-European origins of the Albanian language. After Hahn's pioneering endeavour, there were no new attempts to write about Albania or the Albanians prior to decade leading to the establishment of the independent Albanian state in 1913.

[26]The Dardani were an ancient Thraco-Illyrian tribe, first mentioned in the fourth century BC and which Albanian scholars claim to be the true ancestors of the modern-day Albanians. The evidence of this link is, however, difficult to establish, as there are no archaeological links between the Dardanians and Albanians. The written sources are also non-existent.

[27]Daily *Zëri Ditor*, published in Priština in Albanian language. The quote in question appeared on 12th March 2003, quoted in Herscher (2010), p. 132. The famous Tower of London was built on the foundations of an old Roman *castrum*, albeit several centuries after the Romans left Britain. If the interpretative methods similar to those applied in the case of Serbian mediaeval monuments in Kosovo and Metohija are applied in the case of English history, it would be equally valid to value William the Conqueror (1028–1087), the founder of the English Regnal order, as the originator of genocidal tendencies later displayed by the English in the American colonies and in India.

[28]The churches did not disappear until their systematic destruction began during the Ottoman period. Their sheer presence in the province should have been the reason enough for the maintenance of the Christian faith among the Albanians, had they been present in the Kosovo and Metohija prior to the Ottoman conquest.

[29]Malcolm (1998), p. 55. Malcolm conveniently never mentions this in his writings. However, he indicates "Serbian ethnic segregation of the Albanians and Vlachs" by selectively quoting the early twentieth-century Romanian author Silviu Dragomir who, actually argued that Vlachs were Romanian and as such, different from the Albanians, and as such came into collision with Albanian narrative of the "Albanian Vlachs".

(2) If the population of Kosovo and Metohija, both Serbian and Albanian, in the centuries before the appearance of Islam in this area was Christianized under the strong influence of Byzantium and was, therefore, using the same church doctrine for observing the religious rites, how did it come that only the Albanian population converted to Islam and the Serbian did not?

The revisionist argument insists that the Albanian Christian population of six or more centuries ago resented the mediaeval Serbian kingdom and, thus, eagerly accepted Islam after the Ottoman takeover as their only road to salvation (Di Lellio 2009).[30] Maintaining the narrative along the same lines, the ethnic and religious antagonisms which grew between the Serbian Christian and Albanian Muslim populations over the centuries understandably resulted in the attacks on the mediaeval Christian Orthodox churches and monasteries. The most dramatic that occurred in March 2004 were, therefore, rightfully justified following the 1999 war in the eyes of the Western democracies (Petiffer and Vickers 2009).[31]

The answer to the first question requires the architectural analysis of the monuments in question themselves. Out of hundreds churches and monasteries under the jurisdiction of the Serbian Orthodox Church in Kosovo and Metohija, it will be sufficient to observe those that ended up on the World Heritage List in Danger in 2006: the Dečani Monastery (built 1327–1335 for the Serbian King Stefan Dečanski as his mausoleum), the Patriarchate of Peć (a group of four domed churches built before 1260–1337), Church of the Holy Virgin of Ljeviška (built in 1306) and the Gračanica Monastery (built in 1321 on the ruins of the sixth century Byzantine basilica). Architectural features which distinguished them from other mediaeval and later churches and monasteries built in Kosovo and Metohija were attained under the Nemanjić dynasty (1166–1371), when the Serbian Kingdom reached its apogee.

Mediaeval practice of erecting monasteries and churches on the foundations of earlier shrines and settlements was well-documented and explained throughout world. The conversion to Christianity and/or Islam did not always result in the destruction of older temples and shrines. On the contrary, elements of old buildings were frequently incorporated into the new architectural pieces.[32] Building on top of the old sent strong messages to the surrounding population; irrespective of whether such practise represented the rejection of old deities which were no longer needed or enforcement of the new system of beliefs with present fear of wrath of the old gods or simply because it was cost-effective to recycle the building materials from the old sites, it did not mean that it bore any chauvinist intentions. Quite often, the populations which built on the top of old sites were repopulating abandoned and/or scarcely inhabited territories after the collapse of the previous cultures and civilizations. In any case, the phenomenon of nationalism as it is defined now is a relatively modern concept, as argued by scholars such were E. Gellner, A. Smith or E. Hobsbawm (Hobsbawm 1983; Gellner 1983; Smith 1991).[33]

2.2.1 The Dečani Monastery (1327–1335)—The Final Monument of the Raška Style

The example of the Dečani Monastery solely questions the validity of revisionist claims. As mentioned earlier, the Dečani Monastery was built by the Serbian King Stefan Uroš III Dečanski (reigned 1322–1331) as his mausoleum. Architecturally, the main church dedicated to Christ Pantocrator is built as a five-nave domed basilica in the manner of the *Raška Style of Architecture* that flourished in Serbia between 1160s and early fourteenth century. As the final monument of the *Raška Style*, the Dečani Monastery followed the cannons of this style, first established by the Studenica Monastery about 150 years earlier north of Kosovo, in western Serbia. The *Raška Style* successfully blended the Romanesque—Roman Catholic and Byzantine —Orthodox building techniques (Fig. 5). No other Orthodox architectural achievements ever incorporated elements of Catholic building traditions.

This was not accidental. Stefan Nemanja (1113–1199, reigned 1166–1196), the founder of the monastery and the dynasty, was born in Ribnica near Podgorica in Montenegro and was baptized according to the Catholic rite. He was later re-baptized in the Orthodox church of St. Peter near the then capital of Ras.[34] The wars taking place between the Byzantine Empire and the Hungarian Kingdom at that time were often followed by swaying religious loyalties and the ruling elites converted from Catholicism to Orthodoxy without prejudice. This societal phenomenon reflected on architecture as the builders from Catholic backgrounds were

[30]Di Lellio (2009), p. 32.

[31]Petiffer and Vickers (2009), Preface. Petiffer openly argues in the same book—and elsewhere in his public speeches—that North Macedonia should be divided and its Western part given to Albania.

[32]The Great Mosque in Kairouan comes to mind with its free usage of older Roman, Byzantine and Early Christian edifices from nearby.

[33]Those giants of research studies of nationalism frequently challenged each other and had their disagreements. However, all of them agree that within modern societies political elites often deliberately impose their own views of the past on the populations they rule. Thus, the argument of mediaeval intention of destroying what Albanian scholars see as their own ethnic territories falls within the framework of anachronistic approach. Hobsbawm and Ranger (1983), p. 6, Gellner (1983), p. 50 and Smith (1991), p. 119.

[34]Overlooking the Raška valley, the mighty fortress of Ras was the capital of Serbia. Its importance was such that it influenced the creation of the exonym *Rascians*, used by Hungarian and German authors for the Serbs.

Fig. 5 Studenica Monastery (1186–1196), the first example of the *Raška Style of Architecture*, UNESCO World Heritage Site, 1986

frequently employed to work on Orthodox churches and monasteries and vice versa. There is no information about the architects who built Studenica (1186–1196), but it is almost certain that they were of Catholic background. The monument itself represents fusion of the Romanesque architecture, which is evident in the base of the church, representing an elongated Latin cross and its exquisite stonework (Fig. 6), as well as the Byzantine tradition, which is evident in the choice of dome and fresco paintings.

The most distinctive feature of the *Raška Style* is the unrestricted usage of carved stone sculptures, both on the façade, as was already shown on Figs. 6, 9 and 11, and within the church interiors (Figs. 7, 12 and 13). Unknown elsewhere in the Orthodox ecclesiastical architecture, the *Raška Style* remains the authentic evidence of the turbulent twelfth and thirteenth centuries in this part of Europe. When Studenica was founded and erected, the Byzantine Empire was ruled by the fierce Komneni dynasty, before seizing to exist for half a century in 1204. When it was resurrected under the Paleologues in 1261, the imperial coffers were so depleted of cash that the economic crisis which lasted until the Ottoman conquest disabled the introduction of any new architecture not only in Constantinople itself, but elsewhere in the Orthodox world (Mango 2002).[35] By the time of the

Fig. 6 Romanesque window, detail from the altar wall, facing East

[35]Mango (2002), pp. 248–307.

Fig. 7 Carved narthex with fountain

Fig. 8 Dečani Monastery (1327–1335) the last example of the *Raška Style of Architecture*, UNESCO World Heritage Site, 2004

construction of Dečani, the *Raška Style* flourished in the territory covered by the River Raška (hence the name of the style), which included parts of Serbia west of the Morava valley, Metohija, Montenegro and extended further west to Bosnia and Herzegovina and Dalmatia. Presently, around ten monasteries belonging to the *Raška Style* and dating back to the period in question, survive on this geographic territory. They all bear the characteristics introduced by the architecture of Studenica. No monuments of such characteristics appear anywhere else in the Byzantine world. It is significant to notice that such building practise never reached the imperial capital—Constantinople (Figs. 8 and 10).

Unlike at Studenica, the erection of the Dečani Monastery was well-documented and the name of its architect is known. The founding charter of the monastery, dated at 1330, miraculously survived deliberate destruction of the National Library in 1941, where it is kept to this day. The architect who built the Church of Christ Pantocrator at Dečani was a Serbian Catholic priest Fra Vito of Kotor (Cattaro), famous for his artistic skills and a man responsible for a number of catholic edifices on both sides of the south Adriatic (Ковијанић 2016, reprint 1962).[36] The Albanian press asserted in 2015 that Fra Vito, the Franciscan from Kotor, was actually of Albanian origins from the Albanian Kuçi clan (Malović 2016).[37] Without going further into the dispute, Kotor archives evidence the existence of Fra Vito and his family (surname Čučo, of Slavic origins), father Trifun (name of Greek origins). Fra Vito himself left his signature on the relief of the southern portal in the Serbian redaction of Old Church Slavonic (Fig. 10).

Fig. 9 Romanesque window, above the main portal, facing West

2.2.2 The Gračanica Monastery (1315–1321) and the Church of the Holy Virgin of Ljeviška (1306–1307)—The Serbo-Byzantine or the Vardar Style

Unlike Dečani, the monastery of Gračanica (Fig. 14) and the church of Holy Virgin of Ljeviška (Fig. 15) were built in the style that became fashionable in the Serbian Kingdom at the beginning of the fourteenth century: the *Vardar Style*. Just like the *Raška Style* before it, and the *Moravska Style* after, the *Vardar Style* (c.1307–c.1395) was named after the river valley in which it first appeared and where the majority of the surviving churches and monasteries are situated. Beginning in the vicinity of the then-Serbian capital Skopje,[38] it spread along the Vardar River in northern Macedonia and westwards to Kosovo and Metohija. The main promoter of the new style was King Stefan Uroš II Milutin (1253–1321, reigned 1282–1321) whose long reign was marked by the introduction of the court ethics modelled on those in Constantinople (Fine 1994; Ćirković 1969).[39] Both the Gračanica and the Virgin of Ljeviška were paid for by King Milutin, who acquired the Byzantine cultural framework in order to secure peace with Byzantium while dealing with the constant threat from his more Westwards looking brother (Fine 1994).[40] Following victory, Milutin vowed to build one church for each year of his reign, and it seems likely that he kept this promise at least in part, because the number of churches and monasteries that he had founded and/or reconstructed by far surpasses the number of endowments left by other Serbian mediaeval rulers (Curcic 1979).[41] His choice of Byzantine architectural traditions was directed not as much by his aesthetic preferences as it was by his political and economic decisions. In order to be able to pay for numerous construction projects simultaneously, the

[36]Ковијанић (2016), p. 56.

[37]https://arhiva.nedeljnik.rs/nedeljnik/portalnews/vita-kotoranin-neimar-decana/—Accessed on 2 April 2019. An article by Nikola Malović, *Vita Kotoranin, neimar Dečana*, dated 25 November 2016, disputed this assertion in the press in Serbia and Montenegro.

[38]Now the capital of North Macedonia, which concurrently develops its own new national narrative, separate from Serbian and Bulagarian and in collision with Albanian.

[39]Fine (1994), p. 223. Another argument for Milutin's acquisition of the Byzantine ethics was that he saw the attempted *rapprochement* between Rome and Constantinople at the Council of Lyon in 1274, as a threat to Orthodoxy, which resulted in styling himself as a true defender of the faith. Ćirković (1969), p. 41.

[40]Ibid., p. 257.

[41]Curcic (1979), pp. 6–7. Among others, Milutin founded monasteries in Atos, Sinai and Jerusalem, which still exist.

Fig. 10 Founding charter of the Dečani Monastery, dated 1330, now in the National Library of Serbia

King opted to support rebuilding and/or reconstruction of the existing buildings, erected by the Byzantines in the centuries before that part of the Slav populated Balkans became part of the Serbian crown possessions.

It is difficult to determine the first monument to be built in this style, as many of them were executed almost simultaneously. However, together with a few smaller churches in North Macedonia (Fig. 18), the Church of the Holy Virgin of Ljeviška is usually taken as the beginning of the *Vardar Style* within the Serbian kingdom. Built on the ruins of two much older basilicas dating back from the early Byzantine period, it is argued that it had been modelled on the Church of the Holy Apostles in Thessaloniki, with its cross-in-square basis (Fig. 16). If this was the case, the Virgin of Ljeviška represents a firm departure from the building traditions of the *Raška Style* monuments, with its Orthodox cross plan and harmonious usage of horizontal layers of bricks and stones in the façade walls. The choice of layered walls had not only structural, but decorative role, which does not appear in *Raška Style*. Putting the accent on the façade as a whole, the need for carved stonework had faded, while the fresco paintings took over the narrative role (Fig. 17).

The monastery of Gračanica (Fig. 14) displayed the same characteristics as the small churches in North Macedonia and the Holy Virgin of Ljeviška, only on a much more monumental scale. For the majority of the monuments built in the *Vardar Style*, the names of the architects are lost, as they respected the Byzantine tradition of anonymity. However, the name of the architect who built Ljeviška, Nikola, has been preserved on the inscription in the exo-narthex. If this Nikola is the same Nikola who built Gračanica together with his two brothers, Djordje (George) and Dobrosav, then it can be concluded that the architects were local men, of which Dobrosav was undoubtedly Serbian, which is evident in his Slav/Serbian name.[42]

2.2.3 The Patriarchate of Peć (Before 1260–1337)

The complex of the Patriarchate of Peć consists of four surviving domed churches built over the period of seven decades. The earliest monument is the Church of the Holy

[42] Nikola and Djordje (George) are Biblical names used by the Orthodox Christians and can be attributed to the Greeks or Serbs or Albanians. Dobrosav, on the other hand is a Slavic/Serbian name which the Greeks or Albanians did not use.

Fig. 11 Inscription made by the architect, Fra Vito of Kotor with his name and the year of construction. Southern portal. The inscription was made in the Serbian redaction of the Old Church Slavonic. The supporting angels represent transition from Romanesque to Gothic sculptural traditions

Apostles, built by Archbishop Arsenije I a few decades before 1253.[43] Three additional churches followed in a short period at the beginning of the fourteenth century: St Dimitrius was added to the northern wall by Archbishop Nikodim in 1321–1324, while the Holy Virgin and St Nicolas, bordering the southern wall of the Holy Apostles were both built by the Archibishop Danilo I in 1330 and 1337, respectively.

The complex suffered many damages and changes over the centuries, but the surviving architecture clearly indicates that they were built in combined *Raška* and *Vardar Styles*, which corresponds well with periods in which they were built. The Church of the Holy Apostles (Fig. 19) was modelled on the older Monastery of Žiča (1206–1221), near Kraljevo, western Serbia, and the first seat of the Serbian Archbishops following independence from Constantinople in 1217.

[43]The seat of the Serbian Archbishopric was transferred from the Monastery of Žiča (*Raška Style*) to Peć in 1253, following the disturbance of peace. Thus, the Holy Apostles were already built prior to 1253.

The *Raška Style* features of the Holy Apostles are evident in its base which forms the usual elongated cross and Romanesque windows. The *Vardar Style* characteristics are evident in the Churches of St Dimitrius and Holy Virgin, which both have the typical cross-in-square base. The small St Nicolas has layered stones and bricks façade, also characteristic of the *Vardar Style* (Fig. 20).

3 Palaeologian Renaissance—The Architectural or Painting Style?

As indicated earlier, since the outbreak of wars, a number of new national histories introducing completely revised national *grand narratives* were written and published in Yugoslav successor states. Lacking written sources, both historical and modern, the authors of those studies turned to the architectural and archaeological heritage, which were subjected to unashamed political reinterpretation. When the insurgencies in Kosovo and Metohija broke out in 1997 and 1998, the collections and archives from provincial museums were evacuated to more safe institutions, together with the

Fig. 12 Carved lion on internal side of the portal in Dečani

documents of the Institute for the Protection of Monuments in Priština, anticipating the possibility of outbreak of war (Herscher 2010).[44] The churches and monasteries in Kosovo and Metohija had their treasuries evacuated to the church estates less exposed to the hostilities.

However, the 1999 bombing of Serbia proved to be devastating. The Commander of the NATO forces at the time, Wesley Clark, explained that the command and control facilities, which included general infrastructure, industrial and administrative buildings and bridges were all legitimate targets, regardless of their historical value (Herscher 2010).[45] The architectural and archaeological heritage damaged or destroyed was considered "collateral damage". This explanation directly caused a series of insurgency and counterinsurgency actions of both Albanian and Serbian forces against the cultural and religious symbols of the other nation. The three-month campaign put an end to the hostilities in former Yugoslavia and permanently incapacitated Serbia for any national reassertion in the near future. The Serbian army, together with over 200,000 Serbian civilians withdrew from Kosovo and Metohija,[46] which, supported by the EU and the USA, proclaimed independence from Serbia on 17 February 2008.

After the withdrawal of the Serbian state, the systematic destruction of Serbian Orthodox cultural heritage began in Kosovo and Metohija. When the subject occasionally arose as an issue in the western media, it was explained as an "understandable by-product of the processes of revenge for the Serbian misdeeds against the Albanian population before and during the NATO intervention" (Ćurčić 2000).[47] Following this premise, an American architect Andrew Herscher, who served as cultural heritage officer to the Albanian Kosovo government in 2001–2005, wrote a book on the destruction of heritage in Kosovo and Metohija. Aware of the difficulties that his account would face due to his personal involvement in working for Kosovo's independence, Herscher attempted to balance the West-established narrative of the Serbian aggression on Kosovo (and other non-Serbian territories), with justifiable destruction of Serbian heritage as

[44]Herscher (2010), p. 15.
[45]Ibid., p. 100.

[46]The number corresponds to that given by the UNHCR. See footnote 4, above.
[47]Ćurčić (2000), p. 129.

Fig. 13 Capitals inside the Dečani church. Even though they clearly draw on ancient traditions, the capitals display unique carvings representing various animals and leaves, which was usually practised in the Western churches

understandable revenge (Herscher 2010).[48] According to him, Serbian forces "*vandalised, destroyed or damaged 225 out of 600 mosques during their counterinsurgency campaigns in 1998/99*" (Herscher 2010).[49] Against this data, he gives a number of "*151 Serbian churches and other patrimonial sites*" in the weeks following the withdrawal of the Serbian army and the refugees from Kosovo and Metohija in 1999 alone (Herscher 2010; (Raško-Prizrenska 2004).[50] However, the lists of damaged and destroyed monuments in Kosovo and Metohija *since* the withdrawal of the Serbian state compiled and documented by the Serbian heritage institutions and Serbian Orthodox Church reached a limited number of professional organizations in the West, where they were either received with reserved consternation or, more commonly, remained quietly ignored (Ćurčić 2000).[51] Thus, when the 2004 destruction took place in front of the Western audience, it became impossible to justify Kosovo's application for the UNESCO membership.

Eager to present their involvement in the 1990s Yugoslav wars as a success, the European Union and the USA, quietly supported the ethnical cleansing of Serbs from Kosovo and Metohija simultaneously with the promotion of a new national narrative of Kosovo's history and ethnic composition, which excluded the centuries-long Serbian presence there. Since the number of the Serbian Orthodox monuments by far predates and outnumbers the Islamic edifices used by the Albanian population of the province, the Western *apparatchiks* in collusion with Albanian historians from the University of Priština came up with the idea of the acquisition of the Serbian heritage by the ethnic Albanians. The result of this collaboration was the invention of the term "*Palaeologian Renaissance*" for all of the Serbian mediaeval monuments situated outside of the

[48]Herscher 2010, p. 15.

[49]Ibid., p. 87.

[50]Ibid., p. 136, Herscher's number refers to the period 1999–2003 and is similar to that given by the Serbian authorities. However, since 2003, Serbian heritage institutions count over 151 churches, 5.261 tombstones, 60 graveyards and over 10.000 smaller artistic and religious objects, such as icons, paintings, and other church decorations. Source: Serbian Orthodox church, Raško-Prizrenska Eparchy, http://www.eparhija-prizren.org/?p=26657—Accessed 17 April 2019.

[51]Ćurčić 2000, p. 129.

Fig. 14 Monastery of Gračanica (1315–1321)

current Serbian borders, irrespective of their evident differences in architectural styles.

Initially, the term referred to the revival of Orthodox Byzantine paintings under the last Byzantine dynasty after the short period of Latin rule (1204–1261) by the Western authors. Already in 2000, the Oxford scholar Robin Cormack discussed the term "Palaeologian renaissance" as imprecise and unclear: the word "Renaissance" was always most commonly associated with the revival of the classical art and architecture in fifteenth–sixteenth-century Italian states and suggested the use of "r" instead of the capital "R" in the word "renaissance" in order to distinguish it from the Renaissance in Italy (Cormack 2000).[52] However, a number of German, American and British academic institutions redeployed the term for depicting the last period of the Byzantine history as the "cultural renaissance" which spread from Constantinople towards other Orthodox kingdoms founded on the old Byzantine domains (Fryde 2000).[53] Emanating from the arguments relating to the Byzantine paintings and manuscripts, the usage of the term was quickly transferred to the descriptive analysis of architecture. Conveniently, this practice served as the basis for renaming the Serbian heritage, its architectural schools and fresco paintings originating in the territories belonging to the mediaeval Serbian kingdom into the heritage belonging to the Palaeologue period of the Byzantine Empire (1261–1453). The political pressure of adopting this newly coined term resulted in the four most important monuments of Serbian mediaeval heritage being inscribed to the World Heritage List under the name which denies their Serbian origin (UNESCO 2019a).[54] Paradoxically, the Greek cultural institutions and professionals concerned with architectural heritage appear to be

[52]Cormack (2000), p. 131.

[53]Fryde (2000) is completely dedicated to this cause, arguing that from the onset, the Palaeologues were dedicated to the union with Rome.

[54]https://whc.unesco.org/en/list/724/—Accessed on 10 April 2019.

Fig. 15 Church of the Holy Virgin of Ljeviška (1306–1307) in Prizren

unaware of the terminology, as it is not possible to identify its usage for the architectural analysis (ICOMOS 1988; Republic, n.d.).[55]

3.1 Palaeologian Period—Was There Any "Architectural Renaissance" at All?

There is no doubt that the Palaeologues attempted to restore the empire after recovering it from the Latins in 1261. However, these last two centuries of the Byzantine Empire were marked by constant warfare and economic impoverishment (Mango 2002).[56] The Palaeologian emperors were often referred to as "paupers in their own house" (Crowley 2005).[57] Building and restoration works in the territory directly controlled by the Paleologian dynasty were modest: "… *many of the churches were little more than empty shells and…even the Imperial Palace of Blachernae was crumbling*" (Norwich 1997).[58] It is true that a significant number of intellectual and artistic heritage of the period survives in various European institutions, but the architectural heritage produced by the epoch is far less splendid than the current assertions emanating from the Western scholarship suggest. Within the imperial borders—that is, within the territories controlled by the Palaeologue administration—the architectural styles followed traditional Byzantine models evident in the churches and monasteries being built on a much smaller scale in comparison with the monuments belonging to the previous epochs. As such, they were in

[55] https://whc.unesco.org/en/list/456/documents/—Accessed on 10 April 2019—The document produced by ICOMOS in 1987 for the inscription of the Thessaloniki monuments on the UNESCO World Heritage List does mention the term "Palaeologian Renaissance", but it does not refer to it as the architectural style. The wording implies that the term is used for the historical period. Similarly, all the searches related to the term on the web-pages of the Greek Ministry of Culture concerned with the architectural heritage produced no results. https://www.culture.gr/en/SitePages/searchresults.aspx?#k=palaeologan%20renaissance—Accessed 10 April 2019.

[56] Mango (2002), pp. 248–307.

[57] Crowley (2005), p. 35.

[58] Norwich (1997), p. 366.

Fig. 16 Church of the Holy Apostles, Thessaloniki, early fourteenth century, traditionally dated 1310–1314

sharp contrast to the extensive building programme of the Nemanjić kings, particularly the *Raška Style* monuments which predates the Palaeologian period of Byzantium. As indicated above, the Raška Style originated north of Kosovo and directly contradicts the argument that the Palaeologian Renaissance originated in the Palaeologian Constantinople. On the other hand, the Vardar Style appeared when the Nemanjić's kingdom incorporated the northern Macedonia, which occurred at the same time when the Palaeologues recovered Byzantium. Since the *Vardar Style* monuments in Macedonia and Kosovo and Metohija surpass in opulence the similar monuments found in the Palaeologian domains due to the rich expenditure of Serbian kings, the question remains who had influenced who.

The already mentioned Church of the Holy Apostles in Thessaloniki, also a World Heritage Site built during the "Palaeologian Renaissance", which is usually argued to have influenced the Church of the Holy Virgin of Ljeviška, poses a problem of dating. Both churches were built in the early fourteenth century. The Virgin of Ljeviška is dated to 1306–1307 (Curcic 2005)[59] whereas the Holy Apostles is traditionally dated to 1310–1314 (Sport 2019).[60] If the Virgin of Ljeviška predates the Holy Apostles, how is it possible that it was influenced by it? Even if the Holy Apostles (together with other Byzantine and Bulgarian architectural monuments of the same style) were contemporary with the Virgin of Ljeviška—or even built by the architects who belonged to the same school which spread from Byzantium to the Serbian Kingdom (Čausidis 1995)[61]—how could it be possible that the "Palaeologian Renaissance" first spread to Kosovo and Metohija and reached the fresco paintings of the Dečani Monastery, before making an impression over the vast territory of modern North Macedonia which is geographically closer to the Byzantine territory? And this is where the

[59]Curcic (2005), p. 23.

[60]http://odysseus.culture.gr/h/2/gh251.jsp?obj_id=1682—Accessed 10 April 2019—The date is adopted by the Greek Ministry of Culture on their official web-page. Some scholars date the Holy Apostles as late as 1329.

[61]North Macedonian scholar A. Serafimova holds that monuments built in this period were made by "*those close to the imperial workshop of the Emperor Andronicus II (1282–1328)*"—Čausidis et al. (1995), pp. 112–114.

Fig. 17 Church of Staro Nagoričano, near Kumanovo, North Macedonia, erected by King Milutin in 1313

argument of the existence of the "Paleologian Renaissance" in architecture fails to convince in its entirety.

3.2 Can the Eyes of the Beholder Be Wrong?

It has already been pointed out that the UNESCO sites in Kosovo and Metohija are inscribed on the World Heritage List in Danger as monuments displaying *"Palaeologian Renaissance Style"* (UNESCO 2019b).[62] This terminology is scientifically wrong, as the evidence clearly indicates significant differences between the *Raška Style* and *Vardar Style* in architecture. The Monastery of Dečani, with its Romanesque features, could not under any circumstances be described as "Palaeologian" in style. Modelled on the twelfth century Studenica Monastery which was built in the Raška county of Serbia nearly a century *before* the Palaeologues came to power in Byzantium, the architecture of Dečani excludes itself from any association with the Palaeologues (UNESCO 2019b).[63] On the other hand, the Church of the Holy Virgin of Ljeviška and the Gračanica Monastery, even though architecturally completely different from the Dečani monument, are still labelled as the "Palaeologian Renaissance" monuments, despite the fact that they were not commissioned by the economically deprived Palaeologues, but the Nemanjić kings who were at their apogee at the time of the erection.

The clear dichotomy between the monuments in situ and their politically motivated scientific reinterpretation becomes even more obvious in case of the Patriarchate of Peć. The blend of the *Raška* and *Vardar* styles within a single complex of four churches points to the conclusion that the

[62] https://whc.unesco.org/en/list/724/—Accessed 10 April 2019.

[63] Interestingly enough, when Studenica was inscribed on the World Heritage List in 1986, it was described as being built in the *Raška Style* of architecture. The designation still stands on the World Heritage List. https://whc.unesco.org/en/list/389—Accessed on 10 April 2019.

Fig. 18 Monastery of St Nikita, Skopska Crna Gora, North Macedonia, began after April 1299, appears to be the oldest of the *Vardar Style* edifices. By the time King Milutin paid for its rebuilding, the main monastery church was raised to the ground

mediaeval Serbian kingdom was neither the occupying force in Kosovo and Metohija or North Macedonia, nor that it was incapable of incorporating the political, cultural or artistic models into a distinct cultural framework which could not belong to the "Palaeologian Renaissance".

4 Re-appropriation of the Serbian Cultural Heritage as a Tool of Territorial Expansion

The discourse between the Serbian and Albanian scholars and their Western promoters regarding the Serbian mediaeval heritage has clearly become politically motivated. For the Serbs, the heritage which is now under the occupation of the foreign military forces and outside of the protection of the state institutions is a matter of national survival. For the Albanians, the appropriation of the heritage as purely Albanian is a matter of legitimization of the conquered territory. For the Western sponsors of Kosovo's independence, it is a matter of political prestige in the international relations. The 9 November 2015 vote in UNESCO, according to the words of the then-Serbian ambassador in UNESCO, Professor Darko Tanasković, was guided primarily by the political interests, rather than professional and/or scientific criteria. The political pressure that was imposed on the smaller member states before the vote, however, was insufficient to convince many of those who were well aware of the destruction which took place in March 2004 (Tanasković 2019).[64]

The damage done to the Albanian cause in 2004 alerted the Western sponsors of Kosovo's secession. The initial apologetic explanations of the violence against the Serbian Christian mediaeval monuments could not satisfy the norms

[64]https://www.youtube.com/watch?v=b9pxDTfsN6k&t=4s&fbclid=IwAR3O5fmRPRISLeTtsGa7yuG8uIAtIGVbxzbeVXHoRWcigV62ackaxhputTs—Interview with Dr Tanasković given to the Srpska RTV, on 22 February 2019—Accessed 12 April 2019.

Fig. 19 Patriarchate of Peć, before reconstruction

Fig. 20 Romanesque window, south wall

and values promoted by the EU and the USA. Thus, in order to present the Albanian Muslims as caring of the Christian heritage, a new argument was put forward—that of the Albanian Crypto-Catholicism. Unsurprisingly, the notion of Christian values among predominantly Muslim Albanians came not from scholars of Albanian background, but from the West. The term "Crypto-Catholicism", invented by an American scholar Anna Di Lellio plainly argued that the Albanians are actually "*Crypto-Catholics, who only feign to be Muslim*" (Di Lellio 2009).[65] As such, it resonated well with contemporary western supporters of the Albanian national claims. The term "Crypto-Catholic" itself served to obfuscate the lack of logic behind Western support for the growing Islamic fundamentalism among the Balkan Muslims. Therefore, in defending the decision for the proclamation of Kosovo's independence, Anna Di Lellio in 2008 wrote:

> In the 1990s, when Kosovo started to aspire to independence, there was an open discussion on converting **en masse** to Catholicism, the faith of ancestors, in order to correct the error of mass conversion to Islam. The plan for a mass conversion to

[65]Di Lellio (2009), p. 8.

Fig. 21 A new Catholic cathedral in Priština, the capital of Kosovo and Metohija, dedicated to Mother Theresa, an ethnic Albanian, was built in 2005–2010, despite the fact that there less than 4% of overall population of the province is Catholic. The tallest Catholic Church in the Balkans (bell-tower is 75 m tall) was severely criticized by the Albanian Muslim scholars (Sadiku 2010). Fatlum Sadiku is an Albanian politicologist from Priština accused Vatican and Western scholars writing about Kosovo for Catholic proselytism

> Catholicism was shelved till the spring of 2008, when the public adhesions to the Church by an extended family revived the debate about the Christian origin and identity of the Albanian nation. (Di Lellio 2009)[66]

No such arguments ever came from the Albanian scholars. However, the Albanians from Kosovo and Metohija, being predominantly Muslim, had to have this claim reinterpreted in the light of the existing material heritage. Building on Malcolm's initial assertion of the Serbian occupation of Kosovo and Metohija in the twelfth century, they overwhelmingly accepted the idea of the "Palaeologian" features of the Serbian mediaeval monuments by interpreting the Romanesque details of the *Raška Style* as being of the Albanian Catholic origins, and built by Albanian architects.[67] Simultaneously, the surviving founding charters of the churches and monasteries in Kosovo and Metohija, either kept in the Belgrade archives or inscribed in stone in the buildings themselves, which are normally used as the evidence of their Serbian origins, are simply dismissed by the assertion that back in the twelfth–fourteenth centuries Serbian rulers of Kosovo and Metohija simply produced the forged documents.[68]

Since this was not enough, the Albanian authorities of Kosovo and Metohija, approved in 2005 the project for constructing a new Catholic Cathedral in Priština, "the largest in the Balkans, despite the fact that over 90% of the population is Muslim" (Fig. 21) (CIA 2019).[69] According to the Catholic bishop of Kosovo Dode Gjegjiu, the size of the cathedral is not controversial. Its erection aimed to show that

[66]Di Lellio (2009), p. 34.

[67]Thus, the 2015 unfounded claim by the Albanian popular press that Fra Vito of Kotor, the builder of Dečani was ethnic Albanian. For those interested in the subject, the press sources related to the subject are abundant, although frequently unnamed.

[68]The notion of "forged mediaeval documents" against the non-existing documents and twenty-first century claims represent juxtaposition of events distant 600–800 years. As such, it cannot be taken as serious scholarship. See works of Nasim Feri.

[69]https://www.cia.gov/library/publications/the-world-factbook/geos/kv.html—Accessed on 12 April 2019.

the new Kosovo was close to Europe (Cvijić 2011).[70] The size of the cathedral may indeed not be controversial, but the choice of the façade is undoubtedly directed by the new narrative. Clearly modelled on the Romanesque features of the Dečani Monastery, it aims to "institute the truths of Kosovo's heritage and its Palaeologian character". Being promoted first and foremost by the Western Powers, thus, any notion of the term "Palaeologian Renaissance becomes a fabrication supplemented by architecture" (Herscher 2010).[71]

5 Conclusions

"Reading" historical architectural heritage was always a complicated process, frequently affected by contemporary interpretation of national Grand Narratives which influenced and/or inspired its creation in the first place. In Kosovo and Metohija, as well as in other regions of the world prone to inter-ethnic and inter-religious conflicts, architectural heritage, especially the one bearing visible religious characteristics, often falls victim not only to direct exposure to military actions, but also to the politically charged competing national narratives, specifically designed to justify claims to territory, natural and economic resources or strategic alignment. As presented in this paper, deliberate distortion of historical facts through the usage of the unclear academic terminology with aim to introduce new meanings of the historic heritage does not contribute to the peace and reconciliation. The introduction of the scientifically incorrect term of the *Palaeologian Renaissance* for the historical architecture situated in the modern territory of Kosovo and Metohija represents political attempt to disassociate the origins of those monuments, created by the Serbian Nemanjić Dynasty in the twelfth–fourteenth centuries from its national identification.

The evidence presented in this study concluded that the four main monuments under the term *Palaeologian Renaissance* that are currently on the UNESCO World Heritage List in Danger are wrongly classified under this term. It has been pointed out that both architectural styles and—in the case of the Patriarchate of Peć—the actual monument itself—predate the Palaeologue period of the Byzantine Empire. Architecturally, the Monastery of Dečani has nothing in common with any monument built by the Palaeologues, whereas the Holy Virgin of Ljeviška and the Monastery of Gračanica with their Byzantine-influenced architecture represent the shrewd economic solution of King Milutin to build as many churches and monasteries throughout his kingdom as possible. However, on his death, King Milutin had followed traditions of his predecessors and built himself a mausoleum in the *Raška Style* in the Monastery of Banjska, which was not the subject of this inquiry. This indicates that *Raška Style* with its fusion of Orthodox and Catholic elements was indeed a Serbian mediaeval royal architecture, while Byzantine influences evident in the *Vardar Style* point out to the affiliation to the Orthodoxy.

This short study has also shown that ambiguity of Catholic–Orthodox influences on the architecture of those monuments was used to argue the nature of Crypto-Catholicism of the Albanian Muslim population in Kosovo and Metohija. Historic architecture often bears witness to the historical events that decide its survival or demise on a territory where it is situated. The events of the 17 March 2004 clearly indicated that there cannot be any Crypto-Catholicism among the Muslim Albanians in Kosovo and Metohija. It can be argued that in strictly Huntingdon's sense, these monuments cannot be adequately protected if given to the Albanian authorities and NATO which currently control the province.

Finally, the case study of the Serbian Mediaeval heritage in Kosovo and Metohija is politically most often presented as *sui generis* by the promoters of Kosovo's independence. As argued above, it clearly indicates that politically engineered forgeries can only lay foundations for some inevitable future conflicts.

[70]Cvijić (2011).

[71]Herscher (2010), p. 151.

References

AlJazeera Balkans. (2015, November 9). Retrieved from balkans.aljazeera.net: http://balkans.aljazeera.net/vijesti/kosovo-nije-primljeno-u-unesco.

Čausidis, N., et al. (1995). *Macedonia—Cultural heritage*. Skopje: Misla.

CIA. (2019, April 12). *CIA World Factbook*. Retrieved from Kosovo https://www.cia.gov/library/publications/the-world-factbook/geos/kv.html.

Ćirković, S. (1969). *Srpska pravoslavna crkva, 1219–1969*. Beograd: SPC.

Cormack, R. (2000). *Byzantine art*. Oxford: Oxford University Press.

Crowley, R. (2005). *Constantinople—The Last Great Siege 1453*. London: Faber and Faber Limited.

Curcic, S. (1979). *Gracanica, King Milutin's Church and its Place in Late Byzantine Architecture*. London: University Park.

Ćurčić, S. (2000). Destruction of Serbian cultural patrimony in Kosovo: A world-wide precedent? *Journal of the North American Society for Serbian Studies, 14*(2), 125–131.

Curcic, S. (2005). Renewed from the very foundations: The question of the genesis of the Bogorodica Ljeviska in Prizren. In J. J. Emerick et al. (Eds.), *Renewed from the very foundations: The question archaeology in architecture: studies in honor of Cecil L. Striker*. von Zabern.

Cvijić, Z. (2011, February 3). *Kosovska katedrala, najveća na Balkanu*. Retrieved from www.b92.net: https://www.b92.net/putovanja/arhiva.php?yyyy=2011&mm=02&dd=03&nav_category=823&nav_id=490377.

Di Lellio, A. (2009). *The Kosovo 1389—An Albanian epic*. London: I. B. Tauris.

Fine, J. (1994). *The late medieval Balkans*. Ann Arbor, MI: Michigan University Press.

Fryde, E. (2000). *The Early Palaeologan Renaissance (1261-c. 1360)*. Brill: Brill Academic Publishers.

Gellner, E. (1983). *Nations and nationalism—New perspectives of the past*. Oxford: Blackwell Publishing.

Hellenic Republic. (n.d.). *Hellenic Republic—Ministry of culture and sport*. Retrieved from https://www.culture.gr/en/SitePages/searchresults.aspx?#k=palaeologan%20renaissance.

Herscher, A. (2010). *Violence taking place—The architecture of Kosovo conflict*. Stanford: Stanford University Press.

Hobsbawm, E. A., & Ranger, T. (1983). *The invention of tradition*. Cambridge: Cambridge University Press.

Huntington, S. (1996). Chapter 9—The global politics of civilizations. In S. Huntigton (Ed.), *The clash of civilizations: And the remaking of world order* (pp. 207–238). London: Simon & Schuster UK Ltd.

ICOMOS. (1988). *Paleochristian and Bzyantine Monuments in Thessalonika-No.456*. Retrieved from UNESCO: https://whc.unesco.org/en/list/456/documents/.

Jevtić, M. (2015, November 9). *Kosovo nije primljeno u UNESCO*. Retrieved from RTS http://www.rts.rs/page/stories/sr/story/9/politika/2098204/kosovo-nije-primljeno-u-unesko-spiskovi-kako-su-glasale-drzave-clanice.html.

Ković, M. (2011). Saznanje ili namera: Savremena svetska istoriografija o Srbima u XIX veku. *Sociologija, LIII*, 401–416.

Ковијанић, Р. (2016, re-print 1962). *Вита Которанин, неимар Дечана*. Београд: Magelan Press, re-print Nolit.

Лазић, С. (2019). Насиље над српским светињама неће стати. *Православље*, 24'28.

Malcolm, N. (1994). *Bosnia—A short history*. London: Pan Macmillan, Macmillan Publishers Limited.

Malcolm, N. (1998). *Kosovo—A short history*. London: Macmillan Publishers Ltd.

Malović, N. (2016, November 25). *Vita Kotoranin, neimar Dečana*. Retrieved from Nedeljnik https://arhiva.nedeljnik.rs/nedeljnik/portalnews/vita-kotoranin-neimar-decana/.

Mango, C. (2002). *The Oxford history of Byzantium*. Oxford: Oxford University Press.

Muller, D. (2009). Orientalism and nation: Jews and Muslims as Alterity in Southeastern Europe in the age of nation-states, 1878–1941. *East Central Europe, 1*(36), 70.

Norwich, J. (1997). *A short history of Byzantium*. London: Penguin Books.

Petiffer, J., & Vickers, M. (2009). *The Albanian question—Reshaing the Balkans*. London: I.B. Tauris & Co Ltd.

Raško-Prizrenska, E. (2004, March 17). *Eparhija Raško-Prizrenska*. Retrieved from http://www.eparhija-prizren.org/?p=26657.

Sadiku, F. (2010, November 25). *Muslimani Kosova pored preporoda i diversije politiziranog Katoličanstva*. Retrieved from www.bosnjaci.net: http://www.bosnjaci.net/prilog.php?pid=39696.

Smith, A. (1991). *National identity*. London: Penguin Books Ltd.

Sport, M. O. (2019, April 10). *Church of Agioi Apostoloi, Thessaloniki*. Retrieved from Odysseus, Monuments http://odysseus.culture.gr/h/2/eh251.jsp?obj_id=1682.

Tanasković, D. D. (2019, February 22). *Istina ili Laž*. (S. RTV, Interviewer).

Todorova, M. (1997). *Imagining the Balkans*. Oxford: Oxford University Press.

UNESCO. (2019a, April 10). *Medieval Monuments in Kosovo*. Retrieved from UNESCO https://whc.unesco.org/en/list/724/.

UNESCO. (2019b, April 10). *Studenica Monastery*. Retrieved from UNESCO https://whc.unesco.org/en/list/389.

UNHCR. (2019, March 1). Retrieved from UNHCR-Population Statistics http://popstats.unhcr.org/en/overview#_ga=2.267661582.1431272768.1552401925-809441953.1552401925.

Zulfikarpašić, A. (1998). *The Bosniak*. London: C Hurst & Co Publishers Ltd.

Symbolism and Representation: Utopian Moments and Architectural Movements

The part dives into the relationship between architecture and utopia as one that is longstanding and enduring by drawing upon contemporary and historical geographies. Implicated simultaneously for centuries, utopia and architecture have both gained the interest of geographers, historians, and sociologists, which left room for interpretations and inhabitation.

In "Exploring Utopian Representations in the Architecture of Art Museums in the Middle East" the authors aim to investigate utopian perspectives in the architecture of modern art museums in the Middle East. In describing Utopian approaches and representations of architecture, the authors present a stimulating debate on the basis of two works of Islamic Architecture, the Doha Islamic Art Museum and the Abu Dhabi Louver Museum. After an in-depth review of the case studies, the authors concentrate on the importance of seductive languages with dual meanings in architecture, which have bridged users and architects through the development of stories.

The chapter "Dantesque Allegories in the Symbolic Architecture of Triduo Apparatuses" deals with the Triduo apparatuses, temporary scenes created after the Catholic-Reformation, to strongly affirm the existence of Purgatory and the possibility for the sinner to aspire to the sins' redemption and then ascend to Paradise. Their architectural structures seem to highlight allegories—still little explored—towards the architecture of the situations of Dante's Divina Commedia. The chapter is a significant contribution to the field because considers the installations as real artistic works, one different from the other, consistent with the Baroque style guidelines and explains relevance of their architectural language that symbolizes fictional spaces and, at the same time, is an imagery place.

Moreover, an exploration of traces of intangible heritage, namely a town located in the Campania, Pagani, a region known for the ancient cult of Madonna del Carmine, is provided in the following chapter "Architecture, Territory, Traditions. The Cult of the Madonna Del Carmine Called 'delle Galline' in Pagani". The authors shed a light on the direct relationship between the diffused practice of worship and the structure of a series of public spaces built along nineteenth century. The article is clearly organized through a set of chapters that analyze the origin of the cult from sixteenth century, its influence on the construction of the church of Santa Maria Incoronata del Carmine and the influence on the civic and religious urban spaces of Pagani.

The chapter "Oblique Plane; Layering; Spatial Layering —Pavel Janák's Formal Language Construction" aims to explore Pavel Janák's design techniques through studying his formal expression, which in this paper is conceived as a sort of language. Inspired by Cubist paintings, Janák's building work is composed logically by geometric forms, representing a complex figure. The research method is to analyse the built works of Pavel Janák in different creative stages. Taking his theoretical works as a breakthrough, the paper dissects the Janák's philosophy background, then analyse the formation and transformation of his grammatic structure and of formal expression from a height of aesthetic conception. Following chapters introduce nature lighting, its evocative and spiritual power in sacred spaces and explores lighting strategies, materials used and different technological solutions adopted.

Exploring Utopian Representations in the Architecture of Art Museums in the Middle East

Lamis Mantash and Maged Youssef

Abstract

Utopia is a real tendency that develops out of daily life events, flamed upon realization of what is wanted and what is absent. Throughout history, there have been extensive discussions about utopia and architecture; the clearest current embodiment is through sustainability. However, there is still a mysterious perspective that explores the relationship between the collective and architects who seek to create their own utopia by using implicit representations that conceal ambiguous meanings, which hinders the comprehension of the purpose of design. In light of this problem, the paper attempts to explore the utopian representations embedded in the architecture of contemporary art museums in the Middle East. In Gulf Arab cities, decision makers have recently invited renowned architects to design art museums. Certain critics have pointed out that the architectural language of these museums has received unsatisfactory understanding of design messages due to the exaggerated use of notions as 'Islamic', 'mashrabiya', 'dome', and 'minaret', which claim to emphasize the Arab identity in an attempt to fulfill the collective will. This paper hypothesizes that the architects of these museums have designed utopian environments based on inclusion of different cultural stories, inspired by different religions, rituals, and cults from various nations. Spaces of these museums have turned into narrators of particular stories. The paper adopts an analytical methodology, defining utopia, giving a literature review about the utopian approaches and representations in architecture, and then analyzing two case studies: Museum of Islamic Art in Doha and Louvre Museum in Abu Dhabi. The paper interprets the utopian representations in both projects in order to unveil new meanings and the tendencies of these architects beyond their design language. One of the important conclusions this paper reaches is that the gap between users and architects can be bridged through realization of told stories.

Keywords

Architecture • Utopia • Utopian representation • Art museum • Meaning • Story

1 Introduction

Art museums have been utilized as a strategy that sets out the employment of commissioners by a political, social ,or even economic power that '*seeks to realize its desires*' (Duncan 2005). Carol Duncan listed this phrase, together with the term 'power' in her book Civilizing Rituals in order to show that understanding art museums necessitates understanding and exploring embedded messages and not only historical narration. Epistemologically, the phrase denotes a specialized psychological meaning that further reveals its significance for art museums. It signifies the utopian tendency that might be the main generator and substantive core of the created dramatic field.

The concept of utopia is difficult to be defined because it simultaneously describes a dream that is linked to imagination and reality and refers to time and space; hence, there are always doubts about the feasibility of fulfilling utopia. There are two distinguished approaches to understand utopia: pessimistic (associated to Sir Thomas More treatise on Utopia) and optimistic (associated to philosopher Ernst Bloch in *The Principle of Hope*) (Incorporated 1997). Unlike More who limited utopia to being a pre-existing programmatic state which had to be reached under wise and all-knowing leadership either of a political party or the

L. Mantash (✉) · M. Youssef
Faculty of Architecture-Design and Built Environment, Beirut Arab University, Debbieh Campus, Lebanon
e-mail: lamis.mantash@gmail.com

M. Youssef
e-mail: m.nabilyoussef@bau.edu.lb

C. Gambardella et al. (eds.), *Advances in Utopian Studies and Sacred Architecture*,
Advances in Science, Technology & Innovation,
https://doi.org/10.1007/978-3-030-50765-7_10

church, Bloch highlighted that it is an autopoietic process driven by the laboring, creating, and producing human beings, driven by their material hunger as well as their dreams of overcoming that hunger that can be practiced in various contexts. For More, utopia is nowhere that is either from the past or the future and that cannot be reached. Despite their different thoughts, Bloch and More meet that humanity is the main goal of utopia. Bloch imaged utopia as a journey, and deemed it a necessity in order for humanity to come to itself. He described the journey as following, 'as the more we travel toward the future isle of utopia, the more it will arise from the sea of the possible-out of the present chaos, which the sea represents' (Keller 2016). Bloch describes the role of the artist in relation to utopia, as an artist can represent a skillful essence of the present through mediated thoughts. After all, utopia is a purposive action that starts upon choosing in the present the seeds for the expression of fundamental difference which occurs upon the realization of 'something's missing'. According to Bloch, what is characterized as missed usually has its formentic effect already present through seeds to indicate its strong presence, yet it is difficult to be recognized. For this reason, he linked utopia to reality and the ordinary world, and highlighted the significance of usual daily life experiences as a terrain that carries traces of unrealized potentials. However, it is drawing attention to traces of 'something's missing' that awakens the creative drives that call for actions in order to reach an intended situation and make it real (Zipes 1997). The significance of Bloch's position is that it offers a utopian philosophical framework to understand the meaning of the world. This research paper uses insights from Bloch to understand the meanings of architecture as a communicative art that acts as world-making.

Human dreaming has always reached toward utopia with varying mixes of the abstract and the concrete that produce various expressions of certain states of affairs. The clearest expressions are conceived from the interactions of movements, experiments of utopian thinkers, and architectural theorists and practices through buildings and planning as concretized evidences (Picon 2013; Zipes 1997). In fact, the growing gap between the natural world and the industrial world is the most imminent common threat that utopian thinkers and architectural theorists have tried to resolve. Thus, the relation is in terms of convergent agendas shared by utopia and architecture such as: reconciliation between nature and industrial world, reconciliation of the individual and the collective, and attitude toward time (Picon 2013); this relation is not osmotic. As a deeply communicative art, architecture is about both the individual and the collective. The gap between these two elements has encouraged the proliferation of debatable discussions of architecture that fall within popular terms like: global/local and general/particular. Dutch architect and architectural theorist Rem Koolhaas suggests that utopia is equated or embodies a new ideology that constitutes the unconditional acceptance of architectural star-system that might also be an exceptional opportunity for individualism that communicates receptive messages by digital culture (Picon 2013). In recent years, there has been a considerable growing interest in global practices in which communication is largely based on images and renders that are then compared to the built real projects in order to examine their potential to restore hope for the collective. Here, it is noteworthy that media plays a vital role in culture construction in terms of specifying what the culture ought to accept and construct. The aforementioned comparison between the image and the practice and the unconditional acceptance of architectural star-system highlight the serious problem with communication in architecture, and push us to rethink the utopian concept, be critical of what exists, and understand the constructed social worlds of architecture. There is no doubt that the Middle East has been witnessing increasing turmoil; nonetheless, the Arab region in particular has continued to call for global practices and embrace them (Andraos 2016), especially calls for designing projects of cultural significance such as art museums.

The choice to focus on this region is neither based on ethnic considerations nor on a rejection of foreign architects. It is rooted in the problem of communication in architecture that relies on understanding architecture as images as a method for communicating intended meaning such as the under-construction Guggenheim Museum in Abu Dhabi (Fig. 5.1) which is going to be the largest Guggenheim in the world. Focusing on art museums in the Middle East emphasizes the problem of communication between utopian dreams of the architect and users' perception. The architectural concern for this region is the usage of architectural elements such as the 'dome', 'medina', or 'octagon' as a validation of the authenticity of design and as a mean to market the project for the collective (Andraos 2016). Furthermore, the dominant role of visual presentations (sketches, drawings, renders) often underestimates the power of architecture to be at once enriching and being enriched by people's modes of thinking, and its capacity to convey meaning. While the majority of the Middle East population are Muslims, different religions and sects have coexisted since the dawn of time. Hence, it is a limitation to connect architecture in such rich and vivid region to mere elements that are perceived as presumable fillers of local and global approaches, and to disregard the potential of art museums as a potent engine of ideology. Realization of utopia today as Koolhaas suggested, is through communicating meaning beyond the architecture-image method. Considering that art museums are supposed to mirror their represented communities and that architectural projects are seen as representatives of their respective star architects, this research aims to realize the invisible force moving these architects or individuals which is the utopian tendency and its representations in the two case studies selected for this research. The goal behind exploring utopia in the context of art museums is to propose or conclude a method that could reconcile the gap between the individual and the collective.

Duncan's comparison between the environment of art museum and religious places that was the basis of her theory on ritual settings, incited the authors of this article to further probe into the power of utopian longings and their most ancient representation. These longings are clearly stated in religious and sacred stories that have underpinned the narrative of rituals that are reflected in the architecture of sacred buildings such as churches, synagogues, and mosques. The paper argues that art museums bind communities altogether in a civic body, and hypothsises that the architects in the two selected case studies have represented utopian tendencies that are based on the inclusion of different cultural stories that might be inspired by various religions or rituals, or even might stem from the architect's memories. The study adopts a qualitative analysis approach that considers deductive reasoning in the literature review part and inductive reasoning in the case studies part where also the deductive blurs throughout the interpretative task. Concluding, the paper lists a number of deductions on the basis of selected art museums.

2 Literature Review

The below literature review first defines utopia, then gives background information about convergant agendas of utopia and architecture and its representations. Throughout the literature review, Duncan's theory on art museums as dramatic fields is coorelated to the original role that architect had played in ancient history in order to build a method that clarifies how utopia is related to stories that shape the architect's works. This study approaches the utopian longings and their representations by attempting to understand the narrative structure of explored stories from the architectural cues, architect's statement and interviews, and critics.

2.1 Utopia, Arts, and Art Museums

Historically, Plato defined utopia as the most perfect political state of 'Republic' and he perceived the utopia in laws, ethics, and social norms (Bobonich and Meadows 2018; Bobonich 2004). The architectural design concepts are supposed to respect the collective utopian desires, not to be separated from the collective perception. Koolhaas view on the problem of gap of the 'individual' and the 'collective' on the one hand and the particular potential of meanings communicated through art museums on the other (Winston 2016), drives the research toward Bloch's 'ultimate basic principle of utopian philosophy' that is not separable from the realization of 'something's missing'. This principle includes the interrelationship between the 'self' and the 'we' that controls the manifestation of utopia as a positive relationship with respect to the expanse of human-self presence. For Bloch, the possession of the self is a collective possession through which 'we make of our subjectivity the world'. This process requires a preparation phase that Bloch explained by offering a re-interpretation of humanity's constitutive characteristics, regardless if the person is a producer of any art. He argued that only what is internal in us should be transformed into a 'self-encounter' that enables us to mastermind our objectivity in the external world (Zipes 1997).

Arts are widely considered to be the most important proof of Bloch's utopian philosophy that enquires beholders to understand the whole meaning of the world. For him, utopia is something that constitutes our being from the early childhood phase and enables us to describe a state or place in which we have not yet been there, except we seek to represent it (Zipes 1997). For her part, Duncan treats art museums as secular constructed microcosms that contain 'sites that publicly represent beliefs about the order of the world, its past and present, and the individual's place within it' (Duncan 2005). Her definition of art museums alludes to important terms such as 'humanity', 'cultural heritage', and 'truths', which in turn form intersection points between this typology of architecture in particular and Bloch's thoughts on the significance of arts and their impact on world-making through 'anticipatory illumination' which comprises the listed terms and conditions authenticty. Anticipatory illumination is related to the work of art producer(s) as a coming-to-his/her/their selves and its intersection with contemporary realities, taking a looking-forward direction. As a locale of anticipatory illumination, art is also an experimental site for the utopian process which is known as 'process-fragment' to illustrate the presence of the 'cipher of the authentic' or the 'secret signature' that the maker represents in his/her work (Bloch 1996). The 'process-fragment' is elaborated on later in the paper. Duncan highlighted in her theory that 'ritualistic behavior occurs in secular setting' of the museum by showing the influential relationship between architecture, performance of visitors, and transformative effect (Duncan 2005). Based on Duncan's work and her emphases on understanding the meaning and messages implied by the spatial representation of art museums, this paper stresses the importance of thinking of utopia and realizing its traces on the basis of the role played by the architect to represent other communities.

Fig. 5.1 A 3d model designed by Frank Gehry, Photo courtesy Partners, LLP

2.2 Reconcilation of the Individual and the Collective

Utopia and architecture share an essential objective as a way of seeking a mode of communication to reconcile the individual and the collective. This reconciliation has started taking place in light of the various political and social problems of the industrial age. The social lives of individuals have undergone several changes, which in turn have posed a great danger on the society's social fabric. Many architectural theorists have demonstrated the utopian thinking in the nineteenth and twentieth centuries. For example, Ruskin's theory of Gothic ornament attempts to preserve the collective inspiration by applying a new equilibrium between collective means of production and the individual inventive momentum. The Ruskinian architects Thomas Newenham Deane and Benjamin Woodward design of the museum of Oxford University illuminates the relation between the idea and the wrought iron structure with its sophisticated decor and the stone sculptures carved by hand. A true collective inspiration is also expressed in the same concern that reflects the captivation with the ocean liner that gave authority to the individual life in the industrial age, and also capitulated the individual life to a common destiny that is considered by the utopian writer Victor Considerant in the early 1840s, who is the basic follower and inheritor of the French philosopher Charles Fourier. Considerant illustrated liners as archetypes of the collective living of the future. Further, this was a true model for the Plalanstery collaborative habitation. For his part, French-Swiss architect Le Corbusier introduced in his Unites d'Habitation the impact of the liner as a miniature (Picon 2013).

2.3 Utopia/Genuine Home—from Ritual Functions to Storytelling

Art Museums and religious buildings converge in acting as ritual settings and differ in terms of ideology that Duncan described by considering the former secular sites that convey embedded meanings and the latter sacred sites that are explicit in representing meanings. Meanwhile, philosophers have discussed utopia in relation to ideology and did not constraint it to the static religious realm. Rather, utopia is seen to be flexible, expressing a double ideology (Picon 2013), and that despite its realized seeds of expression, it can be the source of changing continuity. A ritual site of any kind is a place programmed for the enactment of a certain thing. In Western culture, rituals have been always practiced by mimetic performance, with interpretive and influential actions, predominantly dramatic poetry and choral dance, evoked under the title of 'tragedy' (Landrum 2017) that has propelled the development of a past-present literature on storytelling (Mantash 2018). The term 'architect' is present in the Greek drama that defines him/her as someone 'who leads others in ordering action' (Weir 2016), which deems the architect a significant influencer and contributor to communicating and conveying intentional, unintentional, and definitely hidden meanings. In this context, the architect is called upon to play the role of a storyteller. Bloch specially shed the light on stories and their potential of rescuing the function of imagination, considering they are the constitutive content of any cultural heritage. Reciprocally, the imagination of the producer rescues the phenomena or structured story by choosing 'what is significant in it' for him/her. This proves the inclination toward narrating and communicating particular stories. And without 'something's missing', the produced or chosen story lacks the utopian tendency. After all, the mere nature of storytelling is not utopian and utopia is only alive through storytelling. Bloch investigated the causes that constitute the utopian tendency of wonder stories that are famously known for their happy endings. He found that Vladimir Propp limited the absentation and interdiction—which are common aspects in plots—to mere received sufferings that precede the assumed endings. Meanwhile, Bloch explored that the absentation and interdiction are tendentious, by exploring how the protagonist 'purposely' seeks new surroundings to control his/her destiny. Tendentious means is that they tend toward illuminating a path through darkness to a more enlightened world or life situation. Therefore, the protagonist constantly pursues a genuine Home that is commensurate with his/her deep utopian longings. In almost all religious traditions, this Home is associated with the afterlife (Zipes 1997). However, this paper explores how utopia focuses on the present world. It is true that the sources of utopian longings are personal experiences, but this does not contradict the mutual power of both secular/profane and religious/sacred stories with events as seeds of expression.

2.4 Process-Fragment: Thinking as Storytelling

Stories are the means to map the terrain of utopia and make it alive through storytelling (Zipes 1997). Utopian traces or utopian representations are the explored genuine Home, truths, and what is significant for him/her, which all end up showing the 'authenticity' that the architect envisioned with respect to the art museum and its context. While authenticity is a challenging word for architects, it is important to clarify that the process which facilitates authenticity is that of sifting the facts and utilizing them in accordance with their latent tendency of utopian content. Storytelling is to communicate a story deftly through manifestation of narrative in order to engage an audience by inspiring their imagination to construct a new narrative in their mind. Italian architect and architectural theorist Marco Frascari argued that storytelling is achieved by 'confabulation' whereby a story is constitutive, as it gets used, invented, and imagined in a real setting. It is fundamental to note that the epistemological status of a story is that of an enigma, requiring the reader to build bridges between the known and the unknown—the tangible and the intangible (Emmons and Phinney 2017). Story comprises all events which are to be depicted, while narrative is the showing or telling of these events and the mediums selected for this purpose. Hence, storytelling refers to the retelling or circulation of stories in different narratives (Cobley 2014).

In a previous thesis by one of the authors (Mantash 2018), the mutual relationship between storytelling and architecture was investigated as an interpretative methodology in order to disclose meanings. The main contributors of storytelling are the storyteller/author, context, and audience with the story acting as the core (Benjamin 1969). The storyteller is the architect and the producer of art in architecture (Patterson 2000). The context is the storyteller's background against the backdrop of the site of the project. The audience are the visitors whose educational levels vary; hence, they do not all interact with the cue of the spatial representation of the art museum in the same way. For this study, the audience are critics, academics, and researchers from different disciplines who participated in interacting with the selected projects. The fundamental components of every narrative are the theme and the plot of the story/ies. The theme is the central idea and the message (Mckee 1997). The plot is not an isolated set of events; it is a complex entity of events that are bound to sequence, narrative space, and time. As for setting, instead of using the conventional meaning of the term that is constrained to time and place, the paper uses the definition of Paul Cobley, who found that space plays the role of binding effect and is embedded in the plot as it is associated to events (Cobley 2014). In fact, the binding relation between narrative space and event was earlier emphasized by Aristotle and by Ibn Rushd, an Arabic philosopher who did a phenomenological interpretation of this relation and added the value of transcendence instead of relying on the materialistic place or body movement (Hilal 1997; Badawi 1973). The concepts of time and sequence are not explained here because the scale of the paper does not allow enough space to discuss the advanced role of these concepts in clarifying the structure of narrative. However, the researcher structures the narrative based on meanings of the explored stories, and they are categorized later into topics that all revolve along the same theme.

Because the adopted methodology tackles the tangible and intangible medium, the events and narrative space of a story are interpreted in the context of architecture through spatial experience/performance and geometry of being. In fact, this approach is in line with Duncan's theory that uses dramatic field for the architectural situation and performance for visitors. The narrative space of a story emerges highly significant after considering it the source of inspiration and guide toward the genuine Home. The narrative space of a story comprises five levels that are aligned in an ascending order starting from: setting, spatial frames, story space, storyworld, and finally narrative universe (Ryan et al. 2016). Setting is a relatively stable socio-historic-geographic category that embraces the entire story. Spatial frames are the actual surroundings of the characters, and they shift scenes of action that may flow into each other. Story space is the specific space or architectural situation relevant to events of a plot or subplot. It includes the summation of spatial frames related to the plot—the occurring actions. Storyworld is more complicated since it includes the mediation of the reader/art maker in constructing the intended space. In a story that refers to both real and imaginary locations, the storyworld superimposes the locations specific to the plot onto the geography of the actual world in order to ignite the imagination of the readers and call them to contemplate and believe the possibility of erecting this imagined space. Finally, narrative universe includes the total complexity of all of the above, in addition to all the counterfactual worlds constructed by characters as beliefs, wishes, fears, speculations, hypothetical thinking, dreams, fantasies, and imaginative creations (Ryan et al. 2016).

Because an event is associated to a narrative space, the realized narrative space level from the spatial representation of the art museum is at the same time connected to spatial experience that the architect wants the visitors to live. The role of narrative space is to facilitate the exploration of story and consequently the implied meaning. As for architecture, Simon Unwin transcended in his book *Analyzing Architecture*, the limitation of pure geometric analysis of architecture by resorting to the geometry of being. For Unwin, the architectural space is related to the intended experience and lies at the core of conditions of being of space that can be called a place upon its identification (Unwin 2009). Unwin introduced the geometry of being through measure, lines of sight, lines of passage, social geometry, six-directions-plus-center, geometry of making, and circle of presence (major senses). Measure is for order and scale. For example, a grand scale door magnifies the status of the occupant and reduces the status of the visitor. Lines of sight establish contacts and axes between, for example, the source and distant object. For natural phenomena, the lunar eclipse is a significant event of accurate alignment between the sun, moon, and earth. Lines of passage are the ingredients of experience where the same curious situation can be represented by orthogonal lines and curved ones. Social geometry is related to the space of confrontation and to that of togetherness and isolation. Geometry of making is essential to the construction of buildings and conditions the sizes and layout of spaces through its intrinsic qualities and timing of current building practices. The six-directions-plus-center is manifested by establishing resonance between an enclosure and its occupant and also in the world around such as cardinal directions. For the circle of presence, it can be experienced through an extensive circle of visibility or intimate circle of touch ability (Duncan, 2005). Thus, the geometry of being shows the mode of being that is anticipated by the architect, while the narrative space of every story gives glimpses of the imagined Home that is emphasized by the variety of stories. On the other hand, detecting the sources of explored stories allows identifying whether they are secular, sacred or both. This process is proposed to be a process-fragment in order to explore the 'cipher of the authentic' and realize utopia.

3 Methodology

The used method comprises an introduction and listing of overlays (Fig. 5.2). The introduction is a summary of architect-context knowledge and the main explicit message stated in his words which re-appear in the second part as clues depending on the main idea of every overlay. The second part is composed of overlays that are organized in the same order of their corresponding titles in the listing sentence. Every overlay is a definite paragraph that comprises the following: the explored story, narrative space, event/experience, source, and geometry of being. The mutual influence of these elements is ordered differently in every overlay depending on the clues and evidences. The aforementioned allow the exploration of the meaning of 'cipher-authentic' and the realization of the utopia anticipated by the architect in the designed art museum by knowing the represented 'ripened truths'.

The first selected case study is on the Museum of Islamic Art in Doha, Qatar, whose architect claims to represent the 'essence of Islamic architecture', and the second is on the Louvre Abu Dhabi, UAE, whose architect claims to represent 'Humanity in a New Light'. The selection of these is specifically due to: significance of typology; debatable discussions regarding representing 'authenticity'; 'anticipatory illumination' role of the architect; countries of high circles of power; and in order to explore 'what is significant for' international-star architects.

3.1 Museum of Islamic Art, Doha, Qatar, 2008

When Chinese American architect I. M. Pei was asked to design the Museum of Islamic Art, he freely admitted that he could not claim any real familiarity with architecture in Islamic contexts as he began the design process.[1] While working, he took into consideration the objectives of the commissioners and the Al-Thani family ambitions who rule the country today. The architect sought to integrate the community values into the spatial medium of museum and capture the 'essence of Islam', as he identified it. His design

[1]Pei says: *'I chose to do buildings in places I had never been before-I enjoy learning about a place. I knew the East because I was born there, the West because I was educated there, but the Middle East I did not know. My early education never revealed to me Islamic architecture as a major architectural invention,'* (Landin and van Wagenen 2009).

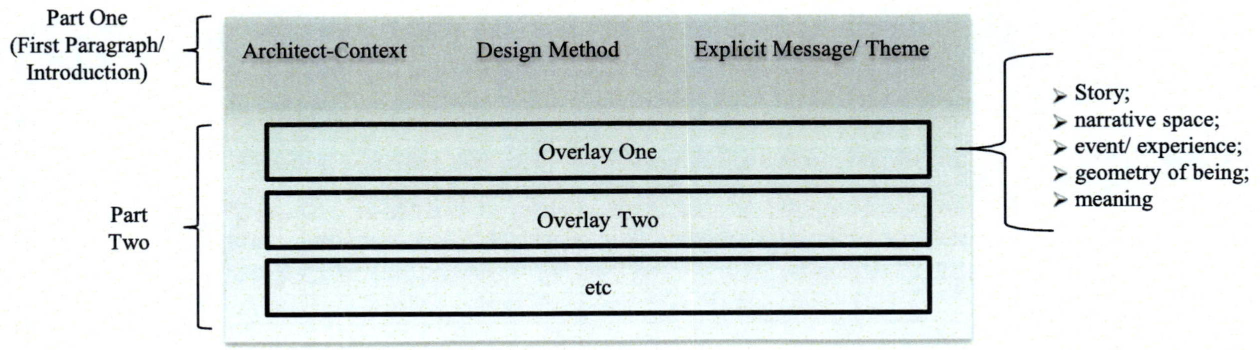

Fig. 5.2 Structure of the method

method necessitated studying the philosophy of Islam and history of Islamic civilizations, and travelling across the regions of Islamic architectural heritage in search of inspiration. He explored throughout the reading of the biography of Prophet Mohammed his narrow realization of Islam (Landin et al. 2009). His search was further complicated when he found that the archetypal examples of Islamic civilizations do not necessarily serve the essence of Islam. He claimed the absence of 'essence of Islam' in Cordoba-Spain, the Jama Masjid in Fatehpur Sikri-India, and the Umayyad Great Mosque in Damascus-Syria. Pei found the essence of Islam in the architecture of Maghreb's and Tunis Fortresses and Egypt's Ibn Tulun Mosque (Jodidio 2008). Pei summarized the response to context as following: 'Architecture is form, space, light, movement, all of that. But more important is the place where you build' (Landin et al. 2009). The central theme implied by Pei is revealing the 'essence of Islam' by evoking the effect of the harsh sunlight through spatial representation (Jodidio 2008). The investigated overlays are listed below:

Ribat—Presence of Allah through Light—Creation of Ibn Tulun Mosque—Buddhist Ritual—Creation—The Search for a Capital.

3.1.1 Overlay One: Ribat

Pei recalled the ribat (fortification) of Tunisia that fascinated him, and considered it as a highlight of Islamic architectural history. He clarified that ribat is a space of life and not simple mosques where worship is practiced (Jodidio 2008). 'Closely connected with the ribat are the concepts of meditation and holy war, for not only did the faithful seek peace and seclusion in them, but they were also meeting places for a warriors that took part in holy wars (al-murabitun)' (Mazot 2013). The battle may as well take the form of an intellectual or a spiritual struggle (ibid). In modern times, these fortresses no longer have the same functional dimension. The storyworld of defense by fighters is clearly reflected spatially in the created context in terms of six-directions-plus-center through involving visitors in the experience of marching to an isolated new reclaimed site that seems impermeable (Fig. 5.3). In terms of geometry of making, the architect reflected the military essence of fear of penetration; he also wanted to give the impression of a strong country by creating a concrete grand solid imposing walls with a solidity effect that is intensified by the dense sunrays. Knowing that this kind of military architecture is very similar to the traditional Romano-Byzantine art of building fortifications, its spatial configuration is designed intentionally to lead to disorientation. It is an imitation of the storyworld of the military ambitions that Islamic civilization utilized many years ago. In terms of circle of presence, the limestone and porphyry massiveness of the museum were used to give the impression of an old monument, which affirms it as a structure dedicated to commemorating a glorious Islamic past in the present. Thus, the desire was to make a powerful presence of the old and still monuments to memorilize military past of Islamic civilization but using different time tense.

3.1.2 Overlay Two: Presence of Allah Through Light

Interviews that sought to shed light on the architect's thoughts and work, alluded to his musing in the expression of light representation in the Noble Qur'an and his correlation of light and presence of Allah as a pathway to reveal the 'essence of Islam' (Jodidio 2008). This paper looks into a Qur'anic verse that illustrates Pei's Qur'anic references on light. In the verse nb (35) in Surah An-noor, the story space reveals the meaning that underlies the spatial representation of stainless steel internal muqarnas dome. The verse says,

> Allah is the Light of the heavens and the earth. The example of His light is like a nich within which is a lamp, the lamp is within a glass, the glass as if it were a pearly [white] star lit from [the oil of] a blessed olive tree, neither of the east nor of the west, whose oil would almost glow even if untouched by fire. Light upon light. Allah guides to His light whom He wills. And Allah presents examples for the people, and Allah is Knowing of all thing.

Fig. 5.3 Context of Ribat. **a** Monastir Tunisian Fortress (Monastir Tunisia 2020), **b**, **c** The main building from Doha corniche (Von Stemberg 2008)

In short, the lamp increased the presence of light by the purity of oil and the purity of the lamp and the narrowness of the alcove of the light of the lamp that protects it from the manipulation of satisfaction (Mughniyah 2009). This verse is an example that Allah has struck on the clarity of evidence of His appearance and His existence in order to accentuate His presense in the created infinite universe. It gives a detailed spatial description to convey how Light (denoted to Allah) is emitted from heavens and communicated to the heart of the faithful.

Paul Goldberger, an American architectural critic, pointed to the wrapping of the stainless muqarnas with four false walls as a contradiction with the honesty of composition and structure that Pei usually used to highlight as a characteristic of historic architecture (Goldberger 2011). Thus, the false wall cover-up and 'dome' scale allow the present study to exclude the purpose of traditional dome that circumscribes a wide breath of space. The combination of stone walls and hidden stainless lamp-like copula is perceived by Goldberger as a dramatic break from Ibn Tulun's fountain. He postulated that the reason behind this is to ensure the scale of the internal muqarnas as deep and narrow space in order to represent the impact of dense sunlight on the stainless facets in terms of geometry of making (Fig. 5.4). The concept of a coffered dome or muqarnas emerged in the Abbasid area, and they were built using stones. It is pierced by a small opening, an oculus, in order to maximize the presence of sunlight in a narrow enclaved space so the sunlight appears as if puncturing this opening (Blair and Bloom 2013). This stainless steel oculus muqarnas can be seen by visitors from all galleries and atriums, and it draws attention to the dense sunlight in this small scale limited space.

3.1.3 Overlay Three: Creation of Ibn Tulun Mosque

Pei's clear declaration and formal association of his discovery of the 'essence of Islamic Architecture' in the fountain of Ahmad Ibn Tulun in Cairo incited Rabbat's reflections after visiting MIA (Rabbat 2011). The fountain dates back to the thirteenth-century Mamluk renovation. Rabbat's reflections transcended architectural analysis to include the role of commissioners and architects in each project and the effect of economical power by retelling the story of Creation of Ibn Tulun Mosque and exploring parallel conditions in this present project (Rabbat 2011). He stated that a source of wealth was unraveled after the horse leg of Egypt's Sultan punched the ground and revealed a hollow where a treasure was hidden in a subterranean cave.

Fig. 5.4 Shimmering enclave (Tahara 2008)

Egypt's Sultan goal used the discovered wealth to build the mosque in order to glorify Allah and make people remember him as a devout and generous ruler. On the other hand, the ruling family of Qatar also has an underground wealth, the subterranean gas and oil, which allowed them to build this museum and buy the artifacts as well. The ruling Prince (amir) of Qatar also wanted to represent the memoralization of Islamic Architectural Heritage in a modern museum. Rabbat's insightful analogy triggered the re-interpretation of the meaning of the discovery of a subterranean cave in Pei's museum. Pei creates a sense of wonder in this space using lines of sight as despite the huge scale of the atrium's space, the oculus appears as a point in relevance to it and the 'mihrab' is the only transparent opening that connects the occupants with the external world. This means that upon entering the museum, one senses transfer from a world to another one, evoking the deep and slightly lit subterranean cave that Ibn Tulun discovered. The soaring scale of the atrium is measured along the discovery of its contrast with the intimate galleries (Ouroussoff 2008) that accentuates the cave-like space and the event of wondering (Fig. 5.5). The displayed artifacts that dwell these galleries in both deep niches and tall glass cases that disappear in the shadowy gallery spaces allowing the works to be highlighted by strategic fiber-optic lighting, evoke the experience of cautious discovery.

Visitors remain ghosts in the shadowy corridors that lead into each other. While social geometry is the primary cause that underlies the galleries spatial representation, the curators in collaboration with the architect emphasized the event of cautious wondering through the material used for the surfaces of galleries that contrasts with stone of the atrium. Wilmotte and Associates used dark gray porphyry and wood tinted with gilt bronze powder to emphasize the darkness and cautious feeling while moving attentively from an individual artifact to another. When the visitor is in the intimate zone of artifact, the strategic fiber-optic lighting turns on, rendering the artifact visible only in case of the proximity of the visitor. The usage of the word mihrab to refer to the only large window that frames the west bay glass towers in the *Museum of Islamic Art* book (Jodidio 2008) was of particular significance because it is usually only associated with the orientation praying in relation to Mecca. Resorting to such term either denotes inaccuracy, or it is intentional, as this paper suggests that Pei wanted to frame the production of towers, which are an erection of petroleum wealth, as another 'Mecca'. Even though the architect caught the essence of Islam by creating a glass mihrab in a museum and not a mosque, he directed it toward emancipations of wealth, which raises many questions.

3.1.4 Overlay Four: Buddhist Ritual

Critics and academics have questioned the insistence of Pei on his interpretation of Ibn Tulun's fountain. Their readings and the research's re-readings, together with the architect's personal experiences and memories, lead to understanding a new yet overlapped meaning of 'defense' that struck Pei in his travels to Tunisia. First, it is important to state Pei's explanation of the unusual sequence of geometric forms within the structure by which the dome is supported. He said: 'a geometric matrix that transforms the dome's descent from circle to octagon, to square, and finally to four triangular flaps that angle back at different heights to become the atrium's column supports' (Jodidio 2008). Goldberger mentioned the unusual sequence of geometric forms within the structure in the context of traditional and modern characteristics (Goldberger 2011), while Ouroussoff criticized this remarkable structure that is not cohesive (Ouroussoff 2008). Here, a logical question is raised if the structural sequence is linked to the exterior cube that could not be seen from the atrium or is an absolute structural solution. O'Kane mentioned the resemblance of stepped order that Pei developed to rotating cubic volumes whose angle of rotation increases to represent the rotational yet stable volumetric composition (O'Kane 2010). However, he critized the difference between Pei's cube and Ibn Tulun fountain's dome. Regrading the cause behind the clear expression of Ibn Tulun dome, literature review states that it is percieved as a final stage of transition from square to octagon that serves as the support of this dome. Some historians explained this sequence as a reflection of the verse nb (17) of Surah Al-Haaqqa, 'And the angels are its edges. And there will bear the Throne of your Lord above them, that Day, eight [of them]' (Dabbour 2012). In fact, Ferwati supported the influence of Ibn Tulun fountain by referring also to a more historic example. He drew a comparison between Jodidio's octagonal description and the octagon plan of the Dome of

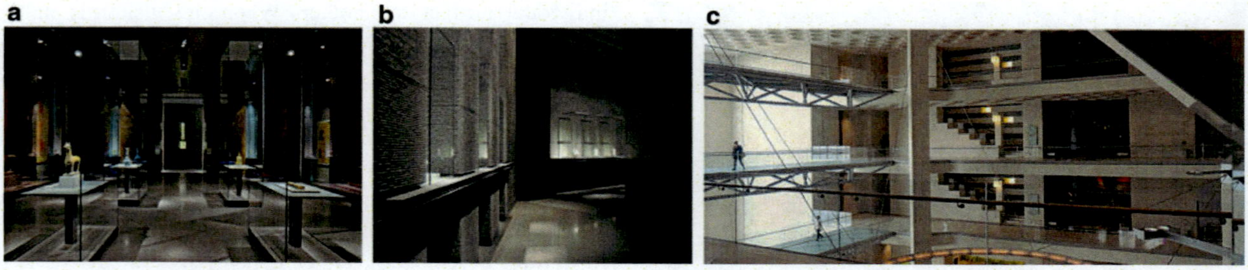

Fig. 5.5 Climax/atrium and galleries contrast. **a**, **b** Cave-like interior *Source* (Wilmotte, 2008), (c) Soaring light atrium (The Architecture of the Museum of Islamic Art, 2016)

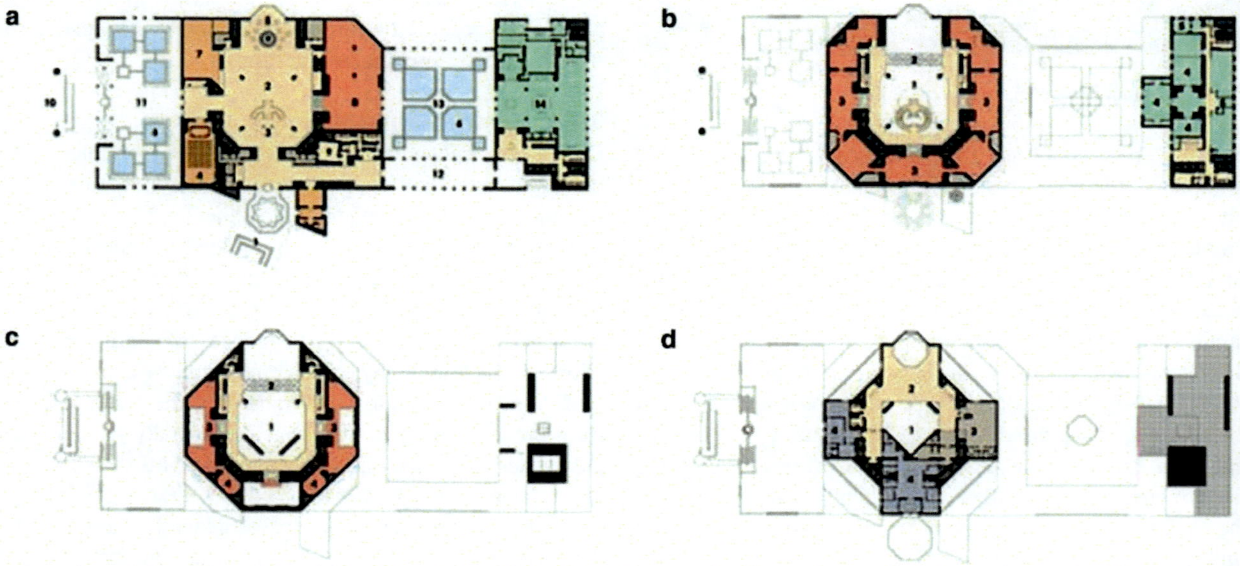

Fig. 5.6 Floor Plans of MIA (Ferwati, 2013). **a** Entrance level, **b** Second and third, **c** Fourth, **d** Fifth

the Rock which is considered the oldest example of an octagon-dome relationship (Ferwati 2013) (Fig. 5.6).

Actually, Jodidio states in his book that 'from the second to the fifth floor, the plans take the shape of an octagon where four of the ribs are distinctly larger than the other four and it gives the impression of a square plan' (Jodidio 2008). Thus, it is a square of slight chamfered corners. The research interest is not to make a pure geometric analysis, but to understand the embedded conveyed meaning. Back to the cube that Pei ended the building's volumetric order with, O'Kane said that what attracted his attention is not the expression of a dome on another cube. However, it is the cube itself that is clearly perceived in terms of six-directions-plus-center and measure. It is a cube of stone walls that is 'alleviated by a lunette on each side that somehow reminded me of the eye on the summit of Nepalese Buddhist stupas' (O'Kane 2010). The geometry of making that triggered previous critics structural questions, Pei's explanation of the triangular flaps in a monumental scale atrium, the vagueness of hidden small scaled dome that is wrapped by a cube, and the stupa insights, all lead the authors to figure out an appealing aspect of Buddhism that Pei used to practice during his childhood. They represent an abstraction for the defensive postures that communicates the meaning of defense and the sense of security that it offers to believeres. For example, at Longmen Grottoes, sculptures of military figures in defensive postures, set on either side of the Buddhist icons, promise a sense of protection (Karetzky 2015). During his early life, Pei had a peaceful and closely connected relationship with his mother Lien Kwun, who was a devout Buddhist believer and a musician who played the flute. Pei was the only member in the family to accompany his mother on her journeys to Buddhist temples and mountain sanctuaries.

Pei held an emotional attachment to these journeys and experiences, especially that he lost his mother when he was

Fig. 5.7 Fighter Posture and imperial scale according to the human body. **a** Guarding figure of Buddha at Longmen Grottoes (Karetzky 2015). **b** Internal structure (Von Stemberg 2008). **c** Three-dimensional section (Pottle and AwadStudio 2003)

still thirteen, so these places would remind him of his mother love and help him commemorate her memory. The narrative universe of this personal experience is essential and clearly traced in the museum. Although it is not a Buddhist temple, the spirituality of the space and spatial configuration reveal evocations of the silence of the mountain that were still alive in Pei's mind. O'Kane mentioned that the scale of the atrium is exaggerated and assumed that it is on purpose. Goldberger also questioned its monumental scale with respect to the tiny hidden dome and its cubic wrapping. Based on these insights, the research explores an evocation of the silence of guarding character of Buddha through this monumental space that Pei had experienced during his early memories through the serenity of Buddhist temples and monumental scale of caves for worship (Fig. 5.7).

3.1.5 Overlay Five: Creation

The previous overlay paves the way toward further understanding and exploration with the help of an insightful confession by the architect himself that was not explicit and popular as the statement on the inspiration from Ibn Tulun fountain. Pei had confessed in a very brief phrase that he was inspired by the Eastern ziggurat architecture (MIA 2006). However, the grandeur scale and stepped order and measure made different critics and scholars from interdisciplinary fields suggest that the design of the museum was based on the ancient Sumer 'ziggurat: of Ur' (Azimi 2016; Warner 2013). Reviewing the intent of constructing ziggurats that were made from mud bricks and were stacked upwards in the direction of the heavens, which tend to communicate the meaning of 'communion with God', helped the research trace the story of Creation or Origin of Beginnings that is found across different cultures. The research assumes that the architect was inspired by the creation story in Buddhism that is historically incarnated through symbols and spatially by 'stupa' (Fig. 5.8), on the basis of the architect's childhood memories.

Buddhism depicts creation through experiencing the journey of spiritual ascent and descent which relies mainly on measure and six-directions-plus-center of the spatial medium or situation. Stupa depicts an image made up of four primary geometrical figures associated with the hierarchy of four natural elements ordered based on the degree of density (Ching et al. 2011; Cavendish 1980). Earth is recognized the densest and represented as a square; water is represented by circular shape, then fire through a triangle and finally air through a crescent. The final element is among the most widely represented and punctuates the end of monuments. It is reinforced when the crescent takes the shape of half moon or half bowl. The present study suggests that limitations identified previously regarding dome-cube relationship and the capstone by O'Kane are solved through the story of Buddhism creation. Ziggurats are stepped terraces where

Fig. 5.8 Creation in various cultures. **a** Massing of the ziggurat (Ching et al. 2011), **b** Buddhism temple (Karetzky 2015), **c** Buddhism universal order (Lotus n.d.), **d** Volumes of the main building (Rabbat 2011), **e** Ablution Fountain of Ibn Tulun Mosque (Curatola 2009)

believers move physically by ascending on stairs; however, Pei did not explicitly bring up the spiritual journey that it is triggered by the absorption and unity of Buddhism's four major elements. Even though a previous study made a comparison between the plan of Dome of the Rock as a popular building of emotional value for Muslims and the plan of the museum (Ferwati 2013), the present study finds that the coherence between universal order and volumes measures and plans is more convincing.

3.1.6 Overlay Six: The Search for a Capital

The confession of Pei about the interest in ziggurats on one hand and the commissioners target to create a capital of culture and education, lead the authors to have a closer look on the influence that the ziggurat had on the Abbasids era. The first step was to search for a capital. When the Abbasid Caliph Al-Mansour wanted to construct the city of Baghdad, he designed a round city surrounded by a double set of walls for defensive reasons (Hattstein, 2013). The walls were pierced at the intercardinal points by four gates from which roads led to the four quarters of the empire. On the other hand, the Great Mosque of al-Mutawakkil in Samarra helps show the size and scale of the Abbasid imperial construction (Fig. 5.9a, b).

The interesting point is the position of the 'mihrab' and the spiral or 'malwiyya' minaret which aligns along the same axis. Pei's design is based extensively on the six-directions-plus-center. There are only four gates that each signposts the major cardinal directions and all are inaccessible except the gate of the main entrance which is the south. Thus, the long approach to reach the building through the green axis leads the visitor to the south gate; by entering it, he/she faces the north 'mihrab' which is directed to the glass towers elevation (Fig. 5.9c, d). The present study finds that this overlay reconciles with the former story of 'defense' and concludes that Pei's perspective of 'Islamic essence' is imperial, meaning it feeds the commissioners' objective to make a statement through this building.

3.2 Louvre Abu Dhabi, Abu Dhabi, UAE, 2017

French architect Jean Nouvel was tasked with the design of the Louvre Abu Dhabi. One of his previous works that incorporated Islamic architectural traditions is a cylindrical tower in Doha, Qatar, which is topped by a cupola as a representation of cosmology to adhere to the essence of a spiritual space. Standing under the cupola gives the impression of being under the sky due to its perforated surface in a manner that the cupola acts as a second sky (Byrnes 2017). Thirty years ago, Nouvel designed the Arab World Institute in Paris through which he sought to evoke Islamic traditions. He saw the intricate ornament and the exquisite decorative patterns that connoted Islamic architecture, and he sought to evoke this intricacy and filigree in an entirely new way by making it the inspiration for an ingenious system of sunscreens. For the Louvre Abu Dhabi, he was inspired by the domes and the common geometrical patterns in Islamic architecture, and also by the city Abu Dhabi itself (Goldberger 2011). Nouvel simultaneously concentrated on universality and the site particularity under the title of shared humanity. In Arab culture, seeing through the filtered light of a mashrabbiya (pierced screen) is very natural (Nouvel 2017a, b; Goldberger 2011), where light and shade punctuate the view creating their own sense of mystery and culture. 'Great Arab architecture is always about geometry and light', according to Nouvel, (Marlowe 2017). He declared that he found inspiration in the context by listing existing basic elements such as sand, sea, water, sky, and the bursting sun (Simpson 2017), and he utilized the latter as the strength-point in his project, being the harshest element among them all. He also highlighted the influence of the 'dome' as a repetitive structure in architecture of Islamic civilizations and described it as a 'cupola' in order for it to be the eternal symbol of almost all civilizations at the same time. Nouvel aimed at representing the universal nature of

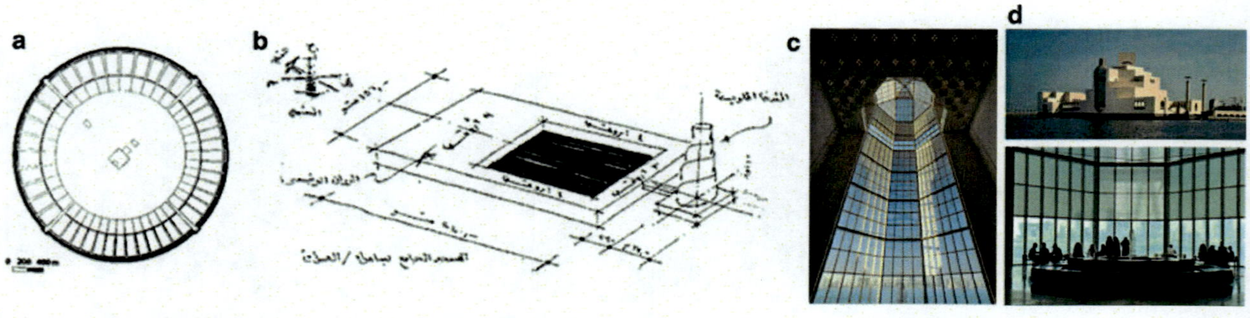

Fig. 5.9 Six-directions-plus-center. **a** Rounded city of Bagdad, **b** Great mosque of Samarra from Lecture of Dr. Tarek Farghaly, **c** Mihrab of MIA and glass Towers (DeHenning 2009), **d** External and internal view of Mihrab (Stemberg 2008)

humanity as a formal association to this art museum (ArchNewsNow 2007). The investigated overlays are listed below:

Man and Cosmos—Foundation of Prophet Mohammad House Mosque—The Rescue of baby Prophet Moses—Deluge—Paradise Lost.

3.2.1 Overlay One: Man and Cosmos

The architect repeatedly identified the hovering structure of the museum as a building that signifies a sky in order to create mystery and let the visitors aimlessly wonder about where the sky building touches the museums blocks (Nouvel 2017a, b). This mystery is a drive toward exploring the 'who', which was apparently the result of curiosity and a belief in the value of cosmos for the sake of humanity. Human interest in cosmology is the predominant reflection in this museum. The architect's intent is to affirm the unity of all humans to this universal existence through creating a cosmic interlace to achieve the essence of spiritual space that is called another sky (Fig. 5.10). In fact, the term cosmos, denotes the idea of 'order' and 'ornament', meaning the universe as an ordered and ornamental whole. Since the building is in an Arab context and the architect described it as Arabic Galagatic and constellations, it is necessary to translate its linguistic meaning. The Arabic equivalent of the word is (kawn) that designates the cosmos as an embodiment of the metaphysical order. Thus, in the cosmic order, cosmos is the mediation between the Creator and Man (Akkach 2005). The sky is the spatial frame which is revealed in a number of verses. For instance, verse nb (64) of Surah AL-Ghaafir and verse nb (22) of Surah Al-Baqara describe the sky as a canopy, while verse nb (2) in Surah Ar-Ra'ad says that heavens are raised without any pillars. Therefore, this hovering structure evokes the presence of a near cosmic hovering volume whose impact is experienced differently during day and night and different times of the day as well (Block 2017). Lines of sight are present extensively through the event of gazing up and down because of the action of rays of light that penetrate the different layers. Therefore, metaphysically and of course not physically or quantitatively, there is a profound correspondence between man and cosmos. This meaning explains the intention of the architect to create the allusion of rain of light. Thus, unity is reflected through these vertical ties that connect humans to their originator through the shared canopy/cosmos that construct the circle of presence. The spiritual significance is by awakening the sense of attraction to the celestial that prevents man from being dragged by gravity to the abysses of subhuman existence. In terms of six-directions-plus-center, it is through representing a remarkable feature of the human state that no matter where and in what condition man may be, he always finds the sky above him and the attraction that pulls him toward the infinite and the Eternal.

3.2.2 Overlay Two: Foundation of Prophet Mohammad House Mosque

The structure of the sky building is popularly identified as the conventional conception of a dome and it was the predominant element in the project that caught the media attention. In the context of Arabic civilizations and especially of Islamic faith, there is a huge body of literature on the development of mosque architecture. Hence, some critics saw the sky building as a re-interpretation of the mosque dome and Nouvel pointed to it as an Arab symbol (Bianchini 2017; Stevens 2017). Nouvel explained that the use of the dome to approach the essence of spiritual protection is not confined to the religious sense (Nouvel 2017a, b), as it was pierced with openings to look like interwoven palm leaves, a traditional roofing material in the United Arab Emirates (The Louvre Abu Dhabi 2017). This has lead the research to a spatial frame in the story of the foundation and construction of Prophet's House that highlights the meaning of protection and directionality. Sources report that during the course of construction, Prophet Mohammed directed his companions, using the following words: 'build it in the form of arish like that of Moses', which reflects the specific miniature he had in his mind. As an indication of shading, 'arish' refers to 'trellis', which literally means the material used for the rooftops of Mecca's houses at that time. Trellis also refer to the 'tabernacle' or 'tent' that God had ordered Moses to build as a worship place. The Prophet's dwelling included a prayer hall that opened onto the courtyard, which represents an elongation of the covered space related to the mosque as an entire place (Akkach 2005). Prayers used to take place in the shaded area back when the direction of qebla was to Jerusalem, and continued to take place there even when the direction was later changed to Mecca. The shading effect is reflected in geometry of making. However, this six-directions-plus-center stands out in terms of the absence of the directionality. Instead, it has a centrifugal direction where no preference direction is emphasized theologically, in order to validate the meaning of universality in which the same spatial area can embrace different dialogues with different destinations since the spiritual action is necessary in itself, regardless of any shape or direction (Fig. 5.11).

3.2.3 Overlay Three: The Rescue of Baby Prophet Moses

The white boxes are repeated components in the museum that are not less necessary than the canopy. The architect offered insightful description of his white cubes. He said: 'My white cubes are the opposite of a typical museum, which are meant to feel blank, like you are nowhere, I want you to feel always like you are somewhere' (Crow 2017). Knowing that these boxes are thematically organized according to chapters of common lives, there is a 'great hall' or 'vestibule' that is identified as the departure space of the

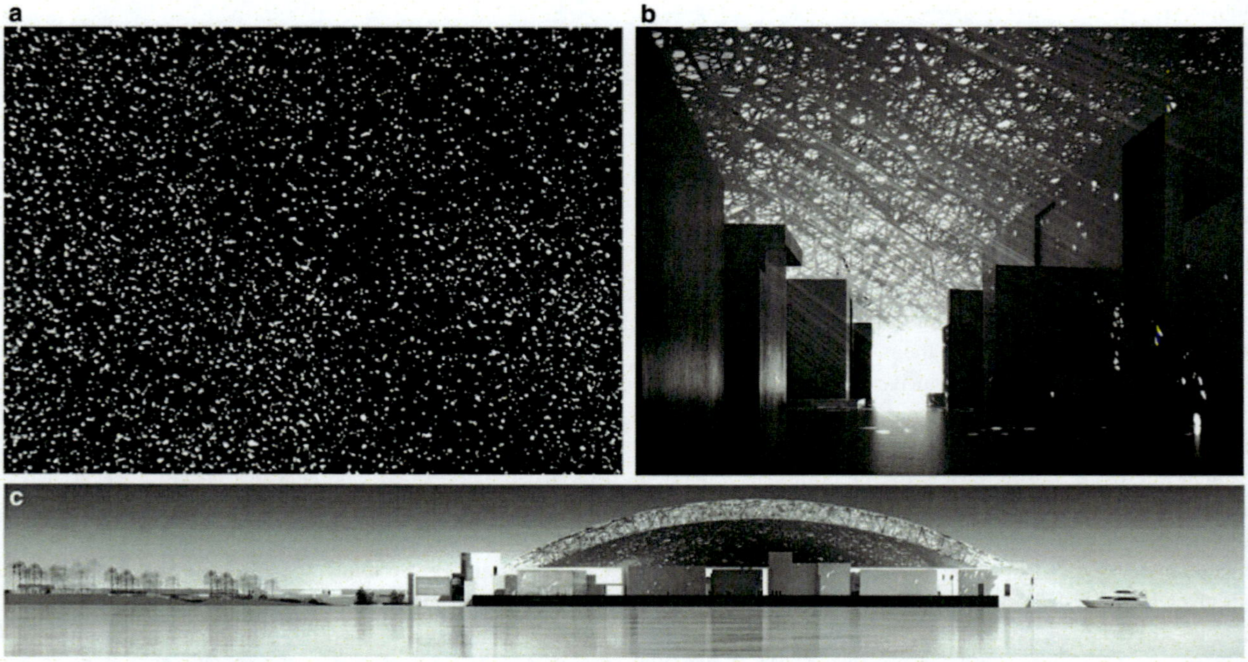

Fig. 5.10 Sky as a Building. **a** Starry sky (Steele 2005), **b** Lines of light (Castro 2016), **c** Canopy as a building in section (Castro 2016)

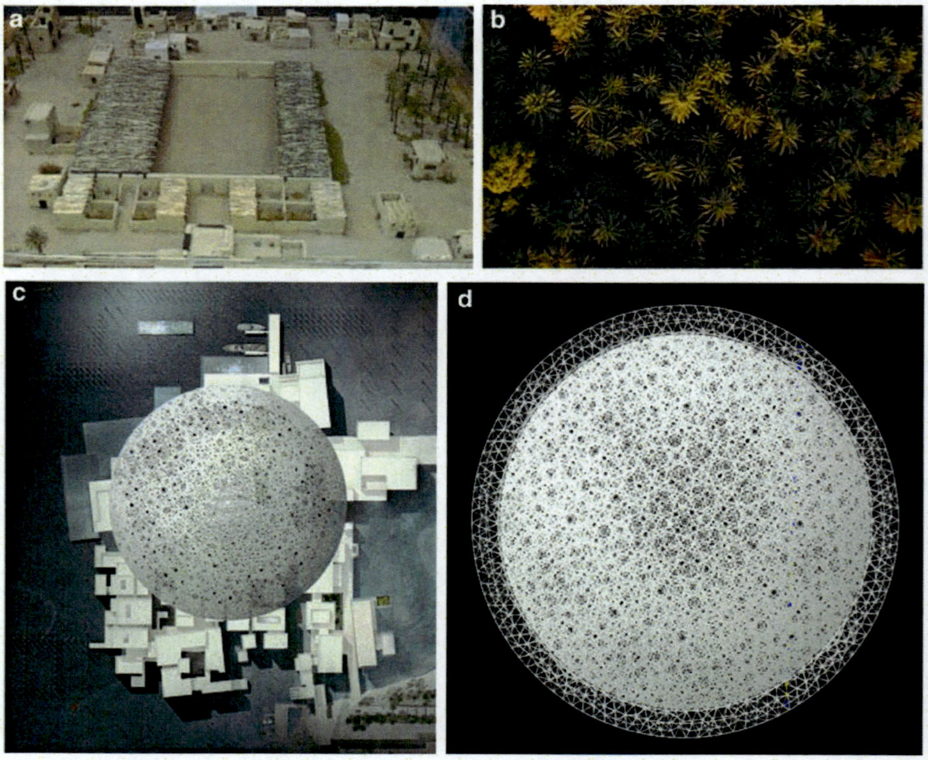

Fig. 5.11 Universality by directionality. **a** Prophet House Mosque model (Conlon 2017), **b** Trellis, **c** Canopy (Castro 2016)

journey in the museum. It is a microcosm of the twelve chapters. Nouvel described it as the representation of the most powerful bond that connects humankind, which is maternity (Leech 2017), by displaying maternity figures primarly. He also said that it represents a common fate that humankind hate but are obliged to experience, and that is death. This allowed the research to uncover an embedded event of rescuing a baby in an ark. It is the story of the baby Prophet Moses that is found in the Holy Books of Abrahamic religions using different narratives, but the event remains the same. When Prophet Moses's mother was deeply worried about the fate of her son, God inspired her to take an unlikely action, and to put him in an ark and cast him in the river away from the sight of the Pharaoh who had commanded the murder of all new-born babies. 'Ark' in Hebrew is tebah, which means the box of life, while in Arabic is tabut. Nouvel had called these boxes in previous interviews, the floating boxes of life, which reveals the storyworld of baby Moses. This event is reflected in social geometry and the circle of presence. For the former, it is through the white boxes of different dimensions that are designed specifically to the displayed items inside and every box protects artifacts that tell stories that are common for many civilizations. This verifies the intention of the architect to make the floating white boxes on water closed surfaces that only have a slight opening in the ceiling to keep in contact with the canopy. As for the circle of the presence, it is the scene of floating white boxes that reflects the rescue and stability effects. Furthermore, it is the 'grand vestibule', which is a room unlike any other building form of all the galleries, that evokes the essence of maritime chart through intersections of the floor compass rose with the dome that is visible through the skylight (Imanova 2017) (Fig. 5.12).

Fig. 5.12 Universal love. **a** mother figures, **b** grand vestibule (Leech 2017)

3.2.4 Overlay Four: Deluge

The museum stands out because of the narratives it utilizes through the exhibited arts, in order to tell the stories of different civilizations and universal themes of humanity (Louvrepress 2017). The architect pushed the geographical limitations of reading and assessing artistic productions that are usually geographically segregated, and wanted visitors to see objects from different civilizations in the same rooms dialoguing together in terms of social geometry. Here emerges the uniqueness of the museum since it is the first to sort displayed artifacts according to themes, which is reflected in the spatial representation, as Nouvel emphasized (The Louvre Abu Dhabi 2017). This sought experience is similar to the event of selective selection of creatures and categorizing them for the purpose of rescue, like in the story of Deluge that like Moses', is found in all civilizations with differences in narratives while the basic plot stays the same (Podany 2014). The embedded meaning of the story is that the flood serves to cleanse the earth from evil and provide rebirth, with the surviving hero functioning as the symbolic seed of the new life (Podany 2014). The verse nb (40) in Surah Hud explains the social arrangement of the creatures who share Prophet Noah the Ark of rescue. The storyworld revealed by the verse indicates that when Allah commanded the earth to blew its streams, a command was given to Prophet Noah to carry in the Ark a male and female pair of each kind of the creatures. In every hall at the museum, one artistic production from every civilization is presented so visitors can recognize similarities and influences between these arts that differ in geography and culture but stand under the same theme.

3.2.5 Overlay Five: Paradise Lost

Nouvel usually offers a scarce clarification; nonetheless, by collecting and re-reading in depth, the authors were able to discover another embedded meaning and new events that are reflected in the narrative of the museum. According to Nouvel, 'the whole territory [of this museum] is envisioned not so much as a nostalgic longing for some remote world, some lost paradise, but as a trigger to question a sense of time' (ArchNewsNow 2007). Mankind hope for regaining heaven is a common longing, although related narratives might differ. Nouvel emphasized this tendency by using terms such as ‘sanctuary’, ‘cathedral’, and ‘medina’, despite his insistence on the spiritual dimension and exclusion of theological dimension. John Milton, an English poet and polemist, is famous for his epic poem ‘Paradise Lost’ written in blank verse. It concentrated on the first fault of Adam and Eve and expanded on the longing aim to what is lost, which is paradise. The introductory verses of Milton poem imply the tendency of Christian and Jewish religions to materially represent the throne of God and place a special focus on arts and buildings for praising God as the source of spiritual power. The Old and New Testament scriptures reveal that God uses earthly things as a mean of communicating with man, whether it be a boat, a container, or a temple (Podany 2014). This reveals the tendency of communion with a transcendent power in order to restore hope. The first event is the communion with God in the tabernacle in the wilderness that is experienced through the sense of cloud protection and invisible approach. The second event is communion with God in the Heavenly City of Jerusalem that is experienced through dancing pillars of light, water rivers, wall inscription, and twelve galleries. The first event refers to the story of Prophet Moses that is present in the final chapters of the Exodus, which describes Prophet Moses second ascent to mount Sinai where he received the Ten Commandments. The story also describes how the Israelites constructed the Tabernacle and the Ark of Testimony, a receptacle for the stone tablets containing God’s laws. By definition, a tabernacle is a portable shelter carried by the Jews in the wilderness; it originates from the Latin tabernaculum, ‘hutment, tent’. The Hebrew word used in Exodus 25:9 is mishkan, which the dictionary translates as ‘habitation, dwelling, hut, tent, tabernacle, sanctuary, temple, grave, haunt’. The word miqdash (sanctuary), literally ‘holy place’, is used in this story almost synonymously (Knappert 1993). The earliest interpreters had no doubt that the importance of the tabernacle lies in its hidden symbolism. The purpose of the tabernacle was to provide a place where God may dwell among men. The tent curtains, and especially the thick veil, served as separator, a dividing barrier, between God and the people. The tabernacle was also a facility which concealed fabulous wealth and beauty. The tabernacle did not only house expensive material goods, but also great works of art. God had commanded Moses to fashion a tabernacle in a way which would involve almost every form of representational art that man had ever known. Once the Tabernacle was completed, the pillar of cloud which God had used to protect the Israelites on their journey from Egypt, would descent above it and the glory of the Lord would fill the place. As long as the cloud would be on top, they would stay in their place. However, as soon as the cloud would float back up, the Israelites would dismantle the camp and resume traveling. Because Moses did not uphold God’s holiness when the Israelites had no water, God forbade Moses from entering the Promised land. Re-evoking the representation of the lost paradise in the museum through the sanctuary (tabernacle), Nouvel re-asserted that the museum is ‘a form that matches its function as a sanctuary for the most valuable works of art’ (The Louvre Abu Dhabi 2017). All these meanings are relevant in explaining that some galleries are divided according to partitions that veil a more important artistic production that mimic the movement toward the Holy of Holies and give the illusion of hovering.

Nouvel also said that the project represents a medina, a term he used to refer to the Islamic Arabic city (Bianchini 2017; Wainwright 2017). The museum complex is composed of several structurally independent low-rise buildings that are arranged in an apparently random sequence to recall the urban structure of medina. Cities in the Islamic world were once the perfect grist for the Orientalists mill (Steele 2005). Rasem Badran did a number of research studies to rediscover the Islamic City by taking a deep look at social, cultural, religious, economic, and environmental issues. He found out that the layout of the medina is not random in terms of sequence and order. This research, however, has no interest in stylistic-driven claims. Even though the border lines of the permanent museum spaces are irregular boundaries, they do not allow a definite line of sight or line of passage because the walls are bare. If the exterior lines of the museum are refined, it becomes a cube with sharp edges. The lines of passage passing from a gallery into another is of maze-like pattern but a known path where orientation is clear from a gallery into another (Fig. 5.13c). The museum’s art galleries do not have an open floor plan. Instead, visitors must follow the official path of the exhibition in order to go from chapter one ‘The First Villages’ to chapter two ‘The First Great Powers’, all the way to chapter twelve ‘A Global Stage’, from where they can then exit. The image presented by the author concerning revelation gives the starting point for finding that the sequence of museum galleries is also an image of heaven as it is described in revelation. This heaven is adorned with jewels and a river runs through it; it is also walled with the names of the twelve tribes of Israel. The boxes also have foundations and walls made out of jasper.[2] This intersects with scarce clarifications that Nouvel gave in various interviews in which he compared the museum to a cathedral (Marlowe 2017). In terms of the circle of presence, the role of water is crucial in the project as a place of abundance and joy, where all people are in direct communion with the Creator, seeing His reflection (light) in the water (Fig. 5.13b).

[2]The Books of Revelation describes the Heavenly city of Jerusalem using these sentences.

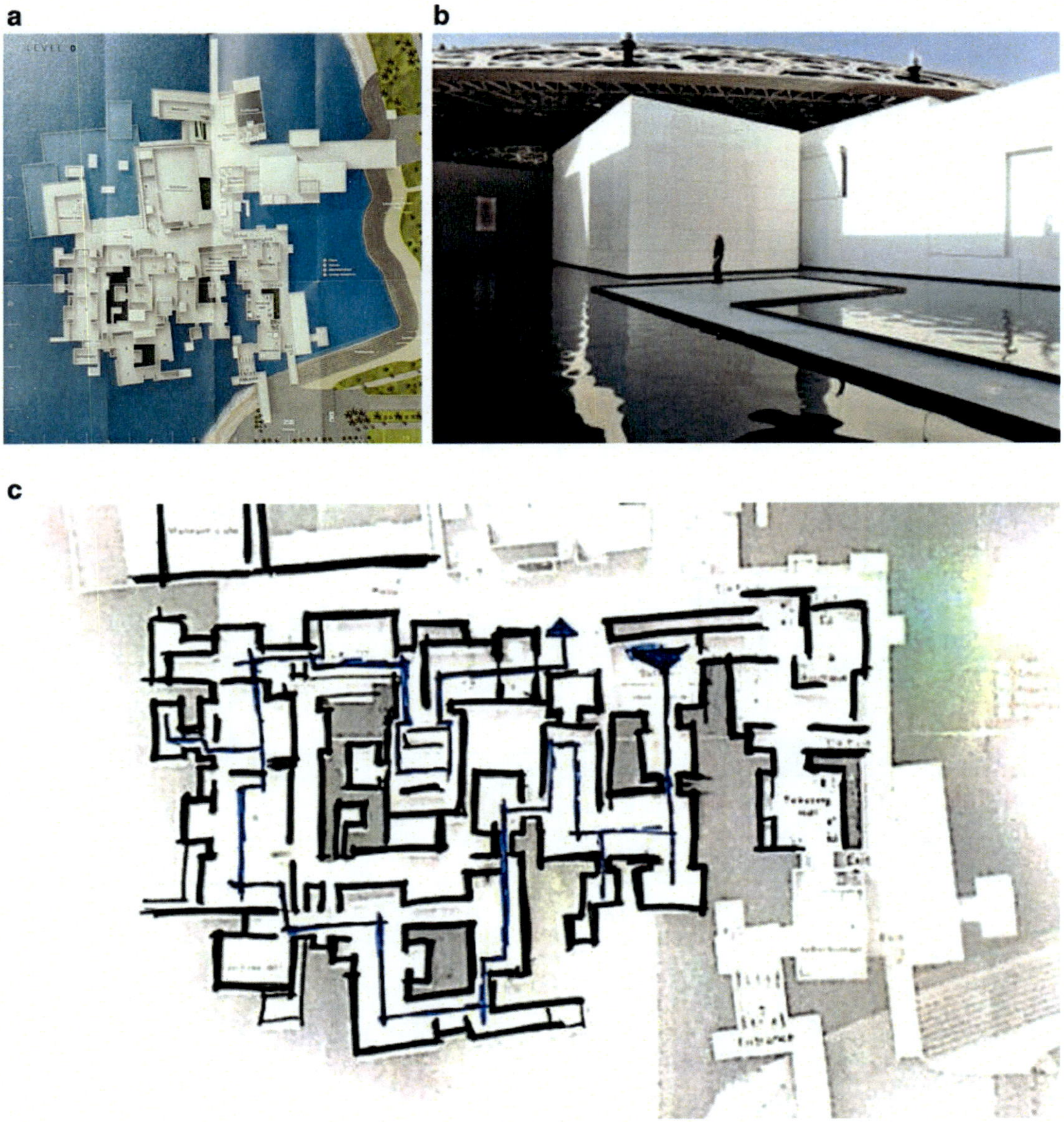

Fig. 5.13 Prescribed path through the twelve themes. **a** Brochure of museum, **b** White-inscribed walls and water (Ghinitoiu n.d.), **c** Line of passage in floor plan of maze-like galleries

4 Discussion

Using the suggested method, realization of utopia in the context of selected art museums allowed the authors to get a further understanding of meanings through the aforementioned overlays as shown in Table 5.1. The themes stated by the architects are listed here in order to show the evolved stories so that they are further understood with respect to the concluded topics; accordingly the implied meanings are disclosed.

Table 5.1 shows that the genuine Home for the Museum of Islamic Art is revealed in the various levels of narrative space: storyworld of Ribat and Creation of Ibn Tulun Mosque, story space for Allah is Light, and spatial frame for Buddhist Creation story, which are all interpreted according to the museum's geometry of being. On the level of six-directions-plus-center, there is an overlap between Ribat, Communion with Buddha, and Buddhist Creation. The common topic among these three stories is Defense and Immortality. On the level of geometry of making, there is an overlap between Ribat, Presence of Allah through Light, and

Table 5.1 Outlining utopian representations by mapping stories and exploring genuine Home

(■ = shaded cell)

		Theme	Topic	Story	Storyline / Event	Genuine Home: Narrative Space: Setting	Spatial Frame	Story Space	Storyworld	Narrative Universe	Geometry of Being: Measure	Line of Sight	Line of Passage	Social Geometry	Six-directions-plus-center	Geometry of Making	Circle of Presence	Physical Space
Art Museums	Museum of Islamic Art in Doha	'Essence of Islam'	Defense	Ribat	impermeability				■						■	■	■	The approach to the site
				Buddhist Ritual	Guarded by grand scale structure					■	■			■	■	■		Triangular Structure
			Divine Presence	Presence of Allah through Light	Illusion of point			■								■		Point Dome
				Creation	Perceiving shape hierarchy					■	■				■			Space of Atrium
			Wealth	Creation of Ibn Tulun Mosque	Discovery				■		■	■		■			■	Oculus and artifacts galleries
				The Search for Capital	Imperial Approach	■									■			Cardinal points, mihrab (glass towers)
	Louvre Abu Dhabi Museum	'Humanity'	Protection	Man and Cosmos	Seeing illusion of hovering		■				■	■			■	■	■	Canopy
				Foundation of Prophet Mosque-House	No preferable direction		■								■	■		Canopy and Centrifugal Directionality
			Rescue	The rescue of Baby Prophet Moses	Enclosed in stable boxes				■					■			■	Main central gallery
				Deluge	Prescribed path				■			■	■	■				Galleries
			Communion with God	Paradise Lost	Illusion of hovering ,movement as holy of holies		■				■		■			■	■	Canopy protecting Boxes of treasures
					Water rivers and wall inscription		■				■		■					Rain of Light and Twelve Themes

Communion with Buddha. On the level of social geometry, there is the story of Creation of Ibn Tulun that includes the discovery of a cave, falls under the topic of Wealth. This indicates that the main concern of the architect is on the level of six-directions-plus-center. On the level of measure, the stories of Creation of Ibn Tulun and Communion with Buddha are overlaid, which is logical because of the contrasting scale of galleries and the building's atrium. The architect used the story of light in reference to the Noble Qur'an that represented at the level of geometry of making. The predominant topic is Defense and Immortality. Considering the considerable significance of the six-directions-plus-center, the essence of Islam by the architect is represented by the used narrative to reveal immortality and power of money. For the Louvre Abu Dhabi, the used method indicates that the narrative space includes a spatial frame for Man and Cosmos and Foundation of Prophet Mosque-House, and a story space for The rescue of baby Prophet Moses, Deluge and Paradise Lost. On the level of six-directions-plus-center and geometry of making, the two stories—Man and Cosmos; Foundation of Prophet-Mosque-House—are represented by the canopy that is perceived the predominant element under the topic of Protection. On the level of social geometry, the stories of the Rescue of Baby Prophet Moses and Deluge are represented through the box of life and the narrative for exhibiting the artifacts. These two overlays intersect under the topic of Rescue. Regarding measure, line of sight, and line of passage, there is an interplay among these stories: Man and Cosmos, Deluge, and Paradise Lost.

The constitutive contents of utopian thinking are summarized in Table 5.2. Pei's quest to seek the essence of Islamic architecture is consistent with the storyteller's context and particularly his personal experience of faith. Nouvel infused with metaphors and poems of definite terms that the present study investigated further to discover that they are overlays of spiritual and transcendental meanings from collective folk, Holy Books, and existing natural parameters

Table 5.2 Conclusion of constitutive content of Utopian thinking

Constitutive content of utopian thinking		Art Museum (Locale material of anticipatory illumination)	
		Museum of Islamic Art, Qatar	Louvre Abu Dhabi, UAE
Causes of Utopia	Genuine home	Setting, story space, storyworld, narrative universe	Spatial frame, storyworld
Means of utopia	Stories	Stories on the same theme but different topics	Stories on the same theme, different but common for holy religions
Authenticity	Hidden essence	Childhood memories	Religious truths
Sources	Truths	Secular and sacred	Sacred

that coexist in spatial representation with open-ended interpretation. Louvre Abu Dhabi represented separate incomplete facts and truths of universal stories in order to trigger the visitor to imagine the experienced situations, relate to them, and color them based on his/her background.

5 Conclusions

Utopia is the genuine Home that the character usually seeks to imagine, and so is the case when the architect thinks of a design for a art museum that is considered a macrocosm that represents the highest truths, values and concerns where the 'cipher of the authentic' and meaning is often embedded. This research primarily argued that art museums are ritual sites that incarnate utopian longings of the respective architects, especially for star architects whose designs are embraced willingly by the wealth patrons and communicated to the public through digital media. Design languages of star architects, entrusted to design art museums in the Middle East Region, imply multiple layers of ambiguity, mystery, and secrecy, even if they claim authenticity in order to influence people by manipulating the tension/excitement connection between them as architects and the collective. The two main architects discussed in this paper, used concise words and phrases to communicate their claims, which incite different interpretations. For example, Pei said he wanted to represent the 'essence of Islam', while Nouvel disclosed he intended to represent 'Humanity in a New Light'. Along these statements, explicit method of communication for the particular Arab context is highlighted by references to the fountain, dome, and medina, in order to validate the relevance of the projects to the context and gain collective acceptance, which in fact highlights the problem of gap of communication between the architect/individual and users' perception/collective. One of the important conclusions is that the gap of communication between users and architects can be bridged through realization of told stories that might be the source of inspiration to a great extent; this is possible even if the architects do not show their process-fragment but only the fruit of work, which this research tried to explore in the context of utopia and its possible means. Just as it is important to explore meanings and understand stories that seem relevant, it is also significant to record their sources, especially when we recall Duncan's treatment of art museums as secular public spaces and not religious sites. Since utopia is based on 'something's missing' and is founded on a selected truth but in a mediated now, the explored meanings were categorized under Truths that are relevant to the author, the context, or the audience through a story in order to specific its significance for a particular party.

The methodology adopted to explore the 'cipher of the authentic' and meanings embedded in art museums is through approaching thinking as 'storytelling', through which stories are the means to realize utopia in various narratives that are inspired from the intangible (oral and written stories) and represented spatially as utopian representations. It was key to take into consideration the 'authenticity' and 'cultural heritage' because these are the claims used to engage the receptivity of the collective by referencing previous architectural elements. To be more precise, the concept of authenticity was clarified with respect to the producer of the art, the architect. Bloch's philosophy on arts, especially the significant role of the art producer, lead to understanding a subtle concept of 'authenticity' and offered insights that fostered awareness of understanding that the possession of the self is a collective possession. For this reason, this research reasons that Bloch's philosophy is in line with the points that Koolhaas made on the unconditional acceptance of architectural star-system that might be also an exceptional opportunity for individualism, and the necessity of thinking of communication beyond the architecture-image method inorder to reconcile the gap between the 'individual' and the 'collective'. The hypothesis of realizing utopia through storytelling is validated by the interpretive task that comprises the narrative space as a source and the geometry of being as an architectural representation, along the performance which depends on events/experiences that are derived from various evidences. Throughout this paper, explored utopian representations of selected art museums proved many unraveled truths that are closely related to the sacred and religious realms and are meaningful for the architect primarily in the first project and to the diversity of civilizations in the second. For example, the narrative universe of Pei's personal experience is implied through his implication of a truth in parallel with his perspective on light and Ibn Tulun fountain; meanwhile, the Louvre Abu Dhabi was itself a dramatic field in which stories of all civilizations and religions were being told at once.

Restoring hope for the collective is the objective of utopia in architectural projects of cultural significance; therefore, investigating utopian representations by thinking as a storyteller is one way to mitigate the communication gap, and may be an answer to the question about what it means to say that utopia is realized. On this ground, the gap between users and architects can be bridged through the realization of told stories. This paper intends to direct a serious comprehensive message to the decision makers and clients, asking them to think deeply about the embedded message/intention of the architect. It also asks the researchers, curators, and critics to comprehend truths beyond art museums because spreading messages that have dual meanings plays a critical role in culture construction in terms of specifying what to accept unconditionally.

References

Akkach, S. (2005). *Cosmology and architecture in premodern Islam—an architectural reading on mystical ideas*. Albany: State University of New York Press.

Andraos, A. (2016, May). The Arab City. *Places Journal*. Retrieved April 15, 2018, from https://placesjournal.org/article/the-arab-city/#0.

ArchNewsNow. (2007, February 2). *In their own words: Abu Dhabi cultural district*. Retrieved from ArchNewsNow.com: http://www.archnewsnow.com.

Azimi, N. (2016, December 19). *The gulf art war*. Retrieved from The NewYorker: https://www.newyorker.com/magazine/2016/12/19/the-gulf-art-war.

Badawi, A. R. (Trans.) (1973). *The art of poetry with Ancient arabic translation, Annotations of Al-Farabi, Ibn Sina, Ibn Rushd, Aristotle Talis*. Beirut: Dar al-Thaqafa (دار الثقافة).

Benjamin, W. (1969). *Illuminations: Essays and reflections*. In H. Arendt (Ed.) & H. Zohn (Trans.). New York: Schocken Books.

Bianchini, R. (2017, December 9). *My museum/Louvre Abu Dhabi*. Retrieved from inexhibit: https://www.inexhibit.com/mymuseum/louvre-abu-dhabi-jean-nouvel/.

Blair, S., & Bloom, J. (2013). Architecture. In M. Hattstein & P. Pelius (Eds.), *Islam art and architecture* (pp. 94–117). Potsdam: h.f. ullman publishing GmbH.

Bloch, E. (1996). *The Utopian function of art and literature*. In J. Zipes, & F. Mecklenburg (Trans.). Cambridge: MIT Press.

Block, I. (2017, December 11). *New photographs reveal the feat of engineering behind the Louvre Abu Dhabi's dome*. Retrieved from dezeen: https://www.dezeen.com/2017/12/11/new-photographs-reveal-engineering-behind-louvre-abu-dhabi-dome-architecture/.

Bobonich, C. (2004). *Palto's Utopia recast: His later ethics and politics*. Oxford: Clarendon Press.

Bobonich, C., & Meadows, K. (2018). *plato.stanford.edu*. Retrieved from Stanford Encyclopedia of Philosophy: https://plato.stanford.edu/entries/plato-utopia/.

Byrnes, M. (Director). (2017). *Seeing the light with Jean Nouvel* [Motion Picture].

Castro, F. (2016, August 12). *In progress: Louvre Abu Dhabi/Jean Nouvel*. Retrieved March 6, 2018, from archdaily: https://www.archdaily.com/793182/in-progress-louvre-abu-dhabi-jean-nouvel.

Cavendish, R. (Ed.). (1980). *Mythology an illustrated encyclopedia*. NewYork: Rizzoli International Publications.

Ching, F. D., Jarzombek, M., & Prakash, V. (2011). *A global history of architecture* (2nd ed.). Hoboken: Wiley.

Cobley, P. (2014). *Narrative(The new critical idiom)*. Abingdon: Routledge (Taylor& Francis).

Conlon, I. S. (2017, January 16). *Ibrahim S Conlon architect*. Retrieved from LinkedIn: https://www.linkedin.com/pulse/true-islamic-architecture-authentic-minimalist-house-ibrahim/.

Crow, K. (2017, November 7). *Louvre Abu Dhabi prepares for its closeup*. Retrieved from The Wall Street Journal: https://www.wsj.com/articles/louvre-abu-dhabi-prepares-for-its-closeup-1510038060.

Curatola, G. (2009). *Visual encyclopedia of art-Islam*. Italy: Scala.

Dabbour, L. (2012). Geometric proportions: The underlying structure of design process for Islamic geometric patterns. *Frontiers of Architectural Research, 1*(4), 380–391.

DeHenning, A. (2009). *Aga Khan award for architecture/museum of Islamic Art, Doha, Qatar*. Retrieved from Archnet: https://archnet.org/sites/6495/media_contents/67673.

Duncan, C. (2005). *Civilizing rituals inside public art museums*. London: Routledge.

Emmons, P., & Phinney, L. (2017). Homo Fabula. In P. Emmons, M. Feuerstein, & C. Dayer (Eds.), *Confabulations-storytelling in architecture*. Oxon: Routledge.

Ferwati, S. (2013). The Museum of Islamic Art: Form, perception, and environment. *The International Journal of the Inclusive Museum, 6*.

Ghinitoiu, L. (n.d.). Photographs of Louvre Abu Dhabi. *Critical Round-Up: The Louvre Abu Dhabi by Jean Nouvel*. Retrieved from https://www.archdaily.com/883911/critical-round-up-the-louvre-abu-dhabi-by-jean-nouvel.

Goldberger, P. (2011, October 29). *Lectures-Islamic architecture, modernism, and I.M. Pei: The challenge of the Museum of Islamic Art*. Retrieved from paulgoldberger: http://www.paulgoldberger.com/lectures/islamic-architecture-modernism-and-i-m-pei-the-challenge-of-the-museum-of-islamic-art/.

Hattstein, M. (2013). Islam-World religion and cultural power. In M. Hattstein & P. Delius (Eds.), *Islam art and architecture* (pp. 9–33). Potsdam, Germany: h.f.ullmann publishing GmbH.

Hilal, M. G. (1997). *Modern literary critisizm*. Cairo: Dar Nahdat Misr.

Imanova, A. (2017, December 4). *Louvre Abu Dhabi's geometric dome has informed its interiors, says Nouvel*. Retrieved from design MENA: https://www.commercialinteriordesign.com/insight/louvre-abu-dhabis-geometric-dome-has-informed-its-interiors-says-nouvel.

Incorporated, B. B. (1997). *Storytelling encyclopedia: Historical, cultural, and multiethnic approaches to oral traditions around the world*. In D. A. Leeming, & M. Sader (Eds.). Canada, Arizona: The Oryx Press.

Jodidio, P. (2008). *Museum of Islamic Art Doha, Qatar*. Munich: Prestel.

Karetzky, P. E. (2015). *Making sense of Buddhist art and architecture*. London: Thames & Hudson.

Keller, C. (2016, April 27). *Utopian dreaming—Ernst Bloch—The Utopian function of art and literature*. Retrieved 2019, from Krustelkram: https://www.krustelkram.com/utopian-dreaming/2016/4/22/ernst-bloch-utopian-function-of-art-and-literature.

Knappert, J. (1993). *The encyclopedia of middle eastern mythology and religion*. Shaftesbury, Dorest, Great Britian: Element Books Limited.

Landin, B., Landin, B., & Van Wagenen, S. (2009). *Learning from light: The vision of I.M. Pei* [Motion Picture]. USA. Retrieved from http://www.imdb.com/title/tt1516559/.

Landrum, L. (2017). Miming a manner of architectural theory-Eudaimonia-A Pantomime Dream Play. In P. Emmons, M. Feuerstein, & C. Dayer (Eds.), *Confabulations: Storytelling in architecture* (pp. 160–170). Oxon: Routledge.

Leech, N. (2017, November 7). *Art and culture/Louvre Abu Dhabi: the long-awaited first look inside*. Retrieved from The National: https://www.thenational.ae/arts-culture/louvre-abu-dhabi-the-long-awaited-first-look-inside-1.673668.

Lotus. (n.d.). *At the Heart of Symbolism*. Retrieved April 8, 2018, from Lotus: http://users.skynet.be/lotus/cross/cres0-en.htm.

louvrepress. (2017, December 22). *Louvre Abu Dhabi, a new cultural landmark for the 21st century, opens to the public on 11 November*. Retrieved from Louvre: http://presse.louvre.fr/%E2%80%8Blouvre-abu-dhabi-a-new-cultural-landmark-for-the-21st-century-opens-to-the-public-on-11-november/.

Mantash, L. (2018, May). Investigating Storytelling Approach in Contemporary Architecture in Arab Islamic Cities. *Not Published*. Lebanon.

Marlowe, L. (2017, October 19). *A giant silver dome filled with art: Louvre Abu Dhabi to open*. Dublin, Ireland.

Mazot, S. (2013). The architecture of the Aghlabids. In M. Hattstein & P. Delius (Eds.), *Islam art and architecture* (pp. 132–139). Potsdam: h.f.ullman publishing GmbH.

Mckee, R. (1997). *Story: Substance, structure, style, and the principals of screenwriting*. New York: HarperCollins.

MIA. (2006). A new awaking of the Arabic spirit/Museum of Islamic Art in Doha (Qatar). *Islamic Arts*, 36–73.

Monastir Tunisia. (2020). Retrieved from Encyclopaedia Britannica: https://www.britannica.com/place/Monastir.

Mughniyah, A. M. (2009). *Tafseer Al-Kashif.* Beirut: daralanwar.

Nouvel, J. (2017, September 6). *A vast floating dome sheds dappled light onto an Arab medina* (V. Noce, Interviewer).

Nouvel, J. (2017, November 7). *Jean Nouvel reveals his singular vision behind the Louvre Abu Dhabi*. (A. Digest, Interviewer). Retrieved from https://www.architecturaldigest.com/story/jean-nouvel-louvre-abu-dhabi.

O'Kane, B. (2010). Museum of Islamic Art, Doha, Qatar. Philip Jodidio, New York: Prestel, 2008. 228. *Museum Anthropology Review*, 209–210.

Ouroussoff, N. (2008, December 12). *Art & Design/Architecture/ For I. M. Pei, history is still happening*. Retrieved from The New York Times: http://www.nytimes.com/2008/12/14/arts/design/14ouro.html.

Patterson, R. (2000). The metamorphosis of tragedy. *Architectural Design (AD)*, 36–43.

Picon, A. (2013). Learning from utopia: contemporary architecture and the quest for political and social relevance. *Journal of Architectural Education, 67*(1), 17–23.

Podany, A. H. (2014). *The ancient near east a very short introduction*. New York: Oxford University Press.

Pottle, J., & AwadStudio. (2003). *Aga Khan award for architecture/Museum of Islamic Art, Doha, Qatar*. Retrieved from Archnet: https://archnet.org/sites/6495/media_contents/67610.

Rabbat, N. (2011). MIA and the memory of Ibn Tulun. In A. Soueif (Ed.), *Reflections of Islamic art* (pp. 231–236). Doha: Bloomsbury Qatar Foundation Publishing.

Ryan, M.-L., Foote, K., & Azaryahu, M. (2016). *Narrating space/spatializing narrative: Where narrative theory and geography meet*. Columbus: The Ohio State University Press.

Simpson, V. (2017, December 13). *Pleasure Dome: Louvre Abu Dhabi by Jean Nouvel.* Retrieved from Design Curial: http://www.designcurial.com/news/louvre-abu-dhabi-by-jean-nouvel-5998780.

Steele, J. (2005). *The architecture of Rasem Badran-narratives on people and place*. London: Thames & Hudson.

Stemberg, M. V. (2008). *Aga Khan Award for Architecture/Museum of Islamic Art, Doha, Qatar*. Retrieved April 8, 2018, from Archnet: https://archnet.org/sites/6495/media_contents/67598.

Stevens, P. (2017, November 7). *Architecture- louvre abu dhabi: jean nouvel-designed museum opens in the united arab emirates*. Retrieved from designboom: https://www.designboom.com/architecture/louvre-abu-dhabi-opening-jean-nouvel-uae-11-07-2017/.

Tahara, K. (2008). *Aga Khan award for architecture/Museum of islamic art.* Retrieved March 4, 2018, from archnet: https://archnet.org/sites/6495/media_contents/67618.

The Architecture of the Museum of Islamic Art. (2016). Retrieved from mia.org.qa/en/: http://www.mia.org.qa/en/about/the-museum-building.

The Louvre Abu Dhabi. (2017). Retrieved from Louvre: https://www.louvre.fr/en/louvre-abu-dhabi.

Unwin, S. (2009). *Analyzing architecture*. Oxon: Routledge.

Von Stemberg, M. (2008). *Aga Khan award for architecture/Museum of Islamic Art, Doha, Qatar.* Retrieved March 4, 2018, from Archnet: https://archnet.org/sites/6495/publications/2252.

Wainwright, O. (2017, November 7). *Louvre Abu Dhabi: Jean Nouvel's spectacular palace of culture shimmers in the desert.* Retrieved from The Guardian: https://www.theguardian.com/artanddesign/2017/nov/07/louvre-abu-dhabi-sheikh-chic-throws-controversial-construction-in-relief.

Warner, M. (2013, August 30). *In DOha.* Retrieved from London Review of Books: https://www.lrb.co.uk/blog/2013/08/30/marina-warner/in-doha-2/.

Weir, S. (2016, December 25). On the origin of the architect: Architects and xenía in the ancient Greek theatre. [A. Douglas, S. Hedges, & R. Jenner (Eds.)] *Interstices-A Journal of Architecture and Related Arts,* 17, 9–15. Retrieved from http://interstices.ac.nz/published-journals/interstices-17-return-to-orgins/.

Wilmotte. (2008). *Projects/museum of Islamic Art, Doha, Qatar.* Retrieved from wilmotte: http://www.wilmotte.com/en/project/59/Museum-of-Islamic-Art.

Winston, A. (2016, May 24). *Architecture has a serious problem with communication says Rem Koolhaas.* Retrieved from dezeen: https://www.dezeen.com/2016/05/24/rem-koolhaas-architecture-serious-problem-communication-oma-american-institute-architects-aia-convention/.

Zipes, J. (1997). The Utopian tendency of storytelling-turning the world upside down. In B. B. Incorporated & D. A. Leeming (Eds.), *Storytelling encyclopedia—historical, cultural, and multiethnic approaches to oral traditions around the world* (pp. 27–32). Arizona: Oryx Press.

Dantesque Allegories in the Symbolic Architecture of Triduum Structures

Ivana Passamani

Abstract

The Triduum structures are temporary scenarios created after the Catholic-Reformation to strongly affirm the existence of Purgatory and the possibility for the sinner to aspire to the redemption of sins and then ascend to Paradise. They are set up three days per year in the presbytery of the church, surmounting the main altar integrating into its shape, or completely hiding it. Enhanced by symbolic decorative elements, their architectural structures seem to highlight allegories—not fully examined yet—reflecting situations in Dante's Divine Comedy. The theme of the religious structures can be found in the treatise "*Instructionum fabricae et suppellectilis ecclesiasticae*" by St. Charles Borromeo: as of this date, the Baroque ornamental character develops. As for Dante we mention the recent editorial by Franco Purini, "Dante and Architecture" (2018) where he suggests a very interesting reinterpretation of the "architecture of the situations," meaning Hell, Purgatory and Paradise. After an in-depth historical-bibliographical research, video and photo shootings during the assembly sessions and the celebrations of the Triduum functions, we proceeded by cataloguing. We analyzed architectural, ornate, typological, distributive, ornamental and symbolical aspects. We analyzed topic of "Architecture and Art" considering these installations as real artistic works, one different from the other, consistent with the Baroque style guidelines. The focus of this paper is the innovative relevance of the architectural language which set ups symbolize: they are fictional spaces and, at the same time, they are places of imagery. The imaginary spaces represented (Purgatory, Paradise) refer to the pyramid-step-shaped structure and to the main focusing element: the sunburst. This interpretation is supported by the iconography of Dante's spaces, starting from the one in the Canto, up to various ancient paintings.

Keywords

Divine Comedy • Paradise • Purgatory • Triduum structures, temporary installations

1 Introduction

This research aims to gain insights on how an addition of a symbolic temporary framework may change the design and consequently the space and the meaning of a church, adding the suggestion of fictional spaces to imagery places.

It may be possible to understand and highlight the deep interconnections between the architectural space of the presbitery and the temporary addition of a structure, called the "macchina del Triduo."

At the same time, the innovative, new purpose is to demonstrate a thesis come to light during the research: these architectural structures highlight conceptual connections and possible allegories—not fully explored—with the architecture of the situations described in Dante's Divine Comedy, in particular Purgatory and Paradise.

Using symbols to communicate is certainly one of the first ways used by men to relate with other human beings. The religions have treasured this primitive expressive language, and over the centuries, they have collected a very rich symbolic vocabulary of signs and shapes (Battistini 2004).

If the language of built spaces is mainly physical, the language of churches is always conditioned by religious symbolism and full of symbolical meanings. We can find them in the central nave or in the transept, but particularly in the presbitery, the area in which the symbols mostly merge.

I. Passamani (✉)
Department of Civil Environmental, Architectural Engineering and Mathematics (D.I.C.A.T.A.M), Università Degli Studi Di Brescia, Via Branze 43, 25123 Brescia, Italy
e-mail: ivana.passamani@unibs.it

C. Gambardella et al. (eds.), *Advances in Utopian Studies and Sacred Architecture*,
Advances in Science, Technology & Innovation,
https://doi.org/10.1007/978-3-030-50765-7_11

2 Object of Study

The Triduum structures are temporary scenarios, created after the Catholic-Reformation to strongly affirm the existence of Purgatory (Le Goff 1981) and, consequently, the possibility for the sinner to aspire to the redemption of sins and then ascend to Paradise. Therefore, their purpose is particularly significant for the community of the faithful; even today they refer back to their countries of origin, to respect the tradition and follow the celebrations.

That custom still occurs: in small towns during the three days of Triduum celebration fairs, markets and other events are organized, and the families are reunited praying for the whole community of the dead.

While other types of functions, like those for "Quarantore" celebrations (that affirm the triumph of the Eucharist), are widespread and used all over Italy, the Triduum ones pertain to a limited geographical area in northern Italy: Dioceses of Bergamo, Brescia and marginally Verona.

The Triduum structures are set up for only three days per year in the church's presbytery, or rather in the most crucial space of the church. Sometimes, they surmount the main altar, by integrating into the altar's shape and resuming the decorative features, or by completely hiding the altar, as you can see in Figs. 1 and 2.

Fig. 1 Lonato (Bs), S. Giovanni Battista: the presbitery without the Triduum structure (in Passamani Bonomi 2009, p. 170)

Fig. 2 Lonato (Bs), S. Giovanni Battista: the presbitery totally transformed by the Triduum structure (photo I. Passamani)

In both cases, these installations and their celebrations are very significant for the community of the faithful.

Enhanced by symbolic decorative elements, their architectural structures seem to highlight allegories—not fully explored yet—toward the architecture of the scenarios in Dante's Divine Comedy: particularly, with Purgatory and Paradise.

3 Main Bibliographic References

In the content analysis of the main references (books, studies, iconographic documents), we propose to stress some keywords that recur in the research. So we would like to suggest some textual or graphic references about the main elements.

3.1 Temporary Installations

The theme of the Baroque religious structures can be found for example in the treatise "*Instructionum fabricae et suppellectilis ecclesiasticae*" by St. Charles Borromeo (1577): as of this date, the Baroque ecclesiastical ornamental character develops, as well as temporary installations.

Andrea Pozzo (Battisti 1996), the very famous Jesuit architect and painter of the XVII century, studied and drew the provisional structures in the first volume of the treatise "*Perspectiva pictorum et architectorum*" (1693-1698), as significantly shown in Fig. 3.

The first systematic study of the Triduum structures is "Il disegno dei Tridui. Il tempo e la memoria nello spazio della chiesa" (Passamani Bonomi 2009). The author allows us to consider these expressions as cultural, artistic, religious and ethnographic documents, bearing in mind the high value of the artistic expressions in religious communication. We would like to point out other contributions of the same author, particularly "The Tridui machines: time relentlessness under the space of the vaults" (Passamani Bonomi 2010).

In this paper, the author anticipates some references and suggestions about the supposed connections between the shape of the frameworks and the structures of Dante's Purgatory and Paradise, like those described in some illustrated books of Divine Comedy, both the ancient and the educational ones (Alighieri 1887).

Fig. 3 Andrea Pozzo "Modo di alzar le macchine che son composte di più ordini di telari" (in Trattato, tomo I)

3.2 Dante, Purgatory and Paradise

About Dante and his poem Divine Comedy,[1] the bibliography is rather extended; there are also several paintings with these subjects. A very exemplifying interpretation of Dante's world is the one by Domenico da Michelino, in Fig. 4.

We mention, however, the recent editorial by Franco Purini "*Dante and Architecture*" (Purini 2018). Recalling how rich Western culture is of images evoked by the three chapters of the Divine Comedy, Purini suggests a reinterpretation of the architecture of the "situations" Hell, Purgatory and Paradise that is consistent with the interpretation of the Triduum, here proposed.

Let us not forget the meaningful illustrations interpreting characters and spaces of the Canto: at the beginning miniatures supplied by the ancient manuscripts, later engravings or drawings arranged for some editions of Divine Comedy. Let us also remember for example that at the end of the fifteenth century, Lorenzo di Pierfrancesco de' Medici commissioned Sandro Botticelli with a hundred parchments to illustrate a manuscript of Divine Comedy.

Moreover, we consider the graphical interpretations of Divine Comedy made by Gustave Dorè (1861) worthwhile, because they give us inspiration and fascination in the design of Triduum equipment.

3.3 "3" Number

According to numerology, the discipline that studies the symbolic meaning of numbers, the number "3" is very meaningful and it is considered the number of perfection (Endres and Schimmel 2006). In every book or study about numerology, it is possible to find the same connections between this number and the divine Trinity (Father, Son, Holy Spirit).

According to Purini, Panofsky states that Dante's vision of the "non-visible" is a symbolic form of thought and of the world of the fantastic (Panofsky 1975); if we do not look at this from the allegorical point of view, we can say that the whole Divine Comedy work is a monumental architecture based on the number "3": 3 parts corresponding to 3

[1]Dante wrote the poem between 1304 and 1321. It is' an allegorical-didactic poem.

Fig. 4 Purgatory iconography in "Allegoria della Divina Commedia," Domenico da Michelino, Firenze, Dome

different places, organized in 33 chapters, written in triplets or hendecasyllables.

We found this number very interesting and significant since in the Divine Comedy declinations and references occur in the shapes and in the allegorical symbolism.

3.4 Light

In the treatise of St. Charles Borromeo, there is particular attention to the role of light. According to Borromeo, it should permeate the church by transfiguring its space. He explains, for example, how positions and shapes of the windows should be.

In this research, the purpose of light described in Dante's Canto is very meaningful; from Hell to Paradise, the light becames stronger and stronger to signify the passage from sin to salvation.

4 The Triduum Structures

In Italy, these structures are called "Macchine dei Tridui": the word "Macchina" ("Machine") is quite suitable, because they are complex systems composed of thousands of pieces. In order to experience the event, it is worth assisting the entire procedure step by step (Figs. 5 and 6).

To clarify, we will distinguish

- the structural parts
- the architectural parts
- the decorative elements.

4.1 The Triduum Structural Components

The assembly lasts some days: some volunteers, who strongly deal with preserving and carrying on this tradition, provide for the raising of the ancient wooden structure or for a more modern and safer metal structure. The vertical high poles are fixed in special slots in the presbytery floor and take up the space between the altar and the choir, to which the various components are connected.

The poles are connected to the horizontal platforms of wooden planks constituting the different floors of the structure and are linked by steep wooden staircases, as you can see in Fig. 7.

It is crucial to point out that people who attend the ceremonies do not see anything of the back structure.

4.2 The Triduum Architectural Components

We can affirm that the visible shape of the structure is mainly made up of three or four parts:

Fig. 5 Borgo Poncarale (Bs), Purificazione di Maria Vergine: assembly phase (photo I. Passamani)

Fig. 6 Erbusco (Bs), S. Maria Assunta: assembly phase (photo I. Passamani)

- the vertical panels
- the candles (real or electrical)
- the impressive sunburst
- the stairway (when visible).

The vertical panels (called flats as well) are made of painted and gold-decorated (or velvet covered) wood, and they are characterized by different and varied shapes and sizes.

Each example reveals different levels of depth, corresponding to several perspectival levels of various shapes and heights.

Many candles are hung on the panels. The two–three hundred candles are crucial during the functions because they indeed introduce the wait, while the volunteers are lighting all the candles, a time of worship or collective singing. Moreover, when they are all lit, they reveal different drawings, and they transform the set ups into light-living machines.

The set of all the components revolves around the focal central element: the large sunburst, surrounded by the candles. The sunburst is always articulate, because it is composed of some concentric rays, usually three. In the center of this component, on a little shelf, the ostensory is exposed.

Many structures have a visible stairway, arranged of flights that go up parallel and/or perpendicular directions in relation to the main side (Fig. 8). In these cases, the faithful can look at the celebrant, ascending with the ostensory to the sunburst.

In several examples, the stairway is on the back side, like a service-staircase. Obviously, the effect and above all the meaning are really different if compared to a visible stairway, as explained below.

4.3 The Triduum Decorative Components

These artifacts are meaningful not only for their appearance but thanks to the many decorations that enhance them:

angels, crosses, hearts, lambs, lions, pelicans, a great variety of flowers and fruits, clouds. Each element has a clear sense and a specific symbolic role in the whole composition: Christ passion or pain, God's goodness, devotion and so on (Ripa, 2012; Van Straten, 2019).

The angels, in particular, take on a meaningful relevance when they are engraved in the wood of the sunbursts as shown in Fig. 9: the reference to the iconography of Paradise is evident.

Let us highlight some other more evident and understandable symbols: the purging souls (the souls that are in Purgatory) and the flames. Both elements clearly symbolize those who benefit from the prayers. In Figs. 9 and 10, there are two different examples.

Fig. 7 Borno (Bs), Ss. Martino e Giovanni Battista: the structure' backstage with the stairway and one of the several platforms (photo I. Passamani)

5 Dantesque Allegories

The survey activity and the following analysis of these artifacts allow us to decipher the most recurrent and unexplored references to the structure of the celestial heavens and the ancient depictions of Purgatory and Paradise, described by Dante in Divine Comedy. Probably, the architects were familiar with some paintings of Dante's poem and of its imagery places and we can imagine that they saw some illustrated books of it as well.

5.1 Interpretative Hypothesis: Purgatory's Allegories

The analysis underway resorts to three concepts:

Fig. 8 Lonato (Bs), S. Giovanni Battista: the central stairway of the equipment during the assembly (photo I. Passamani)

Fig. 9 Borno (Bs), Ss. Martino e Giovanni Battista: S. Raffaele archangel picks up the purging souls from Purgatory's flame (photo I. Passamani)

Fig. 10 Malonno (Bs), Ss. Faustino e Giovita: the purging souls between the flames of Purgatory (photo I. Passamani)

- the pyramidal shape
- the large-step shape
- the framework as a reference to the image of flames.

By looking at a framework from the front, we can inscribe its unique visible front in a pyramidal shape that goes upwards, almost as if it wants to "break through" the enveloping architecture of the presbytery vault.

The same triangular shape characterizes the outline sideways, which is visible only in the section drawings. Moreover, the large-step shape outline is visible, as for example in Fig. 11.

In addiction, drawing its horizontal diagram, we can clearly point out the sequence of flats and the presence of the staircases' flights.

Two allegories emerge from these traits. The first is about the pyramidal shape evoking the sense of the mountain of Purgatory as described in Fig. 12; going up very high to the sky, it embodies the principle of transience. A place of atonement, a path for penance, reflection and repentance, it is described as a climb, a progressive ascent, masterfully described by the artist Dorè (Fig. 13) (Cary 1901).

The metaphor of the ascent as a path of purification and as an ideal spiritual rebirth, allowing us to reach the dimension of the Divine, is taken up by Francesco Petrarca in 1326 in the letter "Salita al Mont Ventoux." In 1496, Pietro Bembo, in "De Aetna," reproposed this theme describing an ascent path on the Etna volcano.

The Triduum frameworks, with their triangular appearance going toward the top of the apsis ceiling, remind us of the precariousness of earthly life and, at the same time, instill hope in the redemption of purging souls thanks to the prayer of the believers (Vovelle 2009). That interesting link depends on their shape, similar to the Dantesque Purgatory; they are the imagery spaces and they link to the fictional space of Purgatory.

The second allegory is about the large-step shape of the profiles of these artifacts. They seem to simulate the cliffs of the mountain of Purgatory, where the suffering that is consumed will make the souls worthy to rise before God.

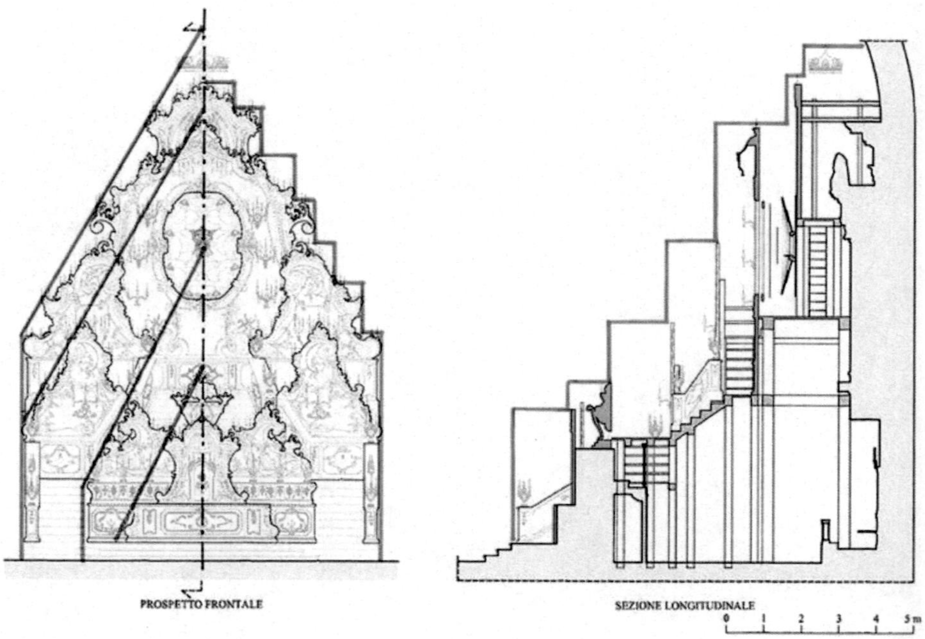

Fig. 11 Castenedolo (Bs), S. Bartolomeo Apostolo: survey drawing of the front and of the sideways section of the structure (drawing I. Passamani)

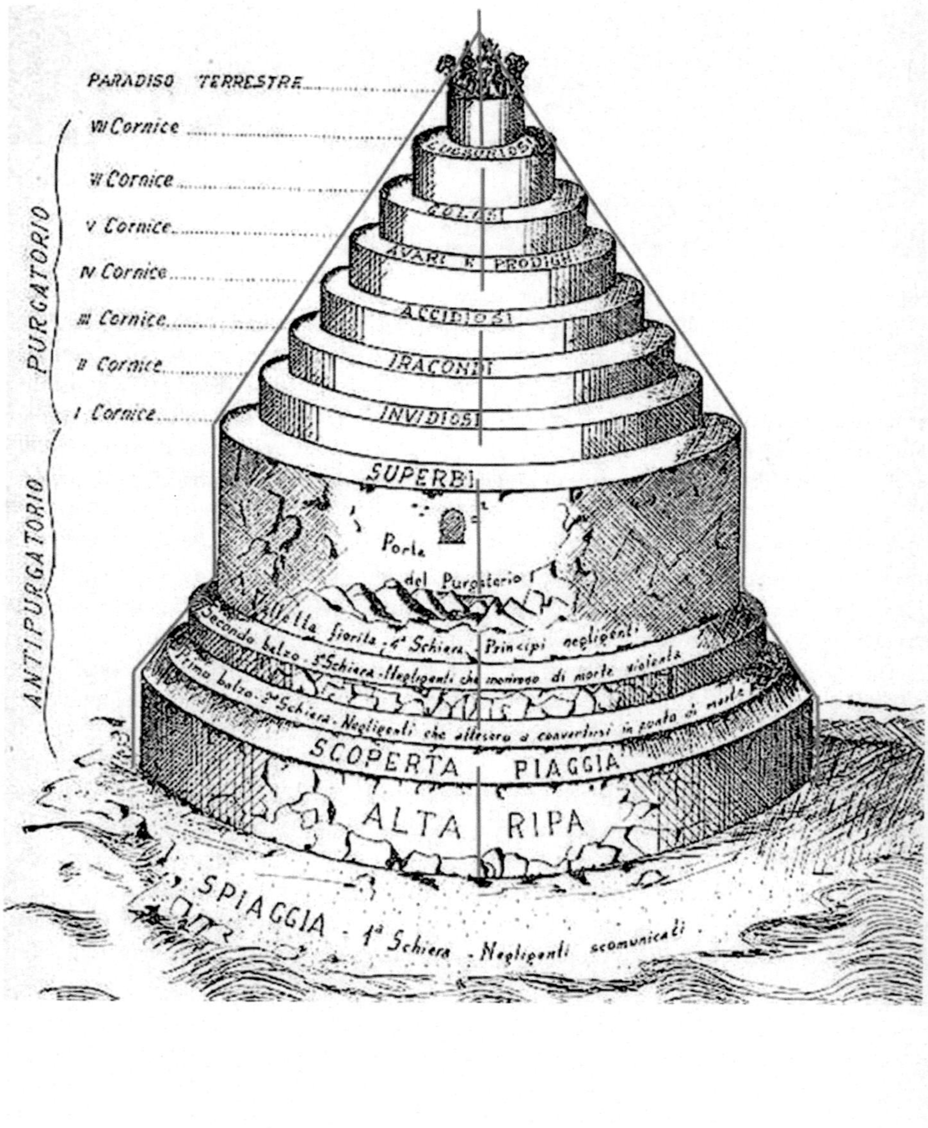

Fig. 12 Purgatory's conic and large-step shape

Fig. 13 Gustave Dorè, Purgatory, Canto IV, rise to the first leap

This reference is further emphasized by the stairway, arranged of flights of steps; the climb of the celebrant up the top of the structure, sometime accompanied by other priests (Fig. 14), reminds the faithful of the chance of salvation.

In the ancient iconographies, both before and after the Divine Comedy, the common transposition of the Purgatory was not a mountain. It was represented in the form of flames, a place of fire, sometimes like a cauldron, other times like a volcano crater.

It is therefore not so overstated to identify another possible allegory. For the faithful, the structure becomes a flaming image that represents the ancient idea of Purgatory as a hostile place not so different from Hell. Figure 15 is really meaningful as it shows how encouraging the participation of the faithful in prayer for the expiation of the purging souls.

Fig. 14 Lonato (Bs), S. Giovanni Battista: the celebrant's climb toward the sunburst to arrange the hostensory (photo I. Passamani)

5.2 Interpretative Hypothesis: Paradise's Allegories

The study and a careful analysis allow other hypotheses:

- the sunburst versus the number "3"
- the sunburst versus Paradise's circular shape (as in Dante)
- the structure as a pulsating light-living machine.

In these installations, the perceptual focus is always represented by the large golden sunburst, which expands in a central and very high position. It presents two different collocations. When it is in the foreground, compared to the panels or to the cloth, it protrudes with its rays and becomes the focal point, like in Fig. 16. In other examples, the panels

Fig. 15 Malcesine (Vr), S. Stefano: the light set up (photo I. Passamani)

are in front of the sunburst: they reveal a large hole (circular or elliptical or oval, like in Fig. 17) where the sunburst is located, accentuating the depth effect, remaning however the main element of the composition.

In both cases, this central element is mainly composed by a sequence of three concentric and consecutive circular rings as in Fig. 18, generally all gilded, often enriched with "putti," cherubs and angels, with rays of different sizes and shapes; one of the rays is frequently better defined as a cloud frame and it is brownish.

Here, the symbolic meaning of number "3" is really evident. In addition to recalling the Holy Trinity, it is a submultiple of number "9" and it can represent a simplification of the structure of the nine concentric skies of Dante's Paradise[2]; they are included by the "Empyrean" or the "pure light sky."

The likeness between the circular shape of the sunburst and the circular structure of Paradise[3] is apparent, as

[2]Sky of the Moon, of Mercury, of Venus, of the Sun, of Mars, of Jupiter, of Saturn, of the Fixed Stars, First Mobile. Different heavenly creatures are referred to each sky: angels, principalities, mayors, virtues, dominations, thrones, archangels, cherubs, seraphim.

[3]The structure of Dante's Paradise takes up the Aristotelian geocentric cosmology of the Ptolemaic system: the Earth in the center of the universe, nine concentric spheres around.

Fig. 16 Gandino (Bg), S. Maria Assunta: the light sunburst (eighteenth century). Photo I. Passamani

Fig. 17 Borgo Poncarale (Bs), Purificazione di Maria Vergine: the sunburst behind the panels (Photo I. Passamani)

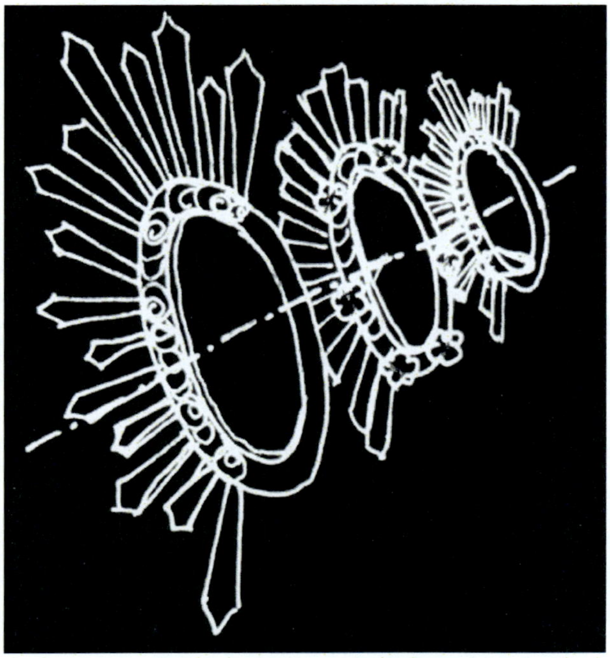

Fig. 18 a sunburst structure model with 3 concentric and circular elements (sketch of I. Passamani)

illustrated by many artists (Fig. 19) and researchers (Battistini 2004).

The last interesting interpretative hypothesis is related to the theme of the light. In Dante's Paradise, the exaltation of the divine creation is strongly expressed by the image of light.

The first representation of God is in Canto XXVIII. God appears as a very bright point in the center of nine concentric circles of shining angels. The hostensory in the sunburst is referred to this evocation. But the most interesting Dorè's illustration rapresents Canto XXXI of the Paradise (Fig. 19): we think that this illustration could have inspired the sunbursts' shapes (Figs. 20 and 21).

The candles' light instills liveliness to the installations as it lights up the hope of redemption in the worshippers. When the structures are lit up in the darkness of the churches (in some cases, a black or red curtain is used to accentuate the effect of surprise, like in Figs. 22 and 23), the experience is always absolutely suggestive and touching: the curvilinear and enveloping surface of the vault of the presbytery, with

Fig. 19 Gustave Dorè, Paradise, Canto XXXI, the multitude of angels

Fig. 20 Breno (Bs), Santissimo Salvatore: the sunburst with angels and cupids (Photo I. Passamani)

its shadow effect, gathers and protects this light-living machine.

The darkness makes the design of simple light more readable: the light becomes one with the architectural shape.

The candles draw original constellations, standing out around their "sun": the sunburst.

It shines thanks to its own light, expanding the effect from the center; the focus of the composition and the overall radiating effect does not really change, both if the sunburst is in front of, or behind the panels. Figures 23 and 24 are illustrative of this special atmosphere.

5.3 Triduum Structures as Purgatory and Paradise's Allegorical Figurations

According to this interpretation, we can consider these temporary installations as symbolic representations of Purgatory and, at the same time, of Paradise.

These installations take on two meanings: memory of the suffering of purging souls (reference to burning fire) on the one hand, hope in redemption (reference to light) on the other.

In conclusion, this type of equipment, assembled to prefigure true architectures strictly integrated into the presbytery, can significantly transfigure both space and perception of it.

6 Methodologies

After an in-depth historical-bibliographical research and many views, video and photo shootings during the assembly sessions and celebrations, we proceeded by cataloguing. We analyzed architectural, ornate, typological, distributive, ornamental, symbolic aspects of many examples.

The next step was the direct metric survey of some of the most significant examples.

Many topics were analyzed through representation: for example, about the "Architecture and Art," we can consider these installations as real artistic works, one different from the other, consistent with the Baroque style guidelines (Norberg Schulz 1979; Terraroli 2004). They continue to be followed during the nineteenth-twentieth century too.

Fig. 21 Ospitaletto (Bs), S. Giacomo Maggiore: the sunburst with angels and cupids

Through research, we have produced different materials and documents:

- photographs of the single components, of the assembly phases and celebrations
- photographs of the church and of the presbitery with and without the temporary installations
- videos of the assembly and of the celebrations phases
- sketches and survey drawings of the frameworks, both in the architectural context (Fig. 25), and as a single element.

The few available assembly schemes have also been collected (Fig. 26). The instructions are still handed down verbally, with a serious risk of losing these craft skills.

A difficulty to overcome was to carry out the analysis and the surveys considering that the installations are mounted and used only three days per year (fortunately not always coinciding). Many photos were taken, and many single components were measured before the assembly work.

But the main complexity found concerns the graphical output of the burning light originated by the multitude of candles.

To graphically fix this special moment of the service, so ephemeral and at the same time meaningful, we recreated through photographic manipulations the conditions of semi-darkness of the environment during the lighting of the candles. We have later abstracted the synthetic design of light highlighting the bright points (Figs. 27 and 28).

Fig. 22 Tremosine Pieve, S. Giovanni Battista. The red curtain opens, the light-living machine slowly appears (Photo I. Passamani)

We tried to simulate what happens when, faced with a disturbing sight, we narrow our eyes to capture the essential features, synthesizing the view.

7 Objectives Achieved

We believe that the purpose of the research about the Triduum structures is achieved.

At the beginning, the declared objectives were:

- to gain insights on how the addition of an equipment changes the sense of a church (of the presbitery in particular), adding the suggestion of fictional spaces to imagery places;

Fig. 23 Magasa (Bs), S. Antonio Abate. The black curtain opens up completely (Photo I. Passamani)

Fig. 24 Rodengo Saiano, S. Nicola abbey. The light-living machine (Photo I. Passamani)

- to demonstrate that these structures highlight conceptual connections and possible allegories with the architecture of the Divine Comedy's Purgatory and Paradise.

The results achieved have demonstrated that the Triduum structures, by transforming the appearance of the church's presbytery for a short time interval (three days per year), introduce a theoretical imagery thought about redemption and salvation that is usually and diffusely visually illustrated in Dante's iconography.

8 Conclusions

When in our research we resort to survey and to drawing, we are sure to translate the peculiarities of an artifact in visible documents and graphic layouts. In this work, the characters, which stand out from the drawings, evidently support the hypothesis that the structures recall the ancient depictions of Purgatory and Paradise.

Moreover, using the drawings, we can simplify complexities of reality, synthesizing the figures excess. For example, the manipulations of the light effects, here proposed, allow to immediately notice the pyramidal shape of many frameworks, and to observe the evident centrality of the sunburst.

Last but not least, we can try to draw the invisible traits.

The very innovative relevance of the architectural pattern of the Triduum structures is to symbolize fictional spaces and, at the same time, places of imagery.

The spaces of imagery represented, Purgatory and Paradise, refer to the pyramid-step-shaped structure and to the main focusing element, the sunburst. This interpretation is supported by the iconography of Dante's described situations.

In conclusion, we can imagine that the artists and architects who designed these temporary structures might have unconsciously used iconographic symbols and images rooted in the Western figurative culture and in its consolidated knowledge.

Fig. 25 Collebeato (Bs), S. Paolo's conversion. Survey drawing of the equipment in the church's section (in Passamani Bonomi 2009, p. 143)

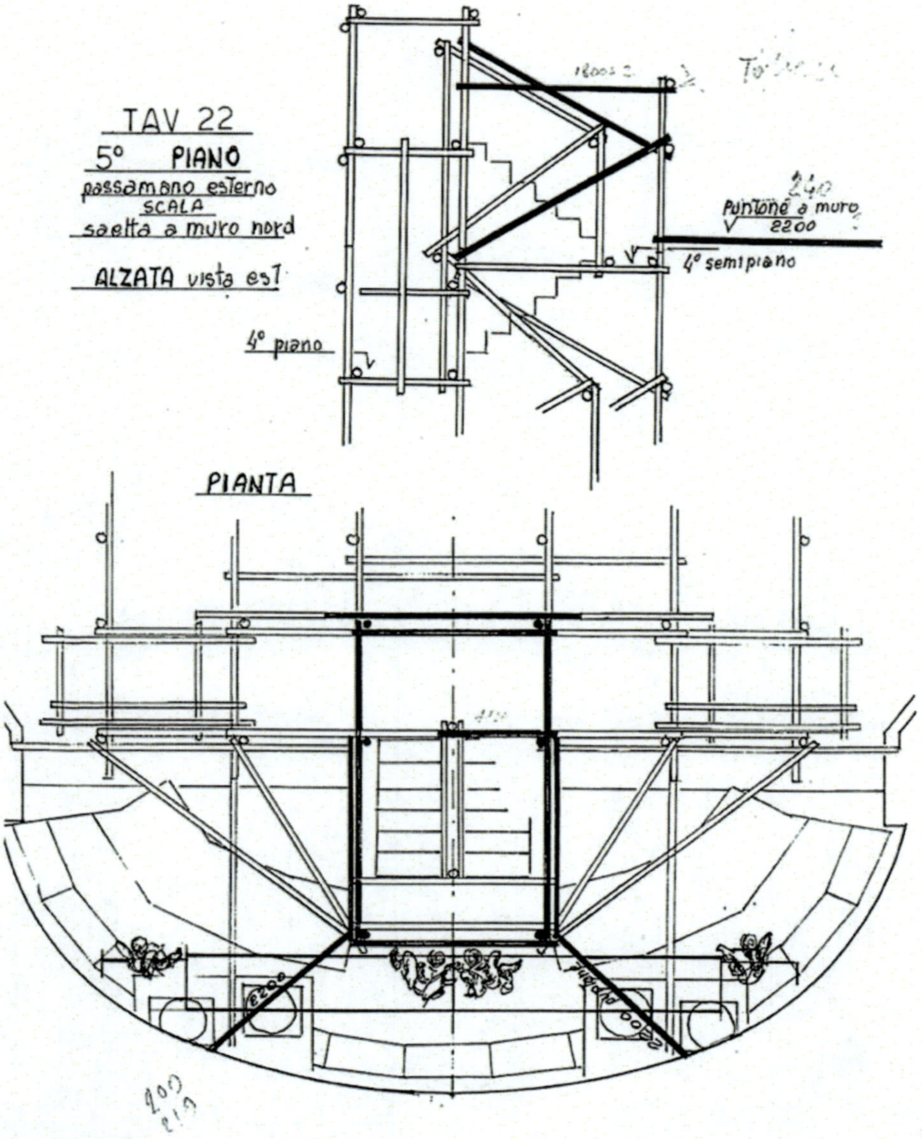

Fig. 26 Collebeato (Bs), S. Paolo's conversion. Assembly sketch (in Passamani Bonomi, 2009, p. 141)

Fig. 27 Malcesine (Vr), S. Stefano: the light's drawing (in Passamani Bonomi 2009, p. 79)

Fig. 28 Siviano (Bs), Ss. Faustino e Giovita: the light's drawing (in Passamani Bonomi 2009, p. 79)

References

Alighieri, D. (1887). *La Divina Commedia* [Illustrated by M. G. Dorè, Ed. E. Camerini]. Edoardo Sonzogno Editore.

Battisti, A. (1996). *Andrea Pozzo*. Luni.

Battistini, M. (2004). *Simboli e Allegorie*. Mondadori Electa.

Cary, H. F. (translate) (1901). *Purgatory and Paradise* [Illustrated by M. G. Dorè]. Thompson & Thomas.

Endres, F. C. & Schimmel, A. (2006). *Dizionario dei numeri. Storia, simbologia, allegoria*. Red Edizioni.

Le Goff, J. (1981). *La naissance du Purgatoire*. Gallimard.

Norberg Schulz, C. (1979). *Architettura barocca*. Mondadori Electa.

Panofsky, E. (1975). *Studi di Iconologia*. I temi umanistici nell'arte del Rinascimento: Einaudi.

Passamani Bonomi, I. (2009). *Il disegno dei Tridui*. Editrice Opera Diocesana di San Francesco di Sales: Il tempo e la memoria nello spazio della chiesa.

Passamani Bonomi, I. (2010). Le macchine dei Tridui nello spazio delle superfici voltate. In E. Mandelli & G. Lavoratti (Eds.), *Disegnare il tempo e l'armonia* (pp. 110–115). Alinea.

Purini, F. (2018). Dante and Architecture. *Area 161*.

Ripa, C. (2012). *Iconologia*. Einaudi.

Terraroli, V. (2004). *Lombardia Barocca e tardobarocca*. Skira: Arte e architettura.

Van Straten, R. (2019). *Introduzione all'iconografia*. Jaca Book.

Vovelle, M. (2009). *La morte e l'Occidente*. Laterza.

Architecture, Territory, Traditions. The Cult of the Madonna del Carmine called "*delle Galline*" in Pagani

M. Martone and A. M. Giugliano

Abstract

Using graphic, photographic, and iconographic documentation, this paper investigates the celebrations of the Madonna del Carmine, "immaterial asset of Italy" according to the Central Institute for Demoetnoanthropology of MIBAC, and their link with the architecture of the town of Pagani, in Southern Italy. The aim is to promote the enhancement and the safeguard of this territory, and its memory consisting of spoken traditions, rituals, worships, festivals, and craftmanship. Intangible assets, as expression of a population's culture and its identity, are currently at risk in today's fast moving globalized society which tends to wipe out local identities despite their important role in the configuration of cultural landscapes. Pagani provides us with a valid example of a place where material and immaterial combine together to create a strong link between the town's architecture and religious cult, local traditions and folkloristic events, that find their maximum representation in the feast of Madonna del Carmine also known as Madonna delle Galline. The painting of the Madonna itself is a replica of an ancient effigy kept underground probably to be saved from iconoclastic destruction, and then randomly found by some hens in a courtyard in the eighteenth century. As memory of this episode, during the festival, the statue of the Madonna, covered with hens and birds, is triumphantly brought in procession around the city, stopping by courtyards of old farmhouses and representative buildings of the town where are the "toselli," special votive altars set up for the occasion. The paper offers a deep dive into this tradition and its architectures.

Keywords

Intangible asset • Madonna delle galline • Local traditions • Tosello • Pagani

1 Introduction

On the basis of the 2003 UNESCO Convention for the Safeguarding of the Intangible Cultural Heritage[1], ratified in Italy with Laws 19/2007 and 167/2007, the Central Institute for Demo-Ethnology promotes initiatives for the protection of the collective identity of the various social groups on the territory and for the expression of cultural diversity, and documents Italian main festivals, listed by regions, through multimedia archives, in order to promote the preservation, and the enhancement of cultural heritage, both material and immaterial.

The institute acknowledges, as intangible asset in Campania, the feast of the Madonna del Carmine, called delle Galline, in the ancient town of Pagani in the agro nocerino-sarnese, developed along the ancient consular road that led from Naples to the Calabrie, passing through Salerno. Pagani, originally linked to the development and the events of nearby Nuceria, in the eighteenth century assumed the name of "Nuceria Paganorum," when the Knights Pagano obtained the territory in feud. The population, consisting of peasants engaged in the lands of feudal lords, lived

[1]2003 UNESCO Convention, Art. 2, Part.1. "The intangible cultural heritage means the practices, representations, expressions, knowledge, skills—as well as the instruments, objects, artefacts and cultural spaces associated therewith—that communities, groups and, in some cases, individuals recognize as part of their cultural heritage. This intangible cultural heritage, transmitted from generation to generation, is constantly recreated by communities and groups in response to their environment, their interaction with nature and their history, and provides them with a sense of identity and continuity, thus promoting respect for cultural diversity and human creativity".

M. Martone (✉) · A. M. Giugliano
Department of History, Drawing and Restoration of Architecture (DSDRA), "Sapienza" University of Rome, Piazza Borghese 9, 00186 Rome, Italy
e-mail: maria.martone@uniroma1.it

C. Gambardella et al. (eds.), *Advances in Utopian Studies and Sacred Architecture*,
Advances in Science, Technology & Innovation,
https://doi.org/10.1007/978-3-030-50765-7_12

in houses gathered around courtyards, where became a practice to live together in solidarity (Istituto Centrale per la Demoetnoantropologia, n.d).

Pagani still preserves, although modified over time, many courtyards where the resident community repeats long-standing daily rituals and hands down ancient traditions. An ancient oral religious tradition, popular in this area, but present with variations throughout the Campania region, links the Madonna delle Galline to other five, or six sisters, worshipped in a deep and heartfelt way in other sanctuaries in the region; these sisters take their name from a distinctive feature or the places they belong to. According to this tradition, we have: the Madonna di Montevergine in the province of Avellino, the Madonna di Piedigrotta, the Madonna dell'Annunziata and the Madonna del Carmine in Naples, the Madonna dell'Arco in Sant'Anastasia, and the Madonna delle Galline in Pagani; added to these six sisters we often find a seventh one, the Madonna della Pace in the town of Giugliano[2] (Canzanella, 2002; Tardio, 2008) (Fig. 1).

The Campanian cult of the sisters, whom, depending on the town, can be three, six, but also seven, belongs to popular religiosity, and it is expressed through pilgrimages and festivals. In particular, festivals, that in some cases have very ancient origins, have survived to these days, although partially freed from original meanings and values, showing a consistent and remarkable capacity for continuity and resistance over time (Canzanella, 2002), preserving their vitality also thanks to new or renewed rituals that slowly integrated with the older ones.

This is the case of the feast of the Madonna delle Galline in Pagani, where religious worship combines and merges with local folk traditions, both old and new, while the city with its urban and architectural context act as a backdrop for the celebrations. The analysis of processional itineraries represents a different methodological approach to the study of cities and, along with archive and cartographic data, represents a useful source for historical reconstruction of urban spaces (Guidoni, 1980). Through the planimetric restitution of the religious processional itinerary and the mapping of the toselli, a particular kind of local votive aedicules, within Pagani urban spaces, this study shows how along the established routes we find the most significant elements of the city and its territory, such as churches, courtyards, palaces, open spaces, and squares.

The documentation of the festival in relation to the territory, such as the analysis of religious rites, the mapping of toselli, the reconstruction of the processional itinerary, and the study of folk traditions, allows an analysis of the urban fabric in order to preserve the existing as a testimony of urban values inherited from the past (Colletta, 2002).

2 Origins of the Cult of the Madonna Delle Galline

The cult of the Madonna delle Galline dates back to the early sixteenth century, in the Octave Day of Easter, when, according to the report of the Bishop Ammirante in 1877, some hens scratching in a courtyard of Pagani brought to light a wooden tablet, of the size of two palms by three, on which was depicted the effigy of a Madonna del Carmine[3] with dark complexion (Sinatore, 2014). After the finding, the tablet was brought and kept for a long time in the nearby oratory of the Annunziatella, also called Spogliaturo, for in that place the members of that Confraternity, before accompanying the dead to the burial, used change their clothes and wear those prescribed from the rite, now kept in the museum inside the Archconfraternity.

According to well-established traditions, the tablet was brought by monks escaped from the East in the eighth –ninth centuries to subtract the sacred images from the iconoclastic destruction (Matrisciano, 2006), and due to its state of decay, it was necessary to reproduce it on canvas, probably after the image became known in the first years of seventeenth century, when it performed its first miracle, the sudden healing of a cripple, to which followed other seven in a short time. The new canvas, depicting the Madonna with the son in her arms, went to cover the ancient tablet (Vassalluzzo, 1987), embellished by the application of two golden crowns and a golden sunburst around the heads and placed on the high altar of the current Santuario di Santa Maria Incoronata del Carmine detta delle Galline, especially built starting from 1610 next to the Oratory. Only at the end of the eighteenth century, we begin to find news about the presence of the wooden half-bust statue that is carried in procession. The statue depicts a Madonna different from the effigy, both in the pose and in the look, showing a light complexion and auburn hair (Fig. 2).

The name Madonna delle Galline seems to have different explanations: According to the local tradition, the name originates from the legendary finding of the effigy, which led to the ancient custom for the population to pay homage to the Madonna during the festival by offering various

[2]According to another tradition the Seven Sisters would be located inside the Vesuvian area and they would be: the Madonna dell'Arco of Sant 'Anastasia and the Madonna Pacchiana of Castello di Somma Vesuviana in the province of Naples; the Madonna delle Galline of Pagani, the Madonna di Bagni of Scafati, the Madonna dell'Avvocata of Maiori, the Madonna di Materdomini of Nocera Superiore in the province of Salerno; the Madonna di Montevergine in the province of Avellino (Tardio, 2008).

[3]Our Lady of Mount Carmel in Italy is also called Madonna del Carmine, from the Spanish correspondent Virgen del Carmen.

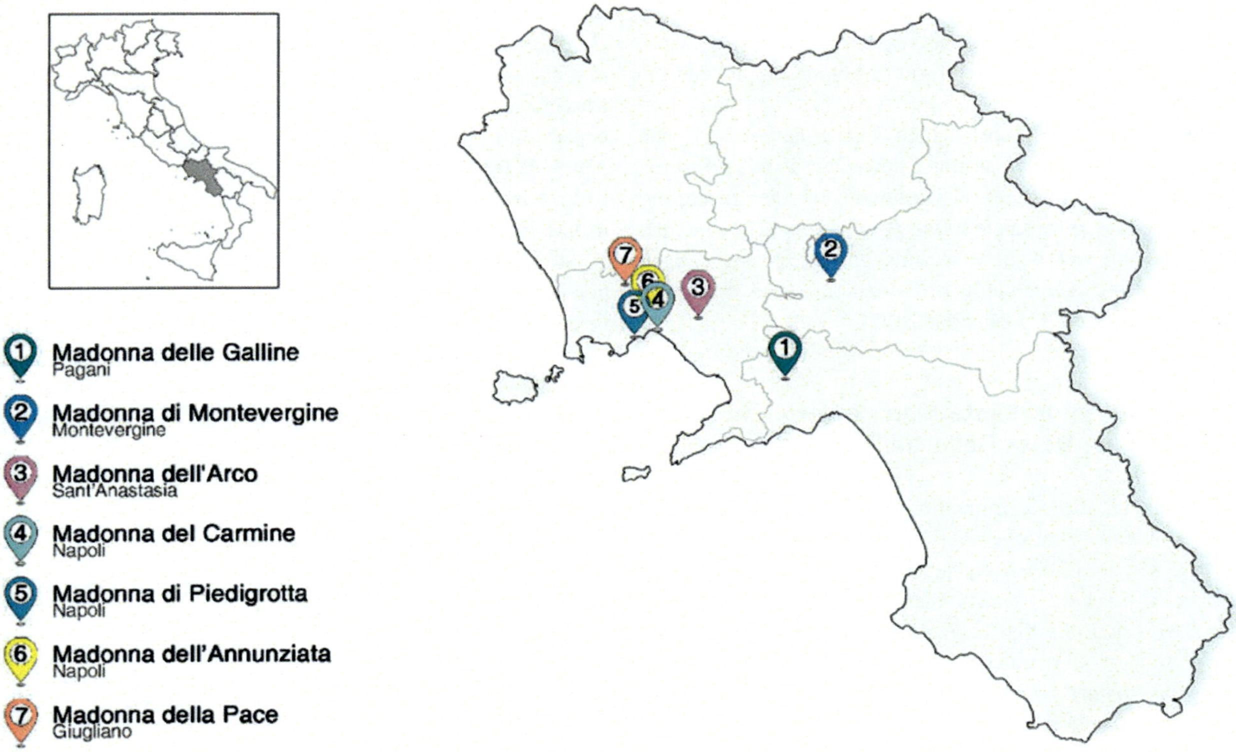

Fig. 1 Worship of the Madonnas sisters in Campania. Location of the Sanctuaries

Fig. 2 From left: Effigy of the Madonna delle Galline; the statue brought in procession; the statue in the niche in the large side chapel during the year, hidden from the view of the faithful by a white cloth

volatiles, mainly hens, as a memory of the ancient discovery (Villani, 1987); another hypothesis, on the other hand, identifies in this practice the origin of both the name and the cult (Montorio, 1715).

The offer of the hens, which began earlier than the eighteenth century,[4] has an important value if related to the original agricultural-rural environment of the sanctuary; considering the unfortunate living conditions of the time, the offer of a hen was a great sacrifice for the peasants, as it represented the renunciation of a precious good that guaranteed subsistence (Canzanella, 2002).

2.1 Sanctuary of Santa Maria Incoronata Del Carmine Detta Delle Galline

As mentioned, after the miraculous events, it was decided to build a more worthy church (Fig. 3) to welcome the worshipper, and in 1610, Monsignor Lunadoro, bishop of the Diocese of Nocera, reports that the construction of the church was possible thanks to the generosity of the people, who gave large sums of money for this to be built. In 1615, during his pastoral visit, Monsignor Stefano Vicari already reports an "ecclesia noviter erecta." In 1665, the prior Mandiello bought some houses located in front of the church in the name of the confraternity and knocked them down to give greater prominence to the façade.

In 1712, the roof was redone and was embellished with the finest craftmanship wooden coffering (Vassalluzzo, 1987). The twenty-one oil paintings, located in the panels of the wooden coffered that decorates the ceiling of the church, depict some episodes from the life of the Virgin (Fig. 4) (Arciconfraternita di Santa Maria Incoronata del Carmine detta delle Galline, 2000). Of the same year are also six frescoes themed with biblical stories. In 1776, began the construction of the Baroque and Altar, completed with various alterations in 1797, along with the first four-side altars. The remaining two are from the nineteenth century; in the biggest one is a niche with the statue of the Madonna del Carmine (Vassalluzzo, 1987).

In August 1786, the chapter of St. Peter in Vatican accepted the plea of the governors of the sanctuary to obtain the coronation of the sacred image of the Madonna delle Galline, which was followed in 1787 by the solemn celebrations so described by canonical theologian Giuseppe Messina in a chronicle of the time: "Si fece un vaghissimo apparato di intorno alla Chiesa e la piazza, sino a settecento palmi di larghezza,[5] di arazzi, festoni, archi trionfali, cappelle e macchine sontuosissime sulle quali vennero messe alcune iscrizioni da me composte (…). Si rinnovò lo stucco dorato della Chiesa e si aggiunse un'altra mezza chiesa posticcia davanti la porta massima per dar comodo alla gente, ed ai forestieri, che vi sarebbero accorsi, poiché la chiesa non poteva essere capace di in tale occasione. Si fecero vaghe illuminazioni per la città e soprattutto intorno alla macchina superba e tutto col disegno del celebre signor Luigi Cartolano[6]" (Vassalluzzo, 1987, pag. 110). In 1790 was introduced the organ with decorated wooden choir.

3 The Celebrations: Customs and Traditions Between Folklore and Sacredness

Arrangements for the celebrations begin on Monday in Albis, Easter Monday. The church, closed to the public for the following five days, gets adorned with blue, red, and white velvet curtains with golden streaks. A rich canopy surmounted by a great golden crown and covered by the same curtains as the church and is mounted on the side of the altar in order to host the statue of the Madonna, usually located in a niche in the large side chapel during the year, hidden from the view of the faithful by a white cloth (Figs. 2 and 3). Nuns are in charge of changing the white dress of the statue with the red one used for the festival. On Friday in Albis, a huge crowd stands in the small square in front of the church, awaiting the Sanctuary Door Opening Ceremony[7] (Fig. 5) and the greetings of the *tammorrari.*[8] When the

[4]In 1515 Serafino Montorio writes: "Fra le offerte, che si fanno alla Vergine del giorno della sua Festa cioè l'ottava di Pasqua di Resurrezione (che per tradizione comune è quello della sua miracolosa invenzione) una se ne fa, che quanto vile, ed impropria apparisce agli occhi del Mondo, tanto è più gradita dalla Madre di Dio, avendolo ella dimostrato con evidenti prodigi. Sin dal principio hanno soluto quei buoni Cittadini, e specialmente le donne, offerire alla Vergine le loro Galline, quale offerta ogni anno è andata crescendo a segno tale che l'anno 1710 arrivarono fino al numero di mille, ed oggi son più, e ciò perché si è osservato, che Maria le gradisce come sperimentosi in tre fatti maravigliosi (…)".

[5]Before 1840 a Neapolitan palm was 0,263,670 m; after 1840 0,264,550 m.

[6]In English: "A very nice system was built around the Church and the square, up to seven hundred palms in width, of tapestries, festoons, triumphal arches, chapels and sumptuous machines on which some inscriptions were composed by me (…). The gilded stucco of the Church was renewed and another half church was added posthumously in front of the maximum door to give comfort to the people, and to the strangers, who would have gone there, since the church could not be capable of on such occasion. Nice illuminations were made for the city and especially around the superb Macchina and all with the design of the famous Mr. Luigi Cartolano".

[7]Outside the church the *tammorrari* invoke the Madonna by calling and knocking on the wooden door of the Sanctuary.

[8]The *tammurriata* is a musical genre, which is expressed through music and dances, closely related to the Marian rites of Agro Nocerino-Sarnese. The music is performed with the tammorra, a frame drum, accompanied by nacchere, or castagnette, two pairs of wood, which, fixed to the middle of the two hands, produce a clear sound.

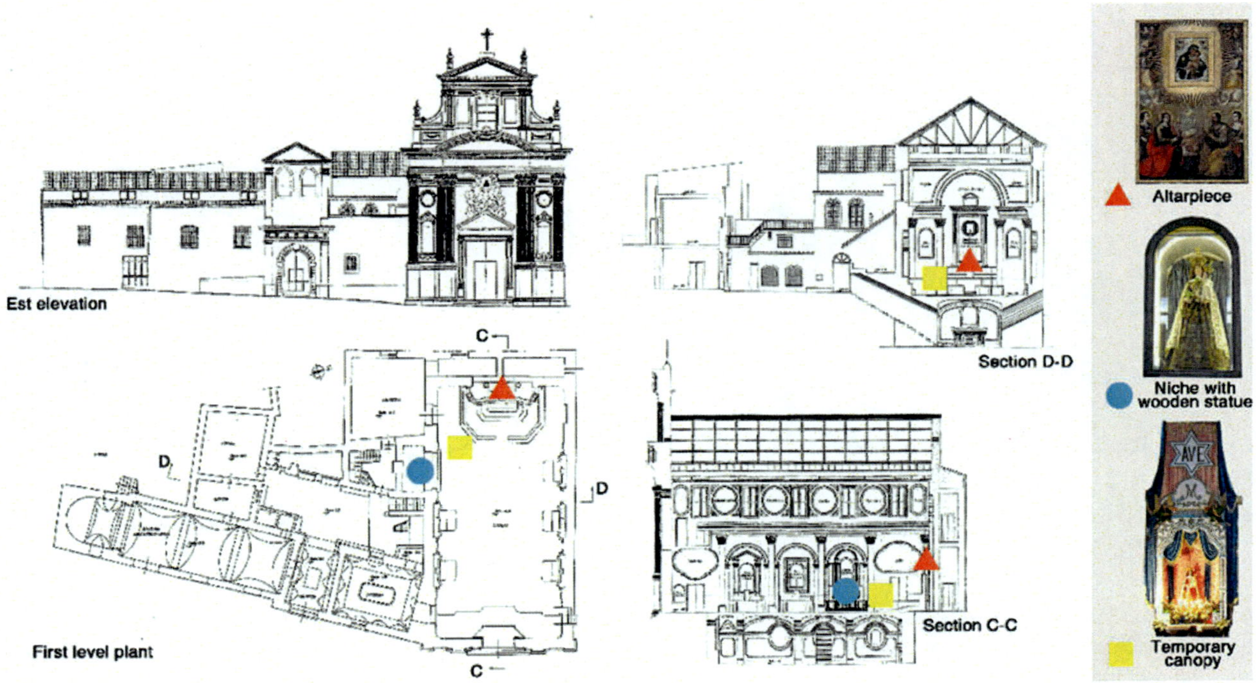

Fig. 3 Drawings of the restoration project finished in 1997, after the damages caused by the 1980 earthquake. Plan, elevation, and sections taken from Arciconfraternita di Santa Maria Incoronata del Carmine detta delle Galline 2000, pp. 46-62. Original drawings 1:100. The table shows the position of the effigy, the statue, and the canopy set up for the feast of the Madonna delle Galline

Fig. 4 On top left: the organ. Bottom left: the Baroque and Altar. On the right: orthorectified image of the wooden coffer obtained from image-based survey with SLR camera and with the software Photoscan; comparison with the drawings of the restoration project finished in 1997 from Arciconfraternita di Santa Maria Incoronata del Carmine detta delle Galline 2000, p. 62

doors open, people release doves, which flutter among the crowd during Holy Mass celebration, after which, outside the church, players and dancers start the vigil of the Madonna to the sound of castanets and *tammorre*.

On Sunday in Albis, the statue of the Madonna is brought through the streets of the city on a wagon, now motorized. People offer mainly hens, but also ducks, doves, turkeys, peacocks, or moorhens (Ugolini, 1987), along with cakes or tortani,[9] while

[9]Rustic cakes stuffed with salami and eggs, which used to be the rich food of the peasants.

Fig. 5 From left: Crowd outside the Sanctuary for the door opening ceremony, photo by Gaetano Del Mauro; Man on the stairs reaching out a child to the Madonna http://www.erodoto108.com/sud-e-magia-la-madonna-delle-galline-nella-citta-di-pagani/, March 2019; A woman trows petals from a roof https://biscobreak.altervista.org/2015/04/madonna-delle-galline/, March 2019

men and women reach out their children to the Virgin, in order for her to protect them, and from the balconies hang nuptial and damask blankets and fall flower petals and confetti, recycled from easter eggs paper (Fig. 5).

The air is thick with the smoke of fireworks and roasted artichokes. Along the procession route worshippers create the toselli, votive aedicules embellished with satin covers, lace, and terracotta molds. In some courtyards, where are the toselli, people gather for exhibitions banquets and for dancing the tammurriata. In front of the pontifical basilica of Sant'Alfonso takes place the exchange of gifts: Redemptorist fathers, according to the tradition started by Sant'Alfonso himself, give a couple of hens to the Madonna, receiving back two doves. After the exchange begins the Soolemn Procession, from Sant'Alfonso to the Sanctuary of the Madonna delle Galline. Then, the procession first stops in front of Villa Comunale gardens in order to pay tribute to players and dancers and, after a last stop in Piazza Corpo di Cristo for the mass, ends at the Sanctuary, where people sing the Magnificat. At dawn on Monday, people depose at the feet of the Madonna the *tammorre* used during the party as sign that the festival is over.

4 The Places of the Celebrations: A Complex and Structured Set of Different Architectural and Urban Subsystems

The analysis of the places of the celebrations starts from the breakdown into different subsystems of the elements composing the festival overall scenario; first analyzed individually and then together in a map that serves as a summary table. Each sub-system is both related to an architectural or urban typology and to a specific action linked to the rituals of the festival. Thus, we find the system of religious buildings with reference to the action of praying; procession route system in relation to movement and city crossing; the Toselli system, for resting, and the courtyards system as extension of city's open spaces, stages of dances, and conviviality moments, to which the system of noble palaces is the backdrop, as well as historical and non-historical buildings.

4.1 Religious Buildings

The city of Pagani is rich in ancient churches, mostly located along the ancient Roman consular road Popilia, called Corso Regio in the Bourbon period, which, starting from Naples, crossed the agro nocerino-sarnese until in Salerno and the Calabrie. The path of this ancient road corresponds to the current via Carmine, Corso Ettore Padovano, via Marconi, and via San Domenico, along which develops the historic buildings of the city, such as palaces, courtyards and churches of artistic, and architectural interest (Belsito & De Pascale, 2014). Walking this urban axis, starting from East, we find:

- *Pontifical Basilica of Sant'Alfonso.* The church was committed by Sant'Alfonso de Liguori, who entrusted the project and the direction of the works to the royal architect Pietro Cimafonte. The construction begun in 1756 and was completed only in 1824. In 1908, Pope Piux X elevated the church to a Pontifical basilica.
- *Church of Santa Maria della Purità, today Sanctuary of Gesù bambino di Praga.* The church was built in 1681 along with the adjacent namesake monastery and stands

on the site of an ancient chapel dedicated to the martyr saints Felice and Costanza.

- *Church of Santissimo Corpo di Cristo.* This church was built in the second half of the sixteenth century, behind the ancient parish church of San Felice, demolished in 1806. Inside is a rich archive.
- *Sanctuary of Santa Maria Incoronata del Carmine detta delle Galline.* (cfr. 2.1).
- *Church of Santissima Addolorata.* The church, annexed to Palazzo San Carlo seat of the Town Hall, was built before 1632 and initially dedicated to San Carlo Borromeo. Since 1814, the church is dedicated to the cult of the Santissima Addolorata.
- *Church of Madonna del Carmine.* It is the last church we find on the Popilia road. Founded in 1491 by Count Francesco Zurlo, it was governed, along with the adjacent convent, by the Discalced Carmelite Fathers.

Other churches scattered in the territory of Pagani are the Church and Conservatory of Carminiello ad Arco in the district called Lamia, as ancient Dea Lamia, Church of Santa Maria delle Grazie in Piazza Cappella, Church of San Francesco da Paola, Church of Madonna di Montevergine, Church of San Sisto, Church and Convent of Sant'Anna, Church of Our Lady of Fatima, and Church of San Domenico di Guzman.

4.2 The Procession Route

In the long-standing tradition,[10] the procession, led by the Confratelli of the ant Arciconfracienternita of Santa Maria Incoronata del Carmine called delle Galline, leaves the sanctuary at nine o'clock in the morning, and in about 14 h covers the 11.98 km^2 of Pagani territory. It almost appears as the Madonna, covered with hens, pigeons, and a peacock, which perches at her feet,[11] is *paying a visit* to all the inhabitants of Pagani, striking out streets, alleys, and courtyards, reaching also the most remote farms outside the urban center (Fig. 6) (Russo, 2001).

The procession reaches by one all the districts of the city through an articulated path, constant over the years except for some small necessary variations, and in some points sometimes retraces its steps for better covering the intricate system of streets and alleys of the city. Randomly along the procession route, the virgin receives by worshippers the offers of gallinaceous, which grow in number as the procession advances and are gradually removed.[12] We find a description of this ancient rite in Swinburne's tales of the journey he made in the Kingdom of the Two Sicilies from 1777 to 1780: "They are also attached to a variety of idle superstition, which, however, are daily losing ground. Of these none is more singular than the ceremony of the feast of the Madonna delle Galline: During the procession, hundreds of hens are placed successively on the poles that support her image,[13] and the miracle consists in the birds sitting quietly. The number of people pressing on every side, and the surrounding noise, make the poor frightened hens remain as still as if perched at roost" (Swinburne, 1790).

Some sources also report that in ancient times "the procession was followed by *gigli*[14] prepared with large quantities of artichokes and oranges then offered to the Madonna" (Di Nola, 1997), a rite unfortunately disappeared. The analysis of the processional route in relation to the location of religious buildings highlights how this verges on all the city churches. From Piazza Sant'Alfonso, where takes place the ritual of the exchange of gifts, to the Sanctuary of the Madonna delle Galline, begins the solemn procession route, which covers seven hundred meters of the ancient via Popilia with great pomp. In general, we can say the processional routes, like the route of the Madonna delle Galline procession, highlights, through the chosen urban itinerary and the stops along the route, all the urban elements and historical places showing cultural and religious values, considered to be a priority within the urban fabric, and which can be a starting point for further exploration and understanding of the organization of the city, from an urban, social, and cultural point of view (Colletta, 2002).

4.3 Courtyards and *Toselli*

The courtyards, which still carry the names of the original families who lived in it, are configured as small semi-independent clusters of houses. In the past, families related to, or economically dependent on, the richer family first established in that place, used to live and work together. On the roadside were shops, accessible from the courtyard as well. Other rooms on the ground floor of the court were used as laboratories and warehouses for products or housed the service areas of a noble palace. The houses were located on the upper

[10]A traditional folk song says: "'A Maronna jesce 'e nnove e và tuorne tuorne e s'arretire 'a calata 'e ll'ora", literally "The Madonna goes out at 9 and goes around and returns when it's dark".

[11]Tradition identifies the meekness of the peacock with a sign of auspiciousness.

[12]At the end of the celebrations the owners 'buy back" the animals offered, commuting the offer to the Sanctuary in cash.

[13]Swimburne seems to refers exclusively to the image and not to the statue, which could have been built after his chronicles.

[14]Wooden structures covered with papier-mâché decorations, stuccos or other materials by local artisans according to religious, historical or topical themes. Particularly famous are the gigli of the city of Nola in Campania.

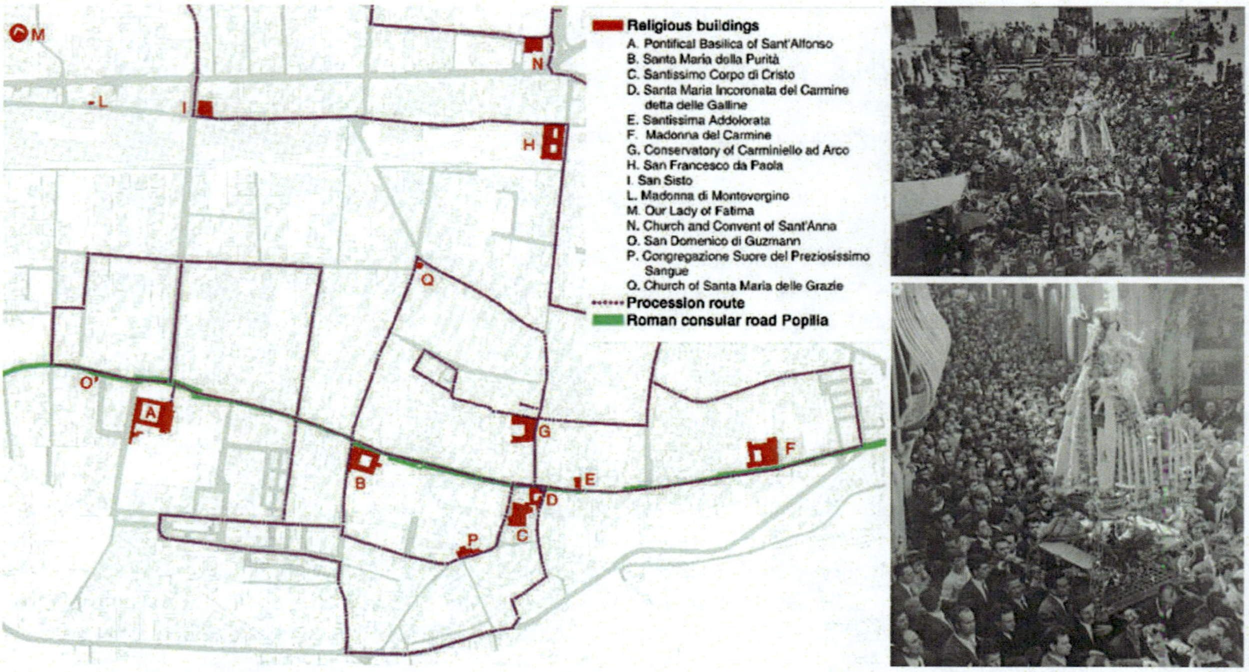

Fig. 6 The processional route is agreed by the authorities, which release it along with religious and secular program. For the creation of this map, we made reference to brochures created by organizers and web material. On the left, map of the procession, cartographic base 1:5000 (Geoportale della provincia di Salerno, n.d), on the right, images of the festival in the nineteenth and twentieth centuries

Fig. 7 On the left: Water well in the Califano courtyard, where the Tosello Madre dell'Africano is set up. On the right: The complexity of the urban space of the courtyard is rendered through a cloud points obtained with photogrammetric survey of the well

floors of the buildings, accessible through open staircases and walkways. Agricultural activities took place in the fields and gardens behind. Many of these courtyards had a common wash house and a well for collecting rainwater (Fig. 7).

Despite the transformations intervening over the last few centuries, Pagani's courtyards retain many of the original elements, which, as a heritage of a population and its culture, should be documented in order to guarantee their transmission to future generations. The survey of the rich heritage of the city of Pagani through Structure from Motion (SfM) could be the key to its critical documentation aiming at cataloging, renovation activities, or reconstruction in case of loss, way up to virtual usability. A particular feature of the feast of the Madonna delle Galline is the setting up, mostly inside these courtyards, of the so-called toselli, votive aedicules to which people offer the fruits of their land. Toselli, whose name arises from the Spanish word *dosel*, canopy, originate from an ancient tradition, still in use not long ago according to contemporary witnesses, when inhabitants of the noble palaces and courtyards used to open the great wooden portals, on which they glued votive images of the Madonna, and prepare small receptions for the porters carrying the statue in procession on their shoulders. After the passage of the procession, inhabitants used to close the portals and continued the celebrations in a more private dimension. The tosello in its modern features, drapes of precious damask blankets, altars, small chicken coops. and objects of popular tradition (Fig. 8), set up in a private space, the courtyard, which for the occasion becomes public, spreads thanks to the folk singer Franco Tiano, who first, in 1971, opened his tosello to outsiders, then in 1975 revolutionized the way of setting up the toselli (Sinatore, 2014).

It is clear that the ancient tradition of toselli evolved thanks to new impulses (Pascariello et al., 2011) and has in fact overturned the concept of public and private, allowing a permeability between the inside and the outside that did not exist in the past, adding a new value to the tradition, for it allows everyone to get to know and experience ancient traditions, which would most likely disappear otherwise and, at the same time, allows the use of an architectural and urban space characterized by precious intangible values. To catalog and map all the toselli has proved to be remarkably complex, mostly for the difficulty to collect fragments of heterogeneous material. The infographic (Fig. 9) offers a reading of the consistency, the trend, and the overall presence of the toselli of the city of Pagani, as well as a classification according to the district.

We were able to create a table providing all the information related to all the toselli documented since the 1970s to present time: name or names, of the toselli through the years; address and correct location on the map; devotees who built it and presence for each year.

4.4 The Ancient Noble Palaces

Pagani historic center is a valuable historical and cultural resource. This urban fabric, extraordinarily rich in exquisitely crafted architectural episodes, is strongly characterized by the presence of numerous noble palaces (Comune di Pagani, 2019). In Pagani and the province of Salerno, the development of this building typology started in 1707 with the end of the Spanish Viceroyalty, when the old nobility and the new aristocracy commissioned new buildings in this area. The noble palace of the city of Pagani is divided into functional areas according to the number of people and main activities: In the courtyard, there are the stables, the carriages, and the access to the cellars below. Upstairs is the noble apartment called *quartino*, with balconies looking outward. The representative space is the gallery, a large hall richly decorated and furnished, which also houses the family collections. The servants stayed on the top floors or in the attics. The main façade represents the symbol of the aristocracy, as well as the portal, on which the owner carves his family symbol. Among the most representative examples of noble palaces in the city of Pagani we find:

- *Tortora degli Scipioni Palace/Tosello Ngopp'o Peron.* Located in via Perone was built in the second half of the seventeenth century the Palatine Count Carlo Pignataro, protomedical of the kingdom, who exercised his profession in Pagani. The building develops around a central courtyard, with the entrance slightly shifted to the right and with rectangular modular rooms overlooking the courtyard, separated by the rear garden by a rectangular building.
- *Tortora Palace/Tosello Palazzo Tortora.* The palace, in via Marconi, is the only example left in Pagani with the original fresco paintings that used to embellish Pagani noble palaces. The building is currently inhabited by various owners, but the dove painted on the three hills under the vault of the hallway confirms the connection to noble Tortora family.
- *San Carlo Palace.* It is located in Piazza D'Arezzo and is currently the Town Hall. It was built as a convent and college of the fathers of the Pious Schools and was later used in part as the seat of the Royal Giudicato and partly a prison. It preserves the façade and the internal staircase in Vanvitellian style.
- *Gatto Palace/Tosello Istituzionale Palazzo Gatto.* The building, located in Piazza Corpo di Cristo, opposite to the church by the same name, was the seat of the Regiment House or Municipality of Nuceria Pagonorum, as attested by the inscription on the keystone of the portal. An integral part of the building is the votive aedicula dedicated to the Madonna delle Galline.
- *Mangarella Palace/Tosello ro' Manganella.* The building, located in Via Astarita, still retains the name of the ancient and noble family of Manganella, now extinct, who had it built between the XVII and eighteenth centuries. Tradition wants the building to be in communication with the Castello di Cortimpiano, a very old area in Pagani, and the Castello del Parco in Nocera Inferiore. The building, although in a poor state of preservation, still retains the architectural features of the seventeenth and eighteenth centuries almost completely intact and presents some elements of great formal refinement especially in the staircase.

Other palaces are: Pagano Palace, Striano Palace, Palazzo Criscuolo Palace, Coscioni Califano Palace, and Fusco Palace, today the seat of the Congregazione delle Suore del Preziosissimo Sangue and of the Tosello Religioso Tommaso Maria Fusco.

Fig. 8 Miscellanea of toselli from past editions

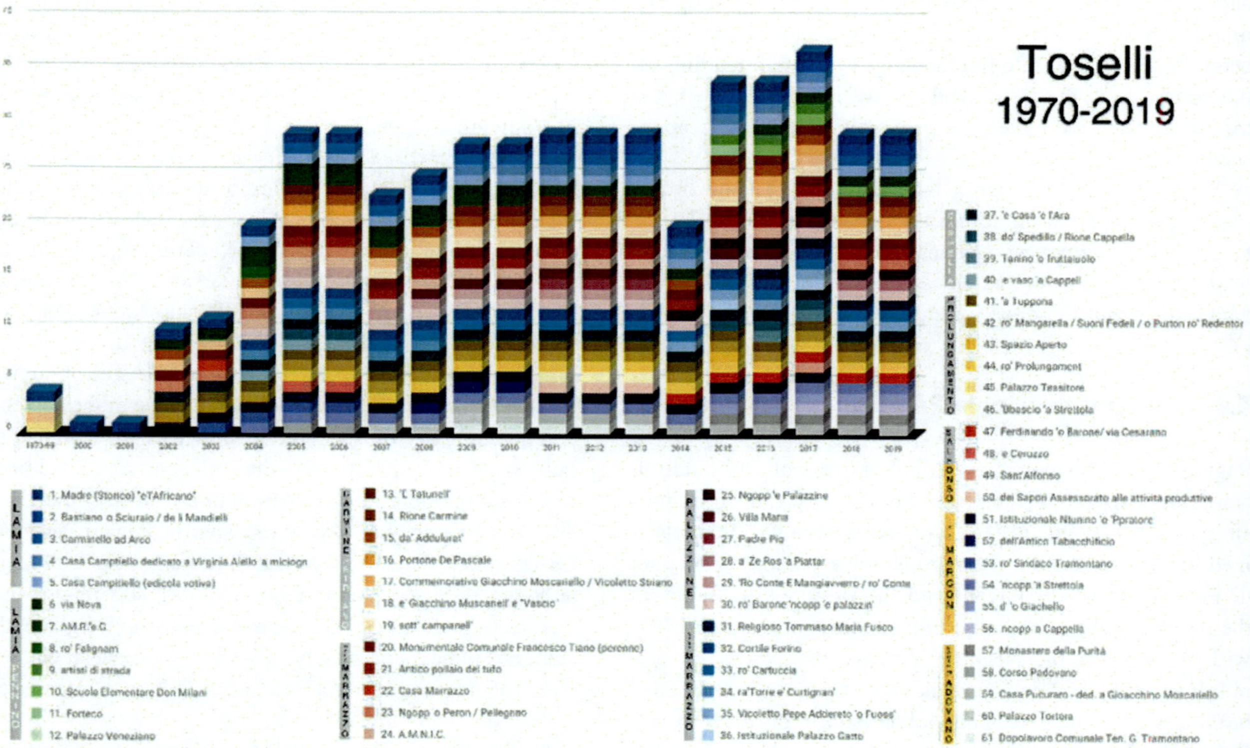

Fig. 9 Infographic: *Toselli of Pagani* documented since the 1970s to present time organized by color families based on the area they belong to and according to the order met by the processional route. In yellow the solemn Procession

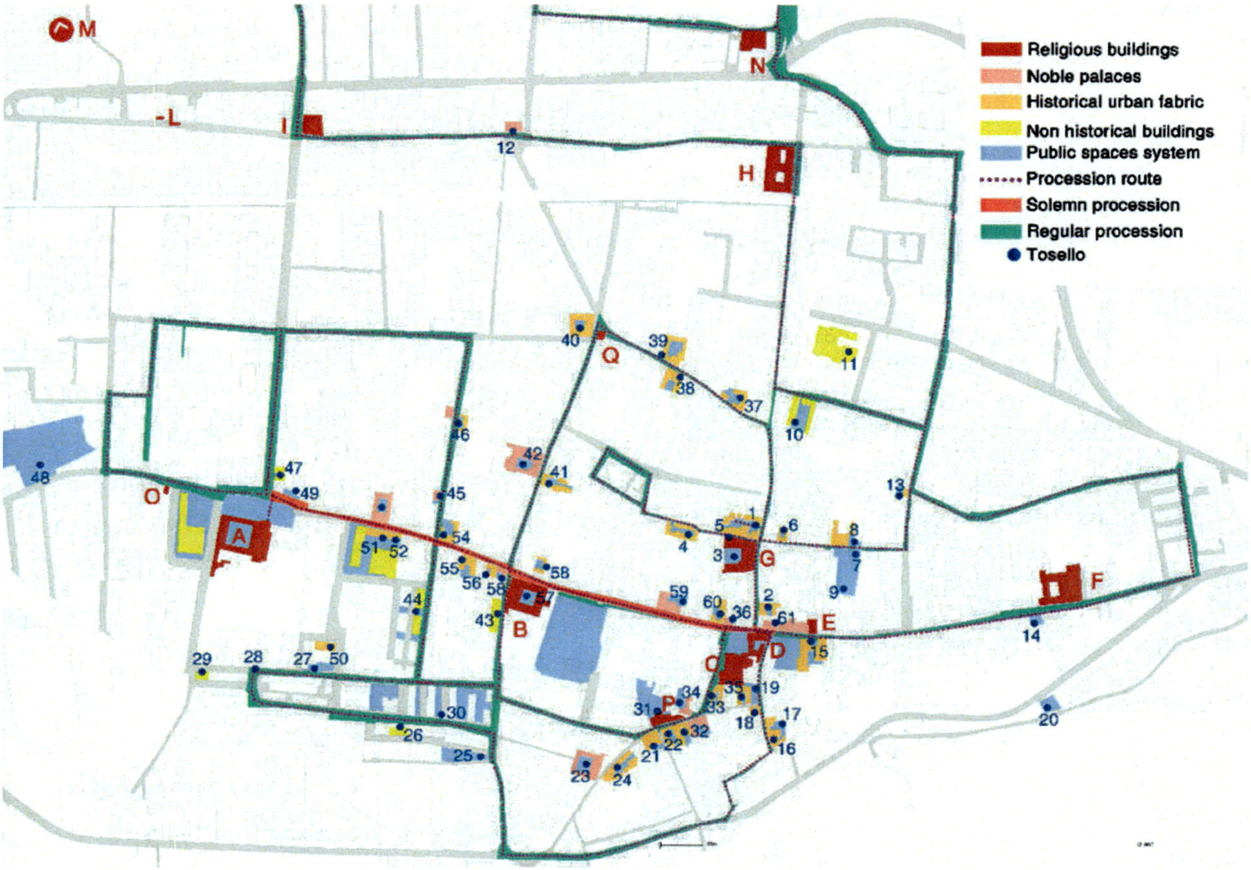

Fig. 10 Cartographic map with the identification of the subsystems previously examined

5 Conclusions

Documenting the feast of the Madonna delle Galline also means to represent a city, Pagani, which during the days of the festival transforms thanks to peculiar and ephemeral setups connected to religious worship and thanks to numerous folkloristic moments of the festival itself; the city opens up to citizens and to all those came to worship the Madonna and celebrate with songs, dances, music, and food. Many private places, such as the courtyards of noble palaces as well as those of historical and non-historical buildings, where the toselli are set up, become accessible to the public, turning into rest areas where it is possible to carry out religious rites or sharing moments of conviviality. The open spaces of the city, such as streets, alleys, squares, gardens, on this occasion of the feast, almost appear as internal environments in which the community is collegially involved in an intimate religious recollection and strongly connected from an emotional point of view.

Although we find evidence of this festival as ancient cult in various bibliographical sources, the main instrument for its documentation remains photography, that portrays the festival in all its aspects, allowing a proper monitoring over time. The feast of the Madonnas delle Galline is widely documented, in fact, by a rich photo collection, also available on the web, of which we propose some in the contribution, portraying the main phases of the festival: The Door Opening Ceremony outside the sanctuary where is the nineteenth-century statue of the virgin, the procession that crosses the main streets, the alleys, the courtyards with the Alzata del Quadro, the raising of the picture representing the effigy of the Madonna, the tammurriata, the return to the sanctuary of the Madonna with the deposition of the tamorre. On the web, it is also possible to watch some interesting videos, including the docufilm "L'Africano," a tribute to the figure of Franco Tiano, presented at the 48th edition of the Giffoni Film Festival in 2018, allowing the people, even those who have never been there, to live the unique atmosphere created the week immediately after Easter in this town of agro nocerino-sarnese.

This paper aims at fixing the image and the story of the feast with a graphic representation; that is, through a cartographic map on which it is possible to overlap the festival and the city identifying the places involved by the procession, the cult, the folklore and food, and wine events (Fig. 10).

In fact, it is the sign to be considered the main instrument for expressing identity, tradition, culture, even of intangible assets, so that these can be kept in the collective memory. It was, therefore, necessary to deny for a moment the immateriality of the festival, in order to highlight the urban paths of the procession to get to know the sites related to the cult of the Madonna delle Galline. The historical center of Pagani is the area of the city interested by the solemn rituality of the festival, the area where are concentrated the main noble palaces, some of which also host workshops or photographic exhibitions that testifies the willingness of the local population to spread a tradition rich in references to the territory of the agro nocerino.

The mapping of the toselli shows a greater concentration in the historical center, mainly in the area around the Sanctuary of the Madonna delle Galline, area in which the devotion to the Madonna is more strongly manifested, almost reinforcing the presence of the Madonna in the church where the effigy and the statue are kept. As for many popular festivals with religious origin, the feast of the Madonna delle Galline in Pagani is also closely linked to food and wine traditions, manifesting through both public and private banquets with local wines and products of the land wisely cooked respecting ancient traditions. It is to be hoped that the dissemination of this festival will be more favored by local and national institutions for the safeguarding of intangible urban values and intangible assets and for a sustainable development of the territory both for increasing tourism and spreading historical and cultural knowledge.

Acknowledgements Based on a collaboration of the authors, paragraphs 1, 4, 4.1, 4.4, 5 are by Maria Martone, paragraphs 2, 2.1, 3, 4.2, 4.3 and pictures are by Alessandra Marina Giugliano.This study would not have been possible without the generous help, the materials and the information provided by the following people: Giuseppe Tortora, current Prior of the Arciconfraternita di Santa Maria Incoronata del Carmine detta delle Galline; Professor and historian Vincenzo Sessa, previous Prior of the Arciconfraternita; Gerardo Sinatore, expert of the traditions related to the feast of the Madonna delle Galline.

References

Arciconfraternita di Santa Maria Incoronata del Carmine detta delle Galline. (2000). *Santuario Madonna delle galline. Un restauro per l'arte, un restauro per la storia.* Edizione a cura dell'Arciconfraternita.

Belsito, F., & De Pascale, C. (2014). *Storia di Pagani in breve. Monumenti, personaggi, tradizioni.* Editrice Gaia.

Canzanella, C. (2002). *I volti di Maria. Miti e riti. Le sei Sorelle. Il culto popolare della Madonna in Campania.* Altrastampa.

Colletta, T. (2002). The routes of the processions as intangible dimension, inalienable in the preservation of south mediterranean towns. In J. Campos (Ed.), *A dimensaso intagìvel na cidade històrica.* Cruarb Patrimonio mundial.

Comune di Pagani. (2019, June 12). *Beni architettonici.* https://www.comunedipagani.it/citta/beni/beni.asp.

Geoportale della provincia di Salerno. (n.d). http://geoportale.provincia.salerno.it/.

Guidoni, E. (1980). Processioni e città. In E. Guidoni (Ed.), *Atlante di storia urbanistica siciliana.* S. F. Flaccovio.

Istituto Centrale per la Demoetnoantropologia. (n.d). http://www.idea.mat.beniculturali.it/.

Matrisciano, A. (A.A. 2005/2006). *La festa della Madonna delle Galline nella città di Pagani.* Master Degree Thesis, University of Naples.

Montorio, S. (1715). *Zodiaco di Maria, ovvero le dodici provincie del Regno di Napoli, come tanti segni, illustrate da questo Sole, etc.* Per Paolo Severini.

Di Nola, A. M. (1997). *Campania Felix.* Editalia.

Pascariello, I., Cilenti, M., Milano, G., & Trincone, M.R. (2011). Per la costruzione di un archivio del patrimonio immateriale. In C. Gambardella (Ed.), *Le Vie dei Mercanti. S.A.V.E. Heritage, Safeguard of Architectural, Visual, Environmental Heritage.* La Scuola di Pitagora s.r.l.

Russo, S. (2001). *Tra fede e folklore: riti e rituali di ieri e di oggi nell'Agro Nocerino-Sarnese.* Edizioni Nord-Sud.

Sinatore, G. (2014). *Madonna de'Pagani detta de'Galline.* Nuova Pro-Loco Pagani.

Swinburne, H. (1790). *Travels in the two sicilies ...: In the Years 1777, 1778, 1779, and 1780,* (Vol. 3). J. Nichols.

Tardio, G. (2008). *Le leggende delle Sette Madonne Sorelle.* Edizioni SMiL.

Ugolini, M. M. (1987). *Animali sacri.* Filelfo: Trilogia.

Vassalluzzo, M. (1987). *La Madonna delle Galline ai raggi X.* Pagani: Edizione a cura dell'Arciconfraternita di Santa Maria Incoronata del Carmine detta delle Galline.

Villani, L. (1987). *Florilegio storico della venerabile effigie di Maria Santissima del Carmelo, "vulgo delle galline", in Pagani,* (Vol. Secondo). Tipografia De Angelis.

Oblique Plane; Layering; Spatial Layering—The Construction of Pavel Janák's Formal Language

Qi Wang and Domenico Chizzoniti

Abstract

The research aims to explore Pavel Janák's design techniques through studying his formal expression, which in this paper is conceived as a sort of language. Inspired by Cubist paintings, Janák's building work is composed logically by geometric forms, representing a complex figure. The techniques he used are considered as the grammar to decode this complexity. So, by means of extracting the geometry archetypes as the solid alphabet from the language, the research will interpret his ever-changing formal expression from Cubism to functionalism and inspect his design techniques and their evolution process. The research method is to analyse the built works of Pavel Janák in different creative stages. Taking his theoretical works as a breakthrough, firstly, the paper will try to dissect the Janák's philosophy background and then analyse the formation and transformation of his grammatic structure of formal expression from a height of aesthetic conception. More profoundly, by studying his personal diary, the paper will inspect his self-reflection and deep thought.

Keywords

Form • Space • Czech Cubism • Architecture • Architectural language • Composition techniques • Functionalism • Layering • Spatial layering

Q. Wang (✉)
Department of Architecture, Built Environment and Construction Engineering, Politecnico Di Milano, Milan, Italy
e-mail: qi.wang@polimi.it

D. Chizzoniti
Polytechnic University of Milan, Milan, Italy
e-mail: domenico.chizzoniti@polimi.it

1 Introduction

Modern architecture originated at the turn of the century in Prague as well as elsewhere in Europe. Two conflicting trends between the past and the present were modernism and historicism. Avant-garde architects and thinkers were starting to question the historicist architecture, which they felt was immoral and anachronistic because it failed to reflect both technological evolution and the social structure and did not correspond anymore to the "*Zeitgeist*". So, they started to promote a universal design principle that would replace the mere reproduction of historical styles.

As Kenneth Frampton justifiably observed that "the country which has always been inadequately represented in any account of the International Style is Czechoslovakia" stated in "Modern Architecture, A Critical History" in 1985, Czech Cubism was always considered as in a prominent position in the modern architecture movement, which had guided the criteria of architecture design towards a scientific direction. It was even once superficially connected to the conception of A. Einstein in 1979 (when he was teaching in German Technical University in Prague at that time) in the first published investigation of Czech Cubism legacy "*Cubism in Architecture and Applied Arts: Bohemia and France 1910–1914*" (Margolius 1979).

Detached from the previous sense of arts and crafts, architecture design activity turned to relay more and more on the material and technology. Not satisfied with this, Cubist architecture was the first tendency that was against modernist spirit in the sense of being opposed to ornament. With Cubist style, a group of Czech avant-garde architects devoted realizing their artistic intention into the design techniques of modern era.

Pavel Janák was one of the foremost representatives of the Czech avant-garde. Cubist art inspired Janák with its essence of abstraction and the kinaesthetic way of perceiving the world. In 1911, he wrote a critical essay, "*The Prism and the Pyramid*" and created a theory of form system using

C. Gambardella et al. (eds.), *Advances in Utopian Studies and Sacred Architecture*,
Advances in Science, Technology & Innovation,
https://doi.org/10.1007/978-3-030-50765-7_13

geometric shapes, which he applied to his sketches, furniture and architectural designs.

Throughout his career, Janák always believed that architecture belonged in the realm of art. Because of this, much of Janák's initial understanding of artistic creation was derived from Riegl's idea "*Kunstwollen*", which means "the artistic will". Riegl saw the aesthetic urge as the key of creation making, not only to what was creatively possible but also artistically necessary (Riegl and Kain 1992). In favour of Riegl's point of view, Janák was against architecture which is too utilitarian, materialist and lacked spiritual beauty. Trying to transpose the principles of Cubist painting into architecture, he claimed that architecture must be more poetic, more expressive and more artistic. Even under the influence of Vienna Secession, Janák had not turned himself into an architect who regards practical functions as the top priority in the architectural design as other extreme functionalists. In the hands of Janák, the functionalist concerns, at last, had become a component of his architectural language as Cubism and other styles of the same validity.

Janák's progress from Cubism towards functionalism was planned and systematic. With the influence of the functionalist movement, he had written a series of complex analytical studies, with mathematical calculations, tables and diagrams to examine the relationship between modern systems of developing urban areas and population density. His architectural work was, however, less orthodox than his theories, and several of his buildings from this period were undoubtedly designed to have an emotive impact. He did not want to lose sight of the artistic aspect of his work. He tended to absorb the substantial character of each period style as source. Historical, folk and functional elements were integrated under the domination of his aesthetic urge and evolved into the forms that the new architecture of the modern era required (Kiesling 2011).

A record of Janák's personal reflections and feelings about architecture was found in his architectural diary. It contains the main ideas from which his key essays of the Cubist period developed. He published three essays between 1911 and 1913 in *Umělecky Měsíčník* the leading forum of the Czech Cubists. "*Prism and Pyramid*", "*Renewal of the Façade*" and "*Of Furniture and Other Issues*" were considered as the most important writings that laid the foundation of Czech Cubism.

In his essay "*Prism and Pyramid*", clearly, the space-generating capability of the triangle is primarily emphasized. He fully linked it to the aim of the abstract expression and had set a spatial law guiding the composition and construction process that he called "spiritual abstraction of matter" by a spiritual arrangement. In the essay "*About Furniture and Other Matters*" Janák spoke about the "spatial expressions of emotion". Later, in the essay "*Renewal of the Façade*", he pursued the idea that the inner three-dimensional space of buildings should provide a place for something far more artistic and spiritual by a "spatial formulation of their surface". Then, what Janák called his concern for "spatial recording of life" was explored both in his published essays and in the raw material of his diary.

By combining Janák's practical and theoretical work, this paper aims to interpret the architectural language of his works and then to extract the specific techniques from this formal language. For this purpose, some architectural examples of his ever-evolving styles will be examined, comparing with his theoretical work in the same period to inspect his deep thoughts in the transition from the classical to the modern era.

2 Formal Language System

Pavel Janák was a protagonist of the Czech Cubism movement. The Cubist architecture, one of the most dramatic phenomena in Czech architectural history, has formed a style of "spiritual abstraction of matter" by his systematic arrangement of geometric forms. It was an attempt to revise the natural, native, technical, material and classical styles, in favour of the Nordic, Expressive, Gothic and Baroque traditions.

From Cubism, on the way to the modernism, Janák never gave up pursuit of the beauty in architecture. His starting point was the view of "*art's autonomy*", proposed by Austrian historian Alois Riegl, which means the artistic will "has no obligations beyond itself" (Ševčík 1986). So, with this obsession of the artistic nature of architecture, he had developed his unique expressive formal language on the base of geometric forms with the method of abstract reality.

3 Form, Order, Style

For Janák, it is the architecture formal system that gives the whole environment a consistency, containing the solid alphabet and the basic rules of composition.

The "Form", which is represented by some geometric archetype, is the solid alphabet of his formal language. However, the term "Style" is a relatively variable value. The distinguishing principle among styles was that of the poetic fiction of matter, rather than form. He extracted the essence of each stylistic principle, translated into his own language, and incorporated it into his own design philosophy. For example, the Dutch elements were taken in by Janák to rule the functionalist works as another language variant. This design ideology can be also seen as a continuity of plastics "Cubism" for its clear, legible and "cubic" composition of masses. The evolution consisted of replacing the acute angles and deformed planes with cubic shapes and orthogonal lines, which conform better to the demands of modern functionalist architecture.

Even the styles are generally abstract, but they should always follow a law, an order, to unify the principles. For Janák, the "order" is the basic rules of composition. They give the whole environment consistency and harmony. They can be seen as the essence of the architectural design, which is a constant property gained by absorbing the nature of present from the history, and then endowing it to the future (Kahn 1944). To achieve the qualities of consistency and harmony, Janák used an architectural grammar to control the geometric unit and drive the detail development. The grammar, which we see as a range of composition techniques, unified the styles by writing the geometric alphabet into a logical structure.

4 The Third Force and the Oblique Plane

4.1 The Third Force

In his essay "The Prism and the Pyramid" in 1912, Janák indicated: "All shapes that occur in inanimate nature and are geometrically complex evolved with the collaboration of a third force" (Janák 1996a). The force could be an external force that broke the equilibrium and could also be an inner force of the crystallization process of substance.

In Janák's opinion, the third, diagonal force is the essence of creative making and creates more complex forms. He has divided the process of formation into two types: firstly, he proposed that gravity, and all other kinds of force that were not opposed by friction, would result in an endless succession of horizontal layers, an infinity of alluvial deposits. The resulting form is orthogonal, bi-planar; secondly, Janák evokes more complex forms originated with the help of a third, diagonal force acting internally or externally: "...snow banks, wash-outs, ravines, caves, earth depressions, volcanoes, are carved out of the matter by a force which deforms matter and displaces it". And he sees crystallization as the most beautiful example of this second process: "... the force of crystallization is so much stronger than gravity that one can almost say that the latter does not influence the former. It seems to be a self-centred inside the mass and so strong that it prevails over other circumstances" (Janák 1996a).

4.2 Oblique Plane and Triangle, Prism and Pyramid

The important third force will occur only when a third plane of an obliquely shaped form joins the naturally balanced bi-plane. With the help of the third plane, the geometry form could be overcome artistically and imbued with spirit. It could represent the artistic intentions and psychologically complex of an architect. In Janak's early Cubist work, the "oblique planes" and "Triangles", appeared as the decorative elements on the gable wall or the chamfer faces that cut into the inner space of the building (Figs. 1 and 2).

Under the "prism" category, Janák attempted to group the principles of Egyptian and Greek architecture. Both of them emphasize in particular poetically transformed matter by endowing it with characteristics, not of this world, as the geometry form. For Janák, the pyramid as shaped by these dramatically ordered diagonal planes is the ultimate form of a spiritually abstracted matter, which is based on the natural prism. It means that a prism is the basis of which we inscribe as a pyramid, and the pyramid is the philosophical replacement of the prism.

An example of Janák's approach to form generation is a series of sketches of interior designs from 1912–1913 (Fig. 3). In the sketches, the interior space presents as a kind of crystalline canopy. We can tell Janák's intention of dramatically decomposing the matter with tapered surfaces and the bevelled angles. For the shaping of the interior space, Janak may have found inspiration in the late Gothic vaults, and probably of the interior of the mervellous Baroque Gothic church that he studied in 1909 (Švácha and Büchler 1995). In all the three versions, the interior space is formed by the connected columns and ceiling, showing us a dynamic image: the rough lines arise from the bottom of columns, extend upward following the direction of the vault and finally combine with the ceiling. Despite an intense feeling of instability, the drawings give the spectators a strong hint that the motions caused by an inserted third force, such as bevelling and carving. It has transported the basic geometric form into a three-dimensional system. Unlike the expressionists, Janák

Fig. 1 Pavel Janák, sketch of a façade design, 1912, NTM

Fig. 2 Modelling of the façade sketch

was not preoccupied with the qualities of material, transparency or reflection, but in the dynamic possibilities of the inner tension between the surfaces and spaces.

The oblique planes, triangle and pyramid are keys to Janák's systematic formal language. This formal language is introduced by his artistic feeling. In this language, Janák suggested a supernatural and idealistic ideology derived from primitive naturalism and pure materialism. The ideology dramatizes the architectural figure with the help of a diagonal plane and makes it no longer belong to the bi-planar system, attempt to revise the natural, native, technical, materialist architecture and the classical tradition. However, as stated by Le Corbusier in his book "After the Cubism", between formal expression and reality, there is an intersection (Le Corbusier and Goodman 2001). This intersection corresponds to these geometric forms found in Janák's built works.

In the Fará House reconstruction project, Janák was asked to rebuild the Baroque building on the square in order to integrate into the neighbourhood context. The idea of restoration was to build a shield to cover the houses of various features around the square so that the square would have a pure image of the old Bohemian character. Janák's task was to remodel the facade of Dr Fará's house, which was also initially built in the Baroque style. During the reconstruction, Janák had cut the corner of the building obliquely in order to soften the right angle and then constructed a shield by adding a triangular gable facing the square with a balcony in front. The original Baroque look of the square was reinforced by the new facade and the balcony over-hanged on the square in the same style, speaking directly a Cubist language. The additional parts are dynamic and excited. The new facade has filled the gap between the two different period styles rather than providing a contrast effect with a modern feature (Figs. 4, 5 and 6).

The realization of the new façade of Fará house validated the technique of Janák's Cubists idea in the manner of imitating the principles of the historical precedent. It achieved a dynamic animation of a feature, and the Baroque appearance was not disturbed by the contemporary addition, but enhanced and complemented by it. Janák's intention seemed to have been both to abstract space into a surface vision and to construct it with planar elements. He explored this pictorial representation of space, on the one hand, through geometrical means, and on the other, through analogies with various art forms that to him closely paralleled the previous architect's concerns.

And in this project, the influence of Baroque and Gothic styles is easily observable. Originating in Cubist painting, Janák used a pure formal expression to interpret the traditional symbols. His attraction to the specific historical styles is found important as a theme in the early Cubist work. Notably, the geometrical structure of Gothic churches in southern Bohemian greatly inspired Janák. He used sharp edges and oblique, constantly touching lines to express the intricate vaulting. Intent on the idea of creative making, he used his spiritual arrangement as a penetrating force to cope with the corporeality of geometry form. He sought equilibrium by hewing corners and edges, by penetrating the depth

Fig. 3 Pavel Janák, sketch of interior design, 1912 (Herbenová and Šlapeta 1984)

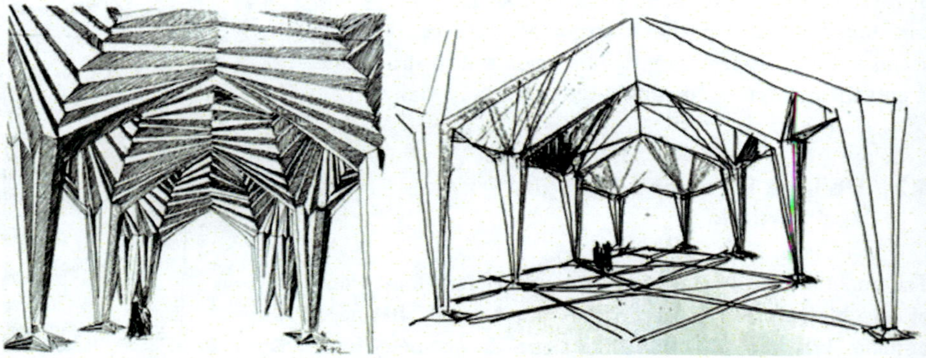

Fig. 4 Historical photo of Pelhřimov Square, 1895

Fig. 5 Historical photo of Pelhřimov Square after the reconstruction of Fará house, 1913

of surfaces. All of these are present in the 1913 remodelling of the Fará house in Pelhřimov, where Cubist features fused harmoniously with the historical style.

5 The Interfaces and the Layers

5.1 The Interface Between Observer and Creator

In his essay of "Renewal of the Façade", Janák stated: "So far, we only know that to make cubic and three-dimensional space does not mean to create space because that merely corresponds to reality. To create always means to extract more than what was already there, i.e. to create volume and space through spatial interpenetration of the surface planes" (Janák 1996b).

For Janák, the façade, in the form of a surface plane, was an interface between the creator and the observer, and the more subjective and abstract thinking we have, the more the facade of architecture asserts itself. The interface, for Janák, was always an essential element in architectural design. This idea was first expressed in the colourful façade design of 1920s factory buildings; then, it was developed with more depth and fewer colours due to the increasing emphasis on functionality. This corresponds with the Cubist style mostly because it allowed Janák to develop a characteristic decorative rhythm with a plastically modelling method.

In Janák's opinion, objects are defined by the light, colour and plastic property of their weight and position. Human

Fig. 6 Details of Fará house, Pelhřimov, 1913–14

beings perceive the world through their eyes and the eyes, with their ability to abstract, help to foster a particular form. This contention is important for Janák and other Czech Cubist architects. The process of visual perception is that the whole world is projected on a spatial plane, to which everything is attached, either materially or in the imagination. Fully oriented towards the eyes of man, it becomes a picture, a relief or a longitudinal facade of architecture. Captured by the figure, that eyes turn to the interesting points and tensions of inner space primarily towards the facade as if the entire content of the building was embodied in its façade.

However, what we see is related to the expectations of the individual and the background of the times. In "The Renewal of the Façade", Janák proposed the visual perception as the way of active participation of the observer. To prove this point, he took series of historical building as examples, e.g. the facade of San Michele in Lucca with its blank arcades that involved our visual perception by suggesting the nature of the space behind; San Marco in Venice and the rock-cut tombs in Petra, represented a sequence of superimposed stories behind; in the expressive Baroque buildings, the different images reliefs of images overlap and penetrate each other, achieving a gradient effect by tilting the plastic elements directly into the visual plane. All these provisions of the spectator, according to Janák, are because the outline of the building comes alive particularly on the façade. Layers are spread out, and all the plastic ledges and arches, otherwise protruded, are articulated into the visual plane, pulling and pushing everything to a higher density.

5.2 Haptic and Optic: Mediator Between Creator and Observer

In Janák's opinion, the role of haptic and optic perception is considered as the mediator between the creator and the observer. They are the bases of creative making. Even in a perspectival world, through the medium of the façade, the depth of a surface can be experienced and expressed differently, conceptually as well as kinaesthetically. The connection with Cubist painting lies here. The spectator can only get the one-sided image by seeing it; by touching it, he can feel the entire body of an object from all sides. So, sometimes by means of seeing, the observer can get one side impression or the illusion of the three dimensions (Janák 1996b).

5.3 Layers

For Janák, the process of visual perception is to abstract space into an optical surface and to construct it with planar elements. The outline of the facade we see in architecture dramatizes the space behind it. It is a starting point of artistic creation. This pictorial representation of space occurs on the one hand, through geometrical means, and on the other, through analogies with various art forms as in painting and sculpture, which to him closely parallel the architect's concerns.

In order to enrich the visual perception, Janák developed a post-Cubist decorative version of style with the technique "layering" which has a tendency of developing all schemes of inner space onto the surface. Using this technique of "Layering", Janák pursued the reality of objects from the illusion of the space. The composition of different materials is transformed into a type of rhythm and the structural elements are converted into a system of rich and lively ornamentation.

6 Techniques

The techniques used by architects are the form giving and form finding geometric operations: linked, separated, attached, detached, interlocking relationships, layering, gradual regularities, etc. Art is above all a matter of conception. Meanwhile, techniques are the instrument of control, of verification of penetration. Technique is only a tool that humbly serves the conception. Thus, the design techniques can be interpreted as a medium through which the creator can transform his artistic intention to the observer.

For most of the observer, the first impression of architecture is the sense of figure, volume, something solid and physical. However, it is not related directly to those conceptual elements of geometry such as the point, the line, the plane and the solid. The impression of a building mainly depends on how it projects in our eyes and then how we perceive it. The visual perception is an active behaviour of observer to participate the creating process. So, the architects, who create, are requested not only to care about the geometry shape, but also to guide the visual perception of observers using some strategies. Unity, balance, contrast, rhythm, proportion and all these tricks (strategies) of architects are achieved by the composition techniques.

For Janák, the techniques are revolutionary. They showed not only the stages that logically were motivated by the previous stage or a sign of his inner struggle but they stand as evidence for a system of plastic shapes through which Janák explored the meaning of modernism.

6.1 Layering: Stratification—Superimposition

In Janák's work, "layering" is realized through the techniques of "Superimposition" and "Stratification".

"Superimposition" repeatedly appears in Janák's diary as a mean of conveying what he calls "the erection of content" as distinct from "growth" (Murray 1990). Optically based, superimposition is a means of making the internal layout visible externally. In the resulting interpretation, he no longer attempts to understand the facade structurally, but rather as "a paraphrase of what is behind, i.e. a transposition of the real content behind the facade into the plane in such a way that each part is superimposed, not "on" but "above the other" and this superimposition signifies perhaps the sequence of things for imagination. In a 1917 drawing of a building's facade in Janák's journal, we can see that the windows and the balcony, these basic elements of the facade, were considered not mainly as the functional elements but a kind of ornament, growing on the surface with a multi-layered dynamic (Figs. 7 and 8).

"Stratification" is the technique by which the substantial and articulate space is constructed. It is the essence of that phenomenal transparency which has been noticed as characteristic of the central post-Cubist tradition. "Flow into one another", "blend into each other" and "wrap around the building" "contribute to that process of loosening up a building which now dominates the architectural scene"

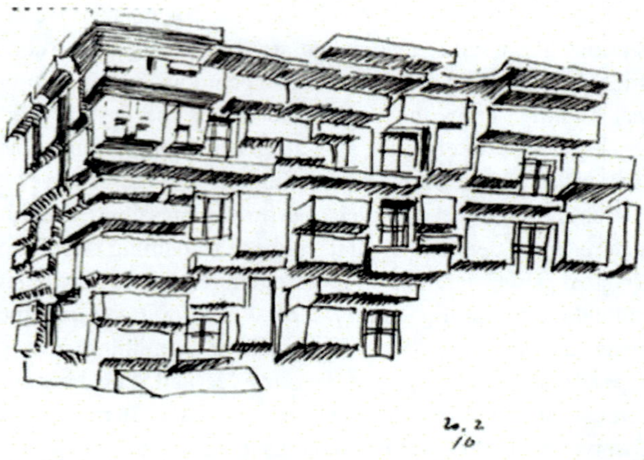

Fig. 7 Pavel Janák, Návrh fasády (Design of Façade), 1917 (Herbenová and Šlapeta 1984)

Fig. 8 Modelling of Pavel Janák's "Design of Façade"

(Rowe and Slutzky 1963), these expression of structural vocabulary consists of a three-dimensional field of elements, in which, the solid elements was enabling the voids to be integral to the whole and equally meaningful, simultaneously allowing one to perceive the whole from a singular viewpoint.

Janák's Pardubice crematorium project speaks directly of Czech national identity through its colours and motifs. In front of the main entrance, a big stair leads to the ceremonial hall on the first floor. Above the entrance, on the triangular gable wall, there is a rose window of 3 metres in diameter. The arches and walls are decorated by the ornamental details. In reference to the Renaissance rustic decoration, they are transformed, flattened and abstracted geometric patterns. The entire interior space, from the vault to the decoration, can be seen as a compositional combination of circular and rectangular shapes. On the dominant arch, the most conspicuous position, there is a mysterious, pantheistic inscription: "I'm alive, and you have lived", which implied the immortality of the soul and the immortality of the nation. A common evaluation of this architecture is a "collection of the national symbol". This prejudiced impression was probably because of the colours with the emphasis on red, white and blue, which are related to the Czech nation and folklore (Fig. 9).

But the layering technique was not merely used for the exterior decoration but also applied, in some sense, spatially. Janák had designed an open gallery around the ceremony hall. Right down below the square-shaped ceremony hall, a similarly shaped space for incineration and all the service rooms are superimposed under the ceremony hall. In the middle of the ceremony hall, a coffin with the hydraulic platform is connected by a system of railed trolleys to the cremation area downstairs. The system could carry the corpse automatically, allowing the furnace to be completely separated from the ceremonial hall. The two spaces of "alive and lived", superimposed on one another, are physically isolated but closely connected with a mechanical system. The main service staircase is hidden behind the ceremony hall. It is indicated by a lower roof, connecting the crematorium keeper's living space, which is composed of a pair of one-story extensions located on each side. Right below it, the coffins depositary with a locker room for the staffs responds to the neat symmetry of the crematorium keeper's apartment.

The entire building is longitudinally stratified into three layers: the entrance stair in front acts as a leading space; the main hall is dominated by the triangular gable roof and surrounded by the semi-open-colonnade gallery; and the service spaces as an attachment behind the service staircase. Meanwhile, horizontally, the building is stratified into two layers of ceremony layer and furnace layer, echoing the meaning "alive and lived". If we consider the entrance stair alone with the first floor slab as a dividing line, the building is cut into two parts with distinct properties: the bottom section with the lower floor height is divided into small, enclosed rooms, mainly for cremation, while the space above is an ample open space covered by a high hanging roof.

However, the stratification of "alive and lived" does not mean the isolation of different functional parts of the building. Besides the hydraulic platform, which is a ceremonial connection, Janák had inserted two interlayers creating a series of openings that allow visual connection between different spaces. On the southern wall, an opening is set over the middle layer where the "choir" is located. It is a place for the performance. Instead of covering it by a screen in a conventional way, in order to prevent the funeral guests from seeing the musical accompaniment during the funeral ceremony, Janák designed it as an open gallery, allowing the musicians to watch the whole ceremony. In this way, the whole interior space achieves a merging of stories and rooms into contiguous and continuous space. This

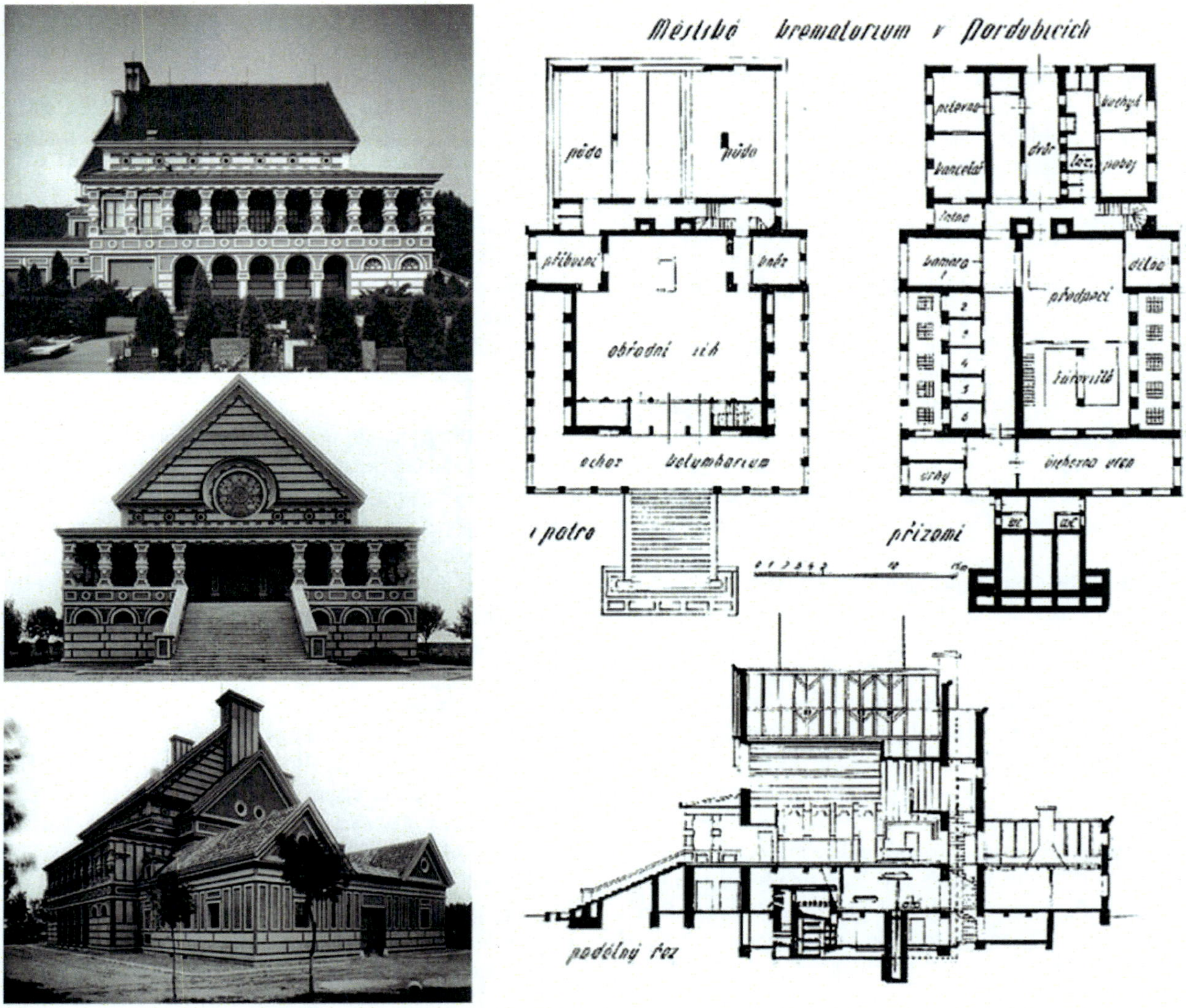

Fig. 9 Pavel Janák, Krematorium v Pardubicích (The Pardubice Crematorium), 1920–1923

connects to Loos's idea of the Raum plan: "Spatial continuity between rooms was created not by omitting walls but by piercing them with wide openings so that views were always framed..." (Risselada 1988). The connection between rooms was not only visual, as through a proscenium, but also mechanical, as the stage changes. Using the technique of stratification, Janák shaped the interface between architectural space and the observers inside, giving these spaces a theatrical quality. As such, the viewer is allowed to "journey" through the space creating a spatial continuum of the layered planes-spaces. For Janák, such a journey might involve penetration as a participant, or merely a gaze as an observer. The spaces are overlapped and different figures interpenetrate causing ambiguity or contradiction of spatial dimensions. The observer may enjoy the sensation of looking through the opening and thus perceiving "here" and "there" simultaneously.

Janák's crematorium is not designed as the temple of the dead or a pure sanctuary space. Its symbolic figure and hybridity figure with dynamic space organization give it the meaning of a "gateway" which everyone must go through. As architect Bohumil Korbel's speech at the opening ceremony: "Death, it is not the end – it is only a continuous sequence – it is only a different form of life". The influence of Eastern philosophy is confirmed by the layering of space, which can be seen as an abstraction of the phase of life, allowing the observer to situate architecture as spatiotemporal rather than planar reality.

6.2 Spatial Layering

Cubism provides an attitude. This attitude reflects Janák's attempt to use formal language to express the hidden meaning of ambiguity in the architecture. This is the most significantly expressed on the surface of the building.

In particular, the concept of layering had a considerable influence in the formation of new techniques of spatial expression of architecture. Techniques such as stratification, superimposition and multiple interpretations found in Cubist layering were incentives for Janák to investigate the application of these concepts in architecture. The concept of layering with that of space resulted in a new concept, "spatial layering" as a continuing paradigm of space in architecture to balance between appearance and concealment, observer and creator. Thus, the definition of spatial layering is the perception of interstitial spaces that advance or recede in reference to the point of view or a direction of motion.

As stated by Colin Rowe in his book "Transparency", the concept of space is fused with the concept of layers. He made the association between space and layer firstly (Rowe and Slutzky 1963). The term "Spatial layering", which means the spatial aspect, dominates the three-dimensional form. A new mode of spatial perception stemmed from the Cubist style and led Janák to engage in a similar manner in the exploration of the concept of layering in architecture. It follows the definition of "layering" with a significant change—spaces replace planes. So, it would be inappropriate to call the quality of the expression involved simply "layering".

A beautiful example is the design of the Hussite cshurch. In this case, the heavy decoration has disappeared; instead, Janák intended to deliberate the geometric factors from the spaces, to stratify the architectural mass making it lighter and slenderer, to move beyond functionalism towards minimalism (Fig. 10).

Thus, we are presented with a composition of a multi-cube-overlapped building whose form suggests the possibility of a reading of space stratified by layers. The whole mass is stratified into segments: the chapter house is sandwiched between two slabs: the entrance porch and the dormitory building. In front of the chapter, a stairway with railings leads to the entrance. The stair is not like the one at the crematorium, which gives perpendicular access right on the central axis of the whole building. Instead, it goes up parallel to the gable, veiled behind the parapet. The vast white wall of the chapter acts as its backdrop. Behind the chapel, the dwelling part with the façade rhythmically divided by the dormitory windows weakens the sense of volume, as a third layer that suppresses the depth of the whole complex. Inside the chapter, a feeling of void and solemnity comes, because of the faint light transmitted by the narrow windows through the glass bricks. The skylight is neatly divided into narrow strips, leaking a soft light into the empty chapter.

We can see that both the plan and section of the Hussite Church depict a diagonal arrow and demonstrate a framing of frames, a penetration through the successive frames of view. The diagonal is essential, as the arrows in both the plan and section refer to the same view and denote the same sequence of framed vistas. Visitors to the building engage in a theatrical voyeuristic gaze passing through the framed spaces. Hence, following the notion of stratification, that is, the sequential layering of frontal planes and spaces, phenomenal or conceptual transparency is achieved.

There is another vital corner, a slender open-work tower standing apart from the entirely rectilinear main building, thrusting a giant copper chalice up towards heaven, filigreed and airy, playing with light and shadow. It is significant that through diagonal views, we can quickly notice that by such a feature as the symbolic element of the church, the translucent corner of the airy bell tower does reduce the heaviness of the observable volume.

Like the crematorium, the church building possesses heterogeneous elements and functions that lead to a multi-layer arrangement, and to the appearance of a further feature which both buildings have in common: the 3-part-layout. But there is also difference: while the crematorium blocks present a symmetrical form of a highly ceremonial manner, the system of symmetry of the Hussite Church is broken by the horizontally held access stair with the attached porch. The whole extension wraps around the bottom of the chapter on the southwest corner. As a result, a new symmetry is formed with the diagonal where the bell tower is situated at the axis. The bell tower, slightly set back along the diagonal, dominates the two major wings of the lateral extension. Between the tower and the main body, an airy gap is formed, re-emphasized by the entrance quay and the lateral corridor, and with its height dominating the entire architectural mass.

7 Conclusion

After going through all the period styles that emerged at the beginning of the twentieth century, Janák had gradually achieved a new and personal style through the use of pure form. His formal language is composed of a geometric alphabet and a systematic grammar of the form and evolving progressively following the ever-changing period style. However, the three characters of his language—geometric, abstract and systematic—are throughout all the evolution process. And his unique techniques of composition dominate all the essential factors of architecture both philosophical

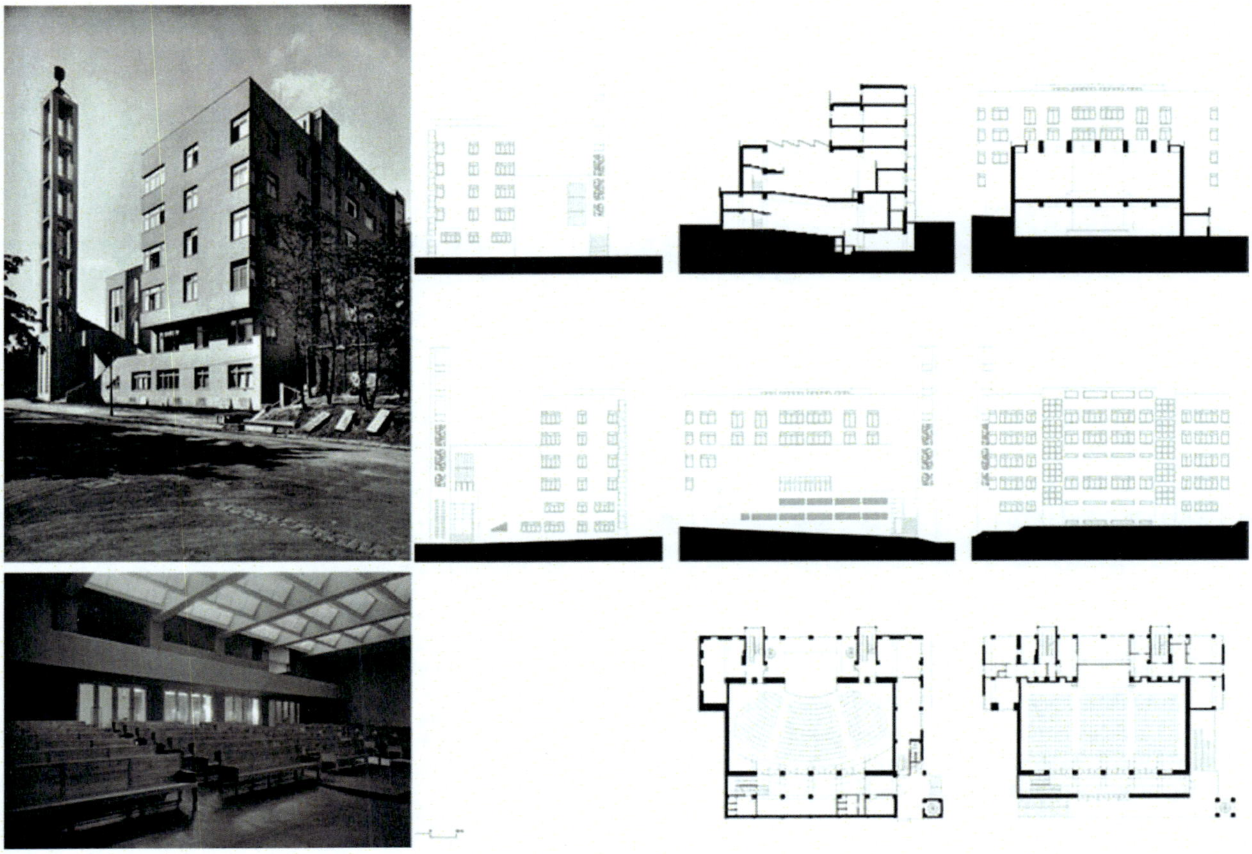

Fig. 10 Pavel Janák, Husuv Sbor (Hussite Church), Vinohrady district, Prague, 1929–1932

and aesthetic. In his reconsideration of each discipline of architecture, form and style are synonyms of starting point and goal of design, and they are connected by his unique techniques of composition.

7.1 Janák's Design Philosophy

The aesthetic urge, which Riegl saw as pivotal not only creatively possible but also artistically necessary, was always the purpose of Janák's building activity. He had always examined the architecture as an art. However, modern architecture is based on material, construction and efficiency, which should be natural, non-speculative, out of historical social life, receiving only the lifestyle of the contemporary world. So, apart from the aesthetic aspect, the problem of how to place the language of architecture into the reality ran through his entire career.

7.2 The Formal Language

Constructivism and functionalism reflect the arrangement of the world and the programme of social life rather than a new art. But modern for Janák is also meant stylistic. In order to satisfy the constructivist and functionalist conditions, he noted, "despite all the programming efforts of the past decades, the order of architecture is not a pure structure but the form system" (Janák 1940).

He absorbed the essence of all the important tendencies of the first half of the nineteenth century to nourish his language. At the same time, with his unique techniques, he unified them and fit them into his design philosophy. During the evolution process, Janák did not accept constructivism and functionalism as a political-ideological social programme. Instead, he has them translated in his own way with an aesthetical and stylistic language.

7.3 The Techniques

This essay is about Pavel Janák's design techniques. The techniques are extracted from the examples chosen in chronological order. In this way, the essay has tried to explain how the design method of Pavel Janák has transformed from "layering" to "spatial layering" in his practical work.

The method used has been first, to study the philosophical background of the architect through inspecting how he perceived the objective world and architectural design. Secondly, through examining the development of his personal design method, the paper has analysed how Janák turned the abstraction matter and kinaesthetic perception of Cubist painting as a reference and introduced the geometric expressions and layers to architecture. Thirdly, the essay has inspected the process by which the geometric shapes grew into a system of forms, and the planar layers on the paintings had approached three dimensions when they entered the realm of architecture.

7.4 The Meaning: Abstract and Empathy: An Observer's Perception-Based Creative Method Approaches from Two Dimensions to the Three Dimensions

These techniques call for the participation of the mind as the main contributor to the perception of the artwork. In the presence of spatial layering, the interplay between opacity and transparency invokes the feeling of ambiguity and hidden meaning. Through opacity, it incites curiosity, and through transparency, it eludes identification. The interaction allows the spectator to ruminate and interpret the space differently from others, which results in a personalized perceptual image of it.

The approach of Janák for design techniques has moved the contradiction and complexity of architecture form from an aesthetic pursuit to the connection between the creator and the observer. In this way, he was trying to erase rhetoric, liberates the language from the historical burdens, in order to purify form, make it neutral and take it beyond art.

References

Herbenová, O., & Šlapeta, V. (1984). *Pavel Janák 1882–1956: Architektur und Kunstgewerbe*. Prague: Umeleckoprumyslove museum.

Janák, P. (1940). Čtyřicet let nové architektury za námi—pohled zpět (40 years of modern architecture behind us—Retrospect). *Architecture, The Work of Czech Architects 1900–1940* (Suppl. of the Czechoslovak Architecture magazine), *5*, 8.

Janák, P. (1996a). The prism and the Pyramid. In A. Vegesack (Ed.), *Czech cubism* (pp. 15–16). Princeton: Princeton Architectural Press.

Janák, P. (1996b). Renewal of the facade. In A. Vegesack (Ed.), *Czech cubism* (pp. 16–17). Princeton: Princeton Architectural Press.

Kahn, L. I. (1944). Monumentality. In P. Zucker (Ed.), *New architecture and city planning: A symposium* (pp. 577–588). New York: Philosophical Library.

Kiesling, N. (2011). *Pavel Janák*. Prague: Arbor vitae.

Le Corbusier, & Goodman, J. (Trans. into Eng.). (2001). After cubism. In A. Ozenfant (Ed.), *L'Esprit Nouveau: Purism in Paris, 1918–1925* (pp. 129–168). Los Angeles: County Museum of Art.

Margolius, I. (1979). *Cubism in architecture and the applied arts: Bohemia and France 1910–1914*. London: David & Charles Inc.

Murray, Ž. (1990). Between plane and the space. In *Sources of cubist architecture in Bohemia: The theories of Pavel Janák (Manuscript)* (p. 31). Montreal.

Riegl, A., & Kain, E. (Transl. into Eng.). (1992). *Problems of style: Foundations for a history of ornament*. Princeton: Princeton University Press.

Risselada, M. (Ed.). (1988). *Raumplan versus Plan libre: Adolf Loos and Le Corbusier 1919–1930*. New York: Rizzoli.

Rowe, C., & Slutzky, R. (1963). Transparency: Literal and phenomenal. *Perspecta, 8*(45), 54.

Ševčík, J. (1986). Šest Kacířských Poznámek K Osobnosti a Dílu Pavla Janáka (Six heretical remarks on the life and work of Pavel Janák). In *Acta UPM XIX Commendations 4, Pavel Janák: Selected Author's Papers and Contributions from the Seminar of the 100th Anniversary of the Architect's Birth*, Uměleckoprůmyslové museum, Praha (pp. 108–109).

Švácha, R., & Büchler, A. (Trans. into Eng.). (1995). *The architecture of new Prague 1895–1945* (pp. 124–125). Cambridge: The MIT Press.

Use of Natural Light for Catholic Sacred Architecture: Technological Strategies and Symbolic Values

Rossella Franchino, Caterina Frettoloso, and Francesca Muzzillo

Abstract

Natural light can filter and penetrate into the spaces through openings and cuts in the envelope going from time to highlight spaces, revealing details and creating atmospheres. The control and diffusion of natural light are strategic to ensure the right light intensity and are, in fact, critical points to manage when it comes to daylighting. One of the first metaphors that we use for an instant understanding of the concept of "holiness" is "light". Since ancient times, sacred architecture has tried to emphasize this intuitive metaphor through a design capable to underline into spaces of darkness the entering of light rays. An emblematic case of creating strategic effect of interplay between light and shadow is possible to be found in the gothic cathedrals. Even in modern and contemporary architecture, a typical design strategy for the project of a mystical atmosphere is guiding people attention with the help of light as a symbolic direction. This strategy let people forget the outside world moving towards a mystical concentration. The work intends to propose some reflections on the role that natural light has in the design of catholic sacred architecture both from the formal-expressive and from the functional-technological point of view through the analysis of some emblematic cases. This role is, moreover, particularly strategic also in relation to the correct lighting of the art world very often present and necessary for the visual enjoyment of the same. In this regards is very important how light sources are set both for the visual enjoyment of the art works and when their valorization is also required. Among the aspects to be considered also, any damage that light can bring to the art works as incorrect illumination can lead to degradation. The work, therefore, will deal with the performance aspects of natural lighting that, in line with the specific functions of the sacred architecture, will concern not only comfort in its various forms but also the relationship between lighting strategy, materials used and technological solution adopted.

Keywords

Natural light • Formal-expressive value • Functional-technological quality

1 Introduction

The control of natural light within buildings is a theme that invests more areas of research, from energy saving to comfort in its different variations. Appropriate use of natural light means reasoning not only in terms of the amount of light intake but above all in light performance (Gherri 2013).

Natural light variability is a particularly significant feature: while it changes according to the weather, season and micro-climatic conditions, on the other, it helps to strengthen the relationship between those who enjoy the internal environment and the outside world also from the point of view of temporal perception.

The project of natural light in sacred architectures is enriched by further ideas for reflection mainly attributable to the symbolic and technological-functional dimension of the project: from metaphor to instrument of space use.

The work, therefore, with the aim of putting in place daylighting strategies, technological solutions and used materials, proposes a series of reflections starting from the selection of emblematic cases of sacred architectures that use passive systems of daylighting.

R. Franchino · C. Frettoloso (✉) · F. Muzzillo
Department of Architecure and Industrial Design, University of Campania "Luigi Vanvitelli", Aversa, Italy
e-mail: caterina.frettoloso@unicampania.it

R. Franchino
e-mail: rossella.franchino@unicampania.it

F. Muzzillo
e-mail: francesca.muzzillo@unicampania.it

C. Gambardella et al. (eds.), *Advances in Utopian Studies and Sacred Architecture*, Advances in Science, Technology & Innovation,
https://doi.org/10.1007/978-3-030-50765-7_14

Daylighting capability of creating a mystical atmosphere inside a building is strictly connected to the shape and the amounts of openings in the envelope. The way in which natural light succeeds in moving people emotions is consequently dependent on design criteria.

Finally, the considerations on the integration of these catch/spread strategies and artificial light respond to the need to have a more complete picture of both the critical issues and the opportunities related to the theme of natural light in the sacred architectures.

2 Light as an Architectural Methaphor

When we think about "holiness", our mind has instinctively the impression of something which is illuminated by natural light. It is probably this impression that has had over time an influence on the use of daylighting in the design of holy buildings (Kieckhefer 2004).

Since ancient times, sacred architecture has tried to emphasize this intuitive metaphor through a design capable of underlining into spaces of darkness the entering of light rays. The concentration of luminous effects into a dark environment, acting in opposition, helps us to remain inward focused towards our spiritual dimension, as being in a mystical atmosphere even in places like sacred buildings, which in the majority of the cases are intended to host a large number of people, even if being in a crowd is not the best situation for concentration.

We specifically wonder how design thinking would frame a strategy in order to create these light effects. And historical references could effectively help us.

Light first of all ties-in with coherent architectural fundamental characteristics, like longitudinal and horizontal section, rhyme sequence of support elements, proportion among parts into a resulting constructive framework. Moreover, if we focus on elements which are the effective medium of illumination vision like windows, skylights, light chimneys, clerestory and loopholes openings, all these elements contribute to design and preview potentialities of light effects. In order to redirect light rays, while entering inside the building, shading devices can then modify sunlight interior reflection also in relation to the different reflectance of various materials, predicting during design phase the daylight distribution inside a close space.

An emblematic case of creating strategic effects of interplay between light and shadow is possible to be found in the gothic cathedrals with the increased area of openings in the facades and large windows with stained glass works which gave new accent of colour to the rays. As Georges Duby has well explained, at the basis of the design for the Basilique de Saint Denis, for example, was the philosophical idea that everything was born in the light and that sacred architecture should directly derive from this idea (Duby 1981).

In Baroque architecture, according to Nick Baker and Koen Steemers, a typical design strategy for the project of a mystical atmosphere consisted not only in guiding people attention with the help of light as a symbolic direction, but also letting people forget the outside world into a spiritual concentration given by light sculpture (Baker and Steemers 2002). Let us think about Borromini's "San Carlo alle Quattro Fontane" and it becomes clear how undulations and curves are capable to design a dramatic contrast between shadow and light, increasing the perception of a complexity geometric configuration and stressing the interaction of different geometric curvilinear elements. The result is a dramatic undulation of the light distribution into a detailed sculptured interior space.

Borromini's capability in sculpturing interior environments with light could provide the appraisal of architecture in terms of metaphoric understanding of signs and symbols among spatial appearances. Consequently, space–time metaphoric mapping in people perception could be experimentally analyzed through the lens of mapped patterns of enlightened and obscured areas.

The enlargement of temporal dimension is obviously related to the perception of space in a spiritual experience, so light could guide new people's feelings with regards to the place in which they are. It is in the essence of the spiritual dimension the translation of the personal reality towards a different eternal dimension. And this is just the goal towards which the design of a holy space naturally moves.

A complexity of previewed futurity and faraway past is typical of the holy experience which is in itself over the limit of present and so holy spaces design should move in this dimension. In order to extend the perception of one's life, space–time architectonic metaphors are useful and it is just in this kind of traditional design experiences that holy contemporary building should be designed, with a composition of enlightened sections between corresponding darkness areas.

The best interpretation of light almost as a real material of architecture, for creating an atmospheric space, seems wonderfully to be found in the words of Juhani Pallasmaa: "*light is the most subtle of the means of architecture; it can express joy and happiness, as well as melancholy and sorrow. We do not see light directly, and architecture must somehow materialize light by means of reflecting surfaces, in order to bring light to our attention*" (Amundsen 2018, p. 2).

3 Natural Light and "Functional" Comfort

"*Daylight can be employed as a diffuse, soft light without shadows or as a strong directed light that casts sharp shadows. Daylight determines how a sacred space and the points of liturgical activities are perceived. The amount and direction of light can be used to focus orientation during the service or to create a place for silent prayer. Daylight can lend sacred spaces a spiritual atmosphere*" (Kreuz 2008, p. 60). The chapel of Notre-Dame du Haut in Ronchamp, the church of the Light by Tadao Ando, the parish church in Riola di Vergato in Bologna province (an Italian work signed by Alvar Alto), are particularly significant examples of the wise and functional use of natural light in sacred environments. Architectures in which natural light is an integral and supporting part of the project certainly combine the power, the symbolic and emotional value of light but also suggest reflections mainly due to its functional role. Analyzing the light component in these sacred architectures, it is possible to introduce a new distinction of comfort. We could define it as the "functional" comfort realized when natural light allows the user to enjoy the liturgical experience or to visit the sacred place in a pleasant way only by illuminating more or less certain elements without causing inconveniences but, above all, emphasizing the spiritual dimension of the religious environment in relation to its specific functions. In a sacred environment, natural light should guarantee performances linked to usability and visual comfort in relation to liturgical needs, both for users and for the presbytery space, or for other spaces functional to ordinary and non-ordinary liturgy. In order to achieve the right natural lighting conditions, it is important not only to allow in the sunlight but also to manage the daytime evolution (and over the seasons) in terms, above all, of light intensity.

Inside the chapel of Notre-Dame du Haut in Ronchamp, the feeling of reflection and silence is evoked by Le Corbusier through the articulated system of openings marked by the Modulor scheme, creating "*a series of spatial dramas in which light animates the space by creating different moods at different times of the day and in different seasons*" (Ramzy 2013, p. 223). The light touches the deep splay of the windows, whose dimensions, depth and inclinations allow the sun rays to enter according to different angles and "*the coloured glass in the splayed windows softens the incoming light and creates coloured shadows in delicate hues on the roughly plastered surface*" (Ramzy 2013, p. 223). The light slides along the walls from the rippled plaster covered with white lime milk and every surface vibrates depending on how it is reached by the light realizing or not bright contrasts (Fig. 1). "*Artistic treatment of light, devices with a shielding or illuminating function are distinguished in Le Corbusier's projects, becoming tools for calibration and modulation of light, which is cleverly concealed or integrated into the shape built and distributed like a veil over surfaces*" (Gherri 2013, p. 46).

A soft and diffused light, with the exception of the cross that lays on the wall on the short side, characterizes the Church of the Light by Tadao Ando: a cut in the wall allows the light to penetrate as long as it is subjected to rigorous constraints, and a radius cuts darkness (Dal Co 1994). What interests to Tadao Ando are the links between light and darkness: the light becomes bright only in presence of a very dark background. The church, like many of its architecture, is therefore a "*eulogy to the shadow*" and to privacy. To emphasize the entrance of the light coming from the cross, Ando uses a 16 mm thick glass directly embedded in the concrete wall that measures about 20 cm. That concrete is not rough like the one used by Le Corbusier but smooth and it is used in a refined way: it dematerializes in the sunlight thanks to its dynamic nature according to weather conditions and time. The only other material used besides the light and the concrete of the walls is the raw wood of the floor and benches: this essentiality of the materials pushes Ando to research an increasingly pure light able to give quality to the architectural space (Dal Co 1994).

The feeling of peace and tranquility that is felt by spending some time inside the church in Riola di Vergato (BO) signed by Alvar Aalto (project of 1966) but built after his death is the result of an everlasting reflected light that comes from above. The Finnish architect devoted great attention to the theme of light as shown by the numerous

Fig. 1 Chapel of Notre-Dame du Haut in Ronchamp by Le Corbusier (picture by the author): natural light and bright contrasts

models made during the design phase in order to understand and control light intensity. This architecture, like other buildings by Aalto, in which light flows over the surfaces making them, in turn, luminous diffusers, is often conceived as large amplifiers of light that in the specific case arrive from the north through curved skylights (Fig. 2). The dynamic shape of natural light that characterizes this sacred space, whose brightness varies according to the time and micro-climatic conditions, is strengthened also thanks to the wise use of artificial light. In addition to the brightness variation during the day, the church is characterized by a luminous progression in the altar thanks to further vertical openings that intensify the luminous field. Even artificial light helps to construct a certain luminous intensity reflecting the functions of natural light. In fact, inside the skylights is also placed the artificial lighting system that allows to integrate natural light, without changing the brightness conditions of the nave.

A massive rammed earth wall delimits the interior of the liturgical hall lit from above by the Chapel of Reconciliation in Berlin opened in 2000. The natural light is intercepted by the external diaphragm, made of wooden strips, and illuminates the ambulatory by projecting an artificial sign on the waxed concrete floor and on the blind wall, accentuated by the striped pattern due to the linen fibres compacted in successive layers (Fig. 3) (Cianfarani 2012). The luminous contrast that is created between the two areas of the Chapel guides and prepares the visitor for the different moments of the spiritual experience that is going to live closely linked to reconciliation and memory. The permeability of the envelope, conceived by the architects Reitermann and Sassenroth as a sunshade, creates a close relationship with the context: from the outside and from a certain distance, in fact, we perceive a compact and austere volume but, as we approach, the view changes and the envelope becomes dematerialized, allowing itself to be crossed by glances as well as by natural light (Fig. 4).

Two corollary are linked to these examples and considerations, the first is related to the binomial "light and atmosphere". In fact, "*hardly the light can be considered in architecture a stand alone element. Besides conditioning the material qualities of the environment, it structure that the atmospheric sense of being immersed or of how one experiences peripherally of a spatial environment that is multisensorial in its different dimensions*" (Mallgrave 2015, p. 206). The sacred architectures examined, through diversified strategies, use light as a tool not only to accentuate or not spatial aspects but, above all, to induce a specific mood

Fig. 2 Church in Riola di Vergato (BO) by Alvar Aalto (picture by the autor)

Fig. 3 Chapel of Reconciliation in Berlin by Rudolf Reitermann and Peter Sassenroth: the ambulatory (picture by the autor)

consistently with the religious thought that the church symbolizes. Light, whether diffuse or punctual, whether it arrives directly or through a more articulated path, guides the user and predisposes him to live a certain spiritual and perceptual experience, "*(…) daylight is still used for spiritual influences more than functional needs in the Modern Period. In these spaces, the harmony of daylighting depends on the ambiance to be attained. Consequently, the spatial organization of the church is directly related to the lighting design*" (Fitoz and Berkin 2007, pp. 8–9).

The second corollary regards the relation "light and constructive strategies". To emphasize the light input, in fact, it is necessary to work on the thickness of the cuts, on the use of diaphragms, on the grain of the finishes. But you also need a clever use of construction technologies to take advantage of every opportunity to capture and channel natural light in order to build real shapes and images. In the Sacristy of the convent of La Tourette (1957—Lyon), defined by the artist and fiolophile Lee Ufan "place of dense, rough and fantastic spirituality", Le Courbusier manages to

Fig. 4 Chapel of Reconciliation in Berlin by Rudolf Reitermann and Peter Sassenroth: the wood envelope (picture by the autor)

create a symbolic effect of light shaping a crucifix skilfully using the formwork. Through an "imperfection" of the wall and thanks to the mitraillettes a lumière, he manages to vibrate the concrete by charging it with a strong symbolic value that contributes to realize that "unspeakable" space mentioned by the famous designer.

4 The Natural Light Technological Use

Worship places lighting is a very challenging theme. Churches not only are home of important artworks but also hold liturgical celebrations and cultural events consonant with the sacredness of the place. In this regard, these places are not just like any other building, but the already mentioned sacredness of the place must be taken into account above all. Worship places do not require constant lighting: it should be differentiated between different parts. On the one hand, the faithful zone must have a sufficient level of illumination in order to allow reading. On the other hand, the presbytery must have a higher level of lighting. Furthermore, the lighting of the exhibited artworks must guarantee their valorisation.

The design of a church's lighting system is therefore a particularly complex intervention. In this regard, the Pastoral Note of the 1996 CEI recommend about the lighting that: "….*We must take into account that the plants are inserted as elements of novelty in a context that did not provide them, and it is therefore necessary to carefully study their physical, formal and functional insertion in the building in order to meet the needs of the celebrations that take place in the church and those of the works it contains… As for the lighting system … it is recommended to take care of its relationship with natural light in order to maintain its characteristics, which vary greatly depending on the eras and architectures. The artificial lighting system is designed in a way that considers, first of all, faithful needs linked with the liturgical celebration, secondly, the conservation needs of the works and the needs of visitors and tourists, avoiding excessive light. Given the delicacy of the problem, it is necessary that the artificial lighting project is studied by specialists of the sector together with experts in liturgy*" (Conferenza Episcopale Italiana 1996, pp. 42–43).

In conclusion, churches lighting other than taking necessarily into account the architectural factors of the sacred building must be sensitive to its religious use. In addition to ensuring the visual enjoyment of places of worship, the lighting must also ensure to highlight any artworks present, trying to exploit as much as possible the natural light possibly in mix with the artificial one when the natural one is not sufficient. Furthermore, the lighting must be chosen also with particular attention to the energy consumption and polluting emissions.

"*The daylighting of churches has usually been dealt with in a satisfactory manner, because of the economic imperative. A church is not occupied for many hours in the day but is visited more often. Therefore, the daylighting must be sufficient for general use, perhaps backed up by artificial lighting when the church is in use*" (Derek 2004, p. 66).

From a technological point of view, it is therefore necessary to manage the lighting of sacred places by mixing natural light with artificial lighting and diversifying the luminous atmosphere for each specific activity.

The first important consideration concerns the type of light used and therefore the choice of natural light or artificial light or a mix of both. Other important factors for proper churches lighting design are the choice of its source and the distance from the light source. Furthermore, the arrangement of the light sources is very important, both to guarantee the right visual enjoyment and for artworks valorisation.

The quality of the light, in fact, is crucial for artworks preservation. The optical radiation affecting the illuminated surfaces is the main cause of their deterioration and depends on the surface quality and its vulnerability to radiation.

In this regard, the following are reported some requirements (Amirante et al. 2013) that must be satisfied for the protection of artworks from light sources:

- adequate lighting of the exposed work;
- uniformity of illumination of the exposed work;
- colour rendering suited to reveal the colour, except the works in monochromatic material;
- light tone, measured by "colour temperature" neared to the prevailing colour of the work illuminated;
- dazzle limitation;
- contrast rendering avoiding the formation of lighting spots on very reflecting surfaces, like those ones of the oil paintings;
- relief effect for three-dimensional objects, such as sculptures and artworks or precious objects;
- luminance distribution, with a higher lighting of the works from damaging lighting radiation.

The choices relating to lighting must necessarily be customized, moreover, also in relation to the type of religious building. In the end, the result is to obtain a balanced light both for historical buildings and the contemporary ones. Among the aspects to be considered, moreover, there are also those related to any damage that the light can bring to the works: an incorrect lighting can cause degradation phenomena. In this regard, the following are reported some criteria of protection from light sources within churches:

- value of the incident radiant flux and related to the illumination;
- exposure duration to light;
- presence of ultraviolet radiation in the spectrum of incident radiation;
- radiant absorption coefficient in the visible surface area.

A high incident radiant flux creates an excessive increase in temperature to the irradiated surface area, especially if the radiant absorption coefficient is high. Such an increase in temperature causes a high heat transmission in the material that can be damaged. Damage also increases as the exposure time increases.

5 Conclusion

This work, also through case studies, deepened the topic of the natural light use in the design of sacred architecture. In particular, the proposed strategies have been addressed not only in relation to optimal technological solutions, but have been related to the important symbolic value and specific functions of sacred architecture.

The centrality of the evocative and spiritual power of natural light, in fact, is linked with the need to ensure and control the right light intensity. At same time, it is also important to think in terms of the potentiality of natural lighting for energy-saving strategies (Lo Verso 2006). It is strategic, mostly, "*in the context of climate change, [where] the use of daylight should also be considered in terms of energy efficiency. Sacred spaces in which services are celebrated in artificially lit surroundings fail to exploit not only the design possibilities but also the energy of sunlight*" (Kreuz 2008, p. 60).

Finally, we could stress the fact that in any case, there is a strict relationship between an adopted technology for sustainable daylight, on the one hand, and a particular spatial organization, on the other one. And the use of natural light in holy spaces confirms our convincement, as it is just in these places that light endorses itself as a material of architecture with a strong relevance in creative design.

In Catholic Church, moving in the light of Christ has the sense of renewing our life, so design tries to emphasize this movement of reunion giving a direction.

Acknowledgements The paper is the result of a common reflection by the authors. Nevertheless, the paragraph: "Light as an architectural methaphor" is edited by Francesca Muzzillo, the paragraph: "Natural light and *functional* comfort" is edited by Caterina Frettoloso and the paragraph: "The natural light technological use" is edited by Rossella Franchino.

References

Amirante, M. I., Franchino, R., & Frettoloso, C. (2013). From the industrial building to the lab museum: fruition and environmental-energetic issues. *Society, Integration, Education, III,* 17–27.

Amundsen, M. (2018) "Q&A with Juhani Pallasmaa on Architecture, Aesthetics of Atmospheres and the Passage of Time". Available at: https://www.researchgate.net/publication/331972704_Ambiances_QA_with_Juhani_Pallasmaa_on_Architecture_Aesthetics_of_Atmospheres_and_the_Passage_of_Time. Accessed April 30, 2019.

Baker, N., & Steemers, K. (2002). *Daylight design of buildings: A handbook for architects and engineers*. London: James & James.

Cianfarani, F. (2012, January–March 3). Rudolf Reitermann e Peter Sassenroth, La Cappella della Riconciliazione a Berlino. *Boundaries*, 122–125.

Conferenza Episcopale Italiana - Commissione Episcopale per la Liturgia. (1996). L'adeguamento delle chiese secondo la riforma liturgica. Nota Pastorale [online]. Available at https://bce.chiesacattolica.it/1996/05/31/ladeguamento-delle-chiese-secondo-la-riforma-liturgica/. Accessed April 16, 2019.

Dal Co, F. (1994) *Tadao Ando. Le opere, gli scritti, la critica.* Milano: Electa.

Derek, P. (2004). *Daylighting: Natural light in architecture*. Oxford: Architectural Press, Elsevier.

Duby, G. (1981). *The age of the cathedrals: Art and society 980-1420.* Chicago: University of Chicago Press.

Fitoz, I., & Berkin, G. (2007). Space, light and beliefs, the use of daylighting in churches and mosques [online]. Available at https://www.researchgate.net/publication/256715298_Space_Light_Beliefs_The_Use_of_Daylighting_in_Churches_and_Mosques. Accessed March 7, 2019.

Gherri, B. (2013). *Daylight assessment. Il ruolo della luce naturale nella definizione dello spazio architettonico e protocolli di calcolo*. Milano: Franco Angeli.

Kieckhefer, R. (2004). *Theology in stone: Church architecture from Byzantium to Berkeley*. Oxford: Oxford University Press.

Kreuz, E. M. (2008). Light in sacred buildings. In R. Stegers (Eds.), *Sacred buildings. Design manuals.* Basel: Birkhäuser.

Lo Verso, V. R. M. (2006). La luce naturale come materiale per l'architettura degli ambienti confinati [online]. Available at https://www.researchgate.net/publication/237028397_La_luce_naturale_come_materiale_per_l'architettura_degli_ambienti_confinati. Accessed March 19, 2019.

Mallgrave, H. F. (2015). *L'empatia degli spazi*. Milano: Raffaello Cortina Editore.

Ramzy, N. (2013). Perceptual dimension of interior daylight in sacred architecture: Analytical study of the lighting programs in five sacred buildings of different styles. *International Journal of Architecture, Engineering and Construction,* 2(4), 219–233.

The Language of Built Space: Sacred and Religious Dimensions

Not merely perceived as a conversational phenomenon, religious transformations continue to leave their mark on urban space and the built environment. This part tackles the changing relationship between contemporary religious architecture and its relation to urban space: from the positioning of buildings to the way urban space is structured.

In "A Unique Space for Different Religions" a reflection on the capacity of modern and contemporary architecture to generate spaces of worship open to all the different religions and community is made. A long-lasting work of research and a design experimentation, carried on during Architecture of Interiors Studios in Politecnico di Milano, explore the possible coexistences, through the material forms of architecture. The argument is important in relation to the current metropolitan situation where different communities inhabit urban spaces which has been able to welcome diversities avoiding any form of conflict. The essay analyses a series of key-studies in Europe, US, and Africa signed by different authors from the post-war to the present situation.

The term "religion", in most European languages, is associated with two controversial meanings through its Latin roots: to knot together the human and divine, on the one hand, and to collect around a mode, chosen to guarantee and verify an identity, on the other. The author of the chapter "Gathering Differences. Sacred Spaces and Diversities", starts from this basic assumption to say that spaces hosting rites live on this ambiguity. The first part of the chapter analyses two case studies recently designed in the North of Europe, which represent interesting examples of an emerging typology for religious spaces in the contemporary city. The second part focuses on a design research carried on with a group of students of a Master Degree of the Politecnico of Milan trying to deepen the research on those hybrid spaces for worship. Other case studies in following chapters are used, such as the Solomon Temple, as relevant reference points for religious building throughout the years, both from a conceptual and egalitarian point of view.

In the chapter "The Convent of La Tourette Between Communitarian Utopia and Religion", the author shows a comprehensive study on the topic in order to draw useful conclusions on Le Corbusier architecture in the Convent of La Tourette. Objective of the work is to demonstrate how the convent is an exemplary case of realized utopia, with all the contradictions inherent in it: according the author La Tourette remains "a beneficiary of his ill-fated urban dreaming" with its open architecture, able to transform itself over time, according to the needs of its users.

In the final chapter titled "Sacred Spaces and Virtual Design. Toward a Digital Prayer Interreligious Architecture", the authors highlight the relationships between the theological issues of Abrahamic religions and a new typology of architectural design. In creating a virtual museum, the authors used software to obtain a database on the user's perception of the pattern of hybrid interreligious space types. Following a thorough study of the steps, the authors stated that they produced a series of useful indicators to map the types and compositional models of hybrid prayer spaces using three Abrahamic confessions.

A Unique Space for Different Religions?

Marta Averna

Abstract

Architecture makes room to life; it hosts the gestures of a single human being or of a group and translates them in material forms, made of space, margins, and equipment. Architecture's hosting ability is clarified and tested in its limits, by its physical features. This ability is stressed in contemporary society, made of moving communities, gathered more around a shared need (study, work, research of wellness conditions) than around a rooted tradition. In them identity-making features, like the beliefs on divine and universal, can be forced to coexist. Men's relationship with transcendent is structured on successive levels: the meditation, silent and private, the prayer, also in small groups, and the rite, collective and organized. If the first is independent from credo, and sharable in quiet rooms, the two following are much more identity related, and they are absolutely difficult to be placed side-by-side. Traditional communities have specific places and modes for all these levels, contrary to newly built communities, looking for new spaces, able not to divide different credo, but to relate them maintaining their differences. Detailed design, the closer to the gestures of dwellers, clarifies the call of this spaces, in the attention paid to materials and finishing, demonstration of their dignity, in the design of equipment, able to suggest with small variations mutual superimpositions or divisions, in the relationship with light, hidden, or direct, but always sign of divine, in the presence of the work of art, often abstract or defined by the light, natural, or artificial, and by the openings on the outer, surrounding world. A long-lasting work of research, on case studies and their theoretical basis, and a design experimentation, carried on during Architecture of Interiors Studios in Politecnico di Milano, explore the possible coexistences, through the material forms of architecture.

M. Averna (✉)
Dastu Department, Politecnico of Milan, Milan, Italy
e-mail: marta.averna@polimi.it

Keywords

Interfaith • Chapel • Light • Interior architecture

1 A Declaration of Intents: MIT Chapel and Interfaith Spaces

The MIT Campus in Boston, built along the banks of the Charles River, is characterized by a stratified fabric with a green square surrounded by trees. On one side of the square, a cylindrical volume of bricks in a circular water basin interrupts the orthogonal grid of the structure (Saarinen and Saarinen 1968).[1] The irregular bricked face rises over the water with a series of flounces, like a gown, leaving the reinforced concrete structure in sight, while a translucent tunnel passes through the basin, creating the entrance (Fig. 10.1). Double light-colored wooden doors along the sides allow to pass from the wide, luminous, treed space of the square to an elongated corridor, closed by translucent or colored large windows, progressively moving away from the whirl of everyday activities. A second series of wooden doors leads to a centric continuous space, closed by an undulating brick face, almost resembling a cloak. There are no windows: The light reverberates in the lower part of the structure through hidden openings that capture the light arriving from the water basin. In the upper part, instead, the light is refracted from the specks of a large sculpture by Harry Bertoia (Kompass Nelson 1970),[2] a *mobile*, onto which it arrives passing through the covering skylight (Fig. 10.2). It is a warm and welcoming space that envelopes those who enter, separating them from their everyday life,

[1]Eero Saarinen, *MIT Chapel*, MIT Campus, Cambridge, Boston, Massachussets, dedicated in 1955 (Saarinen and Saarinen 1968).

[2]Harry Bertoia, *Floating Screen*, Cambridge, Boston, Massachussets, 1955 (Kompass Nelson 1970).

C. Gambardella et al. (eds.), *Advances in Utopian Studies and Sacred Architecture*,
Advances in Science, Technology & Innovation,
https://doi.org/10.1007/978-3-030-50765-7_15

Fig. 10.1 The outer view of the Chapel shows the irregular brick façade rising over the round water pool, letting the concrete structure on view. The cylindric volume is completed by the aluminum bell tower, realized by Theodore Roszak (Leonel Ponce, www.bit.ly/2xaKN6j)

Fig. 10.2 The core of the Chapel is highlighted by a reredos, a mobile designed by Harry Bertoia. It is made of light metal specks, hanged to the ceiling, moving in the air and playing with the light coming from the skylight standing above (Gunnar Klack, www.bit.ly/2xbrwS9)

although keeping them connected. It is a space that allows to look at the sky, without losing contact with the world.

MIT's Chapel was designed in 1955 by Eero Saarinen. It is not a confessional, and it is open to all the beliefs present on Campus. It is currently managed by 25 chaplains representing 40 groups of students, gathered in a religious belief ("Who We Are | Division of Student Life" 2020). A specific institution, the ICEO, Institute Community and Equity Office, takes care of diversity and equity, with the mission to advance a respectful and caring community that embraces diversity and empowers everyone to learn and do their best at MIT ("Institute Community & Equity Office" 2020).

Its architecture and interiors, although allowing everybody to feel welcomed and free to profess their deepest beliefs in total freedom, are not white and silent with the idea to welcome differences. On the contrary, they are connoted in a way that in every language and modality talk about the relationship with what is sacred and transcendent.

Aim of this paper is exploring, through the analysis and the description of contemporary case studies, realized from the second half of last century, the features of architecture and interiors devoted to host people different in their deepest beliefs. Two ideas subtend this text: The first is that architecture has the social responsibility to face the questions posed by contemporaneity, and among them diversity, trying to understand, interpret and make them comprehensible, and the second is that the form is the mean through which this interpretation and donation of sense are told. The paper will first explore the relationship between form and meaning in the specific and complex case of beliefs, and the variations caused to architecture and its interiors by those in the number of users (single person, small group, big group) and in the quality of the gestures carried on (meditation, prayer, rite). Then it will analyze some realized case studies to explore some significant material features: the relations of these spaces with the surrounding environment (insertion/separation), the role of the light and of the works of art, and the possibilities given by the distribution and the flexibility of rooms and spaces.

2 Diversity and Proximity in the Forms of Architecture

The architectural forms play a fundamental role in defining and guaranteeing this sense of freedom to welcome believers. In fact, these forms have the inherent ability to suggest a use, a possible behavior: *in a small room one does not say what one would in a large room*, as Louis Kahn stated in 1971 (Gattamorta et al. 1996).[3] As in every art, the form of things carries their meaning, in a substantial identity between signifier and significant (Ottolini 1997).[4]

Louis Kahn defined architecture as follows: *Architecture is the thoughtful making of spaces. It is the creating of spaces that evoke a feeling of appropriate use* (Kahn 1960). The adjective used, *appropriate*, perfectly describes the ability to make everyone feel at home and have a fruitful reaction toward the world; in other words, to dwell it, recognizing and understanding its own rules and way of working (Norberg-Schulz 1985).

The ability to welcome believers is a challenge for all heterogeneous communities: The presence of different experiences and traditions makes communication and cohabitation complex. Over the history, the closeness between different ways of meaning life often caused tensions, and even wars: We can also remember situations in which the cohabitation was researched and enhanced, as in the so-called *Convivencia* in Spain. In Cordoba, among the IX and the X Century, Christian and Jews were involved in the intellectual life and in the Muslim royal court (Menocal 2002).

The occasions of cohabitation increasingly grew up during the XX Century, because of migrations and of major possibilities to move through the planet for different reasons. They concern both everyday life (families migrating for work reasons, aiming at establishing new roots in a new country) and unusual and temporary occasions (people traveling, or moving for study reasons, taking part to temporary events related to sport, trade, politics, or even obliged to a long stay in a hospital, or a prison) (Rizzi 2019).

Complexity can reach such high levels to make cohabitation impossible, if one of the cultures deems itself overriding, more correct, or dominant compared to the others; or it can make cohabitation difficult, if the non-shared traits are deeply related to identity, such as in the case of religion, and if the will and ability to dialog are scarce or not much cultivated. It is necessary to establish a dialog, codifying a grammar (Sorrentino 2007) that allows heterogeneities to be open and willing to see and understand what is new and different, overcoming places and ways codified by history in order to weave a fabric of fruitful relationships also in communal life.

As mentioned, the characteristics of the place in which dialog takes place play a fundamental role in fostering said dialog. Indeed, this place becomes the symbol of communication, a bridge connecting differences, in cultures, or in worlds, as in the case of man and the divine. All the elements of architecture act a role in this play: that is *margins* (horizontal or vertical, natural or built by men, with their features related to construction and finishing materials and techniques), giving a shape to the *space* (considered not as a void, but as a substance with its material features, such temperature, color, light, …) and equipped by *furniture* (built in or movable, integrated or recognizable, always with their material features) (Ottolini 1997).

Some sacred spaces deeply changed their use during the centuries, proving their capability to speak of God in different languages and in different religions. The current Cathedral of Cordoba was built from the 784 as a Mosque and reconverted as a Church only in 1236, after the *Reconquista*. In 2004 Mansur Escudero, Secretary General of the Spanish Islamic Council asked the Pontifical Council for interreligious dialog the permission to make Christian and Muslim pray together, in this potentially sharable space: The still existing *mihrab* is the peripheral to the main axis of the Cathedral, so the two groups of worshippers would not bother each other. The request, even if considered reasonable, did not lead to any change in the use of the Cathedral, due mainly to practical reasons, as the conservation of an existing balance between the different religious communities (Russo 2007).

As it would be in this, never realized example, space is to be understood *not as a separation that divides, but as a delimitation that symbolically joins diversities*: (Tagliaferri 1995) This is the only way in which *we will be able to experience dwelling as "the way in which mortals are on the earth." As highlighted by Heidegger, the crisis depends on the lack of roots, due to which men "first have to learn how to live."* (Tagliaferri 1995), and architecture can become the medium through which this gets visible and understandable.

3 Gestures and Users: From Individual to Groups, from Silence to Rite

The relationship between man and God is translated in different ways: There is the silent and meditative way of the single individual; there is the organized prayer for small

[3] Louis Kahn (1971).

[4] "However, in any other intersubjective communication, the so-called two faces of any sign, the signifier or material vehicle of the signified, and the meaning or cognitive content associated with it are not of the same nature, one being physical reality and the other mental reality; although it is true that in art the signified does not only presentify in the mind of the subject producing or "listening to" the work, but is inscribed in the signifier itself by specific constitutional means." (Ottolini 1997).

groups; and lastly, the deeply structured acts of collective rites. In traditional communities, members recognize themselves in a single religion, and the entire community knows the rules and correct places for each moment. Moreover, each place is so structured that it is extremely complex for it to open up to experiences different from those for which it was designed.

However, silent and private meditation can be carried out more easily in shared environments, not specific to a given religion: This is evident in the case of *quiet rooms* present in airports, train and bus stations, places characterized by heavy transits connected to the transport system, as well as in hospitals and places devoted to care services. Placed along the main paths to enter and move in these centers, the *quiet rooms* are detached owing to their introvert form and to the design of their access paths that move away from the fast movement of the means of transport or the suffering in hospitals. Ettore Sottsass designed a *Chapel for the Malpensa*[5] airport that has a separate access, although near the hall of the departure floor, reachable through a narrow corridor characterized by blue mosaic walls. It opens to a rounded anteroom, giving access to the higher space of the Chapel. This space, hidden and totally introvert, points at a light scenery flat, always blue (Fig. 10.3), causing to leave the self behind and go beyond the physical space.[6]

However, the continuous movement of recently founded communities due to work, study, participation in events of excellence, such as sport events, has led to the need to face a new challenge: that is, to share places that are more structured than those destined exclusively to silence and a personal experience. During Olympic Games of 1992, the Archbishop of Barcelona promoted the construction of an ecumenical and interfaith center,[7] following the indications of the International Olympic Committee, who recognizes five faiths (Protestantism and Catholicism, Islam, Buddhism, and Judaism) and provides them houses for worship. The *Abraham Center* has an elongated plan, connected by a large stair, leading from the level of the square to the one of the Chapel (Fig. 10.4). The elliptic Chapel alternate full height and lower spaces, as entrances and balconies, naturally lit by big windows. The core is identified by a wooden elliptic false ceiling and points at a work of art, three golden bowls on a white background, symbolizing the three monotheistic faiths.

As soon as we discuss of sharing spaces devoted also to rites, the ground becomes suddenly steep. To define religion is a complex task, as often it involves aspects that are not strictly religious in themselves. This is the case, for example, when having to define what is authentic and what not, or religion's relationship with ethics, laicity, or power. As stated by Altan, *the deep symbolism of a peoples' identity is rooted in religion* (Altan 2002). Besides, religions often describe themselves as the founding element of a new civilization, overcoming a previous unpopular and unfair order.

A first step toward recognizing that all citizens of a same state are equal consists in accepting the equal dignity of the different religious beliefs. Surprisingly, one of the first bodies that promoted the construction of interdenominational or ecumenical Chapels was the US army, in the mid 1900s. In fact, between 1958 and 1968, SOM designed the United States Air Force Academy Cadet Chapel (Powers 1974),[8] in Colorado Springs (Fig. 10.5). Drawing inspiration from the Parisian Sainte Chapelle, the Chapel's design provided for a double-pitched roof, with an extremely steep inclination (Fig. 10.6). The overall structure is organized as follows: The Protestant Chapel is placed on the upper floor, the Catholic one on the ground floor, while the synagogue is on the underground floor. The three deeply different spaces stay side-by-side, and they are not interconnected.[9]

When designing a sacred space, the architectural challenge is to understand how to reinvent some characteristics deemed unavoidable, with the aim to translate them into new forms. Moreover, the new forms have to be equally dense in meaning, and at the same time sharable, thus capable of giving up any symbols that are too evident, or a figurative decoration that can be misunderstood. It is necessary to learn how to design a space which, in its most intimate essence, symbolizes the relationship between man and God. Therefore, we investigated this theme through research and didactics, with the aid of the students of the *School of Architecture* of the *Politecnico of Milan*, in the *Architecture of Interiors Design Studios*. Already existing solutions were

[5]In the original project for the first big enlargement and adaptation to international standards of Malpensa airport, Mario Botta designed an ecumenical Chapel, never realized. Ettore Sottsass designed the interfaith Chapel here described, newly set up in a short time as a Catholic Chapel, the of the Madonna di Loreto of Malpensa Airport.

[6]The glass wall has been realized by the artist Laura Viale, overlapping a picture of a barely veiled by clouds sky (Cielo n. 42, 1997) to light sources, creating a feeling of big peacefulness.

[7]Benedito and Mateos (1992). After the end of the Olympic games, the center has been transformed into a Catholic Parish.

[8]Walter Netsch, SOM (Skidmore, Owings & Merrill), *United States Air Force Academy Cadet Chapel*, Colorado Springs, Colorado 1958–68. (Powers 1974).

[9]In 2007, a Buddhist Chapel was realized as freestanding hall within the Cadet Chapel; in 2011, the Falcon Crest, an open air space defined by an enclosure of big stones, was designed for "The Earth Centered Spirituality" group. The All-Faiths Rooms, worship areas for smaller religious groups, purposely devoid of religious symbolism.

Fig. 10.3 Light scenery flat of the Malpensa Airport Chapel, with his drawing of clouds on a blue sky background, reflects on the pavement, changing its color and its perception (Laura Viale, Glass wall of Malpensa 2000 Airport multi-faith Chapel, Milan, 1998, digital print on transparent film on backlit glass, 3 × 10 m, cCapel design by Ettore Sottsass)

explored and new buildings were designed, often imagined within the University Campus.[10]

In interior architecture, it is particularly interesting to pay attention to gesture, as it is a function dense of meaning that is not only operational—of man dwelling the space—to the formal quality of architecture understood as a complex structure (Ottolini 1997),[11] organized as a whole and as a detail in a very close-up scale. Indeed, the fundamental idea is the Hahnian one according to which *nothing works, until everything works*, (Gattamorta et al. 1996) in a continuous transition among scales and their mutual fallouts.

4 The Context: Insertion and Separation

One of the themes that needs to be faced and translated into form when designing these buildings concerns the relationship with the world, taking into consideration both insertion and separation. The new sacred places have to be recognizable and become part of an urban or natural context. At the same time, they have to abstract themselves from their context, in order to recall something that innerves and at the same time transcends them. When they are recognized as the sacred space shared by a close community, as the SOM Chapel, they become a center on which citizen can orient themselves through the urban fabric, able to give it a deeper

[10]Bibliographical researches, analysis, and projects were developed by the students of the *Laboratorio di Progettazione dell'Architettura degli Interni*, during the academic years between 2005–2006 and 2008–2009, held by me and professors Roberto Rizzi, Stefano Levi Della Torre, and Lukas Janisch, with the help of the architects Aurelia Belotti, Ilaria Guarino and Fabiano La Rocca, and by several graduands.

[11]"In line with the abovementioned laws of Gestalt, Biagio Garzena has spoken of form as an "integral" (in the mathematical sense); that is, as a fusion and not merely the summation of multiple elements or characteristics of constitutional aspects." (Ottolini 1997).

Fig. 10.4 Main façade with the long stair leading to the *Centro Ecuménico Abraham*, in Barcelona, showing the high white walls protecting the shared Chapel (Robert, www.bit.ly/320g1eJ)

meaning. The Cadet Chapel catalyzes the academy space, unique high, vertical volume in the horizontality of the school, and unique one without windows on its sides; the Kresge Chapel stands alone, protected by water, in the middle of a square, clearly with a different call compared to the rest of the M.I.T. Campus: collective, public, and common.

Some other case studies suggest a double possibility of continuity and donation of a sense to the built landscape, and of separation from everyday life difficulties, earning a meaning in themselves. The *quiet room* realized in the Salam Centre for Cardiac Surgery in Khartoum (Serrazanetti and TAMassociati 2017)[12]—designed by the studio TAM Associati in 2007 for Emergency—is formed by two staggered rooms enveloped, with a continuous movement, by a high white wall. Acting as a hinge, a walkway in-between the two rooms passes through—barely above the water—a squared water basin giving it access. The prayer room is hidden by walls, protected by water, screened through its rush matting covering. The room develops on a side of a vertical cut that gives it access. It is protected from sight, and it is lined with a reed matting. In a corner of this room—a place created to mitigate pain and waiting time—life is emphasized through the presence of a tree, sign of the relationship with heaven and the divine.

> A sacred place […] is a symbolic bridge joining heaven and earth. It materializes when the sacred architecture creates the appearance of the other world that is the counterpart of the I and us and a place where the familiarity inherent a house has to coexist with expropriation and non-involvement in order to sense the presence of the other. (Tagliaferri 1995)

[12]TAMassociati, Prayer and meditation pavilion for Salam Centre for Cardiac Surgery, Khartoum, Sudan 2007, Aga Khan Award for Architecture2013 winner. (Serrazanetti and TAMassociati 2017).

Fig. 10.5 Interior view of the Cadet Chapel, showing the different effects given bay natural light passing through the colored glass walls: the white, neutral light behind the main altar, and the colored one, red, blue, and yellow, over the benches for the worshippers (Wally Gobetz, www.bit.ly/320ilCt)

5 Man and the Divine: Light and Works of Art

Indeed, the use of light is never neutral: It is used as the fundamental material of a piece of art, straddled between architecture and the environmental installation. It exploits different qualities and variation possibilities through the use of natural and artificial sources.

Some elements of architecture have the responsibility more than others to build a relationship between two presences, a visible one and a slightly veiled one beyond the ordinary perception. One of these elements is light, as it symbolizes the attempt to tend toward God, emphasizing the heart of the rite. The Cadet Chapel in Colorado Springs is characterized by windows along the ribs of the steep covering, through which a warm blue light in the lower part and a red light in the higher one, that tends to the vertical of the axis mundi, filters in directed toward the lower part of the Chapel. All this is opposed to the white milky light coming from the opal windows behind the altar, a neutral basis for the celebration.[13]

This is what occurs in James Turrel's installations designed for sacred spaces, often confessionals, where the light emphasizes the color and meaning of a space outside of time, putting it by contrast in relation to the time, told by the quality of the light with the passing of hours and seasons.[14] The author said he was *interested in the sense of presence of space; that is space where you feel a presence, almost an entity* (Haskell and Wortz 1980), the relationship between something contingent and something absolute. In the *Live Oak Friends Meeting House*, Turrel realized in 2000 *One Accord*, a skyspace, that is a specifically proportioned chamber with an opening in the ceiling open to the sky. The rectangular opening in the barrel vault of the Chapel can be opened to show the sky, or can simulate it with the use of blue, pink, and red neon tubes, put around it and on the abutments of the vault (Fig. 10.7).

As light, works of art play an outstanding role in defining the identity of sacred spaces, as historical examples show in precise and diverse ways: In confessional spaces, they can be figurative or not, but in any case, they highlight the cores of the rites, the pace of celebrations, and the separation/connection with everyday life.

This way can be analyzed alongside the *Rothko Chapel* in Houston (De Menil 1971),[15] realized around 14 monumental canvases by commission of John and Dominique de Menil. The interior of the building, a brick irregular octagon inscribed in a Greek cross, is a void space, surrounded by the works of art, receiving light only from a baffled skylight: A suspended space, equipped only by eight movable benches, opens to host all believers and support them in prayer.

The connection between ground, human, ordinary and sky, divine, extraordinary is often highlighted by inner and outer work of art: The MIT Chapel is completed both inside by the sparkling Harry Bertoia's reredos, and outside from the aluminum Bell Tower, realized by Theodore Roszak in 1955 (M.I.T. Chapel and Spire 1955), and the *Centro Ecuménico Abraham* on the one hand by the bell tower and on the other

[13]At the ground floor, the Catholithic Chapel is lit by amber windows, and in the underground the Synagogue is closed by colored glass slabs on cypress frames.

[14]Among the installations for sacred spaces realized by James Turrel, we can remind, apart from the one for the Houston's *Live Oak Friends Meeting House*, built in 1995, the one for the Church *Santa Maria Annunziata in Chiesa Rossa* in Milan. This was designed in 1996, the year of the death of the artist, and ended after his death. Neon tubes follow the contour of the barrel vaults and of the apsidal conch: green, blue, pink, and golden lights change deeply the perception of the space.

[15]Philip Johnson, Mark Rothko (paintings), *Rothko Chapel*, Houston, Texas 1962 (De Menil 1971).

Fig. 10.6 Outer volume of the Cadet Chapel is defined by 17 extremely steep ribs, folded by a white metal plate. Between them, not so visible from the outer square, stay the colored glass walls. The inner part of the Chapel, and all its levels, id (Wally Gobetz, www.bit.ly/2XcreoB)

Fig. 10.7 James Turrel, Meeting 1989. One of the skyspaces realized by James Turrel, in which artificial and natural light work together to change the perception of the space (Ray Weitzenberg, www. bit. ly/2XwZGPe)

one by the white panel supporting three golden bowls, recalling the traditional Jewish history of the three golden rings, symbol of the three Abrahamic religions, included in several Italian medieval collections of short histories and especially in Boccaccio's Decameron (1470, 1855).[16]

[16] *I remember often to have heard of a great and rich man, who among his most rare and precious jewels, had a ring of exceeding beauty and value. Being proud of possessing a thing of such worth, and desirous that it should continue for ever in his family, he declared, by will, that to whichsoever of his sons he should give this ring, him he designed for his heir, and that he should be respected as the head of his family. That son to whom the ring was given, made the same law with respect to his descendants, and the ring passed from one to another in long succession, till it came to a person who had three sons, all virtuous and dutiful to their father, and all equally beloved by him. Now the young men, knowing what depended upon the ring, and ambitious of superiority, began to entreat their father, who was now grown old, every one for himself, that he would give the ring to him. The good man, equally fond of all, was at a loss which to prefer; and as he had promised all, and wished to satisfy all, he privately got an artist to make two other rings, which were so like the first, that he himself scarcely knew the true one. When he found his end approaching, he secretly gave one ring to each of his sons; and they, after his death, all claimed the honour and estate, each disputing with his brothers, and producing 24 his ring; and the rings were found so much alike, that the true one could not be distinguished. To law then they went, as to which should succeed, nor is that question yet decided. And thus, it has happened, my Lord, with regard to the three laws given by God the Father, concerning which you proposed your question: everyone believes he is the true heir of God, has his law, and obeys his commandments; but which is in the right is uncertain, in like manner with the rings* (Boccaccio 1855).

Fig. 10.8 Inner space of the *Tag&NachtRaum* designed by Werner Mally for the Städtisches Klinikum of München-Harlaching, in which light and finishing materials define the quality of the space (Tobias Schneider, http://bit.ly/3cNpuLG)

In year 2000, the Czech artist Werner Mally designed *Tag&NachtRaum*, providing it with a luminous diagonal cross that passes through the walls and the ceiling, straddled between an extended field of luminous yellow stucco, and a corner in-between two walls made of dark blue stucco (Fig. 10.8) (Zahner 2004).[17]

6 Adapting Spaces: The Role of Flexibility

Flexibility can be another key to design a sharable sacred space. The possibility to welcome different communities that gather to pray and carry out different rites requires the presence of an equipment that must not be too pervasive and at the same time needs to be flexible in its use, easily adaptable to different users and expectations.

The *Interfaith Spiritual Center* in Boston, a rectangular and elongated space, is designed on the number three (Fig. 10.9). The space, artificially lit on three sides, is phased in three following spans, characterized by three concentric-designed hanging canopies. They highlight the three monotheistic religions to which the space refers. There are few pieces of furniture that can be moved or closed, enabling the space to be used while sitting, standing, looking in various directions, and in variously organized liturgies (El-Khoury and Riera Ojeda 2000).[18]

Furthermore, partitions can be rotated, folded, or moved to create new configuration of spaces, smaller or bigger, inner or outer, private or shared, answering to the need of different groups of users, and to the exigences of prayer or rites.

The ecumenical *Maria-Magdalena-Kirche*,[19] built in Freiburg between 1998 and 2006, is a concrete monolithic volume, formed by three naves. The two side ones host two different churches: the western nave a Catholic church and the eastern one a Protestant church (Fig. 10.10). Their mood is different. If the first is contemplative, lit by few openings, the second is bright and lively, thanks to a big opening facing east. Among the two, a central foyer in which the baptismal font stays. Eight-meter-high sliding concrete walls allow the three part to be joined together for the ecumenical services, creating a unique hall unified by the timber beams of the ceiling (Fig. 10.11).

7 In the End

Rooted in a desire of sharing and understanding diversities, all these elements allow architecture to host different people, and people to feel welcomed: the relationship, of separation and inclusion together, with the outer world, the

[17]Mally (2000), Zahner (2004).

[18]Office dA, *Interfaith Spiritual Center*, Northeastern University, Boston, Massachusetts, USA 1999 (El-Khoury and Riera Ojeda 2000).

[19]Kister Scheithauer Gross architekten und stadtplaner, *Maria-Magdalena-Kirche*, Freiburg, Germany 1998/2005.

Fig. 10.9 Inside view of the *Interfaith Spiritual Center* of the Northeastern University in Boston, with the three concentric canopies, the artificial light hidden behind the side glass walls, colored with different shades of blue, and the continuous wooden pavement (Yarian Gomez, www.bit.ly/324YH8n)

Fig. 10.10 Essential and high interior of the Catholic Chapel inside the *Maria-Magdalena-Kirche* is lit only by the skylights opening over the level of the timber beams of the ceiling (grebrov, www.bit.ly/2NhKIbK)

organization and distribution of the inner space, based also on a reasoned flexibility, the quality of building and finishing materials glimpsed in all the discussed case studies, the role played by natural and artificial light, and the one of works of art, traditional or not, as paintings, statues, light installations and videos. Somehow, the same words used in the vocabulary of confessional spaces, but declined in a new way.

Design culture should reason on these subjects and develop proper forms for them, looking at realized case studies and imaging new spaces, for at least two reasons. First of all, because diversity is one of the traits of contemporary society, people and traditions get close, boundaries weaken leaving space to thresholds, connecting differences. Then, because openness is a basic feature of architecture itself, able to create a significant relationship with the world, so that it can be lived as welcoming and life can be abundantly full of activities and meaning. This can happen only if designers understand and try to represent the identity and the aims of the society in which they live.

None of these spaces is white and neutral, a background receiving an identity from life interacting with it: All of them are perfectly designed as a hole and in detail, dense in meaning, and they are able to symbolize in universal words the relationship between man and God. They are real architecture, that is a material form arise from the complex relationship between several components (Ottolini 1997), in

Fig. 10.11 *Maria-Magdalena-Kirche* appears as a monolithic folded volume, with a few openings and the entrance in a high bay, leading to the shared inner space (grebrov, www.bit.ly/2J33kXK)

other word an integral, in the mathematical sense (Ottolini 1997).[20]

In a 2003 interview, Ettore Sottsass told Anna Detheridge *What really interests me is regaining an almost childish sense of wonder. I recently read some ancient Vedic texts where there appears to be a "pre-religious" sense of the world, and in which the divine plays an important role. We are inside this divine space. Day and night we are amazed by what happens, because we have no explanations—we simply open our eyes and wander through phenomena like children* (Sottsass 2017).

Sacred architecture has to be built around this sense of wonder, deriving from the idea world and things within it, stay in the divine space, and among them stays human existence. The material forms of sacred architecture should become the medium through which this idea becomes clear and understandable, through which the rules underlying world becomes sharable and life rediscover a meaning, going beyond the ordinary and everyday life.

Non-confessional, shared, interfaith spaces more than many others have the specific duty to discuss this topic in their designed and built forms, in their entrances, inviting communities to share experiences and mutual knowledge, in their rooms, variously declined depending on the different uses, and finally through their meaningful beauty.

[20]As stated by Biagio Garzena (Ottolini 1997).

References

Boccaccio, G. (1470). *Decameron.* Napoli.

Boccaccio, G. (1855). *Decameron or ten days' entertainment* (W. K. Kelly, A cura di). London: Henry G. Bohn.

MIT Chapel and Spire. (1955). *Time Magazine.*

Kahn, L. I. (1960, November 21). Structure and form. *Voice of America Forum Lectures, Architecture Series 6.* Washington D.C.

Kompass Nelson, J. (1970). *Harry Bertoia Sculptor.* Detroit: Wayne State University.

De Menil, D. (1971). The Rothko Chapel. *Art Journal, 30*(3), 249/251.

Powers, R. G. (1974, Summer). The cold war in the rockies: American ideology and the air force academy design. *Art Journal, 33*(4), 304–313.

Norberg-Schulz, C. (1985). *The concept of dwelling: On the way to figurative architecture.* New York: Rizzoli International Publications.

Tagliaferri, R. (1995). Lo spazio architettonico nell'esperienza religiosa. La mediazione spaziale della fede nel quadro della ritualità. In A. Piva (Ed.), *La città multietnica: lo spazio sacro* (p. 50). Venezia: Marsilio Editori.

Ottolini, G. (1997). *Forma e significato in architettura.* Roma Bari: Laterza.

Altan, T. (2002). *Le grandi religioni a confronto. L'età della globalizzazione.* Milano: Feltrinelli.

Menocal, M. R. (2002). *The Ornament of the World: How Muslims, Jews, and Christians created a culture of tolerance in medieval Spain.* Boston: Little, Brown.

Russo, A. (2007). Lo spazio sacro nella città interetnica. *Rassegna di Teologia, 155,* 403–420.

Sorrentino, S. (2007). La grammatica del dialogo interreligioso. Per un approccio critico al rapporto fra le religioni. In S. Sorrentino, & F. S. Festa (Eds.), *Le ragioni del dialogo. Grammatica del rapporto fra le religioni* (pp. 179–206). Troina: Città Aperta Edizioni.

Sottsass, E. (2017). *There is a planet*. Milano: Electa - Triennale Design Museum.
Rizzi, R. (2019, June 12). Gathering differences. Sacred space and diversities. *Utopian and Sacred Architecture Studies*. Aversa, CE, Italy.
Institute Community & Equity Office. (2020). Retrieved April 13, 2020 from http://diversity.mit.edu/.
Who We Are | Division of Student Life. (2020). Retrieved April 13, 2020 from http://studentlife.mit.edu/rl/who-we-are.
El-Khoury, R., & Riera Ojeda, O. (2000). *Office dA: Works by Monica Ponce de Leon and Nader Tehrani*. Gloucester, MA: Rockport.
Saarinen, E., & Saarinen, A. B. (1968). *Eero Saarinen on his work: A selection of buildings dating from 1947 to 1964 with statements by the architect*. New Haven: Yale University Press.
Serrazanetti, F., & Massociati, T. A. (2017). *Tamassociati: Taking care: architetture con emergency*. Milano: Electa.
Haskell, B., & Wortz, M. (1980). *James Turrell: Light and space*. New York: Whitney Museum of American Art.
Gattamorta, G., Rivalta, L., & Savio, A. (1996). *Louis I. Kahn. Itinerari*. Roma: Officina Edizioni.
Louis Kahn, I. (1971). *Architecture comes from the making of a room*. Philadelphia Museum of Art, Gift of the Artist.
Benedito, J., Mateos, A. (1992). *Centro Ecuménico Abraham*. Barcelona, Spain.
Mally, W. (2000). *Tag&NachtRaum*. Städtisches Klinikum München--Harlaching, Münich.
Zahner, W. (2004, September). Christian ecclesiatical architecture-Developments from the 1920s to the present day. *Detail*, (4), 932–938.

Gathering Differences. Sacred Spaces and Diversities

Roberto Rizzi

Abstract

Tensions between different religious beliefs are often on the agenda with their ferocious degenerations, discriminations, and radicalizations: "universalistic" visions clash with the need of every credo to be alternative to the others, and so in contrast. This ambiguity is contained in the word "religion" itself. In most European languages two meanings can be associated to this word, through its Latin roots: first, to tie, knotting together human and divine (*religare*: to connect), and second to collect around a mode (a rite), chosen to guarantee and verify an identity (*religere*: to distinguish). Also, spaces hosting rites live on this ambiguity: they are dedicated to gather the broadest and inclusive community, and at the same time they become the sign of difference and limit for every different behavior, in an idea of sacred as secluded and forbidden. Contemporary society amplifies these problems: the mobility of things, ideas, and people intensify potential tensions, and at the same time it opens to places and possibilities to verify new ways of living the religious experience. For this reason, the need of finding new open configurations for every place of worship is accompanied by the urgency of experimenting with sharing places for religious practices in a situation of diversity. We think above all to places of culture and higher education, or infrastructures of transports, but also to community based on the communion of the religious experience. Architecture, especially in a declination careful to people and its gestures, can answer this question through places able to synthesize undeniable specificities and possible sharing. We propose a dissertation developing in parallel theoretical questions, historical case studies, and design evaluation carried on through the university design didactics.

R. Rizzi (✉)
Dastu Department, Politecnico of Milan, Milan, Italy
e-mail: roberto.rizzi@polimi.it

Keywords

Interfaith • Interior architecture • Identies

Tensions between different religious beliefs are a regular occurrence, often fierce and exacerbated by economic difficulties or social marginalization. The mass media document their bloody manifestations and degenerations, highlighting the discrimination and exclusion of people who in that specific context result to be a minority. These radicalized positions intensify differences, transforming them into inequality gaps.

Worldviews with "universalistic" ambitions aim to embrace every aspect of life (and "afterlife") in a single broad and harmonious view. They clash, though, with the tendency, or the need, of every single religious belief to be an alternative to all others, causing the mentioned contrasts.

In an increasingly secularized world, where religions are followed much less than in the past, there is who exploits religious beliefs to reach powerful positions or who has gone back to killing "in the name of God" (Sloterdijk 2008, pp. 49–79; Palese 2013). Moreover, others consider religious beliefs at the basis of the problems induced by globalization and mass migrations, paying more attention to how said problems occur than to the religious or cultural identity affected (Duttweiler 2017a).

If on the one hand, though, the number of people that observe practices and rites has decreased, more or less consistently depending on cases, on the other hand, a sort of "nostalgia for the absolute" (Steiner 1974) stubbornly continues to make religions play an essential role in the construction of our cultures (Givone 2018). However, they are no longer viewed as an element that directs space (in the dialectics between worship buildings and public buildings) and that gives an order to time (with the alternation of weekdays and holidays), but as a litmus test to estimate belongings/identities and exclusions/differences. This triggers two opposite attitudes: the first leads to pay attention to the fallouts of a phenomenon evaluated and lived only for its

C. Gambardella et al. (eds.), *Advances in Utopian Studies and Sacred Architecture*,
Advances in Science, Technology & Innovation,
https://doi.org/10.1007/978-3-030-50765-7_16

possibility to contrast different identities, and the second countering the world's view of religion as an intimate and private experience linked to a substantially detached devotion.

This ambiguity is even contained in the word "religion", as it can have opposing and complementary meanings, deriving from the Latin noun *religio*. It holds, in fact, an "objective" meaning (involving a relationship established with the gods through worship) and a "subjective" meaning (involving the scrupulous observation of the actual rites) (Filoramo 2004, pp. 81–82). These aspects are highlighted also in the relevant verbs: *religare* means to bind, to tie, to establish a relationship between man and the divine (placing the attention on something that unites), while *religere* means to gather, to choose again, to re-read, in the sense of reiterating a continuous verification of the act of joining a worship system that thus becomes the moment for approaching a separate reality (shifting the attention to something that distinguishes itself and has to be always remembered as something totally different) (Agamben 2005, pp. 85)[1] Fig. 1.

The religious experience holds a series of polarities, such as a close connection among different realities and an emphasis placed on jealously cherished differences; the desire to reach the most extensive unity possible, and the watchful monitoring on insurmountable differences; harmony-consonance-concordance and conflict-opposition-contrast.

Indeed, the spaces that host these rites[2] experience this ambiguity as well. In fact, although they gather a community that is meant to be wide, inclusive, and universal, they end up highlighting the difference and limitation of whatever is not in line with the rite for which the assembly is taking place.

These places are separated, sometimes or in part forbidden and inaccessible, for the autonomous manifestation of the divine, of which they preserve and pass down the memory, reiterating its epiphany. For this reason, they risk to become the places in which the divine is relegated, secluded "outside" the profane world. They are places where people, gathered in a community, reassert their allegiance to a belief and a faith. Therefore, they are open and accessible places, totally usable by the community gathered with the purpose to listen, talk, and make comments. At the same time, they risk to become the places through which people celebrating different rites could be recognized as dangerous for their differences.

Such places can be imagined as driven by a polarity, with the predominance of one or the other extreme. It is a polarity that views these spaces as preordained by a preventive unequivocal divine manifestation, defined for the specific rite to be observed, or subordinate and adequate to the assembly's gathering, to the possibility to carry out specific actions, to the fulfilment of gestures that give value to the space hosting them.

This second aspect is supported by the reiterated encouragement within the various beliefs to pray anywhere: the Jewish house, the oriented and clean Islamic structure, up to the simple meeting of two people which, according to the Christian texts, determines the presence of God (Russo 2007, pp. 416).

This is confirmed by the words that define the places of worship.

In fact, the Christian *ekklesia* evokes, from Greek, an invitation, a call to a common assembly; the Jewish synagogue—also called "assembly house (*bet knesset* in Hebrew) —finds its origin in the Greek word that means "to bring together, to gather"; while the word *masjid jāmi'*—the place for the collective Friday prayer—puts together the sense of prostrating and abandoning oneself to God of the word *masjid*, with the function of reception within the word *jāmi'* (assembly) (Ries 1992).

The contemporary society magnifies these issues: the great mobility of things, ideas, and especially people intensifies potential tensions on the one hand, while on the other hand, it opens up to places and opportunities that allow to experience the different ways of living and the religious beliefs.

All this leads to two areas of research.

On the one hand, it is necessary to identify the worship buildings of the single beliefs, that is, open configurations capable of enhancing activities that can create bonds with the surrounding lives, generating widened and inclusive relationships.[3]

Therefore, the poles of the different religions in the urban and social contexts need to be able to produce—starting from core specificities—a halo of possibilities innervated with connections and mutual relationships.

We don't refer simply to a desirable inclination of communities dwelling confessional sacred spaces to interpret this attitude to openness and to provide spaces that could result appropriate for the functions they will host. We insist on the

[1]For some scholars, this is the truest interpretation: to add an accent on an "attitude of care and attention, shaping the relationship with Gods, the troubled hesitation (the "rereading" [relegere]) in front of forms - and formulas- to be observed to maintain the separation between sacred and profane. Religio is not the thing unifying men and gods, but the one guarding to maintain them distinguished" (Agamben 2005, pp. 85).

[2]We refer here, schematically and in a generalized way, to the three monotheistic religions.

[3]Sometimes, it seems even to miss the precondition for this attention, as the debate that the opening, sometimes only supposed, of Muslim places of worship cyclically opens in occidental cities.

Fig. 1 Michelangelo Pistoletto, *Love differences—Mar Mediterraneo* 2003–2005 cm. 738 × 320 cm. Mirror, wood; seats of different country around the Mediterranean Sea. Galleria Civica di Modena—Palazzina dei Giardini. 16th September 2005–8th January 2006

fact this have to happen on an architectural level, because of its skill of grafting in the urban fabric open and dialoguing forms, refusing any desire of monumentality and assertiveness (Scrinzi 2008).

In other words, it is necessary to think of new forms that are not a simple peremptory statement of an exclusive function-oriented ideologically. These forms need to be able to dialogue with the surrounding physical fabric, rooting themselves in the urban structure, creating specificities, also with exceptional nature, starting from the warp on which the surrounding morphological, material, and figurative structures are weaved (Beinhauer-Köhler et al. 2015).

Other studies and researches tried to imagine that the relations between the different places of worship wouldn't be simply those given by urban genesis and urban planning choices, and that is possible thinking and designing "urban scenes" to make buildings and communities coexist and collaborate. Among the possible references, we can quote, for its utopic contents, University research[4] developing the idea of "clod". « It is an architectural concept, intermediate within the town and country planning scale and the one of new buildings [...] it is built by the combination of public and/or collective building, but also residential, to be qualified with an appropriate use of the existing building stock » (Bisogni 2000, pp. X).

On the basis of this premise, some "clod of the Spirit"[5] (Fig. 2) were designed, in which the buildings devoted to monotheistic worship (church, mosque, synagogue) and potentially another building for counterpoint (interfaith center, nonreligious building) constitute a unitary organism. The clod becomes able to structure and allow relationships between isolated and figuratively identified parts, that can become communicative, as the pieces of a bas relief raising from an equipped basement, giving a form to the ground and opening to mutual relationship with the surrounding urban background.

On the other hand, it is necessary to enhance the relationship between the different religions and their practices within the same place.

Similar situations are evident in contexts with a strong cultural mixture, that is places providing higher education and cultures such as universities (Blümer and Marinkovic 2010; Bucher 2010; Hochgeschwender 2010), places relating to transport infrastructure or mobility (stations, airports, ...) (Bernhardt 2010; Kölbl 2010), or those in which people are obliged to stay under forced conditions (hospitals, prisons, etc.); places with a strong concentration of trade or international sports activities; cemeteries, military bases; parliamentary seats (Kaernbach 2010) or supranational institutions,[6] as well as communities that programmatically adopt practices in order to share religious experiences (Averna 2019; Duttweiler 2017b; Hammerskjöld 1957).

In all these cases, the assumption is that the users are not undefined residents of an urban context, that exploit places at disposal of the single beliefs within the normal urban fabric. On the contrary, they are members of a particular community—that can be more or less structured—whose use of those places is somehow steered toward dialogue, programmatically or out of need, and the desire to meet people that belong to different cultural and religious realities.

For each one of these cases, it is possible to identify architectural examples (Cornoldi 1995; Barbara 2017; Cataluccio 2017; Sakralbau 2004)[7]—often of a high level—designed by masters of architecture in free and Avant-garde experimentation. Their designs are based on theoretical, philosophical, and theological considerations concerning the value, meaning, and therefore the nature of those places as spatial bodies. On the basis of theological studies carried out on religions (Knitter 2005), such considerations have somewhat progressed in recent years (Sorrentino and Festa 2007; Razzano 2007), also due to specific opportunities (Francesco 2019a, February 4) (Francesco 2019b, June 21) or local emergencies.[8]

As examples, we can quote two experiences.

The first, soon to be built, concerns the construction of a building, the "*House of Prayer and Learning*", in which a

[4]Interuniversity research "Forma e figura delle architetture pubbliche e servizi per lo sviluppo sostenibile delle aree metropolitane: Firenze, Milano, Napoli, Mestre" (Form and Figure of public architectures and services for the sustainable development of metropolitan areas: Florence, Milan, Naples, Mestre). Ministry of University and Scientific Research 2000. Coordination Salvatore Bisogni (Naples) with Gian Luigi Maffei (Florence), Guido Canella (Milan) and Franco Purini (Mestre).

[5]Specifically on the subject of the "clod of the spirit" worked Naples research unit, Salvatore Bisogni (masterplan and church), Carolina Cigala (synagogue), Eduardo Demasi (mosque) (Bisogni 2000, p. 243-251, 284-285) and Mestre research unit, Sarah Zezza (masterplan and synagogue), Franco Purini (soul museum), Roberta Albiero, A. Gabbianelli, N. Maaloul (mosque), Raffaella Laezza, M. Moreno, G. Santamari (church) (Bisogni 2000).

[6]Among the first case-studies we can quote the "Meditation room", wanted, by Dag Hammerskjöld, UN secretary-general in 1952, with christian, hebrew, muslim communities. For the opening Hammerskjöld wrote: "We have within us a center of stillness surrounded by silence. This house, dedicated to work and debate in the service of peace should have one room this dedicated to silence in the outward sense and stillness in the inner sense....People of many faiths will meet here and, for that reason none of the symbols to which we are accustomed in our meditation could be used.' (Hammerskjöld, s.d.).

[7]Among the most renowned: Bruce Goff, Mc Cann memorial Chapel, Dublin, California 1943; Eero Saarinen, Kresge Chapel, MIT, Cambridge, Massachussets 1950/1955; SOM (Skidmore, Owings & Merrill), United States Air Force Academy Cadet Chapel, Colorado Springs, Colorado 1958–68; Paul Rudolph, Tuskegee chapel, Tuskegge Institute, Alabama 1960–1969; Philip Johnson, Mark Rothko, Rothko Chapel, Houston, Texas 1962; Josep Benedito, Augustì Mateos, Centro Ecumenico Abraham, Olimpic Village, Barcelona, Spain 1990/1992; Sottsass associati, Interfaith Chapel, Malpensa Airport, Italy 2000; TAMassociati, Prayer and meditation pavilion for Salam Centre for Cardiac Surgery, Khartoum, Sudan 2007.

[8]We can simply quote the Assisi gatherings promoted by Giovanni Paolo II, or the debate caused by the request of letting Christians and Muslim pray together in the Cathedral of Cordoba, former mosque. The current Pope Francis based its pontificate on the openness to the dialogue with all the faiths, as witnessed by several trips, meetings and documents as the one "Human fraternity for world peace and living togheter", prepared for the apostolic journey to the United Arab Emirates of the 4th of Febraury 2019 (Francesco, Apostolic journey of his Holliness Pope Francis to the United Arab Emirates 2019). Another interesting one is the recent speech at the Pontifical Theological Faculty of Southern Italy of the 21st of June 2019 for the congress "Theology after Veritatis gaudium in the context of the Mediterrenean" based on the dialog as the foundation of the relationships between people. (Francesco, Meeting on the theme "Theology after Veritatis Gaudium in the context of the Mediterranean' 2019).

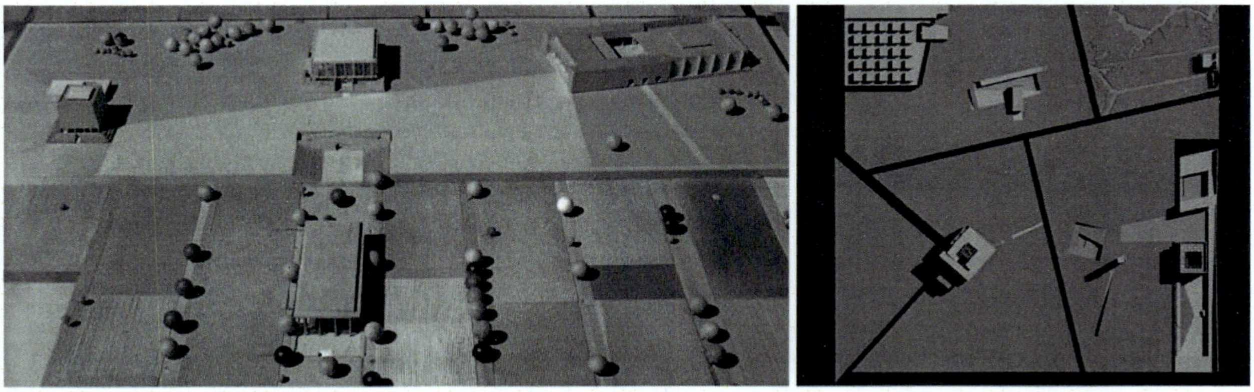

Fig. 2 On the left-side, Salvatore Bisogni (masterplan and church), Carolina Cigala (synagogue), Eduardo Demasi (mosque): *Clod of the Spirit*; maquette. On the right-side Sarah Zezza (masterplan and synagogue), Franco Purini (soul museum), Roberta Albiero, A. Gabbianelli, N. Maaloul (mosque), Raffaella Laezza, M. Moreno, G. Santamari (church): *Clod of the Spirit*; maquette

Lutheran Christian church, a synagogue, and a mosque will find a place (Hohberg and Stolte 2013). It will be built in Berlin, on the site of the demolished St. Petrikirche. The eighteenth-century church, first erected in 1230 and demolished by the DDR government, was the first Berlin sacred building: that is, a historically important and central place.

After the reunification of Germany, a new church had to be built, but the minister of the local community decided to involve also the Jewish and Muslim community, in the idea of designing an interfaith center within the Abrahamitic communities. A long journey of sharing and the writing of a "declaration for the cooperation within Christianity, Judaism, and Islamism", containing the principles for cohabitations and a calendar for the times of use of shared spaces, led to the launch of an international contest won by the Berlin architectural practice Kuehn Malvezzi[9] (Fig. 3).

The call forecasted the realization of three divided spaces, one for every belief, and of a central sharing space, and gave indications on the use of the materials (preferably bricks), on the prevailing role of the space on the structure and technical systems, and on the use of natural light as a principle of symbolic animation of forms.

Almost in all the projects, the three sacred spaces graft on the keystone of the shared spaces. Only in few cases, it becomes able to carry on an active role of mediation between confessional spaces and of connection with the urban fabric, assuming otherwise the features of a distribution emphasized by the dimensions and the use of the light and causing a close and introverted conception of the whole building.[10]

Designed by Bauart Arkitekten und Planer and Urban office Architects, the intervention the Municipality of Bern promoted in 1998 to upgrade a problematic area close to the Hasserholligen railway station, named the "*House of religions—dialogue between religions*", is surely more complex (Sauer 2015). Here eight religions shared a program of collaboration and dialogue in spaces composed of five confessional Chapels for Christians, Muslims, Hindu, Alevites, and Buddhists. Jewish, Baha'i, and Sikh are part of the program, but they don't have a sacred space. Spaces for shared and cultural activities complete the Chapels: cinema, a space for exhibitions and events, a library, classrooms for lessons, a playground for children, and a multifunctional terraced step. These activities, together with a commercial and administrative complex and with underground parking, form an open basement, linked to the urban spaces, from which a 6/8 floors block of apartments rise up as a blade, hosting 19 typologies of apartments for different uses and users. The shared activities, as well as the distribution, the accesses, the opening on the public spaces of the square, and an articulated play of levels, crossed vertically by visual connections, push continuously for meeting and dialogue. The few decorative quotes of the religious traditions are used as an identification of the single places and reabsorbed in a set characterized by the use of simple materials, highlighted by spaces and light (Fig. 5).

[9]First prize: Kuen Malvezzi, second prize: Riepl Riepl Architekten, third prize: Wander Höfer Lorch Architekten, honourable mentions: Joseph Smolenicky (Smolenicky & Partner) and Linazasoro & Sánchez Architecture.

[10]Linazasoro and Francesco Venezia projects, are the only ones able to overpass this locked and para-monumental view in favor of a richness of places and spaces for shared activities, enhanced by light and materials and open to an active integration with the urban context (Fig. 4).

Fig. 3 Kuehn Malvezzi, Das Haus der drei religionen, winner. Views of the wooden maquette, of the inner shared space and perspective section. In the lower floor (underground) the remains of the ancient church are visible

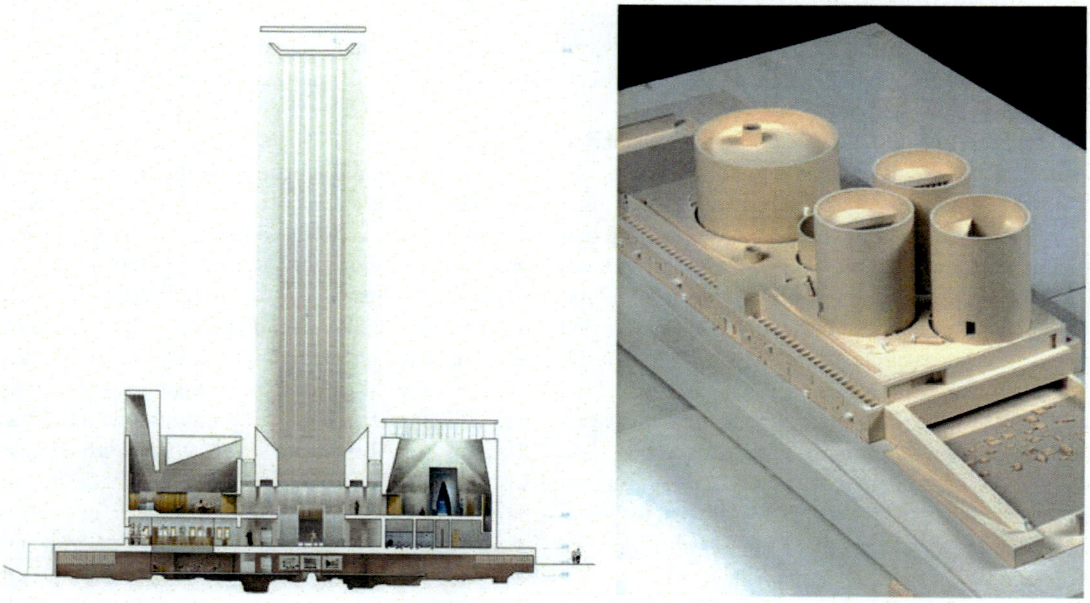

Fig. 4 Das Haus der drei religionen. Project of Linazasoro and Sánchez Architecture (section) and of Francesco Venezia (maquette)

Fig. 5 Bauart Arkitekten und Planer and Urbanoffice Architects, "*House of religions—dialogue between religions*, Berna 2008. Spaces layout and ground floor plan. Views of the worship rooms. Views of the building on the Europaplatz

Similarly to other Universities (Piva 1995; Raffone 2007; Vanacore 2015),[11] these themes were proposed to the Interior Architecture Design Studios of the School of Civil Architecture of the Politecnico of Milan,[12] and analyzed by first-year students and final-year students, also through degree theses. A path was developed based on theoretical considerations that took into account in-depth bibliographic references and on practical analyses of case studies. The latter case kept into consideration both the documentation relating to the simple organization of activities in spaces not defined by a specific architectural project and buildings of architectural quality. The aim was to identify simple organizational implications and their transfer in architectural materials.

The educational context within which this experience was carried out is particularly relevant when considering the theme.

In fact, the "internal space," regardless of its scale, is a founding and genetic element of architecture, considered not as a « *simple logical inversion of the external space, but as the actual genetic interiority of space, its genesis and therefore actual qualification* » (De Carli 1982, pp. 362).

It is not a matter of spatial delimitation, of a "closed inside" countering an "open outside," but of a "restricted equipped open space." In other words, it is a space-delimited by its own margins (the walls, floor, ceiling of a room; the ground and façades of buildings in a square; the state of land, a row of trees of a natural landscape; but also the top, backrest and armrest of a chair) and provided with fixed and movable equipment qualifying its use, available for the "gestures" of those living said space.

[11]Beyond Sandro Raffone, Architectural design studio 2, School of architecture, Naples University "Federico II" (a.y. 200/2001), we can quote: Roberto Vanacore, Architectural design course, School of Engeneering, Salerno University (Vanacore 2015); Antonio Piva, Architectural design course, School of Architecture and Society, Politecnico of Milan (a.y. 1994/1995) (Piva 1995) and Barbara, A. (2016). Design of Interiors workshop, Politecnico of Milan, School of Design, prof. A. Barbara, R. Gilad, P. Gandini; A. Destro, N. Gobini (a. y. 2016–2017).

[12]The classes were carried out in the School of Civil Architecture of Politecnico of Milan, during the Academic Years between 2005–2006 and 2008–2009, held by me and professors Stefano Levi Della Torre, Marta Averna, Lukas Janisch, with the help of the architects Aurelia Belotti, Ilaria Guarino and Fabiano La Rocca.

The form of a place is defined by the system of relationships (the "gestures") existing among people and between people and things. Places are defined by the sense of welcoming and availability that not only allows but also fosters actions, or reveals their new and unexpected modalities. A person as well as groups of people are thus considered in their intimate and inalienable preciousness expressed in their gestural expressiveness crystallized in the material-formal structure of spaces.

The aim of the research was to work on a complex system of spaces used for prayer and cultural comparison. In other words, spaces characterized by a marked functional complexity that host not only moments devoted to prayer, but also cultural and gathering activities (meetings, debates, lessons, shows, study, exhibitions, etc.), organizational and service activities (offices, library/newspaper, listening center, internet point, etc.), and hospitality activities (guestrooms, refectory, etc.). In other words, activities connected to worship services on the one hand, and activities that lead to open up to a relationship with the urban and social context in which they are carried out on the other hand.

With regard more specifically to prayer, the attempt was to consider this aspect both as an interiorized spirituality and an expressed physicality, thus as a contemplative, mental and silent act, but also as an accomplished action, an expressed gesture, musicality. Highlighting the subject's total involvement in this action (AAVV 2000), Novalis commented: "Prayer in religion is the equivalent of the thought in philosophy. The religious sense prays just as the mental organ thinks".

The projects kept into account the different ways of praying for each of the beliefs considered: from the private and personal prayer, carried out in a more informal and meditative way, to the community and collective prayer, with more ritualized forms, structured in space and marked by time.

What varies for every belief is the size of the group (a single individual, a small group, and a more extended group in the form of an assembly). Therefore, spaces must keep into account the need for a permanent form or for flexibility, the times of use, the duration, and alternation.

When evaluating the characteristics of the spaces for each modality mentioned and for the different beliefs, the combination of these variables produces many specificities as to use and space. In fact, they can be grouped, combined or overlapped in different ways. The research carried out in the Studios by the first-year students allowed to identify a series of recurrent configurations that were then further analyzed through more in-depth research projects carried out by the fifth-year students.

The projects were developed on two buildings; in one of them, there was yet the space for a prayer of the Bovisa Campus of Politecnico of Milan. Both the buildings were originally used as warehouses for the industrial complex now occupied by the Campus; at the moment of the projects, they were temporarily occupied by offices and laboratories. Their configuration and dimensions are similar: they develop on two floors, with a narrow rectangular plan and both have a direct relationship with open-air spaces used by the students, close to the library and the classrooms. The first has a flat roof, the second, instead, a vaulted roof.

When trying to identify autonomous spaces for the prayer modalities of each belief, increasingly integrated solutions were identified. Therefore, the aim was to put into relation different situations, with partial overlapping or correspondences depending on the level of unification deemed possible, acceptable, or necessary among the single religious beliefs.

The integration among different prayer modalities within the same belief takes place naturally. The same occurs when considering the same modalities in different beliefs, more frequently in the event of less ritualized modalities (single individuals or small groups).

The spatial solutions identified followed the principles of rigidity and flexibility: they are comparable in the diagram (Fig. 6). In the three superimposed rows, we can find aligned: the solutions designing a unique space shared within different religions on the first row, the intermediate ones on the second row, and on the last row the ones based on autonomous spaces, with intermediate filters and mediation spaces. Along with the columns, we can pass from the left-side, collecting solutions working on rigidity, given by univocal solutions, to the right one working on flexibility, given by sliding and rotating partitions between spaces adjacent in plan or in section.

With regard to rigidity, configurations are blocked and many specialized spaces are connected. In other words, various activities are carried out in undifferentiated common spaces.

With regard to flexibility, undifferentiated, and more anonymous or more specialized spaces are connected each time through spatial and flexible devices typical of the delimiting elements. In other words, specific solutions are identified each time taking into consideration partial closeness or visual connection, also in section.

Next to the space devoted to prayer, a space for private talks was provided, as well as an open space (patio, private garden) connected to the space devoted to prayer, and a small service area for preparing the rites. Therefore, the various places were hinged by the spatial device that

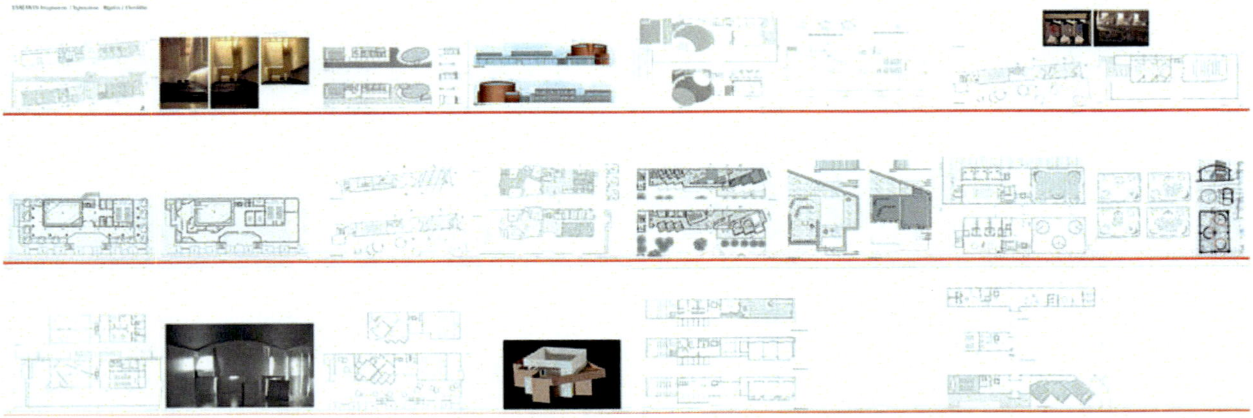

Fig. 6 Comparison between different design solutions: integration/separation–rigidity/flexibility

organizes entrance and distribution, solving and giving meaning to the need to pass from the collective space to the one devoted to prayer with its various structures.

Particular attention was paid to the spatial aspect of the prayer room, especially with regard to the poles of each belief (Khoury 1987)[13] and the orientation of space.[14] The latter variable often led to geometrical complexities in projects, giving life to interesting spatial solutions in which small sizes were alternated with bigger ones, allowing to graduate the different uses.

The projects developed by the students of the last years of the School worked on different areas: a first is adjacent to the entrances of the Bovisa Campus, the other stayed diffused in the city of Milan, characterized by an intercultural call and a significant presence of university students. This variety caused different approaches and results.

The system of support functions frequently introduces the spaces devoted to religious practices. The guest house, the exhibition spaces, the library, the theatre, the open-air spaces for events, forerun the building diagonally closing the plot (Fig. 7). Its three floors are crossed by the three oriented (Mecca, Jerusalem, East) walls, giving birth and order to the worship places, to their activities, and to the natural light–direct or mediated–enlightening them, with increasing levels of privacy in the higher floors. Every chapel is developed on two communicant levels, connected in section, to host properly groups of different dimensions (Rosi 2010).

Often the dialogue within the surrounding urban fabric and the directions of the worship places causes the organizing principle of the project. Two walls, hinged in the connection point with the town, play two complementary roles (Fig. 8). The first, continuous along the high traffic street, is equipped by a shop gallery connecting with its two levels the street level to the one of an inner lowered one. Two ribs, containing the first spaces for meetings, temporary exhibitions and shows (extended outside on terraced steps) and the second a library and some study rooms, graft orthogonally on this system. A high wall diagonally crosses the square and, folding at the end, identifies on different floors the worship places. They found the possibility to establish mutual relationships and sharing moments, in their section and in their system of distribution and access, formed by ramps and free-standing gangway (Mio et al. 2010).

Finally, the enhancement of underground levels allowed both to upgrade the existing ground level as an equipped green, structured on connection systems and equipped for shared activities, and to define spaces with mediated and graduated by following thresholds, relationships with the existing.

A first project, on Campus' boundaries to establish a relationship with the town, is made of a head, containing the sacred spaces, and a body with study rooms, entering in the Campus (Fig. 9). An articulated set of entrances relate the different levels, allowing different ways of entering from the city and from the University. The worship places overlook a shared patio, lit by a glazed winter garden; they can be directly connected by big rotating and sliding panels.

This first patio overlooks a second open-air one. Every chapel is designed on different levels, to be used by the single person or by small groups of people and it's based both on a complex geometry of polygonal chains, following the different liturgical orientations, and on the presence of the water, reflecting the light coming from the splits between walls and roofs (Colzani and Dondi 2010).

The same subject is developed in a second project (Fig. 10). The existing wing of the school was extended on

[13]We are basically referring to the place from which someone reads or speaks (Ambon—in the church, Bimah -in the synagogue, Minbar -in the mosque), to the catholic communion table, to the Aron-Ha-Kodesh for storing the Torah scroll, and to the system Quibla/Mihrab in the mosque.

[14]The Christian East, the Jewish Jerusalem, and the Muslim Mecca.

Fig. 7 Master degree thesis of Patrizia Rosi

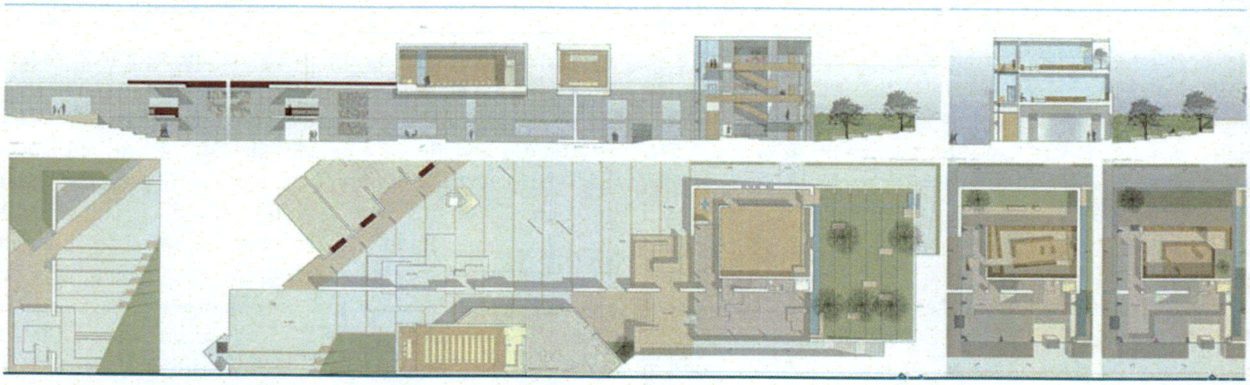

Fig. 8 Master degree thesis of Laura Mio, Sandra Naboni, and Alice Ometto

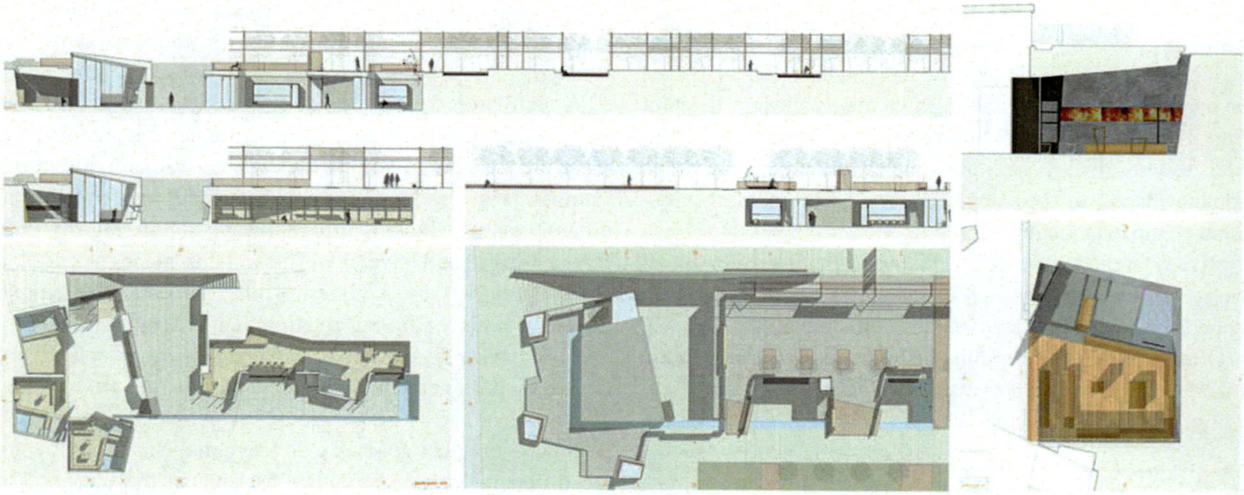

Fig. 9 Master degree thesis of Stefania Colzani and Lavinia Dondi

Fig. 10 Master degree thesis of Giorgio Radojkovic, Nicolò Zanolo and Claudio Zucca

the underground level, to host a library. This new design level hosts also the worship places and a three-foiled auditorium, with extendable stepped classrooms. These spaces are articulated around an open-air patio, following a complex geometry of polygonal chains given by the different worship orientations; they are connected by a path crossing the plot, connecting the different levels and providing protected access for each space. Their underground position suggests the use of natural light sources, enhancing and qualifying the worship places (Radojkovic et al. 2000).

But for the results of the single projects, here synthetically described by samples, what is interesting is the kind of approach and horizon of work.

The main role for identifying solutions is played by the design and construction of the physical space. It is a space that is not to be based on the concept of an already established community but on the idea of triggering paths connected to each other owing to its forms. A space that does not communicate exclusion, but that encourages dialogue between what is different. Therefore, a space is structured "not as a separation that divides, but as a delimitation that unites" (Tagliaferri 1995, pp. 50).

"A space for who never enters, for who has to stay nearby without entering, and for who enters" (Kahn 1957).

Refereces

AAVV. (2000). *La preghiera respiro delle religioni.* Milano: Ancora.

Agamben, G. (2005). *Profanazioni.* Roma: nottetempo.

Averna, M. (2019, June 12). A unique space for different religions? *Utopian and Sacred Architecture Studies.* CE, Italy: Aversa.

Barbara, A. (2017). Interni interreligiosi per l'inclusione sociale. In A. Longo, C. Rabbiosi, & P. Salvadeo (Eds.), *Forme dell'inclusività, pratiche spazi progetti* (pp. 184–195). Milano: Maggioli.

Beinhauer-Köhler, B., Roth, M., & Schwarz-Boenneke, B. (Eds.). (2015). *Viele Religionen—ein Raum?! Analysen, Diuskussionen und Konzepte.* Berlino: Frank & Timme.

Bernhardt, R. (2010). Raumkonzepte als Manifestation von Pluralismusverständnissen. *Kunst und Kirche, 2,* 22–26.

Bisogni, S. (2000). *Ricerche in architettura.* Napoli: ESI—Edizioni Scientifiche Italiane.

Blümer, P., & Marinkovic, P. (2010). Raum der Stille, Haus der Religionen oder Konsequenter Laizismus: Brauchen Hochshulen religiös geprägte Räume? *Kunst und Kirche, 2,* 44–47.

Bucher, R. (2010). Multireligiöse Räume an staatlichen Universitäten. Ein katolisches Plädoyer. *Kunst und Kirche, 2,* 48–51.

Cataluccio, F. (2017). L'astrattezza di Dio: Considerazioni sui luoghi interreligiosi. In A. Longo, C. Rabbiosi, & P. Salvadeo (Eds.), *Forme dell'inclusività, pratiche spazi progetti* (pp. 146–152). Milano: Maggioli.

Colzani, S., & Dondi, L. (2010, Ottobre). Nella città interetnica. Spazi per la cultura e per il culto. Un luogo di riflessione all'interno del campus universitario. *Master degree thesis, supervisor Roberto Rizzi.* Politecnico of Milan: School of Civil Architecture.

Cornoldi, A. (1995). *L'architettura dell'edificio sacro.* Roma: Officina.

De Carli, C. (1982). *Architettura Spazio Primario.* Milano: Hoepli.

Duttweiler, S. (2017a). Entschärfte Säkularisierung—gezähmte Religiosität—(multi-)religiöse Räume als räumliche Materialisierungen der Postsäkularisierung. *Geographica Helvetica, 72,* 283–294 (291).

Duttweiler, S. (2017b). Grenzargeit zwischen Sakralisierung und Profanität. Multireligiöse Räume in nicht-religiösen Kontexten. In U. Karstein, & T. Schmidt-Lux, *Architekturen und Artefakte. Zur Materialität des Religiösen* (pp. 193–211). Wien: Springer.

Filoramo, G. (2004). *Che cos'è la religione. Temi, metodi, problemi.* Torino: Giulio Einaudi.

Francesco. (2019, Febraury 4). *Apostolic journey of his Holliness Pope Francis to the United Arab Emirates.* Retrieved June 25, 2019, from The Holy See: http://w2.vatican.va/content/francesco/en/travels/2019/outside/documents/papa-francesco_20190204_documento-fratellanza-umana.html.

Francesco. (2019, June 21). *Meeting on the theme "Theology after Veritatis Gaudium in the context of the Mediterranean".* Retrieved June 29, 2019, from The Holy See: http://w2.vatican.va/content/francesco/en/speeches/2019/june/documents/papa-francesco_20190621_teologia-napoli.html.

Givone, S. (2018). *Quant'è vero Dio. Perchè non possiamo fare a meno della religione.* Milano: Solferino.

Hammerskjöld, D. (1957). Retrieved from United Nations: https://www.un.org/Depts/dhl/dag/meditationroom.htm.

Hochgeschwender, M. (2010). A secular Faith? Die Geschicte religiöser Räume an US-amerikanischen Universitäten. *Kunst und Kirche, 2,* 54–59.

Hohberg, G., & Stolte, R. (Eds.). (2013). *Das Haus der drei religionen*. Berlin: DOM publishers.

Kaernbach, A. (2010). Der Andachtsraum von Günter Uecker im Reichstagsgebäude in Berlin. *Kunst und Kirche, 2*, 22–25.

Kahn, L. I. (1957). Architecture is the thoughful making of space. The continual renewal of architecture comes fron changing concepts of space. *Perspecta, 4,* 2–3.

Khoury, A. T. (Eds.) (1987). *Lexikon religiöser grundbegriffe, Judentum—Christentum—Islam.* (Italian translation (1991). *Dizionario delle religioni monoteistiche. Islam, Cristianesimo, Ebraismo.* (trans: Benzi, G.)), Casale Monferrato: Piemme.

Knitter, P. F. (2005). Introducing theologies of religions, Orbis Book: Maryknoll/N.Y. (USA). (Italian translation, (2005), *Introduzione alle teologie delle religioni* (trans: Volpe, G.)). Brescia: Queriniana.

Kölbl, A. (2010). Interreligiöse Gebetsräume auf Flughäfen. *Kunst und Kirke*, 27–20.

Mio, L., Naboni, S., & Ometto, A. (2010, Aprile). Nella città interetnica. Spazi per la cultura e per il culto. Architettura e religione a Rubattino. *Tesi di laurea Magistrale, relatore Roberto Rizzi*. Politecnico of Milan: School of Civil Architecture.

Palese, A. (Ed.). (2013). *La guerra dei simboli. Comprendere e gestire i conflitti religiosi nello spazio pubblico*. Lugano: EuropressFtl.

Piva, A. (Ed.). (1995). *La città multietnica: lo spazio sacro*. Venezia: Marsilio.

Radojkovic, G., Zanolo, N., & Zucca, C. (2000, April). Nella città interetnica. Spazi per il culto e la cultura. Progetto di un centro di aggregazione multiculturale a Lampugnano. *Tesi di Laurea Magistrale, relatore Roberto Rizzi*. Politecnico of Milan: School of Civil Architecture.

Raffone, S. (2007). *La casa di Abramo. Aula di preghiera e centro di incontro cristiano islamico a Napoli*. Napoli: Clean.

Razzano, L. (2007, Maggio-Giugno). Ipotesi di un'architettura interreligiosa. Presupposti teologici. *Rassegna di teologia, 3*, 421–449.

Ries, J. (1992). Il sacro nei tre grandi monoteismi. In P. Gennaro (Ed.), *Architettura e spazio sacro nella modernità* (pp. 15–18). Milano: Abitare Segesta.

Rosi, P. (2010, Ottobre 21). Nella città interetnica. Spazi per il culto e la cultura. Progetto di un centro culturale e intereligioso in Via Cosenz. *Tesi di laurea Magistrale, relatore Roberto Rizzi*. Politecnico of Milan: School of Civil Architecture.

Russo, A. (2007, Maggio-Giugno). Lo spazio sacro nella città interetnica. *Rassegna di teologia, 3*, 403–420.

Sakralbau. (2004). *Detail* (9).

Sauer, M. (Ed.). (2015). *Zentrum Europaplatz. Die Verwandlung eines Unorts*. Zürich: espazium.

Scrinzi, F. (2008, Novembre). "Chiese integrate", dialogo religioso, laicità. *Confronti. Mensile di religione, politiche, società*, 1–5.

Sloterdijk, P. (2008). *Gottes eifer. Vom kampf der drei Monotheismen*. (Italian translation (2008), *Il furore di Dio. Sul conflitto dei tre monoteismi* (trans: Quadrelli, P.)). Milano: Raffaello Cortina.

Sorrentino, S., & Festa, F. S. (2007). *Le ragioni del dialogo. Grammatica del rapporto fra le religioni*. Torino: Città aperta.

Steiner, G. (1974). *Nostalgia for the Absolute*, (Massey Lectures, CBC Enterprises: Ottawa. Italian translation (2000), *La nostalgia dell'assoluto* (trans: Cornalba, L.)) Milano: Paravia Bruno Mondadori.

Tagliaferri, R. (1995). Lo spazio architettonico nell'esperienza religiosa. La mediazione spaziale della fede nel quadro della ritualità. In A. Piva, (Ed.), *La città multietnica: lo spazio sacro* (pp. 48–67). Venezia: Marsilio.

Vanacore, R. (2015, Ottobre 22). Architettura e spiritualità. Sperimentazioni progettuali per la "casa dell'Uno". *Convegno*. Salerno: Università degli studi di Salerno, Facoltà di Ingegneria.

Imagined Spaces in Church Architectural Furnishings: Solomon's Temple in Small-Scale Architectural Language

Rita Valenti and Emanuela Paternò

Abstract

Solomon's Temple has been a significant reference point for religious architecture throughout the centuries both from the symbolic and utopian point of view. It is possible to say that in the Judeo-Christian tradition, the inspiration for the temple architecture came directly from God. The intermediary was the powerful King Solomon who built the temple, a model of architectural perfection, obeying God's will. Despite its troubled history characterized by continuous destruction, the temple has been and still is a holy place for Hebrews, Muslims and Christians. The representation of Solomon's Temple in art can actually have a deeper symbolic significance. Probably, it can be associated with the human desire to have a religion and a place where, thanks to architecture as well, different cults could be joined in a single faith. It is known that numerous attempts have been made to rebuild the destroyed temple, a testimony of the utopian efforts to make architecture supplied by God live again for an equally perfect world. Actually, the study deals with imagined and built spaces of religious small-scale architecture such as antependia. In particular, the research focused on a chapel whose altar presents specific furnishings whose space language, both built and painted, is a clear hymn to the Solomonic and to the temple par excellence. Knowledge through innovative technologies, drawing attention to this specific religious symbology, can effectively contribute to the preservation of historical memory and to the inspiration of noble ideals in the next generations.

Keywords

Solomonic order • 3D survey • Photogrammetry • St. Francesco Saverio altar • Cultural heritage

1 Small Sacred Architecture: The Solomonic Motif and Its Symbology

After the earthquake which struck south-eastern Sicily in 1693, an intense reconstruction activity started out determining the planning process of some buildings of extremely high architectural and artistic value. This production determined the development of a new language which led to the distinction between the architecture of the seventeenth century and that of the following century.

The period defined as late Baroque engaged artists and craftsmen, the so-called *lapidum incisores* skilled at stone carving, who determined "the culmination and final flowering of Baroque art in Europe" as the UNESCO criterion for the inclusion of this area in the UNESCO World Heritage List explains.

In this new appeal to reconstruction, the clergy set artistic trends: churches and monasteries extraordinarily embellished the towns in Val di Noto. The study compiled for the research "The re-presentation of Cultural Heritage in Val di Noto: digital instruments for knowledge and dissemination" partially financed by the University of Catania highlighted the importance of the religious theme.

Actually, among the total artistic production, ecclesiastical furnishings arouse considerable interest. They are real jewels, highly expressive and rich in symbolic meanings. Tabernacles, altars, frontals, independently of the context they are set in, convey astonishing scenographic effects, almost dramatic, typical of the Baroque (Fig. 1).

The conducted research on architectural themes makes reference to the Solomonic, formally represented by the spiral column, considered the symbol of the mythic biblical

R. Valenti (✉) · E. Paternò
Special Educational Structure of Architecture, DICAR, University of Catania, Syracuse, Italy
e-mail: ritam.valenti@gmail.com

C. Gambardella et al. (eds.), *Advances in Utopian Studies and Sacred Architecture*,
Advances in Science, Technology & Innovation,
https://doi.org/10.1007/978-3-030-50765-7_17

Fig. 1 Examples of religious furnishings

building. It traditionally represents one of the most evocative architectural details which conjures up the idea of the temple.

According to legend, the ancient spiral columns which decorated Constantine's *pergula*, portico, were brought to Rome as spoils of the temple: eight of them were eventually included in the Bernini's reliquary Balconies, housed in the piers supporting the dome in St. Peter's Basilica.

Solomon's Temple, in religious architecture, is a symbolic and utopian reference point at the same time. In the Judeo-Christian tradition, inspiration for the architectural design of the temple came directly from God.

The powerful King Solomon, a simple intermediary, built the temple, model of architectural perfection, obeying God's will. Despite its turbulent story made up of destruction and reconstruction, the temple is a holy site for Hebrews, Muslims and Christians.

Definitely destroyed in 70 A.D., the temple combined myth and legend, becoming an example of perfection and expression of divine architecture to be used as a frame of reference. About the temple, a large number of treatises have been written, based on the words of the Holy Books, on ancient literary sources, on archaeological remains and on "gerosolimitan" topography.

In religious art, the temple has been represented lots of times, and in several representations, the spiral column is a recurrent theme (Valenti 2003). Baroque art, in particular, consecrates "the mythic image rather than the scientific one" adopting architectural parts and decorative elements which explicitly refers to Solomon's Temple (Tuzi 2002).

1.1 Case Study

In south-eastern Sicily, this artistic attitude is shown on a large scale in the church façades and on a small scale in the holy furniture designed as pieces of architecture. The studied architectural object presented in this paper is perfectly inserted in this context: it is the frontal of St. Francesco Saverio's altar in the Church of St. Francesco Borgia, an annexe to Collegio dei Gesuiti in Catania.

It is a real micro-architecture with scenery inspired by the Baroque canon which the church and the religious orders in particular took into account to spread catholic principles (Ruggieri Tricoli 1992). Indeed, Jesuit work in Sicily had particular relevance (Lima 2001); the order permeated every aspect of society, significantly contributing to spread architectural models which combined religious and artistic concepts. In the specific case, the altar decorated with the frontal is clearly modelled on the Blessed Luigi altar which the Jesuit Friar Andrea Pozzo designed in his famous treatise (Valenti and Paternò 2019). Placed in the right wing of the transept of the Church of St. Francesco Borgia, it was built between 1713 and 1740 and is extremely impressive for the richness of its sculptural elements and for the symbolic meaning of its four spiral columns. The symbolic concept of the Solomonic theme appears again in the frontal: the altar table holds a small-scale masterpiece, witness to a utopian attempt to revive "God infused" architecture. It is a structure whose language of its built and painted space is a clear hymn to the Solomonic theme and to the temple par excellence (Fig. 2).

The conducted research has not come across other similar examples.

Actually, there are a large number of Sicilian baroque frontals made of silver, coral or multi-coloured marble (marmi mischi) which represent with perspective effects, scenes evoking the temple (Rizzo 2008), tough far from the three-dimensional world on the surface (Fig. 3).

The representation of Solomonic columns included in a system of arches is a recurrent artistic figurative element in the Sicilian "architectural" frontals.

In the present case, the representation is expressed in an introverted way, so it is necessary to see through the three entrance porticos to perceive the scene.

Indeed, it is necessary, to feel like going through the entrance arches to have a complete experience of the space.

In detail, the architectural work reproduces a place of worship: groin vaults cover the space, a structure with small spiral columns and small columns with smooth pillars on pedestals, three staircases at the rear next to the apses and balustrades, a floor made of marble inlays: everything in the absolute respect of a rigorous symmetry.

The architectural plan is transversely divided into three parts, longitudinally the scene is partially built and partially painted (scene and background form three parts, two parts are delimited by the axis of the central columns and one is painted on the rear wall).

The inlays on the floor make up a pentalpha, inscribed in circles, near the central arch; near the other two arches, there are two seven-pointed stars.

It is necessary to wait for sun rays which filter from the dome lantern at noon to enjoy the physical and emotional view of all these symbolic signs. The stage machinery in the sunlight provides an ephemeral effect of transcendence capturing the viewer in a context which goes beyond architecture. The purpose seems to be the communication of the power and greatness of God.

Fig. 2 Surveyed micro-architecture of the altar frontal of St. Francesco Saverio

Fig. 3 Examples of Sicilian baroque frontals

Fig. 4 Architectural space lit up by sunrays at noon

"Through art, the message of *Ecclesia Triumphans* had to appeal to the heart, the senses and the intellect of the viewer for a complete spiritual adherence to the values of the Reformed Church" (Trigilia 2017) (Fig. 4).

Unfortunately, this jewel set in the altar of St. Francesco Saverio, even if not easily accessible, is not in good conditions. Several architectural components are missing, others are broken and the inlays are partially detached from the figurative frame.

The research has, therefore, undertaken a syncretic reading of the work of art applying a survey supported by all innovative instruments combined with photogrammetry and dedicated software.

The aim is to make accessible, though virtually, this highly introspective place, contributing to its spiritual preservation in the collective memory and also protecting its characteristics of intimately secluded place.

2 Point Cloud Acquisition Methods

This paper contains some reflections on the relation between religious symbology and digital technologies for the dissemination of information not exclusively on architecture.

The adopted approach aims at the detection of 3D data acquisition, processing and representation about complex small-scale architecture in closed environments. (Rodríguez-González et al. 2015; Caroti et al. 2018)

Nowadays, it is possible to use a vast array of technological instruments for the metric and geometric data surveying that can be precisely retrieved and processed through innovative graphical representation systems.

However, it is advisable to know the characteristics of any technical solution available in order to choose the most appropriate to the specific situations. The latest generations of 3D survey techniques can be classified as range-based techniques (active sensors) and image-based techniques (passive sensors).

Consequently, models obtained through Terrestrial Laser Scanning (TLS) and Structure from Motion (SfM) were compared to consider the advantages and disadvantages of these survey methodologies to find the perfect solution for the complex and unusual studied case.

In particular, two survey campaigns were conducted. During the first one, an instrumental survey through Faro CAM2 Focus laser scanner was carried out for the acquisition of metric data, and during the second campaign, a photogrammetric survey was carried out for the acquisition of colorimetric data connected with the surface materials.

2.1 Data Acquisition Using TLS Technology

In the research, the details of the graphical representation are made clear through the point cloud obtained with laser scanning technology. The adopted instrument, thanks to its reduced weight and small dimension, was appropriate to the needs.

In the specific case, a phase shift laser scanner was employed; here, the distance was measured via the phase difference between the emitted wave and the received signal.

This technique refers to mathematical algorithms to identify spatial coordinates. Phase shift scanners present an extremely high scanning speed and extreme accuracy in point density, thus making it possible to capture details in the mm range, as far as 0.6 mm between each point at a distance of 10 m. However, range limits (0.6–130 m, indoors or outdoors, with vertical incidence on a reflecting surface equal to 90%) caused a few problems in the generation of a satisfying raw point cloud.

The input data acquired from the single scans are characterized not only by x, y, z coordinates but also by values connected to the reflectance of the measured surface. Such values may depend on different factors, for example, the composition of the surface, morphology and the presence of materials which can absorb or reflect incident radiation, humidity, etc.

The considerable proximity of the instrument to surfaces with high reflectance may generate indefinite and imprecise point clouds.

Instruments able to reduce the minimal distance below the present limits are still lacking, so it was necessary to do different tests to establish the appropriate parameters for a positive result.

In order to overcome the problem of the laser scanner proximity to the object of study and taking into account its already mentioned characteristics, resolution parameters were set 1/10 and 8× quality.

It was also necessary to take different scans, five for the interior and three for the exterior of the altar table, so that even the hidden parts could be surveyed.

Data were then processed through cyclone, and it was necessary to use during the survey campaign; three chessboards targets appropriately detected to facilitate, in the post-processing phase, the manual setting of the scans (Fig. 5).

2.2 Data Acquisition Using Photogrammetric Technology

The point cloud obtained from instrumental survey was integrated with that from SfM survey.

Actually, it is very often necessary to integrate different methodologies to gain a highly precise and reliable final model, since, as for the present case, laser scanning survey is insufficient, for different reasons. (Bilis et al. 2017; Pulcrano et al. 2019)

SfM is a technique for the automatic reconstruction of a 3D scene from a set of digital photos, automatically collecting features (correspondences) such as key points recognizable in various images.

The features detected from all the images will then be matched (image matching) (Fig. 6).

Different shots were taken using a Nikon D5300 digital camera especially of the areas where it was almost impossible to place the laser scanner because of the very small dimensions of the altar table.

All shots were taken using a remote control not only to avoid blurred photos but because this micro-piece of architecture is almost inaccessible to the operator. Therefore, it was also possible to survey the vaulted ceiling which is actually hard to explore.

Starting from the key points, through bundle adjustment, the spatial coordinates x, y, z of each key point can be refined describing the scene geometry of the camera position for every single image.

The features are physically processed into a 3D sparse point cloud, that is, a less dense point cloud.

The sparse point cloud is, then, made denser adding an enormous amount of points.

In the specific case, the coordinates of the elements surrounding the key point were detected so to generate a dense point cloud as a final result.

To generate a final, polygonal and coloured 3D model, the point cloud is processed through mesh texturing software. In the specific case, 3D Zephyr Aerial featured the whole digital photogrammetric processing.

It is worth mentioning that in order to detect features, survey must be carried out as rapidly as possible, to avoid

Fig. 5 Laser scanning survey

Fig. 6 Photographic survey

light changing effects. It is also advisable to turn the automatic camera flash off.

However, considering the dim light working conditions, it was necessary to turn it on to focus the pointed objects.

2.3 Data Integration

The dense point cloud generated through photogrammetric survey was aligned with the TLS point cloud, previously imported from 3DF Zephyr Aerial.

This model was then submitted to the next phases of mesh extraction and mesh texturing to obtain a photorealistic 3D model.

In order to contextualize the produced model, it was appropriately connected to a survey previously carried out integrating different techniques and referred to the altar holding the described miniature (Valenti and Paternò 2019) (Fig. 7).

Thus, it was possible to present, through digital rendering software, the complete artefact which has a strong architectural value.

In order to reduce the imperfections due to objective difficulties encountered during the research, a long and meticulous retopology process was executed using the Geomagic Wrap software which recreated a flawless and accurate polygonal model.

However, at the moment, it is impossible to employ a single survey technique which presents a variety of characteristics able to satisfy different needs such as portability, low cost, quality and accuracy of geometric data, data acquisition speed, total automation of 3D model generation, etc (Meschini 2011).

Fig. 7 Final model produced with photogrammetry and TLS technologies: a reconstruction hypothesis with the insertion of the missing columns

That is why, it is necessary to combine different techniques making the most of the advantages to overcome the possible limits.

In particular, knowing the exact performance of a survey system and of other supporting technologies is fundamentally important for data integration and for a high-quality final product.

Comparison with the two approaches shows how the acquired geometric data through laser scanning technology are, no doubt, more precise than those acquired through photogrammetric survey, even if photogrammetric survey provides more accurate colorimetric data.

Thus, the comparative analysis shows that the results from the applied methodologies are similar, but their combination seems to be the most appropriate approach to complex pieces of architecture whose dimensions prevent operations using a single instrument (Fig. 8).

The research has produced a 3D model able to transmit complex information, to analyse and interpret architectural data and above all to visualize formal, physical and colorimetric data otherwise difficult to experience in real life. Moreover, the views of the model on content sharing platforms certainly are a form of culture dissemination.

3 Analysis and Considerations About Reconstruction Hypotheses and Final Products

The 3D rendering of this small architectural structure allows its virtual exploration, stirring a spiritual and emotional involvement similar to a real experience occurring in a place designed in such a way to be commensurate with human existence (Fig. 9).

Fig. 8 Comparison between photogrammetric and TLS models

Fig. 9 Final 3D model: real perception and immersive perception

Taking into account the small dimensions of the altar frontal, this is the only possibility to have an immersive experience of this imaginary and symbolic place, respecting its impalpable characteristics.

Through digital technology, it was possible to achieve a kind of virtual restoration of the structure, recovering in the model the absent architectural components.

After an interpretative study of its forms, the broken portions of the groin vaults, the Solomonic columns and the missing pedestals were added to the model generated from the surveying of the structure.

The final visualization was obtained through a two-level process: in the first one, the added parts are highlighted with colours (Fig. 9); in the second, as previously shown, no

Fig. 10 Model of the altar frontal of St. Francesco Saverio: analysis and perspective perceptions

differentiation is signalled between what was surveyed and what was inserted, in order to give a complete rendering of the structure.

Last, the analysis of the horizontal section through instrumental survey has revealed a slight convergence of the alignment lines of architectural elements.

The resulting effect is that of an accentuation of the real perspective view which gives the scene broader illusory depth than what the human eye can totally perceive from the exterior, including the groin vault, only standing at a height of 50 cm (Fig. 10).

The conducted research, despite the objective difficulties encountered because of the peculiarity of the structure, aims at making visible, through the final 3D model, what is purposely difficult to understand, both for its manufacture and for its meaning.

In particular, the study tested approach methodologies of integrated survey, indoors, on small-scale architecture and through the final product provides a basis for any further determination, useful for the interpretative understanding of the structure.

4 Conclusions

From a general point of view, the present contribution highlights all the possibilities offered by a complex survey implementation on smaller scale architecture with the support of digital instruments.

The 3D model processed in the two versions, both the model of the existing conditions and the model of the virtually restored work (Valenti and Giuliano 2015), has a high scientific value for its geometric and perceptive determinations and a great communicative value making a highly symbolic work of art accessible to a wider audience.

Increasing accessibility to something which was inwardly and intimately designed activates the protection of the work itself allowing the public to discover its sacred space interrupting the mechanism of exclusion in the course of the tour.

Architecture heritage is in fact "the witness of our relationship with time, in particular of the simultaneous relationship between past and future, which represents, when shared, a form of contemporaneity" (Augé 2009, p. 45).

Acknowledgements The authors wish to thank Dr. Gioconda Lamagna, Director of the Museum Centre in Catania, who granted access to the cultural sites and Dr. Luisa Paladino responsible manager of the Regional Interdisciplinary Museum Service in Catania who granted permission to conduct survey on the Church of St. Francesco Borgia in Catania; Dr. Tommaso Palermo for the bibliographical research at the Regional University Library in Catania.This work has partially been financed by the University of Catania within the project "Piano della Ricerca Dipartimentale 2016–2018" of the Department of Civil Engineering and Architecture.Translation by Dr. Melania Grancagnolo.The conducted study within the research programme "Re-presentation of Baroque religious architecture between geometry and symbology. From large to small scale, from façades to altars" (project PON-NEPTIS) is the result of the collaboration of all the authors. In particular, Rita Valenti wrote Sects. 1, 1.1, 3 and 4; Emanuela Paternò wrote Sects. 2, 2.1, 2.2 and 2.3.

References

Augé, M. (2009). *Che fine a fatto il futuro? Dai non luoghi al non tempo*. Como: Elèuthera.

Bilis, T., Kouimtzoglou, T., Magnisali, M., & Tokmakidis, P. (2017). The use of 3D scanning and photogrammetry techniques in the case study of the Roman theatre of Nikopolis. Surveying, virtual reconstruction and restoration study. *The International Archives of the Photogrammetry, Remote Sensing and Spatial Information Sciences*, 97–103.

Caroti, G., Piemonte, A., Martínez-Espejo Zaragoza, I., & Brambilla, G. (2018). Indoor photogrammetry using UAVS with protective structures: issues and precision tests. *The International Archives of the Photogrammetry, Remote Sensing and Spatial Information Sciences*, 137–142.

Lima, A. I. (2001). *Architettura e urbanistica della compagnia di Gesù in Sicilia*. Palermo: Novecento.

Meschini, A. (2011). Tecnologie digitali e comunicazione dei beni culturali. Stato dell'arte e prospettive di sviluppo. *DISEGNARECON, 4*(8), 14–24.

Pulcrano, M., Scandurra, S., Minin, G., & di Luggo, A. (2019). 3D cameras acquisitions for the documentation of cultural heritage. *The International Archives of the Photogrammetry, Remote Sensing and Spatial Information Sciences*, 639–646.

Rizzo, S. (2008). *Architetture barocche in argento e corallo*. (S. Rizzo, A cura di). Catania, Italia: Giuseppe Maimone Editore.

Rodríguez-Gonzálvez, P., Nocerino, E., Menna, F., Minto, S., & Remondino, F. (2015). 3D surveying & modeling of underground passages in WWI fortifications. *The International Archives of the Photogrammetry, Remote Sensing and Spatial Information Sciences*, 17–24.

Ruggieri Tricoli, M. (1992). *Il teatro e l'altare: paliotti "d'architettura" in Sicilia*. Palermo, Italia: Grifo.

Trigilia, L. (2017). Sala ecclesiastica e sala teatrale. L'altare nell'universo barocco. Modelli romani in Sicilia, Malta e America Latina. *Annali del Barocco in Sicilia* (pp. 89–104).

Tuzi, S. (2002). *Le Colonne e il Tempio di Salomone. La storia, la leggenda, la fortuna*. Roma, Italia: Gangemi Editore.

Valenti, R. (2003). Juan Andrea Ricci e il disegno dell'ordine salomonico intero fra pittura ed architettura. *Ikhnos Analisi grafica e storia della rappresentazione*, 11–34.

Valenti, R., & Giuliano, S. (2015). Un percorso diacronico: il restauro virtuale delle mura spagnole in Ortigia. *DISEGNARE CON, 8*(14), 1–13.

Valenti, R., & Paternò, E. (2019). A comparison between TLS and UAV technologies for historical investigation. In *The International Archives of the Photogrammetry, Remote Sensing and Spatial Information Sciences* (Vol. XLII-2/W9, pp. 739–745).

The Convent of La Tourette Between Communitarian Utopia and Religion

Chiara Ingrosso

Abstract

The paper will be focused on one of Le Corbusier's masterpiece: the convent of La Tourette (1952–1960). The convent is an emblematic synthesis of many Le Corbusier's theories about dwellings and common spaces so that through this work, he gave tangible form to his urban imagination. Objective of the essay is to demonstrate how the convent is an exemplary case of realized utopia, with all the contradictions inherent in it: La Tourette remains, indeed, a beneficiary of his ill-fated urban dreaming. Le Corbusier, after years of experimentation, arrived to the to maturity, was able to apply here all his theory. He accepted the religious commission in order to realize his experimentation on housing and on building a community, rather than simply encasing an act of cult. The analogy with the Charterhouse of Ema was fundamental for being the complex similar to a workers settlement, rather that a religious refugee. In La Tourette, the religiosity had to do with the pureness of the shapes and of the volumes, the roughness of the concrete and the minimum sizes of its cells. This approach derives directly from his Calvinist education, which informed his whole *weltanschauung*. As a fulfilled utopia, La Tourette is an open architecture, able to transform itself over time, according to the needs of its users. Today, it is a centre of study on spirituality but also on philosophy, art and architecture, and its spaces have been changing, representing the community occupying it.

Keywords

Housing • Cell • Common spaces • Monastic model • Fulfilled utopia

C. Ingrosso (✉)
Università della Campania "Luigi Vanvitelli", Caserta, Italy
e-mail: chiara.ingrosso@unicampania.it

1 Genesis

During the war, in 1943, the Dominicans decided to buy a vast land called "La Tourette", in Eveux-sur-Arbresle, 30 km from Lyon, in order to found a new convent of studies to train young Dominicans in their priory career. It was an agricultural field, developed on a gentle slope, with an old mansion and a wood. At the beginning, the project had been entrusted to the Savoyard architect Maurice Novarina, author of the Notre-Dame de Toute Grace du Plateau d'Assy (1938–1949). In 1952, the prior of the Province of Lyon, Pierre-Damase Belaud, decided to give the task to Le Corbusier. Father Marie-Alain Couturier was the one who asked Le Corbusier to build the new convent (Aussibal 1988). The friendship between Le Corbusier and Couturier was already consolidated thanks to their exchanges during the Ronchamp designing, when he had the opportunity to see directly his talent. Couturier had a fundamental role in securing the commissions of both La Tourette and the Chapel of Ronchamp, even if, unfortunately, he died in 1954, before the inauguration of Le Corbusier's work: he re-chose Le Corbusier for La Tourette for his modern approach, able to give a radical touch to the project of the convent.

The commission for the convent arrived when Le Corbusier's studio was at a very busy time. After the World War II, the Swiss architect's fame spread throughout the world and permitted him to build on a larger scale, even in the homeland: above all, Unité d'Habitation (1947–1952), Chapel of Notre Dame du Haut (1951–1955), Chandigar (1951–1965) (Cohen and Pare 2018). The French branch of the Dominican order was one of the most important commitments of this period for Le Corbusier, giving him the opportunity to build some the of most famous works of his career, namely, in addition to the aforementioned Rochamp and La Tourette, the unfinished chapel of Firminy (Cocagnac and Capellades 1955, 1960, 1964). The Rochamp project slightly preceded La Tourette and both saw Le Corbusier in close partnership with the religious order. Because of the

C. Gambardella et al. (eds.), *Advances in Utopian Studies and Sacred Architecture*,
Advances in Science, Technology & Innovation,
https://doi.org/10.1007/978-3-030-50765-7_18

scarce economic resources of the Dominicans, the construction took a long time, and for the inauguration of the convent, they had to wait until 1960, only five years before Le Corbusier's death (Fig. 1).

2 In Some Religious Way

Marie-Alain Couturier was an important member of the *Ateliers d'Art sacré*, an artistic movement based in Paris and founded in 1919 by Maurice Denis with George Desvalliéres, aimed at reconciling Catholicism with the secular spirituality (Couturier 1989). Its review *L'Art Sacré*, of which Couturier was co-editor with Pére Pie-Raymond Régamey, was the most effective dissemination tool of the movement (Caussé 2001). In its pages, a systematic operation to modernize sacred art was carried out, opening it up to the contemporary art and architecture, also by atheist authors (Lion 2010). This attitude that characterized the Dominicans of France during the 1950s was so strong that it was noticed even abroad, and the order was strongly criticized also by the Vatican, which considered *the new ideas and tendencies, not only exaggerated, but even erroneous* (O'Meara 1994).

Le Corbusier was not a believer, but Couturier was not interested. He wrote: *We knew very well that some of the artists were not strictly Christians; that some were separated from us by serious divergences of a political as well as of an intellectual order. Trusting in Providence, we told ourselves that a great artist is always a great spiritual being, each in his own manner.* (Couturier 1947, 122). For his part, Le Corbusier was clear: he accepted the project because it was concerned with housing a community, rather than simply encasing an act of cult. In 1961, declining a request for another church commission, he explained: *I built* [...] *the Convent of La Tourette* [...] *because the program (ritual, human scale, space, silence,* etc.*) was favourable, as also were the landscape conditions exceptional. I am not a builder of churches.* [...] *Had you said to me, "Will you create a place open all the year, situated on the hilltops in the calm and the dignity, in the nobleness of the beautiful*

Fig. 1 The convent of La Tourette

Jura site?", the problem could have been considered. It was a problem of psychic nature and, for me, of decisive value. (Purdy 1986, 291).

Despite having received a Calvinist Protestant education, Le Corbusier was an atheist; his spirituality had nothing to share with the Christian religion and, by his own admission, La Tourette interested him for reasons that he was able to filter out from Catholicism (McNamara 1999). He developed an idea of sacrum in a universal way, which over the ages became one with the search for rules to apply to collective life and architecture. A convent was an ideal place to apply architectural ideas about frugality, austerity and individual discipline adapted to industrialization and minimalism. In some way, with la Tourette, the Le Corbusier's monastic dream came true.

The project of the convent gave him the opportunity to imagine a system of monastic cells connected with collective spaces, all in a deep relationship with a natural site. From the Immeubles Villas (1922), to the *Unité d'Habitation* in Marseilles (1947), and then with La Tourette, Le Corbusier's research was seeking a collective enclave in which each cell was a private space in continuous dialogue with the community spaces (Curtis 1986). In Marseille, the study on the cells in a collective building, starting from the Immeubles Villas and the Maison Citrohan (1920), found real conformation. The cells isolated of the *Unité d'habitation* (e.g. the children's rooms) are the most direct precedents of the La Tourette dwellings, especially for its measurements based on the proportions of the Modulor (length 5.92 m, width 1.83 m, height 2.26, for a total of 10.83 m^2) (Fig. 2).

3 Models

If Le Corbusier seems to have accepted the project because it was concerned with housing a community, rather than simply encasing an act of cult (Purdy 1986), the typology of the convent derives from two religious buildings: the Abbey of Le Thoronet and the Charterhouse of Ema. Both architectures were re-interpreted by the architect in order to satisfy the requirements of the Dominicans, applying some of the theories he had been elaborating for a long time.

In 1953, Father Pierre Marie-Alain Couturier alerted the architect to Le Thoronet, a twelfth-century Cistercian abbey located near Provence. He wrote two letters with some sketches, published by Jean Petit in 1961 (Petit 1961), in which he suggested to Le Corbusier the monastery as a reference. At the same time, Couturier gave him a functional program including, according to the traditional plan, three large volumes around the cloister: the church with the refectory in front, the chapter hall on the third side, and two large meeting rooms on the fourth side, the large library, the cells and several common rooms on the first floor. There a lot of analogies between La Tourette and the ancient Abbey of Le Thoronet: *Le Corbusier could hardly be insensitive to the harmonious arrangement of pure forms at Le Thoronet, merely placed next to each other, nor to the blind walls enclosing the inner spaces, nor to the luminous internal harmony* (Erlande-Brandenburg 2000, 69). Le Thoronet inspired him by its roughness and pure geometry, its whiteness, above all in the use of concrete, which was a material largely adopted by the architect at that time. Couturier himself asked Le Corbusier to construct a building with poor materials: hence the choice of raw concrete.

If the Dominicans are neither Cistercians nor an enclosed order, the suggestion from Couturier to look at a Cistercian abbey made the architect free to look for (and to interpret) various historical typologies as references, introducing different levels of autonomy from the original model. In this way, beside Le Thoronet, Le Corbusier chose another Cistercian monastery as a fundamental reference for La Tourette: the Charterhouse of Ema in Galluzzo, near Firenze (Serenyi 1975) (Fig. 3).

Paul Serenyi told us about the references and the means of investigation, with realized and unrealized works, through which Le Corbusier arrived at La Tourette, whose typology sends us back to the Charterhouse of Ema (Serenyi 1967). The charterhouse of Ema had been such a fruitful source of inspiration throughout his career (Talamona 2012), fundamental for his research on dwellings, which, following Serenyi, we could say that Le Corbusier founded the concept of housing on the monastic model. With his worlds: *The beginning of these studies [for dwellings], for me, goes back to my visit to the Carthusian monastery of Ema near Florence, in 1907. In the musical landscape of Tuscany, I saw a modern city crowning a hill. The noblest silhouette in the landscape, an uninterrupted crown of monk's cells; each cell has a view of the plain and opens on a lower level on an entirely enclosed garden. I thought I had never seen such a happy interpretation of dwelling. The back of each cell opens by a door and a wicket on a circular street. This street is covered by an arcade: the cloister. Through this way, the monastery services operate—prayer, visits, food, funerals. This "modern city" dates from the fifteenth century. Its radiant vision has always stayed with me. In 1910, returning from Athens, I again stopped at Ema* (Le Corbusier 1991, 91).

About the charterhouse, Le Corbusier left us some famous sketches, with plants and sections of the monastic dwellings linked to the cloister with the garden, and several letters to his parents. In one of these, he wrote: *I would like to live my whole life in what they call their cells. It is the [perfect] solution to the working man's house type, unique or rather an earthly paradise* (Allen Brooks 1997, 49.) On the basis of his words, we would say the religious complex seems to have been chosen as model for La Tourette not for

Fig. 2 Facade of the convent with the *pilotis* sitting on the hill

being religious but, on the contrary, regardless of its religiosity. In this way, for La Tourette, the reference was the worker's settlement as a rational community, with its rigorous organization, which Le Corbusier had been studying for years and applying it to different projects. Like Purdy noted: *For Le Corbusier, any industrial society must be centrally located, hierarchically organized, administered from above, with the most responsible people in the most responsible position* (Purdy 1986, 289). As model of complex working man's houses, the Charterhouse indicated for La Tourette the functional organization: an efficient program for a convent considered as an *organism* divided in private cells and collective spaces (Sutcliffe 1977, 219).

The silhouette of the religious settlement on the top of the hill, the "modern city", as Le Corbusier himself called the charterhouse, and its relationship with the landscape, influenced the location of La Tourette, but here the structure sits on the hill. During his first visit to the site in 1953, Le Corbusier sketched the precise place where to build the convent, a land with a magnificent view over the Arbresle valley. He drew the landscape: the wood, the fields, the mountain on the horizon, a old sequoia plant, and a ancient mansion called "the castle" to the north, all with sun orientation and the horizon line. He decided to develop the project with this line as a reference, without excavating the earth and resting the building on *pilotis* of different heights.

La Tourette is not an urban convent, but it is isolated from the city where the Dominicans traditionally built their communities. We would say La Tourette is like a new religious citadel on the hill, devoted to the studies. Also, its site was interpreted: *as an attempt to emphasize the building's separateness, and by Couturier's standards, its sacredness* (McNamara 1999, 6).

The biggest difference between Ema, Le Thoronet and La Tourette concerns the different organization of the Dominicans compared with the Carthusians, especially in their communal lives. As Colleman noted: *both sleep in individual cells rather than in dormitories, but whereas Carthusians sleep, eat and work in their cells, Dominicans primarily sleep and study in theirs. Thus, a Dominican cell*

Fig. 3 Facade of the convent with the grid of the cells and the "undulatory glass surface"

is a simple room with a bed, desk, wardrobe and washbasin. Carthusians live in small independent houses. [...] *Dominicans eat in community and are preachers. Carthusians eat in isolation and hardly ever speak. They are hermits who require the solitude that radically claustral settings provide. Dominicans engage the world through preaching and teaching and were, until church and state divided in France, most often housed in convents located in university towns* (Coleman 2005, 142).

Dominican's are a preaching organization, rather than an enclosed order, and the convent is a building based on the dialectical exercise Dominicans play out in their daily life: apostolic and monastic (Rowe 1961). Poverty, up to begging, was another important principle that characterized the order, implemented by the project of pure and cheap material and form.

As regards the cloister, it is inspired by Cistercian cloisters but with a totally new spatiality. The initial idea, elaborated with the engineer Iannis Xenaxis, was to place the cloister on the roof, in order to leave the unevenness of the terrain intact, and provided for a long ramp that would reach the roof. The proposal clashed with the perplexities of the Dominicans, and for lack of money, it was abandoned. The cloister became a cross-shaped path that crosses the central void of the building, an internal architectural promenade that connects the church to the different spaces of the monastery and closed by full-height windows. The introduction of the so called "undulatory glass surface" applied to the articulation of the window frames (which is also present in the Secretariat at Chandigarh) is due to the inventiveness of Xenaxis, still unknown as the great avant-garde musician he would become, and whose most important works "Metastaseis" were composed at that time (Xenakis 1987) (Fig. 4).

4 La Tourette as a Fulfilled Utopia. Conclusion

The convent (Fig. 5) is an emblematic synthesis of many of Le Corbusier's theories about dwellings and public spaces so that through this work, he gave tangible form to his research,

Fig. 4 The cloister with the "undulatory glass surface" applied to the articulation of the window frames

Fig. 5 A view of the convent on the slope of the hill

remaining a beneficiary of his ill-fated architectural and urban dreaming (Coleman 2005).

Paul Serenyi, speaking about the association between La Tourette and the Charterhouse of Ema, refers also to the suggestions that the Phalanstery (probably through Tony Garnier) had on the young Le Corbusier. For Banham, in disagreement with Serenyi, it is not just a matter of typology: it is impossible to reduce Le Corbusier's concept of mass housing to a merger of convent and phalanstery (Banham 1987). His research is about the most complex concept of the *maison des homes,* and the community utopias expressed by Fourier had a huge ideological rule, indicating also a different way of living, in one word a possible utopia. For this reason, speaking about la Tourette, he said: *And one does not have to be much of a cynic to observe that just as the monastery may be seen as a caricature of Corbu's intentions in the reform of urban housing, so La Tourette can be interpreted as a satire on the monastery* [...]. *Here is a building aggressively claustral in form that contains no useful cloister, for its central square is too clogged with miscellaneous structures to be used* (Banham 1987, 115). If La Tourette was a caricature of Le Corbusier's theory about housing or not, it is hard to say. We could say, instead, the convent was a fulfilled utopia, with all the contradictions that ensue. The convent is also an open architecture, able to transform itself over time, according to the needs of its users. Just a few years after its inauguration, indeed, it lost its original function of convent of studies, because of the crisis of the religious vocation and the consequent lack of friars. Today, it is a centre of study on spirituality but also on philosophy, art and architecture and its spaces have been changing, representing the community occupying it.

References

Allen Brooks, H. (Ed.). (1997). *Le Corbusier's formative years* (p. 49). London: University of Chicago Press.

Aussibal, A. (1988). *Saint-Marie de la Tourette. Le counvent Le Corbusier.* L'Arbresle: Edité par Couvent Sainte-Marie.

Banham, R. (1987). La Maison des hommes and La Misère des villes: Le Corbusier and the architecture of mass housing. In Brooks H. Allen (Ed.), *Le Corbusier* (p. 115). Princeton, NJ: Princeton University Press.
Caussé, F. (2001). La critique architecturale dans la revue L'Art Sacré (1937–1968). *Livraisons d'histoire de l'architecture, 2,* 27–36.
Cocagnac, A.M., Capellades, M. R., & Le Corbusier (1955). *Ronchamp. L'Art Sacré, 1–2.*
Cocagnac, A. M., & Capellades, M. R. (1960). Le Corbusier. Un couvent dominicain. *L'Art Sacré 7–8.*
Cocagnac, A. M., & Capellades, M. R. (1964). Un projet d'église paroissiale de Le Corbusier. *L'Art Sacré, 3–4.*
Cohen, J. L., & Pare, R. (2018). *Le Corbusier: The Buildings*. Thames & Hudson.
Coleman, N. (2005). *Utopias and architecture*. USA/Canada: Routledge.
Couturier, M. A. (1947). A modern French Church in the Alps. *Harper's Bazaar, 81,* 122.
Couturier, M. A. (1989). *Sacred Art*, trans. by William Granger Ryan. Austin: University of Texas Press.
Curtis, W. J. R. (1986). *Le Corbusier, ideas and forms*. London: Phaidon.
Erlande-Brandenburg, A. (2000). *Three cistercian sisters of provence: Sénanque, Silvacane, Le Thoronet* (p. 69). In Christopher Thierry (Trans.). Paris: Le Éditions du Huitième Jour.
Le Corbusier. (1991). *Precisions: On the present state of architecture and city planning* (p. 91) (1930), Edith Schreiber Aujame (Trans.). Cambridge, MA: MIT Press.
Lion, A. (2010). Sacred art and modernity in France. The Role of Fr. Couturier. *Revue de l'histoire des religions, 1,* 109–126.
McNamara, D. (1999). *Almost religious: couturier* (p. 2). Sacred Architecture: Le Corbusier and the Monastery of La Tourette.
O'Meara, T. (1994). Raid on the Dominicans: The repression of 1954. *America, 170,* 8–16.
Petit, J. (1961). *Un Couvent de Le Corbusier*. Paris: Les Editions de Minuit.
Purdy, M. (1986). Le Corbusier and the theological program. In *The open hand, essay on Le Corbisier.* Cambridge: MIT Press.
Rowe, C. (1961). Dominican Monastery of La Tourette. Eveux-sur-Arbresle, Lyons. *Architectural Review, 129,* 401–410.
Serenyi, P. (1967). Le Corbusier, Fourier, and the Monastery of Ema. In Serenyi P. (Ed.) (1975). *Le Corbusier in Perspective* (pp. 103–116). New Jersey: Prentice Hal.
Sutcliffe, A. (1977). A vision of Utopia. In Russell Walden (Ed.), *The open hand, essays on Le Corbusier*. Cambridge, MA: MIT Press.
Talamona, M. (2012). *L'Italia di Le Corbusier*. Milano: Electa.
Xenakis, I. (1987). The Monastery of La Tourette. In *The garland essays.* Princeton, NJ: Princeton University Press.

Sacred Spaces and Virtual Design. Toward a Digital Prayer Interreligious Architecture

Mariateresa Giammetti

Abstract

The University of Naples Federico II Department of Architecture is developing research on new types of interreligious space for the three Abrahamic confessions. The research, called *The Shape of Water* (SOW), began in the early 2000s and it has been published in several articles and volumes. To increase the results of the research, the Department of Architecture has promoted another study (called ALEPH) in collaboration with the Federico II University Engineering Department. ALEPH's aim is to design and program software to create a virtual museum related to the three Abrahamic confessions that could be adopted by European museums to encourage intercultural exchange between the three major religious communities in Europe. The software can create a database to collect the point of view of experts and ordinary people about the perception of new types of hybrid interreligious spaces. ALEPH uses digital technologies to build a virtual museum tour into two areas: the first one is a virtual tour through interior spaces of religious buildings particularly significant for each of the three confessions; the second one allows us to go into a virtual hybrid prayer space that was developed by SOW. The strategies produced using the ALEPH digital platform will support ongoing studies of SOW, implementing its databases. So, the visitors of the virtual tour must answer a questionnaire (made in augmented reality) about the perception of the places of the virtual tour and on the space for intercultural prayer. The output of the platform provides a series of useful indicators to map the types and compositing models of the hybrid prayer spaces for the three monotheistic confessions developed by SOW.

Keywords

Preyer Spaces • Virtual • Museum • Interculturalism • Europe • Abrahamic religions

1 Introduction

Recently the US statistical center Pew Research has published a study on the religious geography of the world in 2050 (Pew Research Center. Religion & Public Life 2015). Nowadays, three out of 10 Earth inhabitants are Christians and 2.5 are Muslims; in 2050 the number of Muslims will be almost equal to the number of Christians. In 2010 Muslims were 8.4% of the European population, in 2050 they will be 11% of the population. The number of Muslims would still increase, even if no more migrants arrived in Europe from now on because the percentage is linked to the higher fertility in the Muslim community that has an average age younger than 13 years compared to non-Muslim European citizens. In this intersection of cultures and faiths, the two monotheistic confessions are shaping of today's Europe through their strong demographic presence: surely the future of European cities will be conditioned by the social issues raised by the meeting between local inhabitants and migrants. This new mix of peoples has transformed cities into places of transformation, into threshold spaces. The threshold is a space of transition between two different spaces. The threshold has a temporal dimension. How long it takes to pass from one condition to another? The threshold has a cultural and existential dimension. Today's Europe is on its threshold: beyond there is not only the opportunity to make a new identity subject arisen from the phenomenon of migration but also the opportunity to make a new European national post-identity (Giammetti 2018a).

The European cultural agenda is developing support measures to create a transnational cultural identity, but at the same time, there are still many contradictions and

M. Giammetti (✉)
Department of Architecture of Federico II, University of Naples, Via Forno Vecchio 36, 80134 Naples, Italy
e-mail: mariateresa.giammetti@unina.it

C. Gambardella et al. (eds.), *Advances in Utopian Studies and Sacred Architecture*,
Advances in Science, Technology & Innovation,
https://doi.org/10.1007/978-3-030-50765-7_19

disagreements. There are many political strategies put forth to solve the issue of immigration and the architecture is becoming an important tool for finding new expressions of community and post-national identity (Lawson 2001).

> In Berlin, where once the religious emancipation and the enlightenment became a European project through the friendship of Lessing and Moses Mendelssohn, wherein 1989 the iron curtain fell by the power of words and the power of non-violent resistance, the House of One will be a place where the coexistence of religions is lived peacefully, in great openness and appreciation of diversity (House of One. Three Religions 2020).

This is the concept that opens the website dedicated to the House of One, an architecture that will contain three prayer spaces for Christians, Muslims, and Jews. The House of One, designed by the German-Italian studio Kuenn Malvezzi, will be built in Berlin and its first temporary pavilion was inaugurated not long ago. The experience of Berlin shows that religion represents one of the key points to meet the challenge of integration and to include in a non-oppositional dynamic the communities of migrants who come from Islamic countries. Architectures like the House of One are an experiment to try to decline the relationship with diversity in the name of the meeting, proximity, collaboration, mutuality, and mutual knowledge. An important project on religious dialogue can meet the challenge of integration and it may have a very significant political impact. The political integration may help to reduce the contact between religion and the alarming instrumentalizations that the new myths of terrorism feed and it may help to bring religion back to its natural, spiritual dimension, clearly distinct from the temporal dimension.

SOW and ALEPH can be contextualized in this cultural background. The searches have not focused on the design of a model prayer room, but it has focused on typological research because they have a bottom-up approach (Gleiter 2012; Giammetti 2016b, 2018b). According to the bottom-up approach, the work on the type arises from a natural process of adapting architecture to the cultural changes of the people who inhabit it. The bottom-up approach starts with the needs, fears, knowledge, and points of view on the interreligious issues expressed by the communities of believers. The searches are developing a process of synthesis of knowledge that is gradually acquired and verified by digital spatial models of interreligious prayer places.

2 State of Art

2.1 Functional Hybrids and Hybrid Types

What they mean by type?
Type is not a simple mechanism of reproduction, but it is what underlies the formal structure of space and it is capable of multiple developments.
What do we mean by hybrid building in this research?
It is a building whose shape is structured by a new typology, arisen from the hybridization of existing typologies.
What do they commonly mean by hybrid building?
They don't commonly mean a new hybrid typology, but they mean a functional hybrid, that is a building designed using traditional types, but including a mix of functions.
Analysing the state of the art it is evident that there are many examples of building as a functional hybrid building, but there are no examples of hybrid type.

The research studies and checks the types that can be used to design spaces for prayer dedicated to the three Abrahamic confessions, hybrid spaces, from the Latin *hybridus*, intended as an animal or a vegetable born from a crossbreed of different breeds or varieties. The concept of hybrid, intended as composition in the unity of heterogeneity of different natures, it is also used in architecture to indicate the categories of spaces arisen from the composition of architectures that are usually designed and inhabited separately.

In times of great transformation, architecture has often used the hybrid category, or rather, it has moved into a hybrid *forum*.[1] Since the end of the seventies of the Twentieth century, the attention to the design of hybrid buildings has gradually grown and many architects have ventured into the construction of this kind of buildings. Today they commonly mean hybrid building as a building that contains a mix of very different functions. Contemporary architecture has a tension towards new typology, but the result of this tension is the use of hybrid as a functional hybrid not the research of new hybrid typology. Probably this happens because architecture wants to meet the new needs of our society that moves within a hybrid *forum*, generated by the impact of globalization and of migrations on our physical, cultural and political space. However, there is a difference between the research of hybrid types and the design of functional hybrids: the typological research characterizes structures of the form capable of multiple developments; the functional hybrid deconstructs the bijective correspondence between each building and its specific function (residential, school, office, etc.) and it looks for a new spatial relationship among elements with heterogeneous functions.

In the seventies, Rem Koolhaas, in his Delirious New York, mentions the concept of hybrid in architecture:[…]

> fascinated by the possibilities that the density of the New York metropolis had to offer in terms of hybridization, he identifies in the skyscraper a quality that allows the almost infinite

[1]The hybrid *forum* is a general concept where physical, cultural and political space are *forme a-venire*, that are shapes to be designed through a work on the existing, that is a *mixitè*, because in what exists today there is the seed of what will be tomorrow.

combination of programs that coexist in separate levels (Koolhaas 2000).

Referring to the Downtown Athletic Club in New York, Koolhaas writes:

> This apparent serenity hides the apotheosis of the skyscraper as an instrument of the culture of congestion. […] a machine to generate and to intensify desirable forms of human interrelationships. […] In many respects the history of Manhattanism, intended as a separate and identifiable architecture, is a dialectic between these two forms: between the needle, that wants to become a globe and the globe, that tries, from time to time, to transform itself into a needle. A cross fertilization that results into a series of successful hybrids, in which the ability of the needle to attract attention and its use of a few quantities of soil combine themselves with the great receptivity of the sphere (Koolhaas 2000).

In Koolhaas' writing, there are no more monuments that express greatness and authority, but there are architectures that mix different functions and social realities in a programmed set, in a stratification by a layer that monumentalizes the freedom.

No long after Steven Holl, another contemporary architecture theorist, studies the hybrid building: in 1986 some of his ideas on the hybrid issue were published in Pamphlet Architecture11: Hybrid Buildings (Holl 1985).

During his career, Steven Holl will continue his studies on the hybrid building: Linked Hybrid (Fig. 1) is a complex of buildings built in Beijing in 2009. In the building, there are functional mixes such as residences with cinemas, shops with schools and kindergartens, hotels with exhibition halls, office spaces, restaurants, and sports facilities. Linked Hybrid is a set of towers connected by bridges positioned at the top of the buildings. Steven Holl notes that the walkways are not drying up street life, rather they are generating random relationships between the shops that are at street level and the super-elevated life: as it occurs in the unexpected fullness of metropolitan density.

With Lynked Hybrid, Steven Holl has developed on the urban scale the work started on the hybrid since 1986 with Pamphlet Architecture. In the project for Beijing, the hybrid generates heterotopias, spaces that have the specific characteristic of being connected to all the other spaces, but in order to suspend, neutralize or invert the set of relationships that themselves designate, reflect or mirror.

The Lifted Village (Fig. 2), built in 2011 at the end of the historic Shanghai waterfront designed by MVRDV, it is on the same line as Linked Hybrid. The building combines two types that represent China's past: its urban villages and its future, the tower. The result is a mix of typologies that creates a spatial variety. MVRDV architects have developed a software called *function mixer* to combine parameters such as density or site conditions and to select the optimal configuration chosen from a series of possibilities examined by the program. The program is useful to design functional hybrid buildings that combine functions normally designed in different types of buildings. In MVRDV's architectural research, the hybrid is expressed by the concept of *multiplicity in architecture* that Nathalie de Vries, professor at TU Delft and principal architect-founding partner of MVRDV, defines in this way:

> The majority of buildings should have more productive and transformative qualities. To avoid the creation of architecture that can only be used for a short period of time, attention also needs to focus on a buildings' intrinsic qualities and the ability they have to create value and contribute to our quality of life. The integration in design of these qualities are a new approach towards buildings and building elements that I would like to call multiplicity in architecture (de Vries 2019).

The works of Koolhaas, Hool, and MVRDV didn't produce new types, but they use existing types (such as the tower-skyscraper building, the bridge, etc.) to compose them together. SOW and *ALEPH* try to go beyond the dimension of the hybrid building; the aim of the searches is to know-how and if it is possible to draw a road map that describes a typological behavior pattern of the Abrahamic interreligious prayer spaces, a shape structure that may be developed in design experiments also very different from each other.[2] The type is a tool to transmit information and knowledge through the architectural form, it is a semantic form, therefore, a bearer of shared meanings and values that joins the artistic and applicative practice (Martì Aris 1993; Avecedo et al. 2019). In interreligious prayer spaces, what are the shared values that should join the artistic practice and the experience of inhabiting? To keep together the intangible aspects and the physical components that compose the type, SOW has developed a research method called *extended design approach* used: (a) to design new types of interreligious spaces for Abrahamic confessions, (b) to elaborate impact models on the believers of the three confessions about the project; (c) to have feedback on the project from common people and believers; (d) to elaborate their views on the issue of interreligiousness.

---

[2]This research approach has found a valid methodological support in studies conducted by some European research laboratories such as the Future Cities Laboratory (ETH_Zurigo), the Laboratoire Architecture Anthropologie (École Nationale Supérieure d'Architecture_ Paris-La Villette) and the Master of Science M-ARCH-T (TU_Berlin). These research laboratories work both on pure architectural composition and on sociological and anthropological research. They are analyzing how all social actors carry out to think, plan and imagine in the future.

Fig. 1 Linked Hybrid, designed by Steven Holl, Beijing China

2.2 Hybrid Architectures: Interreligious Spaces for Prayer

In Europe and throughout the world there are many architectures dedicated to the interreligious space, but is there a typology of interreligious prayer space?

Although the globalization and the great migratory phenomena have changed the socio-cultural situation of contemporary cities, there isn't a new typology from which to derive models of interreligious spaces, even if it can be seen that there is a specificity of the project program of interreligious prayer spaces.

Until 50 years ago, European citizens were almost exclusively Christians, Catholics and Protestants; today the religious situation has changed and the society is becoming much more pluralistic. There are many spaces that cannot be entrusted to only one religion and there are many spaces that cannot be defined with a single name (church, mosque, synagogue, Buddhist Temple, etc.), but they are open to different religious groups. These spaces are mainly located in airports, hospitals, universities, parks, government buildings, schools, places open to the new multicultural communities. Rarely, instead, there are religious institutions that promote the construction of interreligious spaces and when this happens, the projects have a completely different approach. Projects such as the House of Religions of Bern, the Abrahamic Family House in Abu Dhabi or the House of One in Berlin (Fig. 3) gather different religions under one roof, but there isn't only one prayer space, every religious community has its own place of worship and every prayer room is positioned around a common space (Somers Cocks 2019; Haus der religionen 2020; House of One. Three Religions 2020). These architectures are functional hybrids that compose buildings with different functions, keeping their typological characteristics intact.

Usually, the common space is a covered or uncovered square; it could be useful to study if this connective space could become a new typology, but right now it is a large indistinct space, much less important than the spaces of prayer that are around it.

We may classify the interreligious spaces into four categories: (a) interreligious celebratory rooms; (b) interreligious rooms for prayer and meditation; (c) rooms of silence; (d) functional hybrids (such as House of One already described in the previous point). Some examples of architecture belonging to these categories will be described below.

Fig. 2 Lifted Village, designed by MVRDV, Shanghai China

2.2.1 Rooms of Silence

The rooms of silence are non-confessional spaces that may be used by anyone who is looking for a quiet place. Everyone can enter and remain silent or simply he/she may be in relax to acquire strength for daily life, to meditate, to pray. An example is the Rothko Chapel in Huston (Fig. 4) or the *Raum der Stille* in the Parliament of North Rhine-Westphalia in Düsseldorf or the Meditation Room at the UN headquarters in New York. The rooms of silence are inspired by the extreme shapes of semantic neutrality. Surrounding (very strong and explicit) to any religious interpretation becomes the compositional tool to characterize the space.

2.2.2 Interreligious Rooms for Prayer and Meditation

The House of Silence in the Westend Campus of the Goethe-Universität in Frankfurt (Fig. 5) is an interreligious room for prayer and meditation dedicated to different religions.

The House of Silence is used to reflect and meditate, it is open to people who want to rethink their relationship with God and it is dedicated to whoever is looking for a quiet place.

There are interreligious spaces even in airports: an example is the *Raum für Gebet und Stille* of Munich airport or the interreligious prayer hall of Köln Bonn airport. The interreligious rooms for prayer and meditation are holistic spaces that have no explicit religious reference, but they are open to a spiritual characterization of space in the name of an infinite aesthetic that may favor different religious interpretations.

2.2.3 Interreligious Celebratory Rooms

Celebration rooms are interreligious spaces used for individual prayer and to celebrate rites, they are different from the interreligious rooms for prayer and meditation, because here it is possible to celebrate rites of different faiths setting up the room, from time to time, with the liturgical and iconic apparatuses of each religion. Without specialist

Fig. 3 House of One, designed by Kuenn Malvezzi, Berlin Germany

characterizations, space can be used by anyone as a meditation room.[3] An example is a Chapel designed by Eero Saarinen at MIT in

Boston (Fig. 6)[4] or the celebratory space in Ingelheim cemetery. Interreligious celebratory rooms don't often express *in toto* the character of interreligious space. In fact, these spaces take on spatial models that are too calibrated on a specific typology of prayer space (such as a church with a basilica plan or a central-plan mosque of the Ottoman type), because each different Assembly needs a flexible space to gather in prayer according to the tradition of each one. Although the globalization and the great migratory phenomena have changed the socio-cultural situation of contemporary cities, there isn't a new typology from which to derive models of interreligious spaces, even if it can be seen that there is a specificity of the project program of interreligious prayer spaces. However, the interreligious rooms for prayer and meditation may offer valid ideas for designing a new hybrid typology, because these spaces focus their attention more on persons than on liturgical equipment, conceptually moving the sacred character of space from physical space to interior space.

2.3 Digital Technologies Applied to Prayer Spaces

Digital technologies can be a good tool to know and compare the cultural differences of the communities that compose the *hybrid forum* of contemporary Europe.

Are there digital platforms to help develop mutual knowledge between believers of different confessions?

Can apps help develop an interreligious dimension of prayer spaces?

Are there non-religious museums that offer virtual tours of prayer spaces of the three monotheistic confessions?

The Replicate App, designed by the Technologies of Vision (TEV) of FBK-ICT, creates a 3D digital copy of any liturgical object or religious icon that can be shared on the network through a viewer or through the virtual reality (SIR 2018). You can take a series of photographs of an object with the smartphone by rotating around it; the app selects about fifteen of these shots and it sends them to a central data server where the object is reconstructed in 3D through images. The result obtained is further improved by the application of a series of textures. The object, transformed into 3D virtual reality, is sent back to the smartphone that can share it with other people in any part of the world. The researchers who designed the app interviewed believers of different religions to ask if a sacred object that becomes virtual remains sacred. The believers responded in the affirmative: the sacredness of an object or space does not depend on the object itself, but it depends on the way to use the object and on the way of feeling a space or an object during its use. The interviews highlighted the following points of view: a) the connection with the sacred depends on the way of feeling and sensing the object or space, it is not important if the experience is physical or virtual; b)

[3]The celebration implies the presence of the equipment used for the rites: the iconic apparatus, the furniture, etc. Commonly, the presence of the equipment used for the rite is strictly connected to a specific typological characterization of the space, designed to best accommodate the ritual equipment. The relationship between liturgy, ritual equipment and space still keeps the typologies of prayer space of each religion firmly and clearly distinct from each other, because in the history of religions this relationship has been overloaded with precise symbolic values that have transformed the prayer space from space to accommodate the Assemblies and their liturgical equipment during the rite into an a space that is sacred in itself. The symbolic value attributed to the relationship between rite, liturgical equipment and sacredness of space is supported by liturgical-theological approach that bases the celebratory aspects on an inconsistent relationship between form and function. This is the reason why liturgists mistakenly continue to hypostatise the identity liturgy – function, suggesting shaping the architecture with the ritual data. This approach is not very correct: in fact, the liturgical celebration is not possible without Assembly, but you can see liturgical equipments without celebration.

[4]The chapel of Saarinen is a very suggestive space, but it is probably too calibrated on a type of prayer room inspired by the traditional *meeting house*, heart of the civil and religious life of the Protestant communities of New England.The podium, with the stone cube/altar, blocks the space and reduces the possibility for the Assemblies to freely position themself. This pattern-space imposes an arrangement of the Assemblies in the shape of the eschatological path, typical of Christian basilica schemes.

Fig. 4 Rothko Chapel, designed by Mark Rothko, Huston Texas

technology is also a useful tool for the spiritual sphere, but not when is controlled by technology.

The 3ALOG digital platform aims the development of the interreligious dialogue on the web and it uses videos that explain different religious contents to ensure mutual understanding (3ALOG 2020). In the team of the platform, there are people who combine their technological know-how with their interreligious and intercultural background. The aim of their work is to integrate different perspectives and to offer a space for the development of new ideas. The platform has developed different thematic areas: (a) 3ALOG & hospitality to organize meetings, interreligious and intercultural festivals; (b) 3ALOG & Media to produce and network interviews, short videos, the design of websites or social media; (c) 3ALOG & Gaming to develop online games to promote democracy, tolerance, and diversity.

There are websites to take a virtual tour of religious places in virtual reality (VR), but they don't use digital technologies to collect information about the impact of the contemporary interreligious dimension. Platforms such as 3ALOG have a role as a cultural mediator, but it could be very important to use virtual museums to measure the impact of the interreligious dimension. There are not non-religious museums that are interested in the experiences of spirituality and that offer virtual tours of prayer spaces of the three monotheistic confessions to encourage the intercultural exchange and the knowledge mutual, but this activity could be very useful in museums. Mass tourism is bringing an increasing number of visitors to museums and could be a useful strategy to reach an increasing number of users with different profiles to collect data on the interreligious issue.

3 Documents and Materials

3.1 Space and Liturgy_ Form and Reforms in the Three Abrahamic Confessions

Since the end of the nineteenth century, many changes have transversally united the three Abrahamic confessions; many reforms have changed their history and changed the way of inhabiting the space of prayer and the liturgy.

The relationship between the nineteenth century and the religion is well expressed by Kàroly Kerényi who said that the nineteenth century had been an era of final palingenesis, where all humanity is embraced in a single, rational and reasonable faith, the ultimate synthesis of its progress (Yerushalmi 1991; Filoramo 2004, 2009; Giammetti 2018a).

The nineteenth century is known as the era of secularization, the age of disenchantment often described as the beginning of the loss of faith. The secularization appeared in European cultures with heavy criticism of some aspects of the Christian religion and subsequently to all other confessions. This long and laborious process of political

Fig. 5 House of Silence in Westend Campus of the Goethe-Universität, Frankfurt on Main, Germany

experimentation transformed the concept of religion: the transformation of religious systems reflected the change in the structures of institutional power.

The centrality of science has been an essential element for understanding the new idea of religion in the contemporary age. The impact of modern science arose in different ways: on one side the science influenced the individual personality, the individual psyche, and its religious symbols; on the other side the modern scientific tools offered methods for analyzing the sacred texts that revealed a different past from the one transmitted by the tradition. The faith in science provided a new worldview, a new device to create the individual identity of modern man. Sigmund Freud, who was the most coherent expression of this faith, worked just to eliminate all those behaviors that belonged to inexplicable aspects of religion. The science struck everybody like a revelation, it could arouse frustration and depression, it could be resolved into an existential conflict which could be resolved in religious faith.

«Goodbye God of my youth! Maybe you'll go back to my bedside. Goodbye. Despite your betrayal, I still love you » , wrote Ernest Renan in his *L'avenir de la science* (Renan 1890). Renan's thought is considered a programmatic platform of the positivist faith, a strong position in favor of the superiority of science. However, Renan's thought is not an obliteration of the presence of God, but an attempt to make woman/man active in the divine presence through her/his own awareness.

3.1.1 The Liturgical Movement

The modern Catholic liturgy conceptually moves the place of celebration from the Temple to the Ecclesia, it moves the Eucharist and the other symbols into a dimension of event and no longer of constant presence, giving more value to the liturgical celebration than to the custody of the symbols. This transformation process is the result of a long work of research conducted by movements, like the Liturgical Movement, that renewed the liturgy and its spaces. Liturgical Movement was born in France during the Restoration and it will lead to the Second Vatican Council and to the

Fig. 6 Interreligious Chapel, designed by Eero Saarinen at MIT, Boston Massachusetts

Liturgical Reform (Richter 2002; Luzzatti 2007; Bauman 2017) (Fig. 7).

3.1.2 Nahda and the Islamic Modernists

Between the end of the nineteenth and the beginning of the Twentieth century, just as for Catholics, there were many reform movements in the Islamic world. The reformed Islam was the Islam of modern man and also his prayer spaces had to have a new architectural shape. The Islamic Reformist Movement questioned the essential points of the religious and cultural Muslim tradition. Until then, only the old jurisconsults *(ulama)*, figures that were halfway between policy and religion, had indicated the political, religious, and moral line of the Islamic world. If the tradition had not adapted itself to Modernity, the Modernity would have become totally a prerogative of Westerners (Gabrieli 1958; Hourani 1970; Filoramo 2004, 2009) (Fig. 8).

For this reason, Islam tried to transform itself into a modern religion, but after the reformist phase of the first half of the Twentieth century, starting in the 1960s with the crisis of secular and socialist Arab nationalism, the Muslim world fell prey to the fundamentalism that today foments international terrorism. However, in recent years, especially in Europe, Islam is promoting a new reformist debate that is beginning to have an impact on the most recent prayer space projects and not only in Europe.

3.1.3 Jewish Emancipation and Reformism

In 1810, Israel Jacobson decided to build a Jewish Temple in Seesen, Saxony. Jacobson used the word Temple, not synagogue and in his thought, the contrast between Temple and synagogue was not only lexical but substantial. Jacobson built a new place of worship, explicitly founded so that the Israelites would abandon the idea of diaspora and they would choose the country where they lived as their homeland where to build a Temple with the same spirit that had animated Solomon in the construction of the Temple of Jerusalem.

In the years 1844–46, in Germany, there were three Rabbinic Conferences that addressed the issue of a decisive reform of Jewish worship to achieve a better emotional involvement of the believers during the rite. One of the proposals to improve the involvement was the abandonment of Hebrew and the use of the language used locally during the prayer (Reinharz 1980; Filoramo 2009) (Fig. 9).

Fig. 7 Rothenfels Chapel designed by Rudolf Schwarz, Rothenfels Germany

Fig. 8 Salman Mosque designed by Achmad Noeman in Bandung, Indonesia

3.2 Style and Character

Speaking of architecture Zygmunt Bauman said that if we believe that designing and planning makes sense because not everything in the existing world is as it should be, we may recognize that architecture has a capacity of transformation and adaptation, in a close correspondence between the impulses that come out of the historical changes that are happening and the capability of space to adapt to these same changes (Bauman 2017). Bauman's words also bring out tension/hope, that goes beyond the simple links cause and effect, supply-demand, function, and space. If designing architecture makes sense because not everything is as it should be, it means that the design can also reflect upon how space should or could be. This is what happened in the history of prayer spaces, whenever they had been transformed or adapted, reusing buildings that originally had another use. This is what has happened to Kuenn Malvezzi

Fig. 9 Synagogue in the desert designed by Zvi Hecker, Negev desert, Israel

in the House of One; this is what has happened to the Roman civil basilicas, like the Basilica of Maxentius, when they were transformed into Christian basilicas. In these examples, the same space has been deprived of a specific symbolic value system and has been loaded with other symbolic values. Therefore, the issue is not what symbolically represents that space in itself, but what it represents for those who live in it. This subtle but substantial difference shifts the capacity to give meaning to space from the *object-space* to the *subject man*. There is a clear distinction between the spatiality of the object itself and the meanings that people give to space when they live in it. Towards the end of the last century, the research on the autonomy of the shape of space, intended as autonomy to have its own meaning, has invested all the fields of architecture, including the religious one. Peter Eisenman has been one of the strongest supporters of the shape autonomy from the historical contingencies.[5]

Since the end of nineteenth century, most theological/architectural research on prayer space for the community[6] has worked on the autonomy of the space shape and on the shift of the signification capacity from the *object-space* to the *subject man*, conceptually moving the prayer space from Temple to *Ecclesia*.[7] According to the tradition that comes from ancient Egypt, the Temple is the home of the God, where the God is kept alive by the rites of priests who care for his image or symbols. The Ecclesia is the shape of the first church, the church of the origins, more Emmaus house that monument, more Assembly house than God house, the house inhabited by the pre-Constantinian community.

This ecclesiological idea projected into the architecture field, it can be translated into the Assembly who, through the experience of inhabiting the prayer space, gives *otherness* to it. The Assembly plays its significant role in prayer space through a *sensitive intelligence* (Autiero 2016) that asks human being a new balance between what is beside and inside her/himself, a balance whereby the sensible sphere, which is outside, is no longer subordinate to the inner spiritual dimension. The *sensitive intelligence* is a tool of mediation between the human being and its spirituality and it uses the experience of the rite to get started.

This new idea of prayer space is arisen by the role attributed during Twentieth century to hermeneutics and to the dynamics established between hermeneutics and experience, a dynamic that intellectualism has eroded, creating an opposition between intellect and experience, that is weakening the sensitive and experiential sphere and it is strengthening the intellect. However, the time is re-evaluating just that sensitive sphere that seemed set aside.

L'aula, the prayer space, is no longer an autonomous and metahistorical religious symbol, its symbolic value is linked to a process of aesthetic reception and production that is realized in the experience of the Assemblies to inhabit that space that can have an ever-new resonance in the function of whom use it. From this point of view, as religious symbols, the spaces of prayer belong to the *aesthetics of reception* (Jauss 1987, 1988; Derrida 2011; Rispoli 2016), so that the experience of inhabiting combines a passivity and an activity that together allow us to define the real act of inhabiting as aesthetic reception of space.

To affirm that architecture, as well as community prayer buildings,[8] concerns the aesthetics of reception and that the shape of space is autonomous with respect to its meaning, it means to separate the syntactic component of the writing of space from the semantic one, with a strong prevalence and an essential autonomy of the first with respect to the second.

Dividing the syntactic component from the semantic one of space for prayer opens substantial problems of the signification of that space, especially if we continue to look for a solution exclusively in architectural writing and if we persist in not taking into consideration the active role played by the experience of inhabiting of the Assembly.

There's a change of *direction* (from space to man) concerning the process of the signification of architecture. The signs of this change are still latent, but they are already in place. Probably the prayer spaces of the communities express more clearly the complexity and contradictions of this change of direction, representing an excellent gymnasium for design experimentation. Entrusting the significant role of space to the form of the community in prayer, with a new balance between *word* and *sacrifice*, represents a high degree of theoretical reflection that the sacred space of contemporaneity could offer. Sustaining this distinction, as well as being strongly innovative, is also important for what is the primary responsibility of those who research in the field of architectural composition: architecture (Galantino 2016).

[5]There are *prayer spaces for communities of believers* and *sacred places*, *places of otherness* or *other places*, that cannot follow the issue of *semantic autonomy* of shape. The *other places* are related to the memory of the sacred and they are not ordinary places and spaces, but they are the Place, the Space.Some places are so full of pure sacred sense, that they transfer it to the architectures they accommodate, so thus, they become the instrument by which you can express the union between God, Nature, Man and his Art. Here, the link between place (natural space) and architecture (artificial space) becomes a vehicle for the irruption of the sacred in the architectural space. The *other places* and their archetypal essence have an evocative force of shape that derives the sense of absoluteness and sacrality of their form. These spaces are ancestral traditions turned to stone, in Modernity's own words, they are expression of the collective unconscious of Jungian philosophy, which surfaces in those places.

[6]The research studies the prayer spaces for the community of believers of the three Abrahamic confessions. The community of the believers who meet in prayer gives shape to the Assembly.

[7]*Ecclesia* is intended as Assembly, as the community of believers gathered during the rite to pray.

[8]To clarify what is meant by community prayer buildings, it is useful to refer to distinction between Temple and Ecclesia, where community prayer buildings are related to the idea of Ecclesia. The attention to community prayer buildings, rather than to *Other Places*, was born from the change in the dynamics of formation of the modern city and its public spaces after the Second World War, when the image of city became the sum of many individual images and where all those people together needed a space that could accommodate them at the time of prayer.

The investigation and the manipulation of structural relationships between the architectural elements that form the space of prayer and its interpreters, that is the members of the Assemblies of the three Abrahamic religions, suggest the possibility of structuring the research starting from two working hypotheses: as a condition for mutual acceptance, the research assumes to disperse the iconographic symbols traditionally entrusted to painting and sculpture and to use only light and empty space to express the idea of the irruption of the sacred. The research does not want to reduce church, mosque, and synagogue *to one*, but its aim is to express the differences not through the *aula* shape, but through the different positions that each Assembly assumes during prayer. The two hypotheses are not in opposition, rather they intertwine and support each other, trying to have an innovative approach in the relationship between architecture and liturgy through the relation between the shape of space and the symbolic shape of the aggregative structure of the Assemblies during prayer.

The first hypothesis, the dispersion of iconographic symbols, is based on the separation between semantic and syntactic components of space. The second hypothesis, the non-reduction *to one* of the three spaces, is based on the significant role attributed to the Assemblies that entrusts to the shape of each community gathered for prayer the task of producing the character of the space.

Is it possible to silence the traditional decorative-symbolic apparatus of the three confessions and to use that absence to increase the expressive power of light, allowing the light itself to animate the shape of space to suggest the irruption of the sacred? Silence, light, and empty space become the protagonists of this different idea of understanding spaces for prayer and their composition.

We may recognize the specific character of each of the three spaces if we center their respective identities, not on the *style* of space, that is the art of transforming matter into a symbol, but on *character,* that is the ability of space to host Assemblies during prayer, whose disposition is the tool to relate that space with its purpose: the relationship with the divine.

The relationship between the shape of space and the symbolic form of the Assemblies finds its theoretical basis on the distinction between *style* and *character* that falls within the definition used by Auguste Perret earlier last century:

> In an architecture we need to consider character and style. Character is the way in which the building responds to its purpose, it's the connection, the relationship between the object and its purpose. Style is the art of using matter, of which architecture is a testament, and of obtaining the highest degree of harmony (Gargiani 1993).

The distinction of Perret between style and character will be recovered later by Le Corbusier, who will theorize another distinction between style and styles: styles are identified by the type of ornament, that is something added to space; the style is indeed the unitary principle of the architecture in every specific historical period. The modern gives up the way of seeing ornament, inherited from the past and it searches its own style with a different critical awareness.

Such an approach has also had effects on the architecture of prayer spaces: whereas in other architectural types this change towards the ornament has been a complex process but, all in all, accepted with less resistance, in religious architecture this change has required and will still require different time of metabolization. This new approach to ornament could also mean renouncing the traditional iconographic symbolic apparatus, de facto linked to the aesthetic perception in relation to the sensibility and to spiritual values.

In modern religious architecture, the compositional and symbolic force of pure shape has grown to the detriment of the traditional iconic decorative apparatus, until to decree not still its end, but certainly a radical reduction. So, the reduction of the traditional ornamental apparatus entrusts mainly to the conformative capacity of the empty space and of the light the potential condition of being a space to inhabit praying. When each Assembly will meet up, it establishes a dialectical relationship with light and space and it gives them a spiritual character. The structural relations among Assemblies, empty space, and light represent a reduction to the zero degrees of architectural writing of the worship houses.

If the reduction of iconographic symbols is part of a transformation of ornament, a cultural process that is typical of the modern architectural research, the verb used to describe the first research hypothesis, however, is not to reduce, but to disperse. To disperse describes an experience that is activated through movement, a movement towards other spaces, spaces not yet given, to be discovered, where the movement is the spatial and existential dynamism of a single or of a group of people, that move themselves in full self-awareness towards new spaces to inhabit, using their full capacity to be simply and beautifully human being.

The research of the greatest modern architects founds a way to reduce the architectural writing of prayer spaces to the *zero degree*, not only through the reduction of the decorative apparatus but also through the use of archetypal elements of the architecture. The archetypal forms use light and empty space to increase their expressive power, becoming evocative of a feeling of the sacred even in the absence of religious symbols, just for their simplicity and just for being universally recognizable to everyone. From this point of view, the modern, with an operation typical of the avant-gardes, has made a linguistic jump in order of accelerating and overcoming the time that had marked the rhythm of the transformation of the prayer space until that historic moment; abandoning all that could be defined traditional, the modern has opened the way to the development of the contemporary space (Figs. 10 and 11).

3.3 Light

The light is matter and gives shape to the material: it reveals the shape, lets the geometries of space be perceived and it underlines its character through the empathic role of the contrast between light and shadow. Through light, space manifests its presence and through the contrasting light—shadow manifests its character. The light not only shapes the space but also its positions and orients who inhabit the space. In the prayer spaces founded on an ontological dimension of architecture, the directional use of light has often been a valid tool to load the space with an eschatological sense, referring to a strong symbolism of the relationship among God-Celebrant-Community-Single. In the architecture of prayer spaces, the use of the semantic potential of light has produced results of great expressive value: for example, Bernini experimented with a light composition technique that produced dramatic and theatrical effects of his architecture. In the Berninian spaces, the architect designed a basic environment with diffused light and subsequently, centered the space geometrically, letting light enter from a point, opportunely hidden, to bring the space in eschatological tension (Campo Baeza 1994) (Fig. 12).

The powerful of anthropocentrism of the contemporary era has worked for leaving the eschatological symbolism deriving from the use of directional light. There is no more directional and eschatological light. The light positions the Assembly, emphasizing the place where it meets, and it directs the Assembly towards itself. This kind of light leaves you alone, it presupposes the awareness of participation in prayer and this is the reason why it becomes widespread, indirect. The light does not frame anything and it does not emphasize directions, it does not establish hierarchies, but it refers to itself, it requests responsibility and awareness for praying and for relating with the divine.

The light indicates an immanent dimension of the experience of inhabiting prayer spaces, but its character is such as

Fig. 10 Gesù Redentore Church, designed by Mauro Galantino, Modena Italy

Fig. 11 Baitur Rauf Jame Mosque, designed by Marina Tabassum, Dhaka Bangladesh

Fig. 12 Ecstasy of Saint Teresa sculpture, Bernini, Santa Maria della Vittoria Church, Rome Italy

to suggest that transcendence is only radically immanent, or rather, it is a transcendence that is unthinkable to the out of immanence (Autiero 2016).

This kind of light remembers the quality of the light that Kubrick uses to illuminate the room of one of the last scenes of the 2001 Odyssey in Space film, certainly, it is very evocative, but at the same time, it is disturbing. There are many architectures where you can see this kind of light: in the Chapel for the students designed by Ottokar Uhl in the Benedictine College of Melk in Austria; in the Chapel for the students designed by Glauco Gresleri in Pordenone. The light falls into space, it rains on persons, it wraps them and it allows them to get in touch with themself or with the other people gathered in Assembly.

The light of the modern and contemporary prayer spaces represents a good tool for the research on interreligious places, because it moves the axis of the signification of space from the ontological dimension to the relational one, an essential condition for designing the character of a prayer space that can adapt itself to accommodate three different religions.

Starting from these hypotheses, the research is studying:

- how light can transform itself to give a different meaning to space according to the Assembly that inhabits it;
- how light can be useful to position and to orientate the Assemblies in the space (Figs. 13 and 14).

3.4 Ornament

The investigation into the manipulation of structural relations between the signs that compose the prayer space and its interpreters suggests the possibility of working on some signs that have played a central role in the semantisation of the prayer spaces: ornament and light. Usually, the identities of the three spaces are centered on the style of space, the art of transforming matter into a symbol. Symbol can be the form of space, that is the typology of architecture: you can remind the basilicas or the mosques with four iwans. Symbol is the plastic elements of architecture that have been translated through its ornamental apparatus and the sculptural and pictorial ones (statues, friezes, frescoes, paintings): each of these elements has been declined in the tradition of each confession and they belong to the decorative apparatus as traditionally intended. The tradition has generated a close relationship between ornament and signification of space, in fact, in the history of architecture, the ornament and the liturgical furnishings have ever had the role of characterizing the preyer spaces. The ornament has built the identity brands that have distanced the church, mosque, and synagogue, building the identity structure of the prayer spaces.

The modern searches its style by renouncing ornament as it was transmitted from the past. In Modernity, the power of pure form has grown to the detriment of the traditional iconic decorative apparatus that has been radically reduced. This idea has also influenced the design of the spaces for prayer, even if with not a few resistances, precisely because the zeroing of the ornament has been equated with the loss of the identity marks of the religious space, entrusted to the traditional iconographic symbolic apparatus.

More than a century after the birth of the reformist theological movements and fifty years after the Second Vatican Ecumenical Council, the dispute between tradition and innovation is still very strong and ornament is one of the most controversial topics.

With regard to the issues that are still contentious, we should work on the research hypothesis concerning the aesthetics of reception[9] to decline the topic of the ornament in the interreligious space.

To better understand this theoretical approach it is useful the classification of the concept of sign proposed by Charles Sanders Peirce, where the signs can be classified as icons, indices, and symbols (Peirce 1980). The icons have a direct relationship with the object; the symbols have a meaning connected to the object by convention; the index alludes to the object, with it it has some qualities in common, but its semiotic result is open because it is allusive, indicative and not directly connected by analogy or convention to a specific meaning. To search for a new typology of interreligious prayer spaces, it is useful to return to the ornament as an index rather than an icon. In this way the topic of identity is not abolished, but, at the same time, it is not stigmatized.

In the era of the transition from machines to digital communication, the debate on ornament reappears and lights up. Contemporary architecture works on the ornament as an index. Ornament manifests itself in the most disparate forms, freeing itself from the concept of ornament as an icon and as a symbol, freeing itself from walls with friezes, to penetrate into the material and the structure of architecture (Gleiter 2012). More than an ornament, it could speak of performative models that are at the same time ornamental. Under the influence of the new digital design procedures, material, structure, and ornament enter into a new interaction, the result of which is a kind of ornament closer to the idea of an index than an icon.

[9]About the aesthetics of reception in the interreligious space, see paragraph 2.2 Style and character. According to the aesthetics of reception the space is its interpretation, the architecture defines itself through the relationship with those who live there, for this reason the space can be recognized from time to time as a church, a mosque or a synagogue.

Fig. 13 Scenes of the 2001 Odyssey in Space film, Stanley Kubrick 1968

Fig. 14 Students' Chapel, designed by Ottokar Uhl, Benedictine College, Melk Austria

4 Research Hypothesis and Method

4.1 Studies of Types for Interreligious Prayer Space of the Three Abrahamic Confessions

The research aim is the study of types for an interreligious prayer space of the three Abrahamic confessions. The study of types has developed three typological categories: (a) directionality; (b) light; (c) ornament.

4.1.1 Directionality

The research of the great architects of the Twentieth century on prayer spaces has provided valid tools for studying some typological aspects. In Germany, a Protestant country, in the early Twentieth century, a group of Catholic theologians/liturgists and architects began to work on a change in the spatiality of Catholic prayer places, focusing their studies on the centrality of the Assembly of the believers in the liturgical space. Theologians and architects such as Guardini, Schwarz, Böhm, Steffann, redesigned the spatiality of the church, even before Vatican II they questioned the basilical scheme with the traditional position of the priest *ad deum.*[10] Rudolf Schwarz hypothesized several aggregative structures of the assembly in prayer that he defined *archetypal images* (Fig. 15) and to which he associated different models of space: (a) *sacred inwardness, the ring,* a closed shape that favors the circular aggregation of the believers around the altar; (b) *sacred departure,* a three-quarter circle or T arrangement; (c) *sacred journey,* the path, the columned arrangement inspired by the eschatological sense of the basilicas (Pehnt and Strohl 2000).

The archetypal images explain that it is not the shape of the prayer room that defines the character of the prayer space, but it is the Assembly, with its way of arranging it that gives meaning to the inner space of the room. Schwarz's archetypal images explain that the prevailing direction of the community of the believers, the direction of the Assembly formed by the position that each believer assumes respect to the other, is a fundamental element to load the space with a symbolic sense. Just as one of the most common archetypal images in Catholic prayer is the sacred journey, where the longitudinal direction that joins the door of the prayer room to the apse is prevalent for its eschatological symbolic value, in Islamic prayer one of the most common archetypal images is the *horizon.* (Figure 16) According to this archetypal image, the Assembly is positioned parallel to the

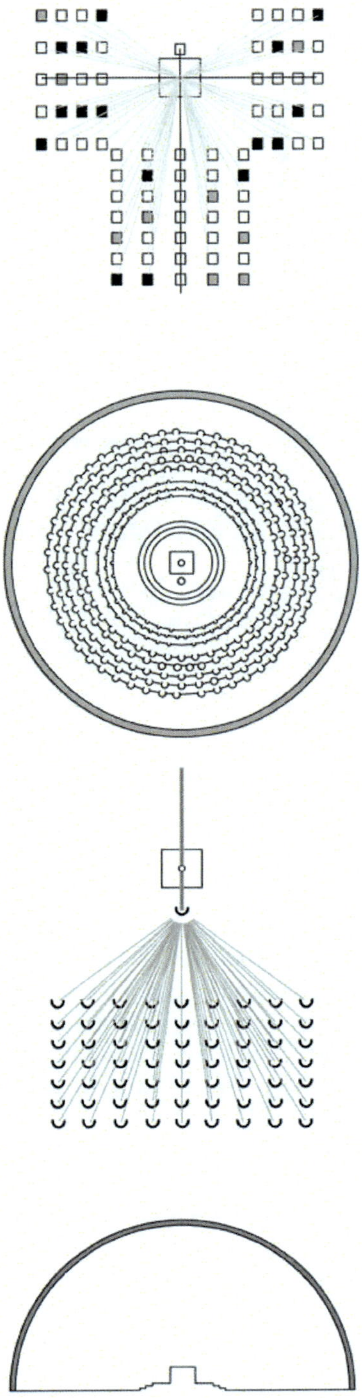

Fig. 15 Rudolf Schwarz, archetypal images

qibla, the wall that contains the *mihrab,* the niche that indicates the direction of Mecca towards which the prayer must be oriented. In this archetypal image the transverse direction is prevalent (Giammetti 2016b). Often, at the origins of Islam, many Christian buildings of worship, mainly

[10]Schwarz and Guardini's research have gone beyond the archetypal image of the columned Assembly typical of the basilical plants that Schwarz defines as *sacred journey.* According to Schwarz and Guardini, each believer had to actively participate in the rite, for this reason in the new churches it was necessary to avoid the arrangement of the believers lined up behind the celebrant.

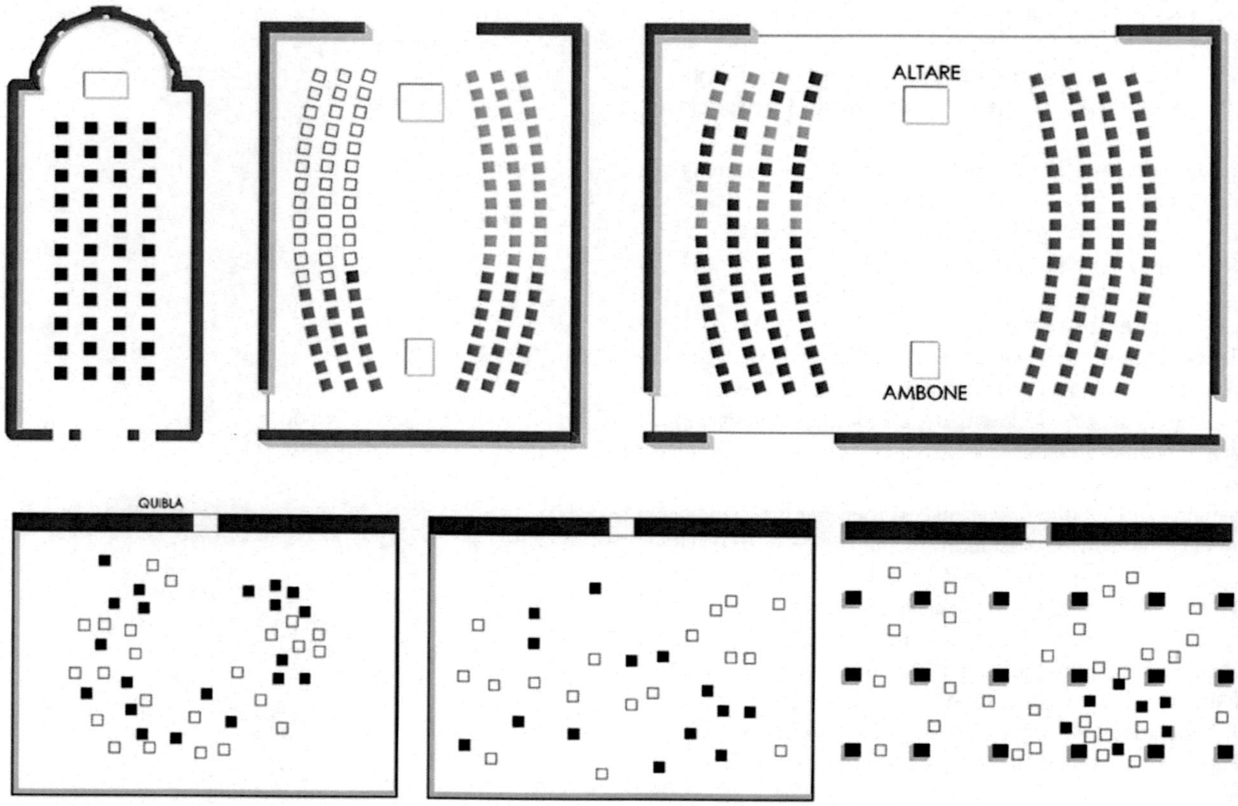

Fig. 16 Archetypal images in Catholic prayer with several cases of longitudinal directions. Archetypal images in Muslim prayer with several cases of transversal directions

basilicas, were transformed from church to mosque and one of the substantial transformations was to rotate the sense of the use of space, from longitudinal to transversal. The rotation canceled the eschatological force of the apse/presbytery system and it strengthened the long walls that would have hosted the *mihrab*.

In the synagogue, the choral shape of the Assembly has assumed symbolic value over time, as in other religions. However, the profound meaning of the synagogue, that is not a Temple but a place of reading and reflecting on the Torah, tells us that the archetypal image of being inside the synagogue is the man who studies God through the Book.

Studying, reading presupposes an individual recollection, in fact, some synagogues have a fragmented space that protects the individual concentration, almost like a library where the protagonists are not the books, but the Book.

It is evident that in each of the three confessions the issue of the prevalent direction of Assembly is main, therefore in the definition of an interreligious building type, it is advisable not to choose in-line schemes that can give life to a space coinciding with one of the directions commonly used for the construction of churches, mosques, and synagogues. The in-line (longitudinal or transverse) schemes strongly influence the Assembly's freedom of aggregation. For this reason, the research has studied a central space type, slightly deformed. The research came to a typological scheme that arises from the hybridization of a central pattern with a diagonal pattern that superimposes the idea of center in a diagonal direction, different from the transverse and longitudinal directions. The hybridization between central and diagonal patterns is necessary because working on the diagonal pattern you may favor the dynamic character of the space and you may introduce a feeling of disorientation because space works on ponderal symmetries and not on classic geometric symmetries. The disorientation does not favor the recognition of the space and induces the believers to recognize themselves in the composition of the Assembly. The character of the space is very dynamic, a large empty space that can be inhabited by the Assembly in its various dispositions, Catholic, Jewish, and Muslim.

The research worked on the hybridization of space both on the x, y plane, and along the z-direction. The research is studying two hypotheses: (a) a vertical space where the development along the z-direction is prevalent; (b) space where the relationship between base and height is in balance so that space does not become excessively monumental and maintains the centrality of the Assemblies.

4.1.2 Vertical Type

Often the character of prayer spaces has been studied according to its different geometries and to the hierarchies between its directions (Giammetti 2016b). Usually, space arises from the translation of a flat geometric figure along one or more straight lines, generally, a function of a rational number, oriented according to the directions to be emphasized.[11]

To study the vertical type, the research resumes Le Corbusier's typological studies on the church. The Swiss architect's research to unhinge the geometric identity between the main axis of the prayer room and the direction of the access paths leads to the project for the church of St. Pierre de Firminy, where Le Corbusier abandoned the line in favor of spiral to move away from the x, y plane and to conquer the space. Like Firminy, the research proposes a spiral scheme for the vertical development of the prayer space: along the z-axis space moves along a spiral, a curve that starts from the access points and bends on itself to return along the same direction of the entry (Fig. 17). Space is a large oriented void, an *objet ä reaction poetique* surrounded by light. The great void is plural and singular at the same time, composed of many areas, which nevertheless participate in the spatial unity, precisely because they are held together by the regulating path of the spiral and not of the line.

4.1.3 Flat Type

The second typological hypothesis renounces vertical development in favor of a greater balance of the proportional relationship between base and height, a useful choice to express an almost domestic character of the space. The domestic character of the space arises from the idea of prayer space as the home of each Assembly (Fig. 18). The renunciation of verticality is a renunciation of monumentality intended as a property to be found not in the space but in the sensitive experience of living in it attending the event of the rite. This hypothesis studies a type of space centered on the human and not on the idea of transcendence, such as the vertically developed spaces. Here the means for the irruption of the sacred is the centrality of human beings.

[11]For example, the basilical pattern arises from the translation of a rectangle along the main axis of the nave. The Greek cross layout arises from the translation of four rectangular figures along as many axes of equal size oriented towards a single point. The central plant is born from the translation of a symmetrical geometric figure along the zeta axis placed in correspondence with its canter of gravity.

4.2 Light and Ornament

The vertical type and the flat type may be characterized by two different types of light. The hypothesis is that the quality of light must not suggest an eschatological character of space, but it must support a strongly immanent dimension of the empty space inhabited by the Assemblies.

An immanent light must be centered on the Assemblies to support the gathering for the group of prayer, a gathering that is inside human beings, not in things or in the space outside. This type of light is produced by artificial lighting or by using natural light indirectly to obtain white light. We can easily compose this type of light with the flat type, but its declination in the vertical type is more difficult, because here the centring of the space on the Assembly is technically more difficult, in fact, the development in height produces a jump in scale between the size of a person and the dimension of space. In the verification phase of the models designed according to research, it will be possible to study the tow type of light and their effects on the models, because the type of light also influences the relationship between spatiality and ornament. The light becomes the only tool to substitute the traditional iconic ornamental apparatus if you use the liturgical furnishings only to celebrate the rites.

4.3 Extended Design Approach

The Actor-Network-Theory (ANT) is a theoretical model developed by Bruno Latour and other French sociologists (Latour 2007). This model says that no subject or object is an isolated entity, but everything must be inserted in a complex network of relationships in which human and non-human, material, and immaterial entities interact. Latour's network is always contingent because it changes every time the space-time condition changes.

The research method of ALEPH and SOW hybridizes the sociological model of ANT with the typological project of interreligious spaces. During the hybridization process, the architectural project is subjected to increase in depth and extension, because at the same time it works to build the space and its communities, a work that can be explained through the term of *construction of co-immunity*, introduced by Peter Sloterdijk—philosopher to whom Bruno Latour makes explicit reference (Sandercock 2004; Sloterdijk 2004). The research defines this idea of the increased project as *extended design approach* (EDA). The extended design approach (EDA) works to study how to promote an intercultural and post-national European lifestyle, as a project to build new communities. Starting from the project on interreligious prayer spaces, EDA's actions encourage the comparison between European communities and migrants: the

Fig. 17 Study of vertical type: space development along z-axis in a spiral movement. Le Corbusier, St. Pierre in Firminy

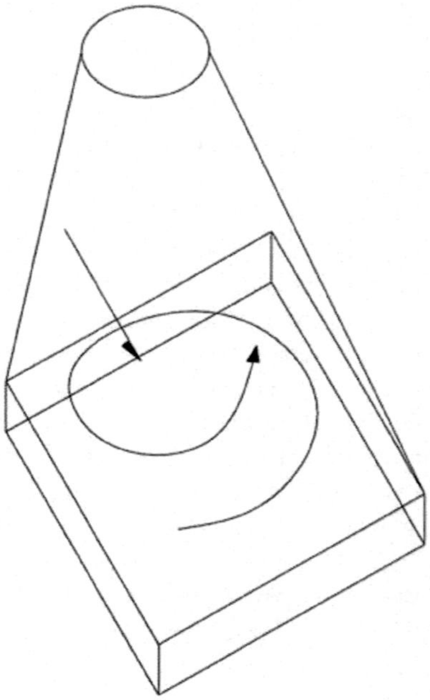

aim is to activate a mechanism to favor the mutual knowledge and to make sure to build new intercultural communities.[12]

4.4 Virtual Tour

Autors: Mariateresa Giammetti and Pierpaolo D'Agostino

The typological hypotheses have been translated into models in the form of spaces built-in immersive virtual reality (RVI). The RVI is useful for involving people in an interactive experience that allows visitors to inhabit the models, that involve their senses, which allows them to move in space, to perceive the quality of its light and its material. RVI can be used with some peripheral devices: (a) a viewer with a system to detect movements, so that by turning the head the same action is obtained even in the virtual environment; (b) earphones to transfer sounds to the visitors; (c) cyber suit that can be used for movements, to issue commands such as typing on virtual keyboards, etc. A virtual tour has been developed in ALEPH with a series of RVI scenarios to be proposed to museums to set up small spaces dedicated to experimentation and verification of the research project:

Scenario1_Individual immersive virtual reality with viewer.

It allows moving inside the rooms of the virtual tour where you can walk in the interior spaces of religious buildings that are particularly significant for each of the three confessions. The visitors can go in the two models (vertical and flat) designed for the research, Visitors do not have the perception of other people in the virtual spaces and along the tour she/he will meet some interactive points where she/he will have to answer a questionnaire about existing religious buildings and models developed for research.

Scenario 2_ Group immersive virtual reality with cyber suit.

It allows to go in and walk in the two models (vertical and flat). This scenario develops a group experience, in fact, cyber suits allow us to have perception of other visitors in

[12]Regarding the theme of interculturalism in relation to the construction of the physical and cultural space of the communities, the EDA method is based on the approach expressed by Leonie Sandercock regarding the project of an ideal city of the twenty-first century, when she write: "I dream of a city… where citizens snatch new possibilities from space to immerse themselves in their respective cultures and those of their neighbours, collectively forging new cultures and hybrid spaces » .The urban planning theories of Sandercock are within the *hybrid forum* of contemporary society, a general concept in which physical, cultural and political space are *forme a-venire*, to be designed through a work on *mixité* that passes first of all for a comparative/cognitive approach. Which are the countries of origin of migrants and European citizens? What are their habits, their ways of living in the public and private space, their ambitions? What are the motivations that have encouraged migrants to leave their countries? This approach allows us to put the human at the centre and not to consider the phenomenon of migration and of interculturalism only as useful numbers to build statistics.

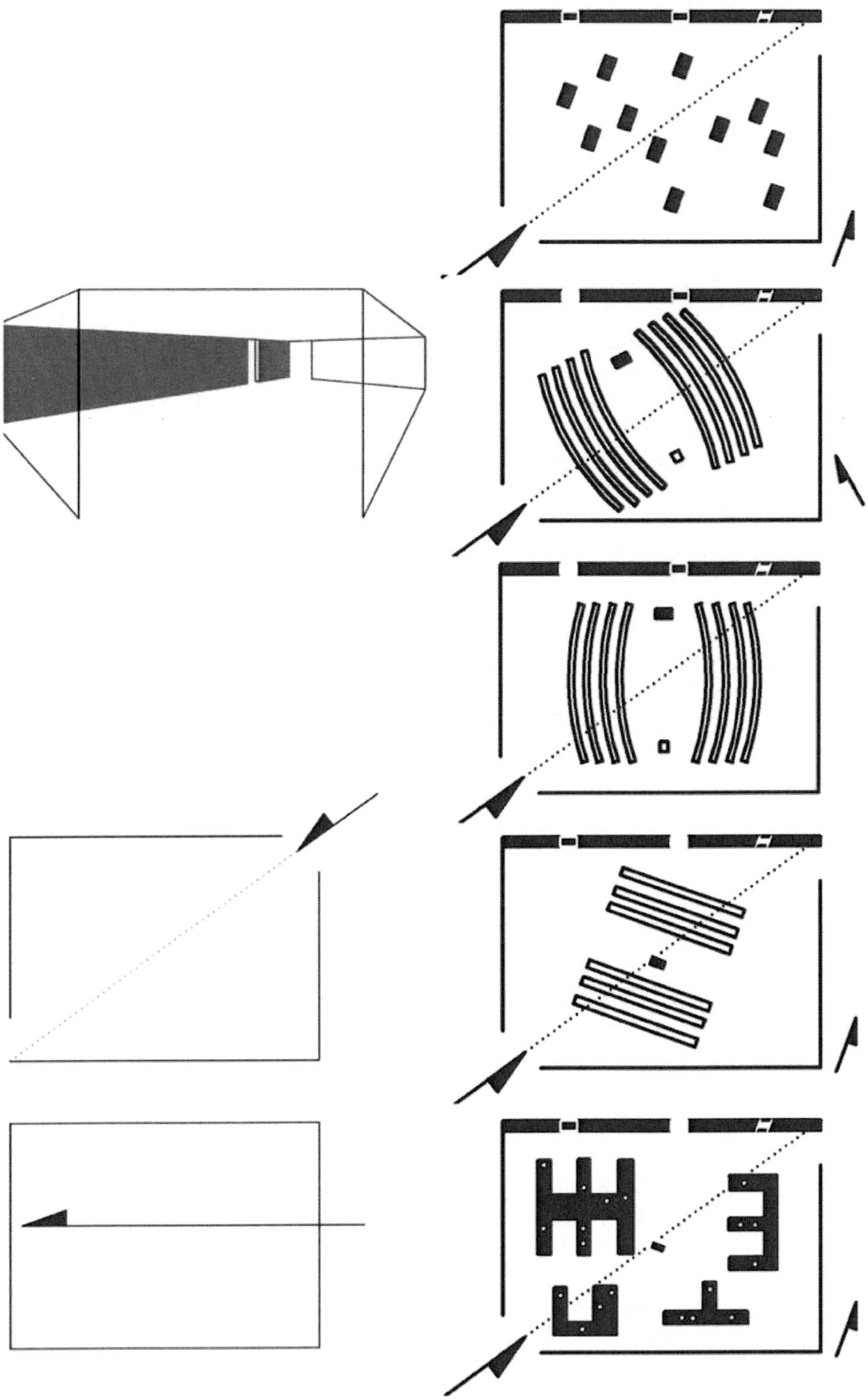

Fig. 18 Study of flat type: central and diagonal space and images of different disposition of assemblies

the virtual space. Each visitor wears a cyber suit that allows him to see other visitors as avatars who move and position themselves in space according to the commands given through the suit. The viewer allows an immersive virtual reality experience and the suit allows us to have a complete sensitive experience that transmits all the tactile sensations on the skin, the touch of other visitors, and the sensation of hot and cold. Group experiences are useful in understanding how visitors position one another in the model of the prayer space. It is also possible to record the data of the immersive experience through a questionnaire that each visitor must complete by clicking on some interactive points.

ALEPH has identified two types of simulations:

mono confessional improvisation: a group of visitors between 20 and 30 people of the same Abrahamic confession goes in the two models. In the models, there are not liturgical furnishings, chairs, and iconographic symbols. The visitors may arrange themselves freely in the space and they may feel the sensitive experience of inhabiting the space.

guided experience: a group of visitors between 20 and 30 people of the same Abrahamic confession goes in the two models. In the models, there are liturgical furnishings, but there aren't chairs and iconographic symbols. The visitors may arrange themselves freely in the space to live the sensitive experience of inhabiting the space through the rite whose guide is entrusted to an avatar.

This type of experience can be organized with groups from different backgrounds: school groups, religious groups, museum visitors, organized groups. The prerequisite for organizing the experience and belonging to the same confession, since it allows to obtain feedback on the same model from different religious communities. Furthermore, it allows us to understand how each type of Assembly tends to arrange itself in that space. This type of experience can be organized with groups from different backgrounds: school groups, religious groups, museum visitors, organized groups. To organize the experience, it is necessary to have groups of the same confession, because it allows obtaining feedback on the same model from different religious communities and allows us to understand how each type of Assembly tends to arrange itself in the models.

4.5 Questionnaire

Different types of questionnaires have been developed to structure a database useful for future research developments. Data collection has not yet been started systematically, because we are working to activate the ALEPH platform in some museums. When the database will collect a significant percentage of information, we can begin to examine them by offering answers to research questions. The study of the data will be useful to build descriptive statistics and a semiotic analysis of the content of the comments written in the virtual tour questionnaire.

Questionnaire about religious buildings particularly significant for each of the three confessions.

It is a written interview with pre-established questions that can be asked of anyone lay and believers of any creed.

The questionnaire wants to know the feelings, the reactions, and the level of knowledge of people about places of prayer that belong to persons of Islamic, Catholic, and Jewish culture.

Questionnaire on typological models.

Multiple choice questions about the perception of the orientation of the prayer space, about the spatial perception of space (disorientation, protection, order, etc.), questions about the role and perception of light. The questionnaire includes questions with answers in the form of comments, in particular about the absence of iconographic symbols and about the significant role of the Assembly for the prayer space.

5 Conclusions

Great cultural transformations are taking place in Europe: the challenge concerns the formation of a new identity subject that includes the phenomenon of migrations and new forms of post-national identities.

In these processes, religion continues to be one of the key points to meet the challenge of integration and to frame the strong presence of migrants from countries of Islamic culture in a non-oppositional dynamic. It could be possible to develop a relationship with diversity in the name of the meeting, proximity, collaboration, and mutual knowledge.

The research took up the European challenge on integration policies and it is designing a new typology for the interreligious prayer spaces of the three Abrahamic confessions.

The hypotheses for the new type project are: (a) to use a design method based on the extended design approach; (b) to entrust the significant and symbolic role of space to the shape the Assemblies of the believers gathered in prayer; (c) to reset the traditional iconic symbolic apparatus and to entrust to the expressive power of light the role of conveying the sense of the sacred.

The study focused on the search for possible characteristics of the space that could be shared by the faithful of the three confessions in order to generate a strong and useful sensitive experience in activating the event of the celebratory rite. The research method was founded on the extended design approach to connect pure design studies specific to the field of architectural composition with research by experts from other disciplines and with databases that collect the point of view of people on the interculturalism applied to spaces for prayer.

Until now the research achieved two partial objectives: (a) the study of two types of Abrahamic interreligious space declined in two models created in the form of 3D space usable in immersive virtual reality; (b) the development of ALEPH, a software that manages a virtual museum of the three Abrahamic confessions which also includes the virtual models. ALEPH has a dual objective: encourage intercultural exchange and use interactive questionnaires to collect feedback and views of experts and ordinary people on the perception of new types of hybrid spaces for prayer designed by research.

The expected results in the future are (a) the implementation of the database through the activation of the software in museums and the dissemination of workshops in schools and universities; (b) the processing of data to verify the typological hypotheses developed in this first phase; (c) the implementation of the typological project through the feedback obtained and the development of new models.

References

ALOG. Digital Inter-religious Platform web site. (2020, March 30). https://3alog.net/.

Autiero, A. (2016). Sensus fidelium e Magistero dal Concilio ad oggi. In A. Rovello (Ed.), *La morale ecclesiale tra sensus fidelium e Magistero* (pp. 13-16). Cittadella Editrice.

Avecedo, C., D'Aram de Valda, S., Jones, T., Kowoll, J. H., & Parry, E. (2019). Editorial. *Types of Assembly Architype Magazine, 1*(1), 4–5. TU Berlin Fachgebiet für Entwerfen und Baukonstruktion Raier Hehl.

Bauman, Z. (2017). *Vite di scarto.* Editori Laterza.

Campo Baeza, A. (1994). Architectura sine luce nulla est. *Domus, 760* (1), 86–89.

de Vries, N. (2019, November 22). *Inaugural address Prof. Ir. Nathalie de Vries at TU Delft.* https://www.mvrdv.nl/events/1470/inaugural-address-nathalie-de-vries-at-tu-delft.

Derrida, J. (2011). *Politiques et amitié.* Galilée.

Filoramo, G. (2004). *Che cos'è la religione.* Einaudi.

Filoramo, G. (2009). *Le religioni nel mondo moderno.* Einaudi.

Gabrieli, F. (1958). Il Risorgimento arabo. Einaudi.

Galantino, M. (2016). *La forma dell'acqua. emblemi spaziali ed emblemi dello stare in uno spazio di preghiera comune per le tre religioni abramitiche* (M. Giammetti, Ed.). RiStampa Editore.

Gargiani, R. (1993). *Auguste Perret 1874-1954. Teoria ed opere.* Electa.

Giammetti, M. (2016b). *La forma dell'acqua. Emblemi spaziali ed emblemi dello stare in uno spazio di preghie-ra comune per le tre religioni abramitiche.* RiStampa Editore.

Giammetti, M. (2018a). *Forma e Riforma/e.* Arthinkeditions.

Giammetti, M. (2018b). Silence, light and space aleph, a space for three abrahamic religions. *SMC Magazine, 7*(1), 50–57.

Gleiter, J. (Ed.). (2012). *Ornament Today. Digital, Material, Structural.* Bu, Press.

Haus der religionen (2020, March 30). *History.* https://www.haus-der-religionen.ch/en/history/.

Holl, S. (1985). *Pamphlet Architecture 11: Hybrid Buildings* (J. Fenton, Ed.). Princeton Architectural Press.

Hourani, A. (1970). *Arabic Thought in the Liberal Age.* Oxford Univ. Press.

House of one. Three Religions (2020, March 30). *Architecture.* https://house-of-one.org/en.

Jauss, H. R. (1987). *Esperienza estetica ed ermeneutica letteraria.* Il Mulino.

Jauss, H. R. (1988). *Estetica della ricezione.* Guida Editore.

Koolhaas R. (2000). *Delirious New York.* Electa.

Latour, B. (2007). *Reassembling the Social: An Introduction to Actor-Network-Theory.* Oxford University Press.

Lawson, B. (2001). *The Language of Space.* Architectural Press.

Luzzatti, L. (2007). *Dio nella libertà.* Zanichelli.

Martì Aris, C. (1993). *Le variazioni dell'identità.* Città Studi: Il tipo in Architettura.

Pehnt, W., & Strohl H. (2000). *Rudolf Schwarz 1897-1961.* Electa.

Peirce, C. S. (1980). Semiotica. Einaudi.

Pew Research Center. Religion & Public Life (2015, April 2). *The Future of World Religions: Population Growth Projections, 2010-2050.* https://www.pewforum.org/2015/04/02/religious-projections-2010-2050/.

Reinharz, J. (1980). *The Jew in the Modern World a Documentary History.* Oxford University Press.

Renan, E. (1890). L'avenir de la science. Calmann—Lévy.

Richter, K. (2002). *Spazio Sacro e immagini di chiesa.* Edizioni Dehoniane.

Rispoli, F. (2016). *Forma data e Forma trovata.* Istituto Italiano per gli Studi Filosofici.

Sandercock, L. (2004). *Verso Cosmopolis.* Edizioni Dedalo: Città multiculturali e pianificazione urbana.

SIR (Società per l'informazione religiosa) Information Agency web site. (2018, May 11). *"Religion2Go!", la religione entra nel mondo vir tuale* https://www.agensir.it/mondo/2018/05/11/religion2go-la-religione-entra-nel-mondo-virtuale/.

Sloterdijk, P. (2004). *Sphären III - Schäume.* Suhrkamp: Plurale Sphärologie.

Somers Cocks, A. (2019, December). Le tre religioni insieme ad Abu Dhabi. *Il Giornale dell'Arte: 403.* https://www.ilgiornaledellarte.com/articoli/le-tre-religioni-insieme-ad-abu-dhabi/132426.html.

Yerushalmi, Y. H. (1991). *Freud's Moses.* Judaism Terminable and Interminable: Yale University Press.

Utopian and Sacred Visualization to Culture and Spirituality

This part focuses on visualizing newly contrived methods and approaches aimed at outlining and the possibilities of contemporary construction through shedding a light on the newest technologies. In the chapter "Design Approaches in the Service of Persuasion in Sacred Architecture", the author argues that "sacred architecture has been the playground of experimentation in displaying a supreme reality existing chiefly in the realm of imagination". Intricate visual methods were contrived to generating a synthesis persuasive of its existence. The author uses Gestalt theory of perception but even personal findings and sight for interpreting through this the significance of perception and exploring its impact in sacred spaces constitution.

The chapter "Utopian Sections: Visionary Architectural Drawings from Early Approaches to the Contemporary Era" focuses on the 'utopian sectional architectures' analysing some significant examples and proposing a historical reconstruction of the role of the section in the utopian architectures and drawings, aimed to outline the possibilities of contemporary construction, in the light of the newest technologies.

The chapter "Sacred Architecture and Fashion Drawing: The Late Antique Decorations in Cimitile as Ideational Traces for Fashion Design" deals with an original and certainly interesting topic. Its field of investigation is cross-sectional to Architecture, Fashion, and Communication. The Architecture is the sacred one of the Paleocristian Basilical Complex of Cimitile in Campania regions. The author rediscovers geometry and meaning underlying of the sacred space decorations, through the 3D laser scanning and traditional surveying, transforming them into stylistic motifs. The project of innovative textures for Fashion is actually aimed at promoting the medieval complex of Cimitile through a wider communication project.

The intent of the chapter "Sacredness as Morality: Ledoux, and Chaux's Utopia" is to establish from Ledoux's lessons a significative equivalence between sacredness and morality, starting from a not obvious notion of utopia. By the imaginary designs for the tiny village of Chaux, published in the essay "L'architecture considerée sous le rapport de l'art, des moeurs et de la legislation", Claude Nicolas Ledoux wanted to show a city that could be built, at least in virtual form, and to lend credibility to its textual description and accompanying comments: true literary proof that aimed to show the possibility of a conciliatory utopia, in which social conflict could be subdued by a cult of Nature. The paper expands the knowledge and represents a significant contribution to the field in order to re-meaning in a broader sense the idea of sacred.

Design Approaches in the Service of Persuasion in Sacred Architecture

Iakovos Potamianos

Abstract

Sacred architecture has been the playground of experimentation in displaying a supreme reality existing chiefly in the realm of imagination. These experiments did not simply focus on a representation of another world but worked persistently to generating a synthesis persuasive of its existence. To this end, intricate visual methods were contrived. These varied considerably depending on the tenets and sophistication of each culture. This paper will present perceptual analyses of four such methods attempting to extract significant perceptual factors involved in the promotion of a utopian idea. Perceptual stratagems were devised in order to attain persuasion of the truth of imaginary constructs most effectively. A perceptual analysis is a procedure focusing on revealing the distinct aspects that play a role in achieving a concrete expression and its subtle gradations. The synthesis of such expressive factors results in an intense outcome able to convey a more or less concrete meaning. The analyses will not be simply based on descriptions of evident form but on deeper and more abstract relationships based on methods discovered by the Gestalt theory of perception attempting to derive the ways in which such visual methods may exert an effect on perception at a subliminal level. The first analysis will focus on the abstract visual methods employed in a Greek temple the second in a Byzantine church the third in a Gothic and the fourth in a Baroque church.

Keywords

Perception • Sacred architecture • Greek • Byzantine • Gothic • Baroque • Light • Shadow

1 Introduction

Each religion is founded on a narrative of a utopian condition of a perfect world, usually attained after death, although the admission of its tenets profoundly affects day-to-day life. According to a quite extensive recent study, the formation of large scale, complex societies preceded the belief in moralizing gods in order to facilitate cooperation among strangers (Whitehouse et al. 2019). Religion was used to effectively regulate societal relationships. Toward this end, it produced extensive belief systems capable of monitoring every aspect of human life. A belief system, once embedded in cultural life, exerts an immense pressure on individuals to comply. One is expected to conform to certain prescribed norms so as to be rendered fit to transfer into an alleged utopian world after death. The key prerequisite, however, is for those norms and conditions to be fully adopted. A person may not choose, relying on his own will and judgment, which tenets to abide by and which to relinquish.

In an attempt to convince him to waive his freedom of volition, potent methods of persuasion ought to be devised. Such methods rely largely on the production of relevant literary works, but also on visual and sonic compositions. These three aspects must be intertwined in order to effectively impress, influence, and ultimately sway the observer. While these aspects are most effective when they operate jointly, this paper will focus on visual factors alone. The discussion of these aspects will be looked at in a specific way focusing on their perceptual operation.

In order for such an exploration to be effective, the researcher must turn inward and observe the process of seeing itself. Only then one may reach an understanding of the manner in which the emergence of expression and meaning occurs and the ways in which these become capable of eventually affecting the psyche. Toward this end, the examination will rely on certain principles and processes studied by Gestalt psychology. Since, however, Gestalt psychology developed in the beginning of the twentieth

I. Potamianos (✉)
Aristotle University of Thessaloniki, Thessaloniki, Greece
e-mail: iakovospotamianos@gmail.com

C. Gambardella et al. (eds.), *Advances in Utopian Studies and Sacred Architecture*,
Advances in Science, Technology & Innovation,
https://doi.org/10.1007/978-3-030-50765-7_20

century, a question may arise regarding its use to interpret artifacts that long predate it. Yet, there is evidence in antique philosophical writings that great interest had been devoted regarding the manner in which the visual "aesthesis" or perception operates (Stovaios 1992; Andriopoulos 2003). Also, developments in the arts exemplified by certain refinements in the geometry and distribution of light and shadow in temples (Potamianos 2017) indicate a quite profound understanding of visual aspects, still not entirely grasped to this day.

It appears that ancient knowledge on such fine design aspects was retained and developed almost seamlessly while passing into the Byzantine era. Even though religious doctrines may have changed to a great extent in the passing from paganism to Christianity, from an engineering point of view the scientific knowledge of subtle visual refinements had been preserved through Latin and Arabic translations of late Greek works (Neugebauer 1975).

2 The Greek Temple

In order to comprehend the development and extent of sophistication of the visual refinements explored in order to reach a profound level of persuasion one ought to certainly consider an Ancient Greek precinct. For the purposes of this paper, the Athenian Acropolis and the Parthenon will be analyzed from a perceptual point of view.

As one enters the Acropolis precinct (Fig. 1) through the Propylaea building (Fig. 2), a considerable distance intervenes between the observer's position and the locus of the Parthenon. This spatial void operates on various perceptual levels. On the one hand, it allows the entire image of the Parthenon to be grasped at once. On the other, as this image is viewed within a large surrounding space the monument emerges from the mostly celestial background particularly reinforced. At the same time, the building appears to stand in isolation within its surroundings. Perceptual isolation is acknowledged by Gestalt psychology of perception as one of the most important factors for making an object stand out, acquire visual power, and occasionally even a sense of uniqueness (Fig. 3).

Another significant fact is that the Parthenon presents two of its elevations to the viewer. This placement, uncommon to modern building standards, has led early scholars to consider the layout of the Acropolis precinct as haphazard. The exposure of two facets of the orthogonal parallelepiped structure, though, has several offshoots serving certain perceptual objectives. The first objective is the presentation of the monument in a form as close as possible to its three-dimensional wholeness. The eye is practically unable to perceive more than three facets of a cuboid at the same time, and this is true only if viewed from above. As the Parthenon greatly exceeds human height, an upright standing man is incapable of seeing more than two facets.

The second perceptual objective is for the temple to be revealed at once in its entirety as one enters the Acropolis precinct. It befalls the viewer as a sudden vision of an almost explosive character. This takes place at the moment one exits from the interior space of the Propylaea while concurrently entering an exterior space. This condition is a contradiction in terms and a visual influx of paradoxical experiences. Perception struggles to cope with a juxtaposition of contradictory complexities between visual apprehension and conceptual meaning. While just becoming accustomed to seeing partial views of an oversized, monumental interior within the Propylaea, one gets prepared for an exiting process during which he impulsively expects to free himself of enclosure and move over to an outdoor unstructured reality. Instead, one enters a spatially finite precinct developing in open air. At this very moment, one is also faced with large structures that present themselves in their entirety. One's perception is in essence bombarded by paradoxical information. The impression provoked is profound and catalytic. The outdoor instead of being free and unstructured it is seen as highly organized although quite differently than the interior just passed up.

As the Parthenon is revealed, its two facets are turning away toward two different vanishing points while presenting to the viewer an edge which invades the perceptual field by directing its sharpness right at his eyes. By the placement of the building so as to view one of its corners, a two-point perspective is established which generates an ensuing peculiar effect. This effect is not focal, that is, it does not draw visual attention toward a single point—as a central perspective would have done—but subdivides it into two directions drawing it out and away toward the horizon. While a tension is generated by the intrusion of the vertical edge into the visual field, this tension is somewhat mild, moderated by the subdivision effected and the unhurried convergence of the perspectival lines.

The moderation is further reinforced by the rhythmical placement of the columns. The building presents its corner to the observer which corresponds to the foremost protrusion into one's visual field. The columns recede away from the point of observation to opposing directions toward the margins of the visual field. In this way the image of the building acquires an obtruding tension. The contrasting tension toward the depth of the visual field becomes subdivided into rhythmical recessions toward two opposite directions. This step-by-step transition while it is leading toward the depth, at the same time, decelerates this motion. This is one means by which it is visually moderated. The concept of moderation or measure, frequently referred to in ancient literature (Homer 2003), achieves an important goal. A higher power is always distinguished by measure.

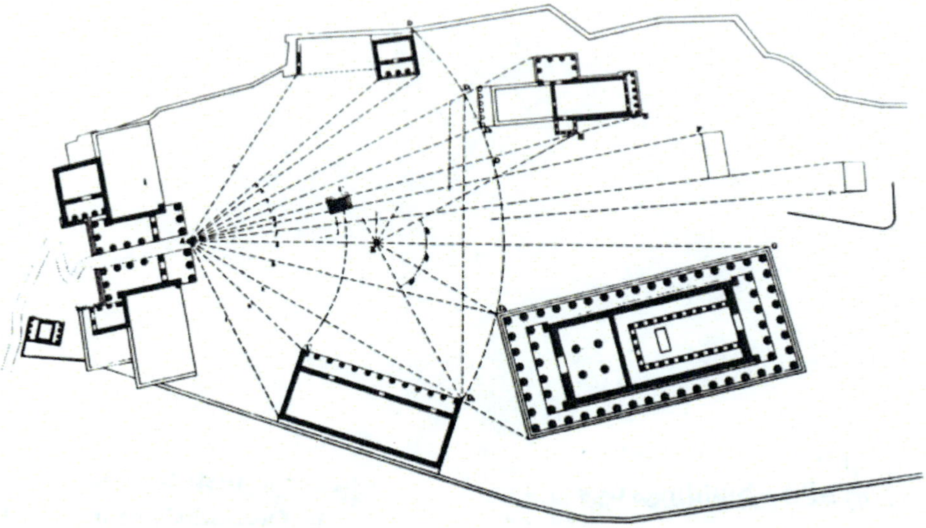

Fig. 1 Acropolis, viewpoint at Propylaea exit, after Doxiadis, 1937 (Doxiadis 1972)

Fig. 2 Propylaea, Acropolis, Athens 5th c. BCE

Consequently, an edifice dedicated to such power should not be placed in such a way as to compel toward more rapid consummation. An unhurried retraction toward depth confirms the sense of remote magnificence.

The planes of the building itself are receding due to the chiaroscuro. The recession generated, at first, by linear perspective is based on the apparent conversion of parallel lines toward the two vanishing points. This recession is augmented by the rhythmical, step-by-step progression of the columns toward these same points. On top of that, a third and subtler type of recession is in effect. The columns are clearly protruding into the light while the wall behind the colonnade remains hidden into darkness. This chiaroscuro effect produces a dramatic contrast which is perceived as a recession from a lighter first plane to a darker second one; as if the building were gradually withdrawing into itself, within the confines of its own body (Fig. 4). Thus, three different visual methods are employed to bolster the same concept. To this is added the ever so slight inclination of the columns inward, meeting at the building's central axial line at 2.2 km

Fig. 3 Parthenon, Acropolis, Athens 5th c. BCE

above the temple base. Although, this may be regarded more as an optical rather than a perceptual refinement, it certainly works in support of the same distancing effect (Fig. 5).

The various cues, outlined above, cooperate, each offering a slight accentuation of the impression. A distancing effect is achieved by the cumulative impact of the visual tensions. Through the Gestalt isomorphic method this effect is partly conceived as an attitude of indifference to reality. The monument appears to keep its distance, to feel remote. A sense of visual detachment is generated which is, in turn, felt as becoming aloof and withdrawn to a transcendental or spiritual level. It is sensed as surpassing common reality, staying uninvolved with it, thereby felt as belonging to a different or higher veritableness or truth. It is present before one's eyes, and yet it convincingly presents itself as belonging to a world beyond the real, or more to the point, visually establishing the presence of an alternative, higher, more spiritual reality. In this case, then, the persuasion of spirituality is effected through two spatial-visual qualities; of distance and poise or reserve.

3 The Byzantine Church

From its first inception, the Byzantine church exhibits a strong emphasis on interior space (Michelis 1955). In some early examples, the interior may present a certain stylistic correspondence to the exterior as in the case of the original design of Hagia Sophia of Istanbul. In most cases, however, though the exterior form may appear pleasing, it does demonstrate a ubiquitous conflict between the rectilinear aesthetic and the curvilinear one (Potamianos 2017). It presents certain paradoxical transitions not withstanding severe aesthetic criticism. As a relevant instance may be considered the curious transitions and incongruities between plastic forms as viewed from the outside (Fig. 6). If the exterior appearance were the principal concern such aesthetic conflicts would be thoroughly prevented.

As one enters into the building a central perspective is established. The eye is directed toward the center which is at the end of the main axis of the church and is marked by the

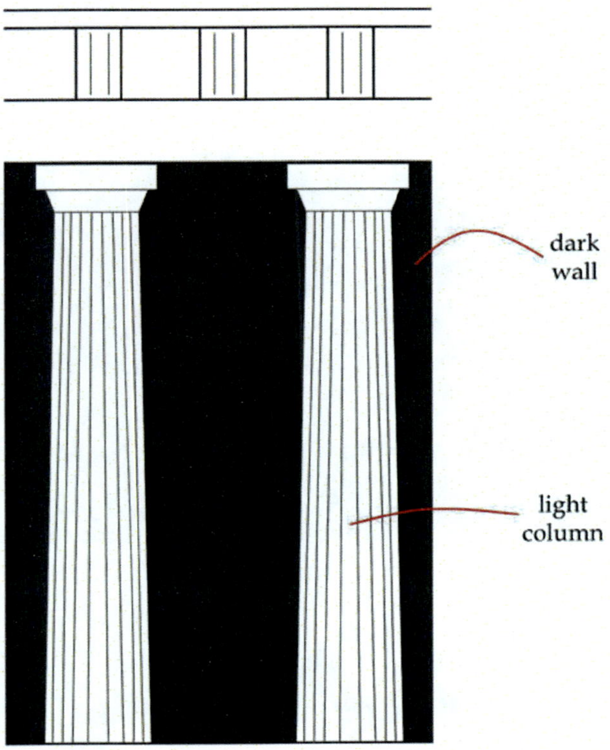

Fig. 4 Contrast of column against wall (Potamianos 2017)

position of the altar. However, this central perspective is not particularly strong because it is moderated by the equality of the arms of the cross in plan. Most Byzantine churches adopt the Greek cross in plan which consists of four equal arms (Fig. 7). Instead, in the Latin cross—found in later churches predominantly in the West—the lower arm is much longer than the rest. While this difference is understandable and perhaps justifiable in plan, it is not as clear to conceive its perceptual consequences when it comes to the third dimension. As one enters the Byzantine church and passes through the doors separating the narthex from the nave the angle of vision expands equally to both plan and section (Figs. 8 and 9). Thus, the entire interior expanse is revealed to the observer.

The focusing on the altar is moderated rendering the altar as an important point but sharing its gravity with the dome. The two are concurrently present in the visual field without any substantial movement of the head. Because of this sharing, a sense of moderation is here induced as well. The space seems to be enveloping and at the same time be of a certain measure. The space, as it expands around the viewer, is not enormous. It is not too high neither too long. It is moderately expanding toward all directions. This occurs in perception for two reasons. There are no cues drawing the eye too intensely toward the depth. As a result, the depth along the principal axis in plan does not become dominant. No cues are drawing the eye sideways too intensely either. The same is true when one looks upward. Although the dome is more intensely lit (Fig. 10), the distance to its apex does not exceed by far the dimensions of the remaining of the church. It appears to be part of an almost evenly distributed spherical space. The perceptual tensions are moderate.

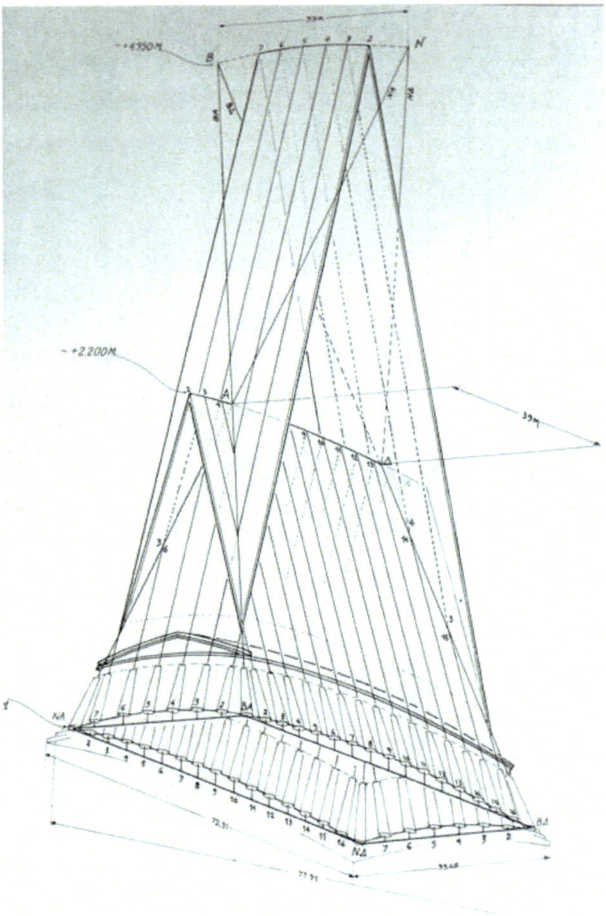

Fig. 5 Optical Refinements exaggerated, after Korres (Korres in Tournikiotis, 1994)

Perceptual intensity is distributed by different means. Toward the altar a reinforcement of perspectival focusing is achieved by the delineation of intermediate arches placed on the barrel vault above the iconostasis. One may notice the peculiar subdivision of the barrel vault that connects the dome to the quarter sphere above the altar in a Greek cross-plan church. In Middle-Byzantine churches two intermediate arches are placed along this barrel vault. The interposition of these arches is not justifiable by structural reasoning. Each time such an arch is interposed, the ensuing

Fig. 6 Hagios Paneleemon, Thessaloniki, 13th c

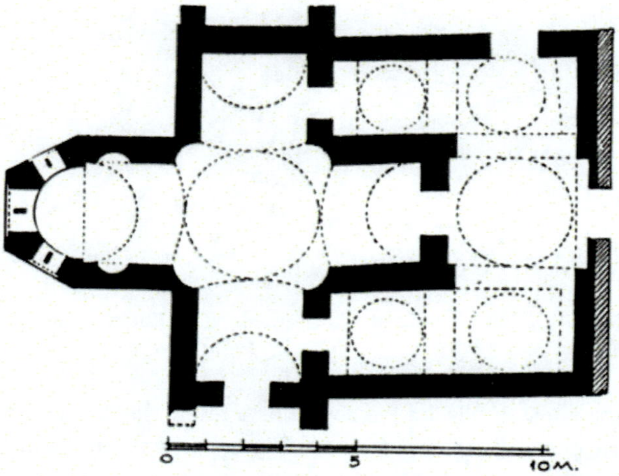

Fig. 7 Paliopanagia, Manolada, Greece, 12th c., in Gkioles (Gkioles 1987)

radius of the barrel vault is reduced (Fig. 11). Viewed from a point along the central axis, however, the sense of convergence toward the focal point is intensified to a certain degree. As a result, the locus of the altar is intensified so as to stand out and it acquires perceptual strength counteracting the overpowering, highly lit dome. No part becomes overly dominant.

The surrounding space is moderate in size. A feel of enfoldment and intimacy ensues. The sense of measure is established by the moderate scale. The distance to any of the surfaces does not exceed by far the size of the human body. Therefore, no great sense of monumentality is established on the basis of size.

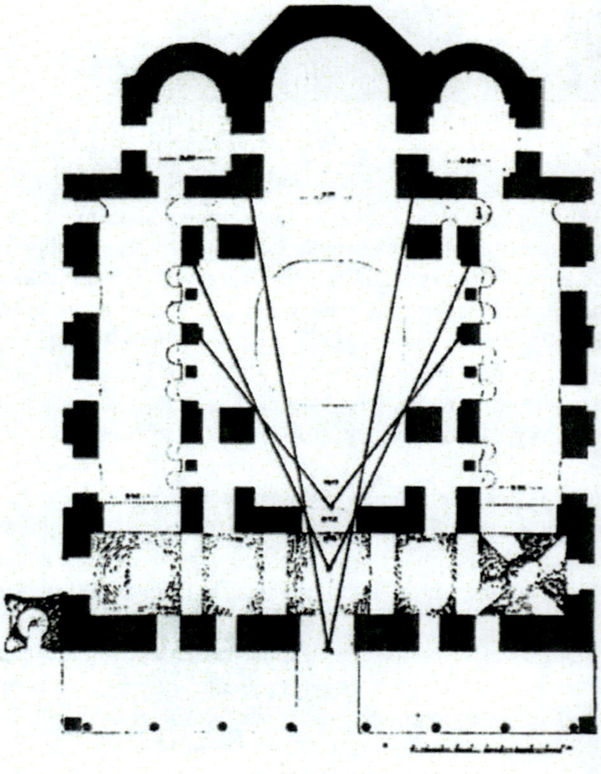

Fig. 8 Hagia Sophia Thessaloniki, 7th c., plan after Kalligas (Kalligas 1946)

Most lines do not last for long in the form of a straight line. They quickly transmute into curves which are austere, rigid semicircular lines. Such curves delineate the apse but also the dome. The moderation of the straight lines' length

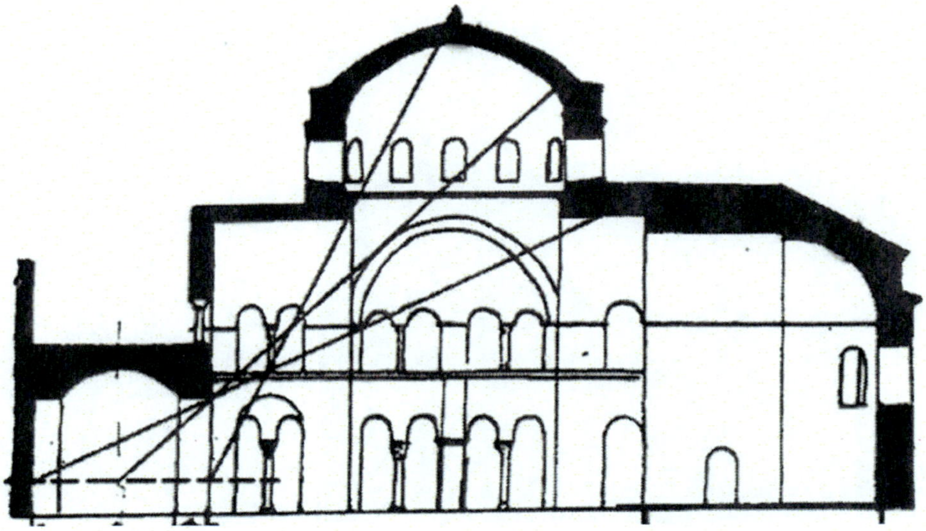

Fig. 9 Hagia Sophia Thessaloniki, 7th c., section after Kalligas (Kalligas 1946)

Fig. 10 Metropolis, Mystras, Greece, 13th c

and their transformation into semicircular curves generates a perceptual motion and effortless leaps from one spatial void to another. Thus, a single dominant direction is not indicated with certainty, and the sense of an enveloping world is strengthened.

A patchwork of light and dark starts developing as this perpetual movement is mounted on perception. Within the void of the apse, a light shaft is generated on specific instances rendering this void dependent on time. This phenomenon occurs unhurriedly and only at times of liturgical importance. On the other hand, the whole of the space appears disinterested in time with its crowning dome acquiring a highly lit surface, remarkably more so than the rest of the surfaces in daytime. Regardless of the time a sense of stability is developed. A certain dramatic conflict like a tug of war appears to be at work.

Fig. 11 H. Sophia, Thessaloniki, 7th c., partial section after Gkioles (Gkioles 1987)

This ambiguity is reinforced by light shafts developing within space making the void appear more energetic. This void is rendered visible through the depth produced by the light shafts. Light and darkness generate a second type of dramatic conflict which often repeats itself at various points concurrently. Thus, perception becomes persuaded of the presence of space as an entity with substance and density; an

Fig. 12 Hagia Sophia, Istanbul, 6th c

entity through which light cuts with effort (Fig. 12). In Byzantine thought, both space and light have been considered as qualities of the divinity.

4 The Gothic Cathedral

In the Gothic cathedral, the emphasis is also placed on the interior. This time a more overt system of persuasion is at work. The church is usually laid out in a Latin cross-plan (Fig. 13). As one enters the church through the main doors a long corridor opens up before him. The straight line tends to impose on perception certain dynamics hard to swerve from. The longer a straight line becomes the greater its imposition on perception. Thus, attention is swiftly dragged toward the final aim located at the apse (Fig. 14). As if this were not adequate, the left such line is still reinforced by a symmetrical identical line on the right-hand side. The two work together as a powerful magnet literally launching perception toward the endpoint. These two lines are formed by the rhythmical pattern of the colonnades but also of the pointed groin vaults in the ceiling (Fig. 15). The central perspective has a dual effect. It tends to ingulf perception into its final destination, almost the same way as a black hole does to light, while concurrently it generates an equally powerful outward force like a point explosion. This dual force acting upon perception leaves no alternative apart from surrender

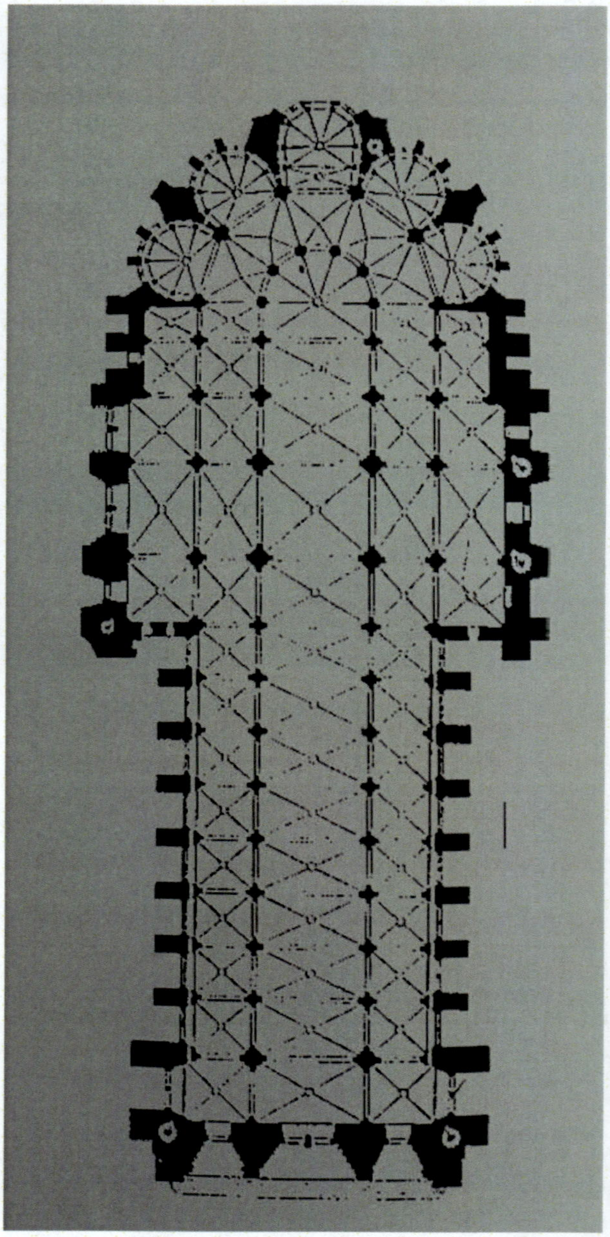

Fig. 13 Reims, France, plan, 1211

and sheer captivation. Attention becomes a prey to line dynamics. No escape is possible. The endpoint becomes its ruler.

In later, Gothic churches or reconstructions of Rayonnant style an ancillary method of coercion is contrived. Each column is subdivided into a number of lines just like stems of plants. The columns that were previously distinct wholes are now transformed into a group of straight lines moving upwards (Fig. 16). Lines until then served the visual goal of delineating the form of a column and other structural or

Fig. 14 Westminster, London, 1258

Fig. 15 Notre Dame, Paris, 1163

decorative forms while now they acquire a new and more abstract one. By teaming together in groups and based on the inherent dynamics of the straight line, they gather perceptual momentum and launch it upward. In doing so, they pass by the previous point of destination, which was the underside of the floor above and continue unbroken to the upper floors and passing by them too, they move without interruption all the way to the apex of the groin vaults. Thus, in case attention is momently sidetracked, a powerful force seizes it and leads it upward, generating on the one hand a sense of uplifting, while on the other, once the eye reaches at the top, the rhythmical pattern of the groin vaults places it back on track toward the focal point (Fig. 17).

This perceptual structure is quite severe. It is yet reinforced by the structure of light and shadow. The windows are of a very large size which constituted the motivation for certain structural innovations. The stained glass although thought as transparent, it is in fact rather translucent. Light does not fully penetrate space through them; as if it were impeded from freely flowing into the church. It appears to be rather arrested on the surface of the tinted glazing. Thus, as the space mostly develops in two distinct directions perpendicular to each other, i.e., the depth and the height, the structure of light and shadow evolves as rhythmical glowing patches in a mostly dark interior, creating a stark contrast. Each glowing surface tends to exceed its own limits generating a corruptive effect on the window outline, adding a sense of unreality or spirituality to the image. This lack of clarity contributes greatly to the otherworldly atmosphere of the interior.

Finally, at the point of the transverse aisle two new directions open up. The one in the vertical direction, toward the internally lit lantern tower (Fig. 18). The other opens up toward the side choirs. The opening up, however, occurs at once as a sudden visual revelation for which the observer is unprepared. As a perceptual event, it acquires once more an explosive character.

5 The Baroque Church

The linear perspective, perfected by Filippo Brunelleschi and Leon Battista Alberti, was so well comprehended by Sebastiano Serlio that he became capable of using it in a very ingenious way, fit for theatrical purposes. He realized that space itself could be visually compressed in order to deceive the eye into seeing it as deeper than it actually was. This compression of perspective was incorporated into a makeshift theater constructed within the atrium of Palazzo Porto (Fig. 19). Later, the method was incorporated in the

Fig. 16 St. Denis, reconstruction, France, 13th c

Fig. 17 St. Denis, rhythmical transition

Teatro Olympico by Vincenzo Scamozzi and was further developed for architectural and religious purposes by Francesco Borromini and Gian Lorenzo Bernini.

Instead of the straight lines used in Gothic, curved lines are often used in Baroque churches. In some cases, such lines fluctuate to a great extent both in the exterior and in the interior, as for instance in the San Carlo alle Quattro Fontane (Fig. 20). In perception this results to an oscillating effect, which draws the viewer into it more slowly but as surely as in the Gothic interior. A wave like sensation is generated through the imposition of an alternating repelling and attracting tension. The eye, in its effort to change focus to adapt to the vacillating distance, it becomes strongly engaged to the forms delineated and are driven from one to the next. In the case of the San Andrea al Quirinale, the elliptical plan (Fig. 21) exerts changing pressures as the glance moves toward different directions, at the moment one enters, coercing oscillating visual, and kinesthetic responses. These responses are due to the manner in which perception operates while they are sensed as emitting some kind of eerie power.

The chiaroscuro structure was developed to an extreme in the baroque style not only in paintings but in the appearance of buildings. Specific light and shadow handling devices were contrived for purposes of dramatic appeal. The best-known examples are the heavy excretions and deep recessions on the building surfaces. These generated a high relief with sharp transitions from light to shadow. Since such transitions are not common in the appearance of objects but are more gradual under earthly lighting conditions, these can be easily associated as an expression of an attitude emerging from within and instinctively attributed to the object itself. For instance, a human face with eyes hidden deeply in heavy shadow will appear as expressive of a threatening attitude. The transition from light to shadow is even more gradual in an interior space because of the multiplicity of reflections in the surrounding surfaces. Therefore, sharp transitions are even more effective in an interior space and become capable of generating unworldly appearances or even reversing sensations. Important examples may be found in the Oratorio dei Filippini in Rome, in San Giovanni di Laterano in Rome, and the Scala Regia in the Vatican (Portoghesi 1968). In the Oratorio dei Filippini (Fig. 22), looking upward one sees various surfaces one above the other in which light does not follow a normal pattern of distribution from light to dark. Instead, abrupt shifts and sudden reversals are seen. Thus, the distribution is not conceived as governed by an environmental rationale but rather by the presence of conscientious volition denoting specific intention. In San Giovani di Laterano, something similar creates a separation between horizontal surfaces that produces a sense of openness to the heavens within an enclosed space (Fig. 23). In the Scala Regia at the Vatican, on the other hand, an entire theatrical

Fig. 18 Ely Cathedral, lantern, England, 1322

effect is set up employing the Serlian compression of linear perspective in conjunction with a simulation of the Da Vincian "atmospheric perspective." A person descending that stair appears gigantic and distant at first while gradually acquires human size. A subliminal impression of a divine being descending to earth to mingle with humans becomes inescapable to unsuspecting eyes. However theatrical, the image is staggering and long-lasting in human imagination (Fig. 24).

6 Conclusions

Different visual methods have been used to persuade perception of the presence of a utopian world and the existence of divinity in different cultures. A concerted effort employing various visual means was founded on a careful study of perceptual and gnostic functions. This quite deep comprehension led, in turn, to the contriving of relevant methods of visual and spatial composition. Ancient knowledge on perceptual functions evolved gradually and was later embraced and developed still further by architects of the Christian era. The ancient spatial designs indicate a high level of sophistication in the engagement of perception followed by Byzantine examples suggesting a comparable degree of compositional refinement. The Gothic churches reveal more straightforward rather than suggestive methods in compelling insight of divine presence. Baroque designs were also based to a certain degree on ancient knowledge but proceeded to explore new ideas and invent new methods. These designs were mostly driven by the Renaissance invention of perspective and the infinite possibilities that opened up by compressed perspective. These chiefly optical rather than perceptual advancements were balanced by strong compositional inventions and individual skills of a theatrical dramatic flair that incorporated a quite novel

Fig. 19 S. Serlio, Wooden Theater, Palazzo Porto, Vicenza, 16th c

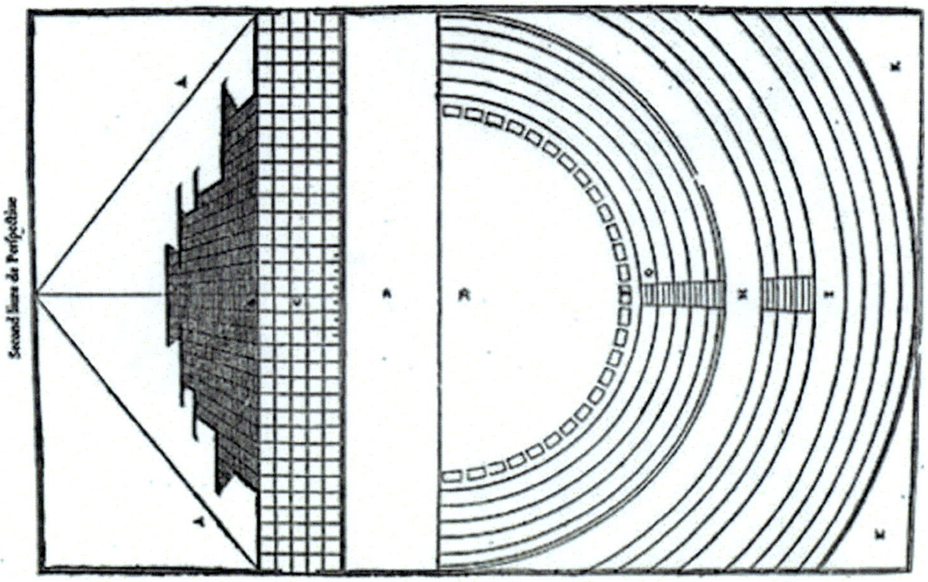

Fig. 20 F. Borromini, San Carlo alle Quattro Fontane, Rome, 1646

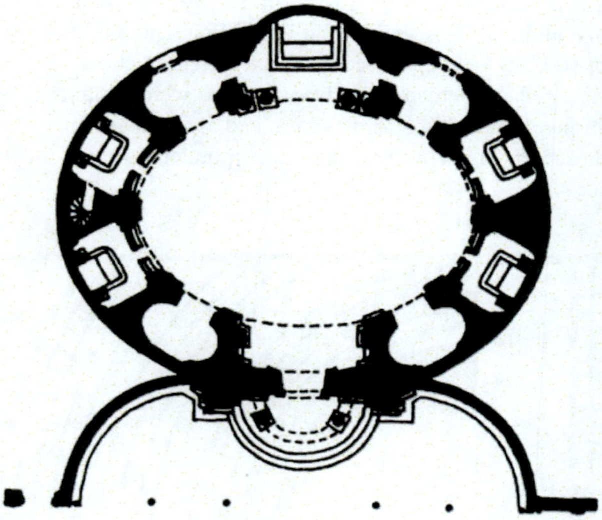

Fig. 21 G. Bernini, San Andrea al Quirinale, Rome, 1670

Fig. 22 F. Borromini, Oratorio dei Filippini, Rome, 1637 (Portoghesi 1968)

Fig. 23 F. Borromini, S. Giovani di Laterano, Rome, reconstruction. 1650's (Portoghesi 1968)

Fig. 24 G. Bernini, Scala Regia, Vatican, 1663

system of handling light. Thus, the attempt to persuade one of the existence of a utopian world in the realm of human imagination has undergone an unceasing development since antiquity generating a wide-ranging and powerful compositional vocabulary greatly enriching the field and the extents of architectural theory and practice.

References

Andriopoulos, D. (2003). *Archaia Ellenike Gnosiotheoria [Ancient Greek Theory of Knowledge]*. Athina, Greece: Papadimas.

Doxiadis, C. (1972). *Architectural space in ancinet Greece (in German 1937)*. Cambridge, MA: MIT Press.

Gkioles, N. (1987). *Vyzantine Naodomia 600-1204 [Byzantine Religious Architecture]*. Athina, Greece: Ekdoseis Kardamitsa.

Homer, (2003). *Odysseia, Rhapsody Zeta*. Athina, Greece: Eksantas.

Kalligas, M. (1946). *He Aisthetike tou Chorou tes Hellenikes Ekklesias sto Mesaiona [The Aesthetics of Space of the Greek Church in the Middle Ages]*. Greece: Athina.

Korres, M. (1994). *He Architektoniki tou Parthenona* [The Architecture of the Parthenon]. In P. Tournikiotis (Ed.). *O Parthenonas*. Athina, Greece: Melissa.

Michelis, P. A. (1955). *An aesthetic approach to byzantine art*. London, UK: B.T. Batsford.

Neugebauer, O. (1975). *A history of ancient mathematical astronomy*. Berlin, Germany: Springer.

Portoghesi, P. (1968). *Francesco Borromini*. London, UK: Thames and Hudson.

Potamianos, I. (2017). The handling of light: Its effect on form and space in the greek temple and the byzantine church. In C. Papadopoulos, H. Moyes, (Eds.). *The oxford handbook of light in archaeology*. Oxford, UK: Oxford University Press.

Stovaios, I. (1992). *Eklogai, Apofthegmata, Ypothikai* [Selections, Quotes, Precepts], (Tomos 4 (5th c.)). Athina, Greece: Kaktos.

Whitehouse, H., Francois, P., Savage, P., Currie, T., Feeney, K., Cioni, E., et al. (2019). Complex societies precede moralizing gods throughout world history. *Nature, 568*, 226–229. https://doi.org/10.1038/s41586-019-1043-4.

Utopian Sections. Visionary Architectural Drawings from Early Approaches to the Contemporary Era

Mickeal Milocco Borlini and Maria Vittoria Santi

Abstract

Starting from the Piranesian visions of the Carceri (1745–1750), through the drawings of Pantheon by Francesco Piranesi (1790) and the fantastic paintings by Nils Ole Lund as The house in the cliff and L'éléphant dell'étoile by Charles-François Ribart (1758) up to the Utopian Urban Machines by Le Corbusier and the incredible perspective sections of Paul Rudolph (Art and architecture building, 1963), it is visible how the section is gradually becoming a fundamental tool for adequate parameterisation of the human being in space and in the city, whether in the real or imaginary world. Thanks to the examples cited and an imaginative vision, which wants to go beyond boundaries of what is achievable, we reach the projects of contemporary visionaries such us Julio Lafuente with the hotel on the cliff in Gozo, Malta (1967), Lebbeus Woods with his 'sectional' hybrid architectures, Gordon Matta Clark with his Home Cuts and Stefano Boeri with his Milano Animal City (MAC). These dreamers produced 'utopian sectional architectures', regarded as such only because they were not built at all, even if they are potentially feasible. Through a careful analysis of some significant examples, it is possible to perform a historical reconstruction of the role in the utopian architectures and drawings section. The final objective of this survey is the reinterpretation of the creative and design value of those seemingly unattainable visions, in the light of the newest technologies and achievable possibilities of contemporary design.

Keywords

Utopian architecture • Drawing section • Urban fabric • Experimentation • Design

M. Milocco Borlini (✉) · M. V. Santi
University of Udine, DPIA, Udine, Italy
e-mail: mickeal.milocco@gmail.com

1 Introduction

«They despise my modernity; I condemn their modesty».
(Giovanni Battista Piranesi)

The following text tries to give value to the utopian dissection tool that practices a method of investigation and research since ancient times. Through the etymological explanation of the term section, passing through its classification as an instrument, measure and 'cut', it is possible to proceed to its definition as a design tool when it is accompanied by visionary ideals. That will show how and what the Utopian Section is and what it has served over the centuries, bearing in mind that the heterotopic and imaginative aspects of some contributors. Finally, it is possible to proceed with an analysis of specific case studies, divided by historical period, and conclude with some considerations on the research work carried out.

2 The Section as a Survey, Research and Design Method

2.1 Definition, Etymology and Meaning

«Section noun, UK /ˈsek.ʃən/US /ˈsek.ʃən/:

Section noun (part): one of the parts that something is divided into

section noun (cut):

- Specialised medical a cut made in part of the body in an operation
- Specialised biology a very thin slice of a part of an animal, plant or other object made in order to see its structure:
- Specialised architecture, geology, engineering a drawing or model that shows the structure of something by cutting part of it away: This vertical section of the soil shows four basic soil layers.

C. Gambardella et al. (eds.), *Advances in Utopian Studies and Sacred Architecture*,
Advances in Science, Technology & Innovation,
https://doi.org/10.1007/978-3-030-50765-7_21

- The shape of a flat surface that is produced when an object is cut into separate pieces
- In section: showing what something would look like if the surface was cut away and you could see inside: The first diagram is a view of the building from the street, and the second shows it in section.» (https://dictionary.cambridge.org/dictionary/english/section)

If in the medical and anatomical field to dissect means to dismember, to divide a given object into parts in order to analyse and discuss it, we can say that the same methodology is applicable to architecture, in particular to the Architecture of Section. The subdivision of an element into several parts facilitates the understanding, the study, through a method that, consequently, links the conditions of the context, of the soil and of the elements which might interfere with it. The use of the dissection technique in the artistic-anatomical field has always focused on the anatomy of the human body to make all the parts examined explicitly and, in particular, their functioning (Lewis et al. 2016; Privitera 2009a, b; Milocco 2018). Thus, we understand that the anatomical technique is not that distant from the 'architectural cut'. The Manual of Section—mentioned several times in this paper thanks to its particular research depth and vision—by Lewis et al. (2016)—shows some emblematic examples of the use of the anatomical practice of dividing the parts of a body; among the most cited ones, there are the studies of Leonardo da Vinci and the De Humani Corporis Fabrica, in the Baroque period, in which there was a real staging of the 'dissected' human bodies being exposed to the public, showing all the anatomical levels of the corpses (Lewis et al. 2016; Privitera 2009a, b).

The act of dissecting gradually increases the cataloguing of what is shown, therefore, what remains is the result of this process. We may report some concepts to fully understand what the theme is truly about. In Giovanni Michelucci: Anatomy of Space by Privitera (2009a, b) the author, through a meticulous study of the art, starting from The Dissection represented by Rembrandt (Anatomy lesson by Dr Tulp) (Fig. 1) and making interpolations with Michelucci's architecture,[1] she composes a critical text with a poetic and illuminating inclination for the theoretical path of this thesis. The theses proposed by the author are exemplified in the methodology of investigation, research and discovery of the concreteness of the artefact or body being examined, considering the functionality and the relationship between the exterior and the interior parts, as for a building or a human being (Milocco 2018). The aspect of the architectural section that attracts interest, for these considerations, is the research for 'presences' between what is interposed between the external and the internal world. The Section is, thus, an effective instrument which is able to show and study the human and urban systems, and where the anatomy of the territory and its natural and anthropised components can be seen in their entirety without neglecting the urban and social relationships that trigger 'between' the spatial composition of the urban and the social elements of the landscapes (Privitera 2009a, b).

To fully understand the methodology of operation of the Section, it is compulsory to observe what the Section normally hides; the Section looks at the architecture (a building or a portion of a city) through a different point of view that allows us to observe the different dynamics among the existing architectures. Thanks to the study of the section we can hypothesise new ways of making architecture and solving urban space problems (Secchi 2006; Allen and McQuade 2011). We can, therefore, consider this text as a "public manifestation of the Section" to celebrate and improve the skills and interests of the designers throughout a journey of disassembly, study and reassembly of the contemporary metropolitan architecture (Milocco 2018).

A Section is a part of an object or body intended as a result of a cutting operation that allows the viewer to see all the parts simultaneously (Lewis et al. 2016); it is also a technique of representation (a figure) of a volume which is crossed by the passage of a plan. In architecture, this tool allows us to parameterise the human figure with the vertical dimensions of the urban environment: «in contemporary architecture, the Section is the generator» (LTL Architects 2016). Finally, in this paper, the Section is intended as a geometric technique, a set of spatial organisation methodologies and as a starting point for a historical, critical and iconographic excursus towards an architectural instrument which, as we shall see, reveals its usefulness in designing the complexity of the city (Chapman 2009).

2.2 The Section: Classification, Cut and Design Tool

> «The work we begin […] must explain the general principles on which all science and art, liberal or mechanical and the most remarkable details that constitute its body and essence.» (Casini 1966)

Among the earliest examples of in-depth analysis and cataloguing of 'things', the Encyclopédie of Diderot and D'Alambert contains many drawings that disassemble and classify entire objects of common use to compose a general compendium of knowledge. It was the eighteenth century and Denis Diderot, who was the director and publisher of the Systematic Dictionary of the Sciences Arts and Crafts,

[1]See Michelucci's drawings of Chiesa di San Giovanni Battista or "dell'Autostrada', Campi Bisenzio (Florence) 1960–1964 at http://db.michelucci.it/archivi/disegni/?opera=P145&project=on&check=on (last retrieved March 2020).

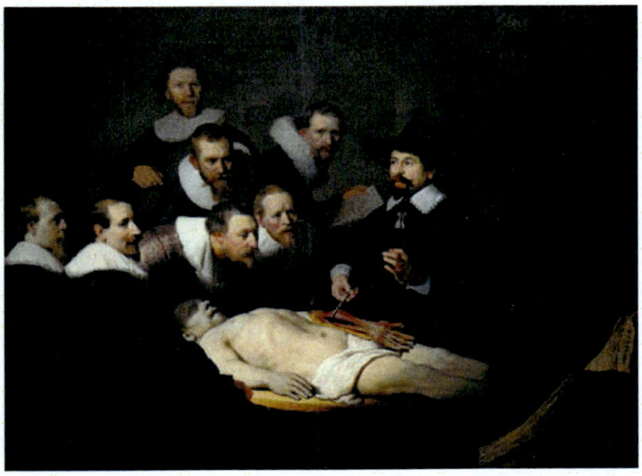

Fig. 1 Rembrandt, the anatomy lecture of Dr Nicolaes Tulp, 1632. *Source* https://it.wikipedia.org/wiki/File:Rembrandt_-_The_Anatomy_Lecture_of_Dr._Nicolaes_Tulp_-_WGA19139.jpg

completed what is now considered the archetype of the modern Encyclopaedia and was a summary of all scientific knowledge. The methodologies used in the Encyclopédie—i.e. disassembly, dissection, classification and assembly (Fig. 2)—represent excellent support to explore in detail the object of this study, the Section (Milocco 2018): «interruption of the continuity of a body, operated by dividing it in all its thickness or by incising it with greater or lesser depth by the means of a sharp instrument. Excision, amputation. Detachment. A part separated from a whole» (Pavia 2016).

The relationships between the section and the analytical knowledge of the built environment (Gregotti 1984, 2011) might be taken into account when it is explained through perceptive, technical and documentation experiences. In fact, according to Lewis et al. (2016), the Section is intended as an 'imagined cut' of a building and an impeccable tool for describing a construction, a body, in its entirety; in this context, the author's words are confirmed assuming that the 'cut' is explicitly associated with the dismemberment of either a body or a volume. Furthermore, the act of dissecting should be an instrument that records the strata of buildings, cities and object giving back new points of view and new tools for studying—through a series of «drawn sequences» (Lewis et al. 2016)—the interaction between the built environment and its users. The Section defines and highlights the spatial organisation of a building and of the different layers of which our cities are composed (Zanni and Trillo 2010), through a vertical cutting view that allows to better understand the dynamics of the city-systems on each scale. With these premises, it is possible to say that the Section is an abstract method of investigation which is current since Renaissance experiments (Ackerman 1954, 1997) in which the task of the cutting was to trace the external and internal profile of the walls of the buildings to report, as previously expressed, the structural skeleton, the heights and volumes of the buildings comparing them with both the human figure and the context (Milocco 2018). According to the studies exposed in The Manual of Section (Lewis et al. 2016), emblematic examples (Fig. 3) are certainly the Phanteon (Adriano 28 AD) and Palladio drawings (Palladio 2002). The first one is a container that encompasses—holds together—large volumes and different historical periods (De Cesaris 2012) which are studied through section reproductions that have been made in all ages. In Palladio's drawings, on the other hand, as P. Lewis recalls (2016), «the alignment of planimetric and sectional information undoubtedly represents the trends of representation of the late sixteenth century» and that strategy allows us to read the structures and the aesthetics of the building with exceptional synchronicity.

Documenting and cataloguing the architectural works, particularly in the historical and archaeological practice, has always been relevant: emblematic testimony to the countless Piranesian drawings of Roman ruins (Purini 2008). The dissection tool which is used to comprehend the material layers and the structural solutions of buildings and ruins of the classical eras (Guillerme and Vérin 1989) has also acquired a methodological value to provide the basis for the design of innovative architectural works (Milocco 2018). In fact, the Manual of Section defines Piranesi drawings as an «illustrative practice, which maximises the visual appeal» (Lewis et al. 2016), because some parts of the architecture, which were difficult to study properly, were often completed in any case. In this context, the section is the instrument capable of representing both the cutting profile (what can be seen inside the building) and the whole extension of the architectural work in a certain environment, describing in a more precise way many of the technical–technological information for the executive construction of an architecture without forgetting the human livability aspects that are coming 'to light' and can be improved through a vertical view that allows to see an entire section of the urban context.

This is the reason why, during the eighteenth century, in the projects of Charles-Francois Ribart, Claude-Nicolas Ledoux and Etienne-Louis Boullèe, the ideal and/or utopian forms in architecture found in the section their most successful representation, which was capable of providing information also of dimensional and constructive features. In the past and nowadays, from a design point of view, the need to elaborate, analyse and operate through the Section would represent the desire to reveal the complexity of a building and its relationship with the outside; this necessarily means coming to terms with the size of the surrounding landscape, with the mobility and with all the social dynamics that are activated in the proximity of the designed building. Additionally, the section shows the structural skeleton of the architectural work and, consequently, it clarifies the optimal

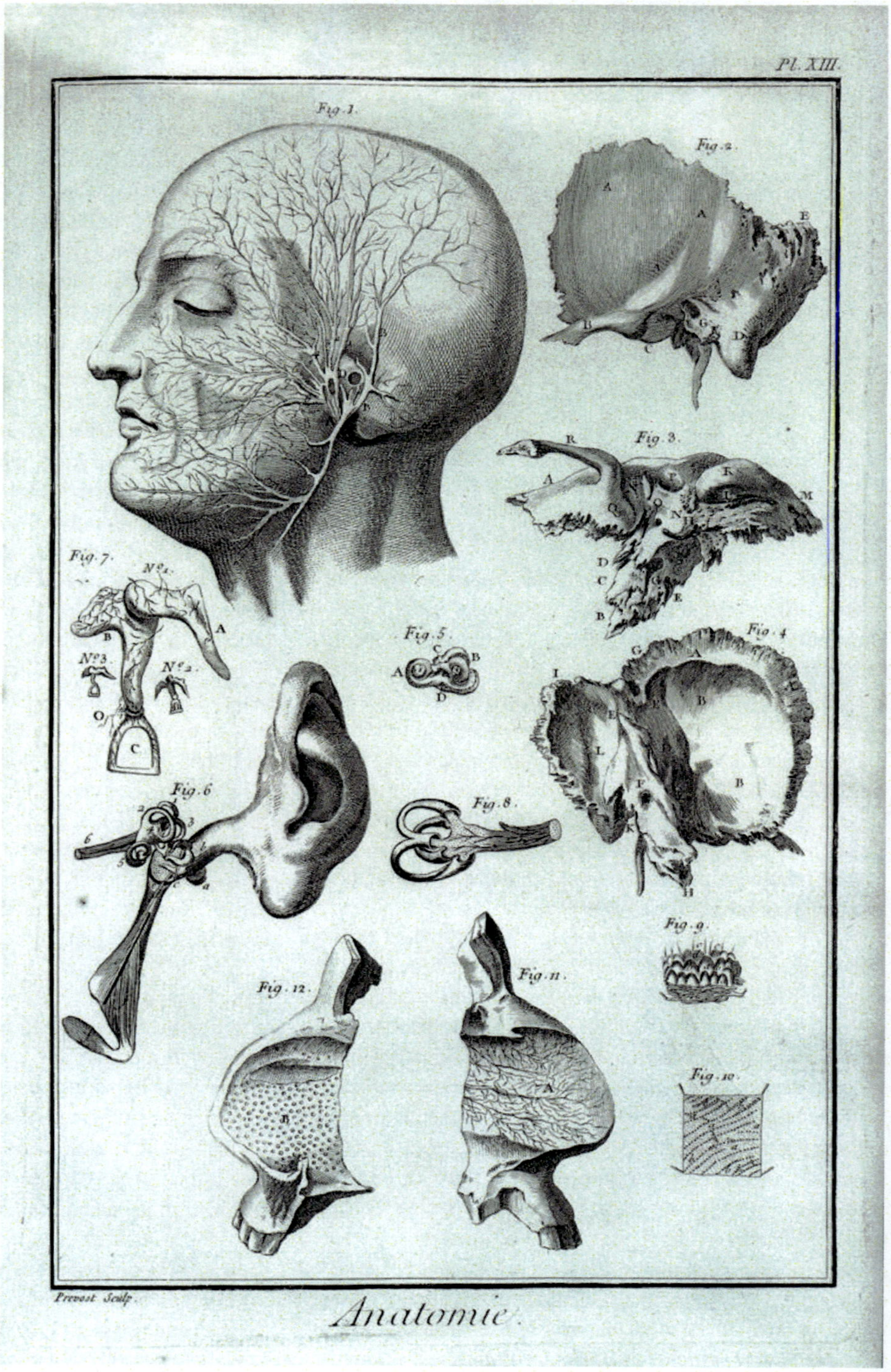

Fig. 2 L0028244 D. Diderot and D'Alembert, "Encyclopedie", 1762. *Source* Wellcome Library, London, https://commons.wikimedia.org/wiki/File:D._Diderot_and_d'Alembert.jpg

position of all the design elements. Clearly, the section, from a compositional point of view, is an 'ideal' visualisation, or «a form of abstraction in restoring the consistency and features of an artefact» (De Cesaris 2012) and for this reason that it was often used, in a utopian way, as a tool for understanding and as an effective practice of representation.

Fig. 3 Rome, Pantheon, drawing. Wilhelm Lübke, Max Semrau: Grundriß der Kunstgeschichte. Paul Neff Verlag, Esslingen, 14. Auflage 1908. *Source* https://es.m.wikipedia.org/wiki/Archivo:Pantheon.drawing.jpg

3 Section and Utopia

3.1 Utopia, Heterotopia, Visions Beyond Utopia

«Utopia, Noun [C or U] UK, /juːˈtəʊ.pi.ə/US /juːˈtoʊ.pi.ə/,

(The idea of) a perfect society in which everyone works well with each other and is happy

Eutopia, Noun UK, /jʊˈtəʊpɪə//juːˈtəʊpɪə/

A perfect (imagined or hypothetical) society or state of existence; a place of supreme happiness.

Also: a literary work describing such a place; a vision of an ideal state of existence.

Dystopia, Noun, UK, /dɪsˈtəʊpɪə/

An imagined state or society in which there is great suffering or injustice, typically one that is totalitarian or post-apocalyptic.

Heterotopia, hetero- + -topia, or heterotopy, Noun [C or U] UK, /ˌhɛtərəʊˈtəʊpɪə/

(Pathology) Normal tissue (or an organ) present at an abnormal part of the body

(Ecology) The occurrence of an organism in a number of different habitats

(Philosophy) A type of utopia that actually exists in a society.» (https://dictionary.cambridge.org/)

The features of the utopian and heterotopic spaces, which are of major interest for this study, are the aspects of hyper-connection and simultaneity with their surrounding environment. It is, in fact, the concept of heterotopia that which «designates open places on other places, places whose function is to make spaces communicate with each other. However, where utopias designate environments lacking actual localization, heterotopias are real places» (Panella and Spena N.d.; Boulland 2003; Rosenfeld 1978; Rohan 2014). Foucault himself highlights the formal elements of heterotopias, considering their innate ability to overlap multiple places in the same point; what makes these places 'different' is precisely the unification, on more dimensions and different *locis* (Foucault and Vaccaro 2011). For our considerations, the section is, at the same time, *eutopia* and *heterotopia*. One of the keywords that emerge from the aforementioned philosophical extracts is certainly 'to connect'.

A brief excursus among the Urbanistic Utopias of the last century (Hénard 1910) is useful to understand the role of the Section in the visions *beyond utopia* (Choay 1973; Tafuri 1977; Coleman 2005). In this context, the sectional operations arose in the production of design ideas which considered the soil as a non-reproducible material, leading to compositional approaches towards solutions for the multiplication and consolidation of residential and working activities in punctual urban nuclei, independent but reconnected to larger territorial transport systems. A remarkable first example was the activity of the architect Edouard Utudjian who, already in the 1930s, founded the research group Gecus (Groupe d'etudes et de coordination de l'urbanisme souterrain) for the exploration of urban solutions based on "more levels", such as the *ville en épaisseur* (Utudjian 1966; Avanza et al. 1991). Additional relevant examples generated from the theories of spatial urbanism where the basic concept was the super-integration of the building with other systems, through vertical and horizontal connections. This is made to overcome the conception of the construction as a single object in the urban fabric and otherwise to assume the idea of the urban mechanism as a *super-infrastructure* (Avanza et al. 1991). Features and purposes of this architecture originated mainly from the desire of reducing soil consumption and optimising its use for the future, careful of natural and territorial resources has been studied by Yona Friedman, the "spatial city" project in 1960. Other proposals must also be mentioned: in the 1960s, Paul Maymont proposes a project to drastically reduce the car congestion of the city centre of Paris; this (utopian) idea involved the insertion of a second urban layer, a metropolitan section below the riverbed, which contained, in addition to new traffic arteries, a parking system connected to the main surface roads. The most interesting aspect of the system was the designer's willingness to accommodate at the same level different kinds of services by transforming the entire project into a multi-use and multi-level architecture: «the historical model of the first mega-structural imaginations was the transatlantic: object of high technology, with an alveolus belly full of life but drastically excluded from the external environment [...] It inspired Le Corbusier to conceive the Unité d'Habitation, and Utudjian benefited from this intuition and, after timid beginnings, ended up conceiving his underground city as a series of linked spaces of various functions» (Nicoletti 1980).

3.2 Utopian Section

In the light of the foregoing considerations, it is imaginable to clearly define the meaning of the Utopian Section within this research. Starting from the Piranesian experiences, we can deduce that the motivations which led designers and visionaries of every age to use the 'cutting tool' are, for instance, the reconstruction of parts of the buildings invisible to the human eye and the historical and archaeological cataloguing of the ruins of the classical eras—even of works of a distant past, no longer existing or inaccessible. The Section, either utopian or real, has the power to parameterise the human being in relation to the structure/building and the surrounding landscape. From an aesthetic point of view, the section has given way to the imagination of dreaming possible futures, in the form of projects and mega-structures that develop into multiple levels, which challenge the gravity or heaviness of the soil with the common aim of improving at local or urban/sub-urban scale the 'boundary conditions' (Crotti 1997), the accessibility, and the quality of life.

Furthermore, the section tends to subvert the redundancy and abuse of exclusively planimetric visions, re-evaluating itself as a drawing and generative tool, especially in the contemporary era. The Utopian Sectional projects—here presented—deconstruct the standardised space through unusual visualisations and iconographies, also allegorical, towards a formal, structural or simply goliardic research.

Finally, the Utopian Section has the aim to give credibility to only imagined architectures (Magrou 2012) which are apparently impossible to be constructed, nevertheless fundamental for an experimental development towards a possible realisation of the works, even partially. The purpose of the Section (Isola 2002) is as precise as it follows; in other words, the main purpose of it is to dare as much as possible in order to proceed towards compositional syntheses that can decree the construction of new urban areas in line with the needs of the contemporary age, of reuse and environmental sustainability.

Through a careful analysis of some significant examples, from the early historical approaches to the contemporary era, we can easily make a historical reconstruction of the Section's role in Utopian design and architecture.

4 The 'Construction' of Utopian Architecture Through the Section

4.1 The Eighteenth and Nineteenth Centuries

During the eighteenth century, the utopianism, which was recognised in the context of the works of the Enlightenment, was applied in the design of 'ideal, visionary and/or utopian' settlements and buildings, which begun to be consistently present in the aesthetic and design research. Dimensional exaltation and geometric purification, which constitute the constants of these projects—focusing not so much about unrealisable dreams but rather on experimental models of a new way of design—took on a concrete meaning (Tafuri 1977) also and above all through the Section.

Emblematic is the case of the Triumphal Elephant of Charles-Francois Ribart de Chamoust, dating back to 1758; Ribart, better known for his architectural writings than for his buildings, did not like conventional solutions; he designed a multi-storey elephant-shaped building comprising internal rooms accessible via a central staircase. As tall as a 5-storey building, it would become the focal point of the Champs-Élysées, nestled among the greenery. An extraordinary structure externally, it is no exception inside (Wilkinson 2017), and the Section describes its complex organisation; the distribution of the rooms on the different levels on the sides of the central staircase, the main rooms on the upper floor to occupy both the body and the head, the decoration of the dining room of naturalistic inspiration, the finishing and furniture of the ballroom and buffet rooms.

Claude-Nicolas Ledoux for the Royal Saline of Arc-et-Senans, in 1770 ca., designed an ideal city which, through the very shape of its complex—which should have been "as pure as the sun during its journey"—united symbolism and rationalism; the choices, however, are not governed by symbolism alone, but functional aspects are present in the project—for example in the location of buildings based on their intended use (production, housing, offices). More than in the distinctly symbolic plan of the Oikema—house of pleasure, in the two spherical-shaped buildings (a house and a cemetery), the Section is the most effective tool for representing intent and possibilities: the sphere as a symbolic form and together with its ability to become a 'functioning' architecture (Fig. 4).

The theme of the sphere is similarly explored in the Cenotaph of Newton, by Étienne-Louis Boullée, dating back to 1784, in which the «pure visual effect and symbolism of the choice of the sphere» (Wilkinson 2017) for a spectacular project are, once again, made 'concrete' by the sectional representation, which introduces dimensional and proportional references (the presence of human figures to account for the 150 metres of diameter, the tallest building in the world at the time), and report the double effect on the visitor inside, from the unnatural experience of a spherical space to the different day and night scenery of the incredible structure (Fig. 5).

And again, for the Bibliothèque Nationale, in 1785, Étienne-Louis Boullée imagines the interior as a large free-space, which can only be optimally represented through the perspective section: a large arched vault—more than 30 metres of span—with coffered ceiling and a large zenithal skylight, multi-level shelving on the sides of the room, with the result of leaving a generous central space for the users; and yet, section gives information also of the original constructive solution for the wide wooden trellis covering which uses the existing buildings as lateral supports.

In 1830, architect Thomas Willson proposed housing London's dead in a gigantic Pyramid Cemetery, designing a metropolitan cemetery on a scale which should be commensurate with the necessities of the largest city in the world so sufficiently capacious to obtain 5,000,000 of dead bodies (Haskell 2015). The main Utopian Section of the project—including an enlargement in which the network of thousands of graves and concentric corridors is evident—represents the building in its 94 storeys which is organised around a single central stairwell, for a total area of 7.3 ha.

4.2 1900–1950

From the first decades of the twentieth century, the prefiguration of utopian architecture and engineering works developed, through the Section, both on the urban-territorial scale and on the more strictly architectural-design scale, starting from the innovative visions of the avant-garde movements to culminate in the experiments of the Modern Movement (Coleman 2005).

In the project for La Città Nuova of Sant'Elia (1914), although the Futurist architecture prefers the dynamism of perspective representation, among the numerous drawings some report elements in section (e.g. elevators). In particular,

Fig. 4 Chaux Cemetery Project, C. N. Ledoux, 1804. *Source* https://commons.wikimedia.org/wiki/File:Chaux_-_Projet_de_cimetière.jpg

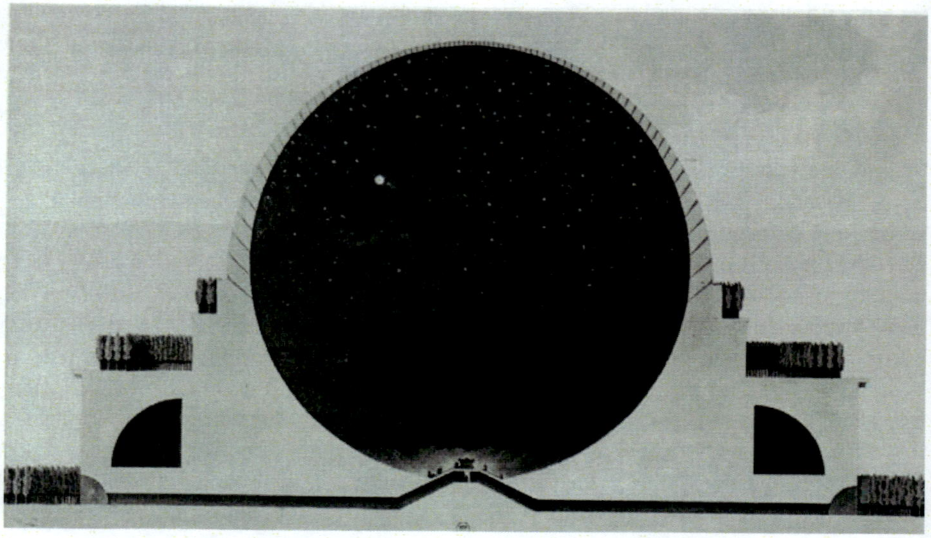

Fig. 5 E. L. Boullée: cross-cut Newton-Memorial. Daylight with artificial starlight, 1784. *Source* https://upload.wikimedia.org/wikipedia/commons/1/15/%C3%89tienne-Louis_Boull%C3%A9e_Memorial_Newton_Day.jpg

its prototypes for the New House introduced a new idea for high-rise buildings with separate towers to house elevators, positioned back-to-back, creating internal corridors, bridges and overpasses—which anticipate many subsequent solutions for the intensive building type.

A similar idea of the organisation of the housing organism through an A-shaped structure, represented in a section which highlights the relationship between structural and spatial elements (Nyilas 2018), is predominantly clear in the drawing section of the project by Walter Gropius and Frank Möller for the Wohnberge (1928); here, a giant A-frame structure supports 32-storeys high-rise building that is organised in lateral living units and central common space with suspended bridges—that is realised probably due to pedestrian circulation.

For the Modern and modernist architectural ideal, the Section became an indispensable representation tool, both for the human parametrisation of architecture and for highlighting its constructive, functional and functionalist aspects.

El Lissitzky produced perspectives and sections for his Horizontal Skyscrapers (1923–1925)—Wolkenbügel, 'cloud-hangers', 'sky-hangers' or 'sky-hooks'. A series of eight structures designed to mark the major intersections of the Boulevard Ring in Moscow. A simple schematic section summarises the unusual idea for the skyscraper: a flat three-story, 180-metre-wide L-shaped slab raised 50 metres above street level; it rested on three pylons, two of which extended underground into the proposed subway station (Fig. 6).

Walter Gropius, in the visionary project for the Total Theatre (1926), conceived a variable theatre space which would have been modified in accordance with the staging needs and intentions of the director—Erwin Piscator. Convertibility, flexibility and anonymity in architectural design are effectively portrayed, in the scale-model and in the drawing sections of the theatre,[2] as three basic versions were designed: an arena arrangement, a proscenium stage and a classic depth perspective scene.

4.3 1950–1990

The Utopian Sections of the late twentieth century have in common the ideals of 'the machine', the movement, the soil-saving, given the growing industrial and technological revolution. The general intent of the experiences mentioned below was meant to multiply the territorial surfaces as much as possible through actions such as overlapping or extending soils; the common denominator was the desire to increase building areas, trying to contribute to the growing need for space given by the exponential growth of the population.

In 1958, Walter Jonas designed the Intrapolis[3]; macrostructures in the shape of an inverted cone which follows an introversion principle that turns the accommodations inside of structure by limiting its communication with the outside. Each of these mega-infrastructures could contain many housing units that comprise industrial production areas, stations, offices and shops on the lower floors, which are made clear just by the section drawings (Badalucco 2017). The project for Infinite House by Frederick Kiesler (1960, NYC MoMA, Visionary Architecture show), which is perceived as unattainable, has the power to bring to

[2]See W. Gropius drawings, Total Theatre, 1926, Berlin at https://www.harvardartmuseums.org/collections/object/50995?position=22 (last retrieved March 2020).

[3]See drawings of W. Jonas, Intrapolis, 1958 at www.walterjonas.ch (last retrieved March 2020).

Fig. 6 E. Lissitzky, Wolkenbügel, 1926. *Source* http://gallery.ru/watch?ph=yge-cJ6ny

light linear compositional strategies, without caesuras, with a spatial succession of different environments without any interruption. There are no corners, only organic, sinuous, rocky spatialities as if the whole project is closer to a sculptural rather than architectural work; and hence, the research is based on pursuing infinite dynamic forms. (Archiscultura 2017). Ron Herron's project[4] for the Walking city (1964), is a set of mega-structures that can move thanks to the mechanical legs that are capable of supporting them; the architecture moves in any direction and in any part of the world trying to favour the ideals of the following decades whose «anticipate the hope […], that the profession can finally project itself towards utopia» (Presstletter 2012).

Peter Eisenman and Michael Graves in the Jersey Corridor project of 1965 devised a linear city (with an urban character) that connects two major cities (Boston and Washington), in which the same architecture moves, engages, winds between the existing territorial infrastructures (Aureli et al. 2007; Architectmagazine 2015; Alter 2014); the theoretical aspect resides in the experimentation and decoding of the 'right scale' between infrastructure, environment and human being (Zuliani 2019). The buildings become part of a double complex and parallel system of horizontal communication, evident in section (whilst crossing the central street), differentiated from the destinations of use: work and private/commercial life (Alter 2014). Peter Cook, in the ideal project of the Plug-in City—Maximum Pressure Area of 1965, composed a series of supporting infrastructure, that contains all the urban systems, on which volumetric units of different functions (residential, commercial and industrial) can be moved, translated and reconnected (Presstletter 2012). Cedric Price was always interested in a visionary architectural and social future. In The City of the Future (1965), he tried to point out how technology, as well as the social freedom of the human being, represents the fundamental element of his research (MoMA 2019). The City of the Future is a spelling book of a possible city that must interact with the primary user (the inhabitant) in an autonomous and rapid manner by responding to his needs—in line with contemporary times with smartphones and home devices that efficiently respond to the requests of their owners (MoMA 2019).

New York Habitat of 1967 by Moshe Safdie shows the need to condense housing units as little as possible in the smallest possible space, given the growing urbanisation of the American metropolis. The housing units, as in HABITAT 67 in Montreal, are wedged together, but in the American case, they are inserted in pyramidal-like structures that increase the possibility of vertical development of the work without exceeding the regulations of the East Side River (McGill 2019). Julio Lafuente, for the hotel on the cliff in Gozo (in Malta from 1967), explores the utopian idea of

[4]See Ron Herron's project for the Walking city (1964), MOMA, NYC at https://www.moma.org/collection/works/814 (last retrieved March 2020).

born from the desire to respond to the growing demands of tourism and, consequently, from landscape needs (the most beautiful view) (Lafuente Studio 1967). The designer subverts the excessive ideals of preservation of the natural environment by inserting and hybridising the project 'between' the cliff towards the seascape.[5] The composition of the project is manifested on the vertical axis (*Y*) cut 'flush' with respect to the present rock face. It develops in height for over 100 m, and the distribution is guaranteed by a full-height lift. The greatest feature of this text is undoubtedly the desire of Lafuente to reconvert natural landscapes into totally artificial landscapes perfectly integrated with context whilst demonstrating the theatricality of the structure with the intention of trying to assign to it a new local identity (Pastor 2016; Farrugia 2012). Following the same line of thought as the aforementioned author, we cannot fail to mention Richard Buckminster Fuller, with his Dome over Manhattan (1968) or Justus Dahinden, Arkopolis Leisure City (ca. 1970). We may find liaisons between dissection practice and the history of contemporary art especially in the works of the American artist, Gordon Matta-Clark who has been analysed here for some macro-structural interventions; the building-cuts.[6] To simplify the comprehension, we can borrow some words from Minuto (2017) in Gordon Matta-Clark. Dissection as an Allegorical Process; here the author describes the 'extreme' interventions of the artist with surgical detail, the same 'sharp' meticulousness that Matta-Clark uses in his creative process where he tries to subvert the orthogonal dynamics of the buildings and return to the user a new architectural environment (Minuto 2017). The central idea of the artistic-architectural productions of Matta-Clark is expressed through the social and the interactional dimension between individuals and the new dimension of space which is given by the 'breaking' of artefacts which lead to creating new views and unexpected urban perspectives (Minuto 2017). The artist's architectural dissections are 'gestures' that may be synthesised in a volumetric action that transcends the canonical metropolitan fabric.

4.4 1990 to Today

The 'Utopian Sections' of the last 30 years have been influenced mainly by historical events of global importance and by the growing development of new technologies. This happened due to the raising of population awareness towards climate change, re-use, and reconversion of abandoned or degraded areas of today's metropolises.

Lebbeus Woods, Berlin Free-Zone 3-2 (1990),[7] is a proposal for an abandoned government building in reunified Berlin. Lebbeus Woods is remembered mostly for works that have never been built (Chayka and Vartanian 2012). The architect, through his imagined sections, has investigated the socio-cultural interpolations of post-wall Berlin; a significant example is Berlin Free-Zone which is the emblem of an expressiveness that is free from geopolitical, physical and social limitations in which almost alien structures are hybridised (in section) with buildings restored after the destruction of the wall: «the interiors were designed to be difficult to inhabit—a strategy to shield the typical bourgeois» (Ouroussoff 2008). It can certainly be said that Woods' creative thinking wants to show 'the ruin', the historical presence of a political conflict breaking with the traditional housing units, creating spatiality—with unusual rules—to be shared with the community (Strathshepard 2013). His parasitic architectures are considered as «sanctuaries for the most vulnerable of society: marginalised, rebels, heretics and dreamers [...] [in] a world in the constant struggle for survival» (Ouroussoff 2008). The Danish architect Nils Ole Lund (Lucarelli 2016), Maestro of Collage, combines in post-modern philosophies with Pop/satirical notions; he wants to emphasise the devastating effects that some contemporary architecture can have on the landscape and The House in the Cliff[8] is a case in point (Lucarelli 2016; Designlevelzero 2012). Rem Koolhaas whose Les Halles project in Paris (Koolhaas and Bruce 1995), the non-winning OMA project for the Forum (hypogeum shopping centre and transport exchange hub) of the Paris' Halles, is emblematic for this paper because the concept was conceived in a sectional manner. Large, translucent volumes rise from the surface of the ground and penetrate it by showing all levels and possible commercial environments; the section of the buildings, in relation to the ground line, seems to rise and fall continuously, almost in a sinusoidal way, guaranteeing adequate public space at the 0 Level and facilitating access to the railway stations (Professionearchitetto 2004). The deformation of the architectural elements of the project of Zaha Hadid for Leicester Square,[9] that was never built, takes place through the fragmentation and alteration of hierarchies and spatial priorities

[5]See Gozo Hotel, by Studio Lafuente at http://www.studiolafuente.it/portfolio/albergo-e-ville-a-gozo/ (last retrieved March 2020).

[6]See G. Matta-Clark Building cuts at https://www.tate.org.uk/research/publications/tate-papers/07/towards-anarchitecture-gordon-matta-clark-and-le-corbusier (last retrieved March 2020).

[7]See L. Woods sectional drawings at https://www.architakes.com/?p=4140 (last retrieved March 2020).

[8]See N. Ole Lund. The House in the Cliff at http://sorbusaucuparius.blogspot.com/2012/08/images-of-month-august-20125.html (last retrieved March 2020).

[9]See Z. Hadid concepts for Leicester Square at https://www.architectural-review.com/essays/folio/folio-zaha-hadids-reimagined-leicester-square/8638056.article (last retrieved March 2020).

that have the necessity to be subsequently recovered and declined (at the design level) to facilitate the comprehension by the user. In fact, the human figure must be re-parameterised within morphologically unusual spatiality (respecting a standard) in order to benefit from all the available space. By paying attention to the words of Z. Hadid in Remodeling Project for Leicester Square, it can be deduced that a person could design entire sectional spaces in the underground being in direct relation to the urban surface by not setting limits, deforming it through topographic gestures, penetrating the ground up to significant depths (Hadid 1991); these underground movements represent «an urban negative [in which the section project] overturns the logic of historical stratification without affecting it, aiming at the construction of a new urban and cultural identity given by the overlap of more layer. [That of Z. Hadid] is an excessive, exaggeratedly deep, dizzying, unattainable proposal. […] The zero quota […] is corrugated, articulated through folds and cracks, accommodates habitable rooms, buildings solid and transparent […] knife-like between full and empty of an articulated new urban spatiality.» (De Cesaris 2012). Thanks to this last case study the possibilities of interpretation of sectional utopias are now more understandable as in M.A.C. (Milano Animal City).[10] The ideal project stems from research conducted at the Politecnico di Milano University, during a university course conducted by the architect Stefano Boeri. The work is exemplified with visions and re-conversion of abandoned spaces (such as the space between the ground and the inter-floor of an elevated road) by placing various services; the idea studies repopulation of plants and animals within the metropolitan city (Fenini and Boeri 2015).

5 Conclusions

The final objective of this investigation is the reinterpretation of the creative and design value of the seemingly unrealisable visions, in the light of the technologies and the actual possibilities of contemporary design. Through extreme sectional solutions, we can think of solving and reprogramming parts of corrupt cities or manifesting the dissent of soil abuse (compare Nils Ole Und) and re-elaborate an idea of cities based on a human scale.

This text has the virtue of subverting the minority meaning that is generally given to the theories and uses of architectural section (Magnago Lampugnani 1982). As a matter of fact, the aim is to focus attention on the effectiveness of the cutting tool due to the intrinsic characteristics of research and composition; the power of the Section is linked to the possibility of simultaneously seeing several floors, multiple layers, and as a consequence it allows us to recognise the weak points of the buildings and settlements, and additionally all the possible ways of implementing interventions aimed to urban and social improvement—towards a constant interaction of those systems—in a punctual, general and hyperconnected way.

[10]See Stefano Boeri student's project, Animal City MAC, Milan, 2015 at https://milanoanimalcity.wordpress.com; and http://www.affaritaliani.it/milano/mostra-milano-animal-city-130215_mm_463508_mmc_3.html (last retrieved March 2020).

Acknowledgements All Italian quotations have been translated into English by the authors.Thanks to F. Battaglia for English proofreading.

References

Ackerman, J. S. (1954). Architectural practice in the Italian renaissance. *Journal of the Society of Architectural Historians, 13*(3), 3–11.

Ackerman, J. S. (1997). Villard De Honnecourt's drawings of Reims Cathedral. A study in architectural representation. *Artibus et Historiae, 18*(35), 41–49.

Allen, S., & McQuade, M. (Eds.). (2011). *Landform building*. Baden: Lars Müller.

Alter, L. (2014). *Instead of Vertical Cities, should we be thinking about linear cities?* Retrieved January 2019, from https://www.treehugger.com/urban-design/instead-vertical-cities-should-we-be-thinking-about-linear-cities.html

Archiscultura. (2017). *La Casa Infinita di Fredrick Kiesler*. Retrieved January 2019, from http://archiscultura.altervista.org/la-casa-infinita-di-friedrick-kiesler/

Architectmagazine. (2015). *Michael Graves New Jersey Corridor project*. Retrieved January 2019, from https://www.architectmagazine.com/videos/michael-graves-new-jersey-corridor-project

Aureli, P. V., Biraghi, M., & Purini, F. (2007). *Peter Eisenman. Tutte le Opere* (pp. 56–57). Milan: Electa.

Avanza, F., et al. (1991). *Progettare il Sottosuolo* (pp. 19–27). Milan: Franco Angeli.

Badalucco, G. (2017). Le Città Del Futuro. *Edicolaweb*. Retrieved January 2019, from http://www.edicolaweb.net/atlan19a.htm

Boulland, F. (2003). *Michel Foucault, Penseur de l'Espace. Conférence*. Université Charles de Gaulle.

Casini, P. (Ed). (1966). *D'Alembert—Diderot, La Filosofia dell' Encyclopédie* (p. 44). Bari.

Chapman, J. P. (2009). *The planned city in section*. Undergraduate Thesis, Princeton University.

Chayka, K., & Vartanian, H. (2012). *Remembering radical, theoretical architect Lebbeus Woods*. Retrieved January 2019, from https://hyperallergic.com/59590/remembering-radical-theoretical-architect-lebbeus-woods/

Choay, F. (1973). *La Città. Utopie E Realtà*. Turin: Einaudi.

Coleman, N. (2005). *Utopias and architecture*. London, New York: Routledge.

Crotti, S. (1997). Interspazi, Dai Siti Pubblici ai Luoghi Comuni. In P. Caputo (Ed.), *Le Architetture dello Spazio Pubblico. Forme del Passato, Forme del Presente* (p. 39). Electa: Milan.

De Cesaris, A. (2012). *Il Progetto Del Suolo-Sottosuolo*. Rome: Gangemi Editore.

Designlevelzero. (2012). Retrieved January 2019, from https://designlevelzero.wordpress.com/2012/10/20/collagesnils-ole-lund-the-future-of-architecture/

Farrugia, J. J. (2012). *The art and architecture of twentieth century Gozo*. Retrieved January 2019, from https://www.um.edu.mt/library/
Fenini, R., & Boeri, S. (2015). *Animal City, Le Città del Futuro*. Retrieved January 2019, from https://www.panorama.it/cultura/arte-idee/stefano-boeri-animal-city-citta-futuro-rivolte-agli-animali/
Foucault, M., & Vaccaro, S. M. (2011). *Spazi Altri: I Luoghi Delle Eterotopie*. Mimesis.
Gregotti, V. (1984). Modificazione. *Casabella* 498.
Gregotti, V. (2011). *Architettura E Postmetropoli*. Turin: Einaudi.
Guillerme, J., & Vérin, E. (1989). The archeology of section. *Perspecta* (Vol. 25). Cambridge: MIT.
Hadid, Z. (1991). Remodelling project for Leicester Square 1990. *El Croquis, 52,* 130.
Haskell, D. (2015). *London's Pyramid for the dead, Ernest Journal*. Uncharted Press Ltd. Retrieved January 2019, from http://www.ernestjournal.co.uk/blog/2015/7/17/londons-pyramid-for-the-dead
Hénard, E. (1910). The cities of the future. In *Transactions, Town Planning Conference* (pp. 345–367). London.
Isola, A. (2002). *Manuale e Atlante, Forme Insediative E Infrastrutture* (pp. 205–206). di F. Lambertucci, Politecnico di Torino: Marsilio.
Koolhaas, R., & Bruce, M. (1995). *S, M, L, Xl*. New York: Monacelli Press.
Lafuente Studio. (1967). *Albergo e Ville a Gozo*. Retrieved January 2019, from http://www.studiolafuente.it/portfolio/albergo-e-ville-a-gozo/.
Lewis, P., Tsurumaki, M., & Lewis, J. D. (2016). *Manual of section* (pp. 1–49). Hudson, New York: Princeton Architectural Press.
LTL Architects. (2016). *In contemporary architecture, the section is the generator*. Retrieved January 2019, from https://www.metropolismag.com/architecture/contemporary-architecture-section-is-generator/
Lucarelli, F. (2016). *The future of architecture and other collages by Nils-Ole Lund*. Retrieved January 2019, from http://socks-studio.com/2012/08/05/the-future-of-architecture-and-other-collages-by-nils-ole-lund/
Magnago Lampugnani, V. (1982). *La realtà dell';immagine. Disegni di architettura nel ventesimo secolo*. Milano: Edizioni di Comunità.
Magrou, R. (2012). The glories of the architectural section. *Harvard Design Magazine, 35,* 34–39.
McGill. (2019). *Habitat New York I*. Retrieved January 2019, from http://cac.mcgill.ca/moshesafdie/fullrecord.php?ID=10818&d=
Milocco, B. M. (2018). *Dis-Sezioni Metropolitane. Suoli, Sezioni e Interspazi nella Città Contemporanea*. Doctoral Thesis, Sapienza University of Rome.
Minuto, M. E. (2017). Gordon Matta-Clark. La Dissezione come Processo Allegorico. Retrieved January 2019, from https://www.strozzina.org/admin/wp-content/uploads/2012/02/maria_elena_minuto_03.pdf
MoMA. (2019). Cedric Price, city of the future project (perspectives) c. 1965. Retrieved January 2019, from https://www.moma.org/collection/works/104697
Nicoletti, M. (1980). *L'Architettura delle Caverne* (p. 16). Bari: Laterza.
Nyilas, A. (2018, 1970). *Beyond Utopia: Japanese metabolism architecture and the birth of mythopia*. New York: Routledge.
Ouroussoff, N. (2008). *An architect unshackled by limits of the real world*. Retrieved January 2019, from https://www.nytimes.com/2008/08/25/arts/design/25wood.html
Palladio, A. (2002). *The four books on architecture* (R. Schofield, & R. Tavernor, trans.). Cambridge: MIT.
Pastor, M. (2016). Against the ecological prejudice: The specific experience of the built landscape. *REIA (Revista Europea de Investigación en Arquitectura), 5,* 131–151.
Pavia, R. (2016). *Il Suolo come Infrastruttura Ambientale*. Retrieved January 2019, from http://www.casadellacultura.it/419/il-suolo-come-infrastruttura-ambientale
Presstletter. (2012). Retrieved January 2019, from https://www.presstletter.com/2012/09/6-1-17-archigram-gli-inizi-1961-1964/
Privitera, F. (2009a). *Disegnare Dialoghi, Esercizio Della Sezione E Progetto nell'opera Di Giovanni Michelucci*. Pontedera: Bandecchi E Vivaldi Editori.
Privitera, F. (2009). Giovanni Michelucci, Anatomia dello Spazio. *Firenze Architettura, XIII*(1), 94–99.
Professionearchitetto. (2004). Paris: Quattro Visioni sul Futuro di Les Halles. Retrieved January 2019, from https://www.professionearchitetto.it/news/notizie/2820/Paris-Quattro-visioni-sul-futuro-di-Les-Halles
Purini, F. (2008). *Attualità Di Giovanni Battista Piranesi*. Foggia: Libria.
Rohan, T. M. (2014). *The architecture of Paul Rudolph*. New Haven: Yale University Press.
Rosenfeld, M. N. (1978). *Serlio on domestic architecture*. Mineola, New York: Dover Publications.
Secchi, B. (2006). Progetto Di Suolo 2. In A. Aymonino & V. P. Mosco (Eds.), *Spazi Pubblici Contemporanei Architettura A Volume Zero*. Milan: Skira.
Strathshepard. (2013). Retrieved January 2019, from http://strathshepard.tumblr.com/post/48421393675/lebbeus-woods-berlin-free-zone-3-2-1990-a
Tafuri, M. (1977). *Progetto e Utopia: Architettura e Sviluppo Capitalistico*. Rome: Laterza.
Utudjian, É. (1966). *Architecture et Urbanisme Souterrains*. Paris: R. Laffont.
Wilkinson, P. (2017). *Phantom architecture. The fantastical structures the world's great architects really wanted to build*. London: Simon & Schuster.
Panella, G., & Spena, G. (n.d.). *Il Lascito Foucault*. Florence: Clinamen.
Zanni, F., & Trillo, A. (2010). *Abitare La Piega, Piegare, Incidere, Stratificare*. Maggioli: Santarcangelo di Romagna.
Zuliani, G. (2019). *Oppositions 1973–1984*. Retrieved January 2019, from http://www.famagazine.it/index.php/famagazine/article/view/137/688

Sacred Architecture and Fashion Drawing. The Late Antique Decorations in Cimitile as Ideational Traces for Fashion Design

Alessandra Avella

Abstract

The field of investigation of this study is cross-sectional to architecture and fashion. Always a source of mutual inspiration, today architecture and fashion draw the "new forms of living" (Avella 2018) designing spatial configurations animated by the bodies that "inhabit" it (in architecture) or that "wear" it (in fashion). In this study, the architecture is the sacred one of the Paleochristian and early mediaeval Basilical Complex of Cimitile in the Province of Naples and near to Nola, whose historical-cultural value returns to that of Roman Pompeii. The site of Cimitile, despite being an excellent testimony of sacred architecture in Europe, does not enjoy the same notoriety of the Pompeian archaeological site. The aim of this research is to promote and enhance the sacred architectural space of Cimitile through fashion as a cultural and social phenomenon that nowadays more than others is able to reinterpret in a contemporary key the historical testimonies so to make them endure in actuality. The decorative motifs of the Basilical architectures in the Cimitile's site are the signs of the past that become the ideational traces of the fashion design through the drawing of new textures that propose a contemporary narrative about Cimitile's archaeological site able to contribute to building a conscious memory of it.

Keywords

Geometry • Traditional surveying • 3DLS surveying • Textures • Textile design • Cultural heritage • Cultural communication

A. Avella (✉)
Department of Architecture and Industrial Design,
University of Campania Luigi Vanvitelli, Aversa (CE), Italy
e-mail: alessandra.avella@unicampania.it

1 Introduction

The field of investigation of this study is cross-sectional to architecture and fashion. Architecture has always been an important source of inspiration for the most different fields of interest, including fashion design. In the same way fashion, and in particular contemporary fashion, becomes the concept of the most authoritative international architects who shape their artefacts giving rise to interesting formal and spatial solutions that look at the futuristic creations of Fashion Design 4.0.

Architecture and fashion are intertwined, therefore, in a combination of experimental creativity between architectural spaces designed according to real "tailoring" and tailoring models built as "architectural structures" supporting the shapes and surfaces that envelop the body's geometries. It is from the human body, as the highest living proof of symmetry and proportion, that designers and stylists derive models and archetypes for the composition of architectural and stylistic paths, thus orienting the production of all times.

Fashion and architecture, in fact, united by recognizing in the human body a common reference, draw the "new forms of living" designing spatial configurations animated by the bodies that "inhabit" it (in architecture) or that "wear" it (in fashion), depending on a greater or lesser degree of spatial proximity. Starting from antiquity the body, the way of conceiving it and drawing it, the function and the meanings assigned to it, had a decisive role in the definition of the fundamentals of every form of representation, and in particular of the thought Modern Western architectural. The similarities between body and architecture are already present in *Marcus Vitruvius Pollio* (first century BC) that in the third book (dedicated to the temples) of his Treatise *De Architettura* (about architecture) shows that there can be no temple that is not governed by principles of harmony, order and proportion between the various parts of the building, in the relationship between them according to a precise aesthetic canon deduced from the system of human proportions.

C. Gambardella et al. (eds.), *Advances in Utopian Studies and Sacred Architecture*,
Advances in Science, Technology & Innovation,
https://doi.org/10.1007/978-3-030-50765-7_22

This canon described by *Vitruvius* in his studies on the proportions of the human body is represented in Leonardo da Vinci's famous drawing *Uomo Vitruviano* (1490). The work of *Vitruvius*, considered the first and most famous theorist of Architecture of all time, continued with the Renaissance artists engaged in the search for a law of proportionality between the various parts of the body and a reference criterion to interpret the ancient buildings and to design the new constructions.

In this regard, the well-known drawings produced starting from the great Masters of the Renaissance represent the anthropomorphism of Architecture according to the assumption by *Vitruvius* (Fig. 1).

As in past times *Vitruvius* from the body and even women's clothing draws inspiration for the sinuous lines of the Ionic order, as he declares himself writing «*[…] fecero così la colonna alta otto volte il diametro dell'imo scapo, onde avesse un aspetto più slanciato; sottoposero una base a mo' di calzare; a destra e a sinistra del capitello collocarono poi le volute pendenti come crespi cincinni di capigliatura, e ornaron le fronti con cimase ed encarpi disposte a guisa di crini, e lungo il fusto tutto lasciarono andare a basso i listelli delle scanalature, come pieghe di lunghe vesti matronali*», thus, in the architectural culture of the twentieth century Le Corbusier, performing an application of the golden section in its *Modulor* (1945), elaborates an architectural proportional system based on the proportions of the human body, thus, giving life to a "man-sized" architecture.

Even in the multifaceted overview of the latest trends in contemporary architecture, the design language is enriched with the possibilities offered by new digital technologies and innovation in the structural, material, energy, etc., to propose an efficient and sustainable architecture, once again focused on man, aimed at satisfying the housing needs, improving the well-being of life.

In the same way, since the past, the drawing of the human body is known as the first fashion drawing (Fig. 2). The fashion designers, in fact, dedicating themselves to the construction of the clothes, implicitly draw the bodies that "inhabit" them: the shapes and surfaces of the "worn architecture", in some cases they adapt to the geometries of the soma, equiwing it with a second skin, in other cases oppose them, redefining their silhouette and designing a new anatomical topography (Avella 2018).

The garment, therefore, as the first house that is "inhabits", is designed around man, conditioning the "body clad" (Calefato 1986) in posture, gestuality to make it suitable to establish existential relationships with the space around itself, depending on its needs and demands (Giannone and Calefato 2007).

Architecture and fashion, thought around man, are the concept of each other so much to be merged often in common ideative processes and in some cases shared through collaborations that produce the creation of unique capsule collections, such as extravagant footwear with flowing and sinuous lines designed by Zaha Hadid for Charlotte Olympia.

If in architecture field, the groove of Ionic columns is traced back to the folds of women's clothing and contemporary buildings are designed flaking structural plots of the perspective layouts that seem to recreate the textile plots obtained by designers with technologically innovative materials; in the same way, in the fashion field, both the architecture of the past and the contemporary are a source of inspiration for the most famous fashion designers. Look, for example, at the haute couture garments with the rigorously minimalist style designed by the French stylist Madame Grès, evoking the purity of hellenic sculpture and the tradition of ancient civilizations, or the more refined fashion creations taken from world of the built as the bag by Paul Ka that follows the Bibliotheque Alexis De Tocqueville (2017)

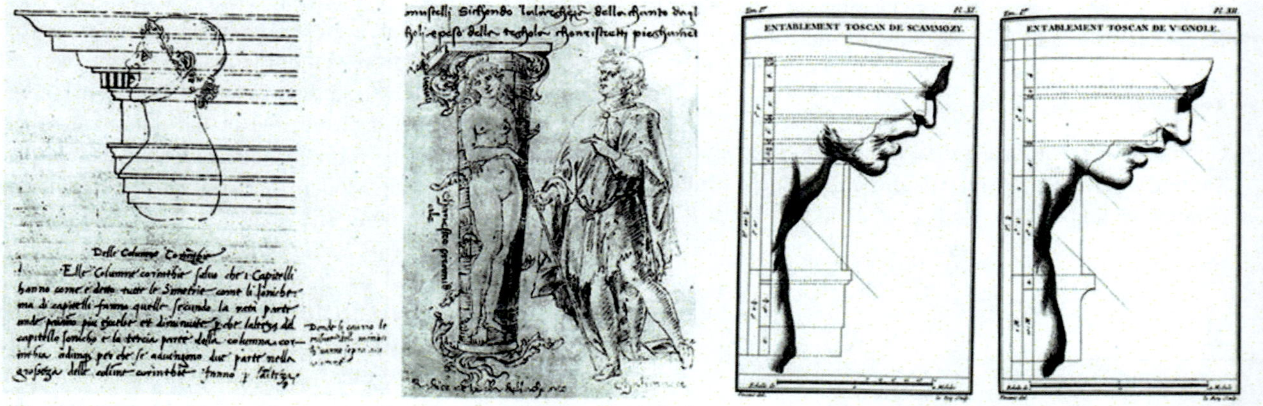

a b

Fig. 1 **a** Francesco di Giorgio Martini, *Codice Saluzziano* 148, ff. 21r, 14; **b** Jacques-François Blondel (1705–74), relationship between physiognomy and Tuscan orders according to the Treatises of Scamozzi and Vignola (Portoghesi 1999)

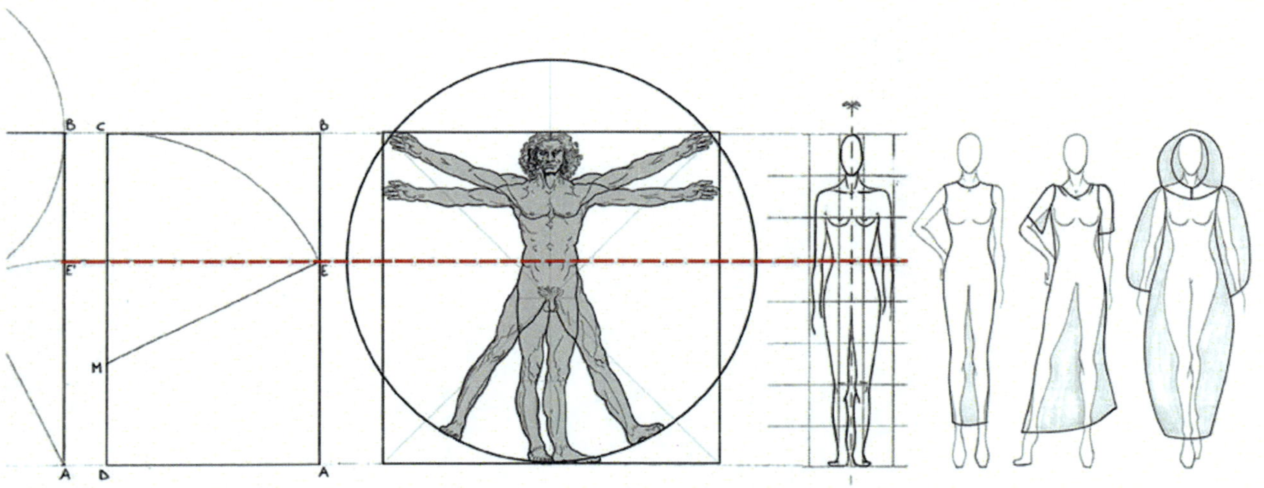

Fig. 2 "Extreme and mean ratio" from the drawing of the human body to the fashion drawing (graphic analysis by A. Avella, 2017)

in Caen in France designed by OMA and the shoes by Marla Marchant inspired by the Zubizuri Bridge (1997) in Bilbao by Santiago Calatrava (Fig. 3).

Fashion designers come into the very body of architecture—both of the past one and of the contemporary one—analysing and interpreting, with the artist's cultured and refined sensibility, its multiple dimensions from the structural to the formal one to the decorative one that thus become the ideational traces of the fashion design.

2 Materials and Methods

In this study, the architecture is the sacred one of the Paleochristian and early mediaeval Basilical Complex of Cimitile (Fig. 4) in the province of Naples and near to Nola, whose historical-cultural value returns to that of Roman Pompeii. It is a sacred enclosure circumscribed to a few square kilometres that has known in the time alternating fortunes. Born as a suburban necropolis of ancient Nola, chosen by the Fathers of the Church for the experimentation of the first forms of coenobitic life, abandoned between the end of the nineteenth century and the mid-twentieth, the site of Cimitile is now a monumental palimpsest of Paleochristian and Byzantine art, still partly to be investigated, consisting of seven basilicas and a material sedimentation documenting an anthropination that has affected more than seventeen centuries (Jacazzi et al. 2011). The restoration and excavation project that involved the archaeological area at the end of the last century gave new important data on its articulated stratification. As evidence of the undisputed religious as well as historical-archaeological value, on 23 May 1992 the Sanctuary welcomed Pope John Paul II as an illustrious pilgrim to the tomb of St. Felix. The creation of the archaeological park and the setting up of the *Antiquarium*, sponsored by the Archaeological Superintendence of Naples on the occasion of the 2000 Jubilee, further enhanced the extraordinary uniqueness and beauty of the Basilical Complex.

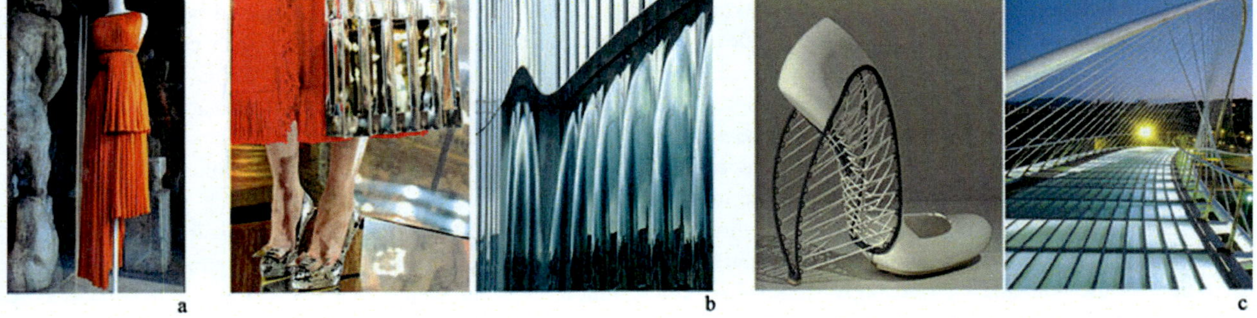

Fig. 3 **a** Madame Grès exhibition at the Bourdelle Museum, Paris 2011; **b** Bag by Paul Ka and the Bibliotheque Alexis De Tocqueville by OMA; **c** Shoes by Marla Marchant and the Zubizuri Bridge by Santiago Calatrava in Bilbao

Fig. 4 Paleochristian and early mediaeval Basilical Complex of Cimitile, Cimitile (Naples) (Photo by the author, 2015)

Despite being an excellent testimony of sacred architecture in Europe, the site of Cimitile does not enjoy the same notoriety of the Pompeian archaeological site, if not at the specialist circuits dedicated to the only insiders (ecclesiastics, scholars, archaeologists, local tourist guides, etc.).

The aim of this research is to promote and enhance the sacred architectural space of Cimitile through fashion, the cultural and social phenomenon that nowadays more than others are able to reinterpret in a contemporary key the historical testimonies so to make them endure in actuality. Through fashion and its products, sharing the logic of the most current market trends according to which the philosophy of branding is more and more an expression of the aesthetic significance and the cultural capital that the product confers, more than the value of the goods in itself, the latent identity of Cimitile's site is promoted from the tangible signs of the time traced in the heterogeneous ornamental *corpus* that connotes the Paleochristian Basilicas. The decorative motifs in the St. Felix's Basilica and in the Saints Martyrs's Chapel are analysed as signs of the past that become the ideational traces of the fashion design through the drawing.

The methodological approach adopted is analogous to that of the aforementioned stylist Germaine Krebs, known as Madame Grès, who applies her sculptural capacities to the world of textiles and tailoring, giving life to timeless artworks, so as to be exhibited in a continuity philological next to the monumental statues of the house-studio of the sculptor French Antoine Bourdelle. The asymmetrical draperies of the dresses modelled directly on the body are the tangible proof of how the past can relive in the present through a contemporary interpretation of the original characters of inspiration with outcomes surprisingly innovative and deliberately timeless, capable of responding to the needs of the market and to meet also the tastes of the current clientele.

Returning to the heterogeneous ornamental *corpus* of Cimitile, the drawing of the decorations that characterize the architectures is in the majority of the cases of geometric or vegetal nature. The geometric patterns, of the type of the rhomboidal lattices, laced circles and peacock tails, underlie, from the point of view of the configurative genesis, respectively, straight links inclined in opposite verse at 30°, and circular meshes. The vegetal patterns, instead, of the type of vine shoots, underlie a geometric system of bilateral symmetry and continuity of the polycentric curve with concave-convex pattern to simulate the curved and spiral pattern of the shoots. The decorative motif composed of intertwined vine shoots is one of the most significant figurative legacies that the ancient world of the Mediterranean has handed down to our contemporaneity both from the ornamental and symbolic point of view.

The measurement of decorative patterns in Cimitile was performed through both traditional and digital survey techniques. The new digital technologies, in fact, have finally burst into the cultural heritage sector study that for a long time had seemed reluctant to be contaminated by the culture and languages of the digital world (Guarnieri et al. 2006).

In the Basilica of St. Felix—erected between the end of the fifth and the beginning of the sixth century around the tomb of the Saints *Felice* and *Paolino*—the ornamental motifs of the parietal mosaics that decorate the inside of the newsstand with a square plan have been surveyed through three-dimensional laser scanning technology (Fig. 5), which is the most widely used today for the study of Cultural Heritage from monuments to archaeological finds. The parietal mosaics of the newsstand—a structure made up of four brick walls, each of which has three arches supported by columns and re-used capitals—are made with tiles with a monochromatic background in thin gold leaf and mostly in blue for the ornamental motifs in order to ensure the traditional function of favouring the expansion of space and light above all through the particular technique of setting the tesserae according to specific angles. In particular, the west wall has an elegant scale pattern, unlike the others, which are each decorated with three pairs of spirals. Few remains of the decorations can be recognized on the outside of the east walls (palms and crosses) and north (vine shoots with large bunches), while in the soffits alternate geometric motifs (bipartite scales, laced circles, rhombus or rectangle lattices) and vegetables. Geometric patterns with connected circles and rhombus lattices are also present in the marble gates of the St. Felix's Tomb worked as a fretwork, according to a reason very common between the fourth and sixth centuries (De Matteis and Ebanista 2008).

The laser acquisition of the newsstand of the St. Felix's Basilica give a 3D point cloud model. This model is a digital three-dimensional cast in true shape and measure of the real object, from which it was possible to extract two-dimensional images for the analysis of the decorations.

The traditional survey of the decorative motifs in St. Felix's Basilica (Fig. 6), such as those in the entrance columns of the Chapel of the Saints Martyrs (Fig. 7), was performed through the technique of drawing known as *frottage*, based on the principle of rubbing. This technique consists in overlapping a certain support, like a sheet of paper or a canvas, to a surface that has more or less marked reliefs and in rubbing the support with a soft pencil so as to bring out the reliefs of the underlying surface.

Each survey was digitized through the vectorialization in dedicated software of the drawings obtained through the *frottage* technique. The metric data of the ornamental motifs surveying was verified through comparison with the measurements detected by traditional technique and represented in the survey's drawings (*eidotypes*).

The vast decorative heritage represented by the surveying activities has turned into an extraordinary activator of creativity that, freeing the imaginary, has allowed the design of new textures of fabrics all freely inspired by the drawings of the late antique motifs of Cimitile.

3 From the Drawing of the Decorations in Cimitile to the Fashion Design of New Textures

The activity of acquisition and visual reinterpretation of the ornamental motifs in Cimitile through simple or complex geometric transformations of the underlying module has allowed to capture the sensations of great suggestion that the decorations of Cimitile they are still able to transmit today, in order to convert them back into a visual heritage with a strong identity. The new textures are a contemporary narrative through which Cimitile enters into real dialogue with the reality lived by the visitor and the local communities and contributes to build a conscious memory of it.

The method of this research is aimed at explaining the role of the project as a place of synthesis of identity and

Fig. 5 3D point cloud model of the newsstand of the St. Felix's Basilica in Cimitile's site from 3DLS surveying (3D laser scenner surveying and data post-processing by A. Avella, 2016)

Fig. 6 Drawing of the geometric patterns with laced circles and rhomboidal lattices in the St. Felix's Tomb and in parietal mosaics that decorate the inside of the newsstand erected around the Tomb. From the survey's drawings performed through the frottage technique, it is possible to read on the slats of the marble gates of the Tomb the inscriptions in Latin which show biblical maxims of a moral nature. The frottage's drawing of the southern marble gate shows on the upper strip the inscription "QVI PARCIT BACVLO ODIT FILIVM SVV", while on the lower one the sentence "RADIX OMNIVM MAL-ORVM CVPIDITAS"

innovation in which the processes of innovation, however, give positive results only if they are able to interpret, with contemporary solutions, the complexity of a tradition that has ancient origins. In the case study as illustrated in Fig. 8, the ornamental motifs measured through the traditional surveying are those of the entrance columns to the Chapel of the Saints Martyrs. The two square section columns have geometric and vegetal decorations. The surveying activity of the perimeter surfaces of the columns is followed by the identification of their geometric derivations.

In particular, Fig. 8 illustrates the pattern analysis of leaves that decorate one of the four sides of the two columns and the geometric transformation of the hexagonal-based module tracked. The figure of the irregular hexagon was

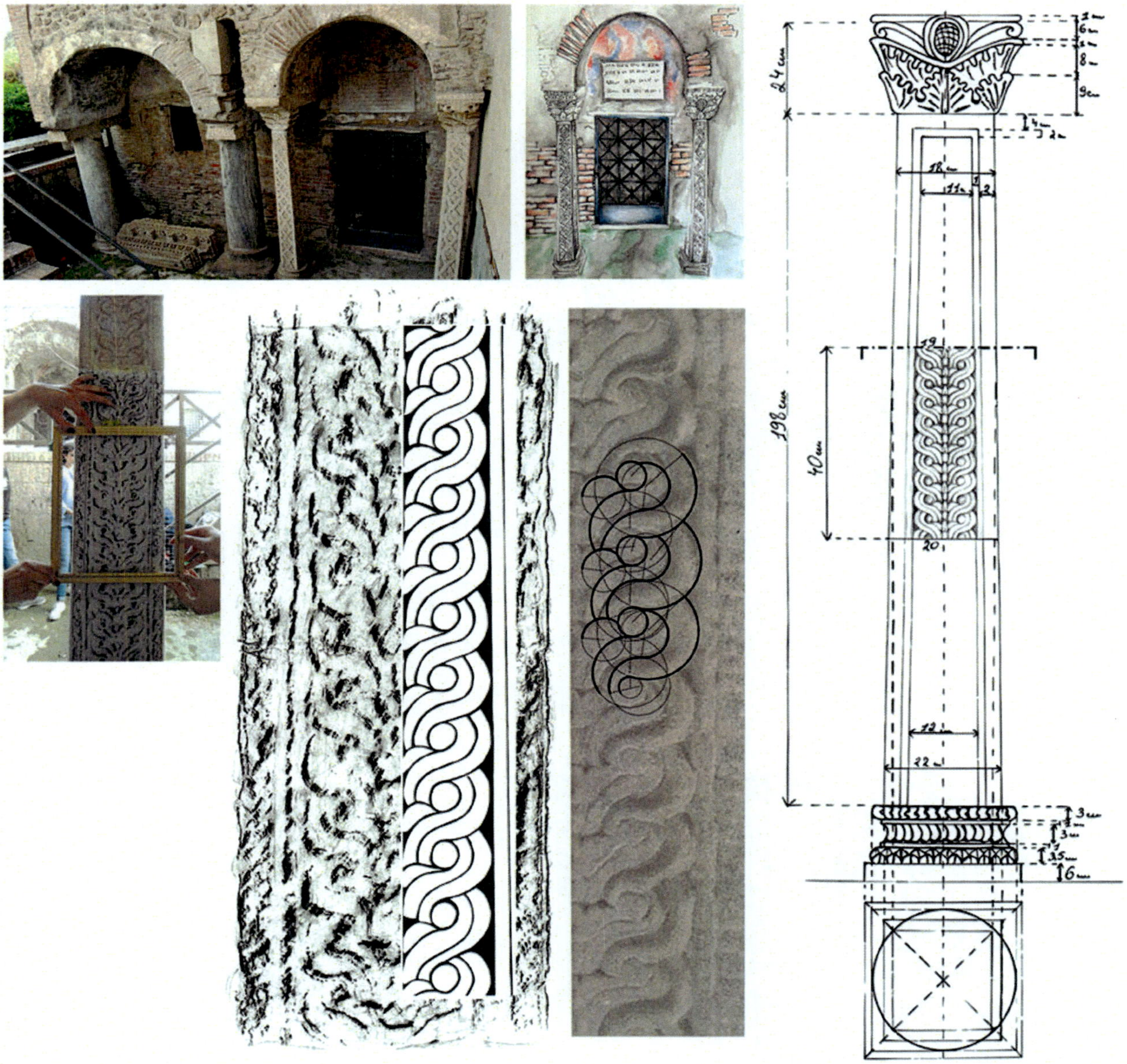

Fig. 7 Traditional surveying of the geometric patterns of the entrance columns to the Chapel of the Saints Martyrs in Cimitile's site

alternated with that of the hexagram (six-pointed star) drawn inside it. Through symmetrical derivations (translatory, bilateral and roundabout) of the geometric pattern thus obtained, various textures and different tailoring models were designed.These textures and models, although innovative, are inspired by tradition (Fig. 9).

Fig. 8 From the traditional survey of the vegetal decorations of the entry columns to the Saints Martyrs's Chapel to the design of new textures

Fig. 9 Sacred architecture and fashion drawing in Cimitile's archaeological site: from the idea to the project

4 Conclusions

The analysis of the geometries underlying the Cimitile's decorations through the 3D laser scanning and traditional surveying is the knowledge activity propaedeutic to the interpretation of the signs through which the Cimitile's site tells itself in architecture and re-proposes itself in fashion through the design project of the new "forms of inhabiting" the human body.

This methodological approach shows the manifold chances provided by geometry not only in the representation of the design project, but also in the conscious creative process.

Geometry, in fact, can be interpreted either as the geometric thought that rules the codification of the unitary *corpus* of the different methods of representation through the control of shape and space, or as a structure of thought that, with a morphogenetic role, causes and stimulates the invention of shape and space, thus speeding up the creative process. According to this meaning, geometry or rather "geometrization" as in a well-known quotation by William S. Huff (Huff 1984), follows a heuristic process a priori (Quici 2004) according to which inventing means to design the invention, which clashes with a heuristic process *a posteriori* according to which inventing means to rely on one's inspiration, thus associating the creative act to a sort of innate ability, to a "gift" as Huff would say.

In conclusion, with this study, we want to show that the heuristic nature of the geometric thought seen as a generator of the creative process also resides in the process of configuring the fashion event, in which the geometry is not only an aid to the representation of dresses and accessories, but also, according to a more authoritative meaning, an aid to both the analysis of the existing object and the genesis of what does not exist yet, that is the project.

The project proposal of innovative textures that refer to the decorative motifs of Cimitile's site is aimed at enhancing the cultural heritage of the archaeological site through a communication project aimed at promoting and disseminating the results of a project activity that interests and involves both local users to stimulate the sense of belonging of the local communities, and an external user to attract potential users according to the local tourism authorities. Therefore, the main result of this research involves the adoption of a model of communication that must impose an interactive dialogue with the cultural heritage understood as a common value on which establish collective identity processes.

Acknowledgements Figures 6, 7, 8 and 9 are graphic re-elaborations by Alessandra Avella from the exam drawings of her Courses in Fashion Drawing 1 and Fashion Drawing 2 (Bachelor Degree in Fashion Design) of the students L. Cianchetta, A. Bettino, A. Mottola, Y. Babychuk, A. Trasacco, L. Ardia, V. Maisto, T. Di Biase (2015/16 a.y.) and from the thesis of A. Dell'Aversano (2014/2015 a.y.) of which Alessandra Avella is mentor.

References

Avella, A. (2018). Disegno di Moda. Corpo|abito|illustrazione. *Fashion drawing. Body|clothing|illustration.* Aracne Editrice, Roma.

Calefato, P. (1986). *Il corpo rivestito*. Bari: Edizioni Dal Sud.

De Matteis, M., & Ebanista, C. (Eds.). (2008). *Il Complesso Basilicale di Cimitile*. Napoli: Arte Tipografica Editrice.

Giannone, A., & Calefato, P. (2007). *Manuale di comunicazione, sociologia e cultura della moda: Performance*. Roma: Meltemi.

Guarnieri, A., Remondino, F., & Vettore, A. (2006). Digital photogrammetry and TLS data fusion applied to cultural heritage 3D modelling. In H. G. Maas & D. Schneider (Eds.), *ISPRS—Commission V symposium, image engineering and vision metrology* (Vol. XXXVI). Dresden: ISPRS.

Huff, W. (1984). Geometrizzare e percettualizzare. *Rassegna, 19*.

Jacazzi, D., Carillo, S., & Petillo, P. (Eds.). (2011). *Materia Cimitile: memoria di segno*. Napoli: Misura di storia. La scuola di Pitagora editrice.

Portoghesi, P. (1999). *Natura e Architettura*. Milano: Skira.

Quici, F. (2004). *Tracciati di invenzione*. Utet, Torino: Euristica e disegno di architettura.

Sacredness as Morality: Ledoux and Chaux's Utopia

Nicola Delledonne

Abstract

In 1804, Claude Nicolas Ledoux (1736–1806) published *L'architecture considerée sous le rapport de l'art, des mœurs et de la législation* (*Architecture Considered in Relation to Art, Morals, and Legislation*). Despite the travails it took to complete the work—which was in the making for almost twenty years and was to have been released as an opus of several volumes, but was finally published as a single large book—the architect's editorial choice was quite focused: He wanted to tell of an ideal city, to be built in the countryside and to be modeled after the emerging physiocratic ideas. The selected location was Chaux, a tiny village in the Franche–Comté region near the Swiss border where, between 1775 and 1779, Ledoux had built the Royal Saltworks (listed since 1892 as a World Heritage site by UNESCO, for its architectural value). While the Saltworks were the result of a specific professional engagement, all subsequent designs for Chaux came from Ledoux's imagination; in fact, they exist solely on paper. Their purpose was to show their audience a city that could be built, at least in virtual form, and to lend credibility to its textual description and accompanying comments: true literary proof that aimed to show the possibility of a conciliatory utopia, in which social conflict could be subdued by a cult of Nature that everybody would subscribe to. Nonetheless, Ledoux was not content with a simple cult. Instead, he imagined secular rituals that would revolve around the budding public institutions, and that he later attempted to extend to community life. In so doing, he made a fundamental contribution to the efforts of the Enlightenment—which were almost universally frustrated—to transform sacredness into morality.

Keywords

Architecture • Utopia • Ledoux • Chaux • Sacredness • Morality

1 Introduction

After René Girard—the anthropologist, literary critic, and philosopher—published *La violence et le sacré* in 1972, every new discussion about sacredness has sooner or later led back to his fundamental work (Girard 1972). In this paper, however, we will follow a different path: We will endeavor to analyze this specific term not from an etymological perspective but in its metaphorical meaning. Our goal is to understand whether, and under what conditions, a relationship between sacredness and morality can be established in the context of French culture during the second half of the eighteenth century, when philosophers like Montesquieu and Rousseau began to rethink the meaning of religion by conferring a sacred connotation upon morality, in the name of secular terms like "common good" and "general will."[1] It was at about the same time that La Font de Saint-Yenne, regarded by many as the originator of art criticism in France, encouraged history painters to provide "une école des mœurs" [lessons on morals] by portraying the virtuous deeds of great men (Honour 1968).[2] Exaltation of virtue and condemnation of vice, two themes that often exist side-by-side, formed the most important chapter of the philosophy of the Enlightenment.

Ledoux, who was influenced by the aforementioned thinkers both as an architect and as a scholar of architecture,

N. Delledonne (✉)
Architecture Department, College of Environmental Design, King Fahd University of Petroleum and Minerals, Dhahran, Saudi Arabia
e-mail: arch.nicola.delledonne@gmail.com

[1]The main works referenced by Ledoux are *L'esprit de lois,* which was published anonymously by Montesquieu in 1748, and Rousseau's *Contrat social*, which was printed in 1762.

[2]See *Sentiments sur quelques ouvrages de peinture, sculptures et gravure*, 1754.

C. Gambardella et al. (eds.), *Advances in Utopian Studies and Sacred Architecture*,
Advances in Science, Technology & Innovation,
https://doi.org/10.1007/978-3-030-50765-7_23

placed morality at the center of his theoretical analysis and used it as the basis for his personal utopia. He worked in France—both in the city of Paris and in the provinces—beginning in 1770, and he was quite successful for at least fifteen years in his professional endeavors, which included both public and private buildings. After the French Revolution broke out, Ledoux fell into disgrace and risked the guillotine due to his friendships with several clients who were members of the aristocracy. At the end of the two-year period ranging from 1793 to 1794, the so-called *Terreur* (the Terror), he was released from prison and he resumed a normal life, but he was never able to regain the public consideration that he had previously enjoyed as an architect, despite his efforts. His book (Ledoux 1804), *L'architecture considérée sous le rapport de l'art, de mœurs et de la législation* (henceforth *L'architecture…*) was meant to be not only the legacy of an individual who had devoted his entire life to architecture, but first and foremost an instrument aimed at restoring him to the front of his profession. Unfortunately, as we know, his wish did not come true. Nevertheless, in the span of a little over a century the book came to be regarded, for better or worse (Kaufmann 1933, 1952, 1955; Sedlmayr 1948, 1955), as a decisive factor in the evolution of the theory of architecture, and in architectural production as well.

2 Morality and Character

In the second half of the eighteenth century, both within the *Académie d'architecture* and among knowledgeable professionals, a debate took place on the idea of character, which required that buildings should manifest their purpose via their form. This idea was deployed to different ends: in residential buildings, to represent the hierarchical structure of society through a hierarchy of aesthetics; in public buildings, to reveal the civic meaning of public institutions; and, in the temporary structures and projects that were staged in the context of the French Revolution, to garner popular support. Architects and academics alike, including Boffrand (1745), Blondel (1771), and Le Camus de Mézières (1780) began to interrogate themselves about the role of sentiment as a tool for the mediation between architectural forms and their societal and civic meaning. Quatremère de Quincy (1788, 1832), the architectural theorist, analyzed the idea of character in reference to different civilizations and the historic and geographic conditions in which they had developed. Boullée (1790), himself a professor and architect, went even further, maintaining that character had to be considered as a theoretical notion capable of describing the general architectural concept of a building. On this basis, he interpreted old building types in new ways, for example, by imagining theaters as temples of good taste dedicated to beautiful ladies, and libraries as basilicas for scholars. Ledoux took this premise one step further when, in the beginning of *L'Architecture…*, he defined character as the specific nature of a monument that contributes to the diffusion and the purification of morals. Thus, character became a pedagogical and moralizing feature that was used to determine how each type of building should be regarded in the social order, by virtue of the specificity of its architectural forms. In other words, Ledoux attributed a particular moral meaning to civic and social functions. The resulting sacralization should be viewed as a secular form of religion that rediscovered in architecture a powerful form of expression.

In addition, in his definition of the character of buildings, Ledoux postulated a comparison with an architectural and literary heritage that had become established over the span of several centuries. Beginning with the years of his education in the classrooms of the *Collège de Beauvais* in Paris, an institution that was famous for its prestigious courses in rhetoric (Vidler 1990), Ledoux chose a referential world that was populated by ancient myths, which he interpreted and revisited freely. It is not by accident that in *L'Architecture…* he cited ancient writers such as Homer, Virgil, and Horatius, as well as the more recent Jean de la Fontaine, who could himself be counted as an "ancient writer" for the position he took in debate over *La querelle des anciens et des modernes* (The Quarrel of the Ancients and the Moderns). Through his reasoned manipulation of the historical universe, Ledoux gave the classical world a new meaning and, by blending it with the philosophical ambitions of the Enlightenment, he outlined an encyclopedic, educational, and moral ideal that he attempted to translate into architectural forms. In his "translation," which from a stylistic standpoint resulted in an *antique modernisé* (a modernized antiquity) and in a *classicism éclairé* (an enlightened classicism) (Rabreau 2000), Ledoux conferred upon the idea of character, i.e., the evocative power of architecture, an inalienable role that was an exact parallel—to use Ledoux's own expression—to the role that manliness played as an essential attribute of love. Indeed, from his perspective, the character of a building had to be unequivocally *lisible* (intelligible). For this reason, he tried to create an *architecture parlante* (a "speaking" architecture), although he did not always succeed in making it *éloquente* (expressive, or eloquent).

Ledoux was truly convinced that architecture could be conceived as a language capable of bringing the people of a community, for example, the inhabitants of the village of Chaux, to a state of good civil coexistence. However, for this to happen, the language of architecture needed to be written down, in order to clearly explain its utopian content.

3 The Narration of Utopia

Ledoux never defined the Chaux project as "utopia." Doing so would have been against his own interest, as he was looking to establish a new professional space in a rural setting after the failures he experienced in the big city.[3] Yet, many scholars (Ozouf 1966; Gallet 1980; Vidler 1990; Santangelo 1998; Rabreau 2000) agree in considering Chaux as the outcome of a utopian belief, if only for the kind of life that the architect thought would spread as a result of it. After all, the way he spoke of Chaux in *L'Architecture...* does support this interpretation. We also need to keep in mind that the book can hardly be considered a treatise on architecture, except in a very broad sense. Ledoux, who not coincidentally was granted the titles of *architect-ecrivain* (writer-architect) (Kaufmann 1933; Didier 1986) and *aéde moderne* (modern bard) (Rabreau 2000), had intended to create an imaginary *récit de voyage* (a travelogue) in which the journey was both the pretext for describing a certain idea of the city and for reviewing its architecture.[4] Indeed, this device allowed Ledoux not only to portray the traveler as he visited different buildings, but also while he observed the activities that took place inside them. It is from this perspective of direct experience, if only as an imaginary ploy, that Ledoux developed a theoretical idea of narration through which he could depict his utopia. His theory established the architect as a political player and forced him to address not only the relationship between architecture and art, but between architecture on one hand and morality and law-making on the other. At the time he wrote the book, however, these relationships were not well-established; and even if they were, they were quite open to critique. Therefore, Ledoux attempted to transform them by imagining a social order that his lengthy story should bear out as feasible, at least on the level of a simulation. From this perspective, the projects included in his collection represented the fragments of a possible utopia. As such, they were not based on actual work orders, but on wholly invented programs that, thanks to the rhetorical flair of their author, produced a result that was very different from the very dry aggregate of functions they were subsequently reduced to, first by engineering culture and then by the architectural functionalist movement. Indeed, before analyzing some of Ledoux's projects, we need to clear the field of one possible misunderstanding: His desire to establish a link between architectural forms and people's social behaviors should not be mistaken for a generic search for function. In fact, his aim was set on the evocative power of architecture rather than on the proper functioning of the building. Moreover, the aseptic and falsely objective concept of function implies the existence of spaces that should be optimized on the basis of consolidated habits, whereas the social life proposed by the architect—one in which moments of rewarding work should alternate with moments of joyous celebration and fruitful study (Ozouf 1966)—and also the allegorical characters that bring it to life, is entirely a product of his own imagination. After all, in his book Ledoux did not intend to speak of actual reality, but of one to be desired: an actual utopia. For this reason, he abandoned the *style de la définition* used by his teacher, Blondel, in *Cours d'architecture*,[5] in favor of a *style figuré*,[6] which was better suited for describing his plan as a fragment of moral philosophy to be written in stone.[7] That same philosophy had to become the true object of sacralization, and the imaginary buildings described in *L'Architecture...* were destined to bear witness to it.

4 Civic Monuments and Dwellings for Communal Living

Ledoux, who was drawn to physiocratic ideas, envisioned a rural utopia in a small town that—due to its proximity to the French-Swiss border—would base its economy on the commerce of produce and crafts, as well on salt extraction. The life that Ledoux imagined for Chaux (Fig. 1) was primarily agrarian and represented the ideal context for the exaltation of Nature, which he considered to be the basis for every moral principle: an idea that epitomized the philosophy of the Enlightenment. At the same time, he expressed the contrast between the city and the countryside, disparaging the former and extolling the latter.[8]

To prove the economic feasibility of his utopia, Ledoux conceived his project for Chaux's marketplace (Fig. 2) as a small town that could exist only in a rural setting. That project, however, did not represent the moral essence of his ideas, despite the fact that he had attributed a symbolic significance to it. Indeed, in each of the four courts located at the corners of the building, which were smaller than the roofed central square, Ledoux placed a *columna lactaria* (a

[3]These failures had begun well before the Revolution, with the design and the partial construction of the so-called *barrières*, i.e., the toll gates around Paris.

[4]Among the possible literary models that might have inspired Ledoux's book are *Les Aventures de Télémaque*, published by Fenelon in 1699, and *Voyage du jeune Anacharsis en Grèce*, published by Abbé Barthélemy in 1788.

[5]JACQUES FRANCOIS BLONDEL, *Course d'architecture*, I, 374.

[6]CLAUDE-NICOLAS LEDOUX, *L'Architecture*..., 16.

[7]This writing style can seem obscure at times. Mona Ozouf used the expression *text illisible* (unintelligible text) for *L'Architecture*...

[8]Ledoux loved the countryside not only for his physiocratic beliefs, but also for autobiographical reasons, since he hailed from the bucolic hamlet of Dormans. At the same time, his "pastoral sensibility" was influenced by his friendship with a man of the letters, Jacques Dellile, known as *l'abbé*, who had translated Virgil's *Georgics*.

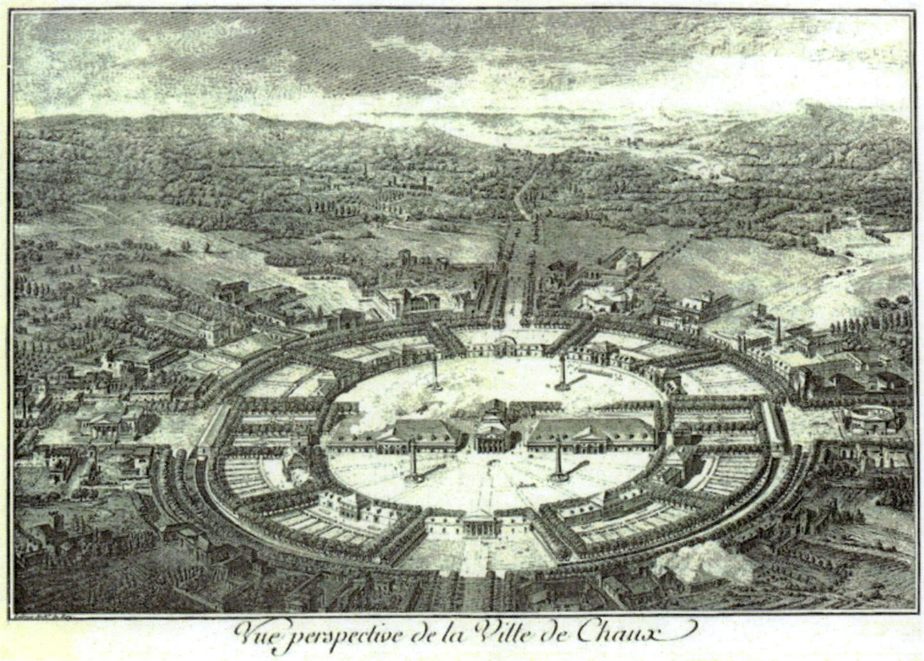

Fig. 1 Chaux, Royal Saltworks, overall view

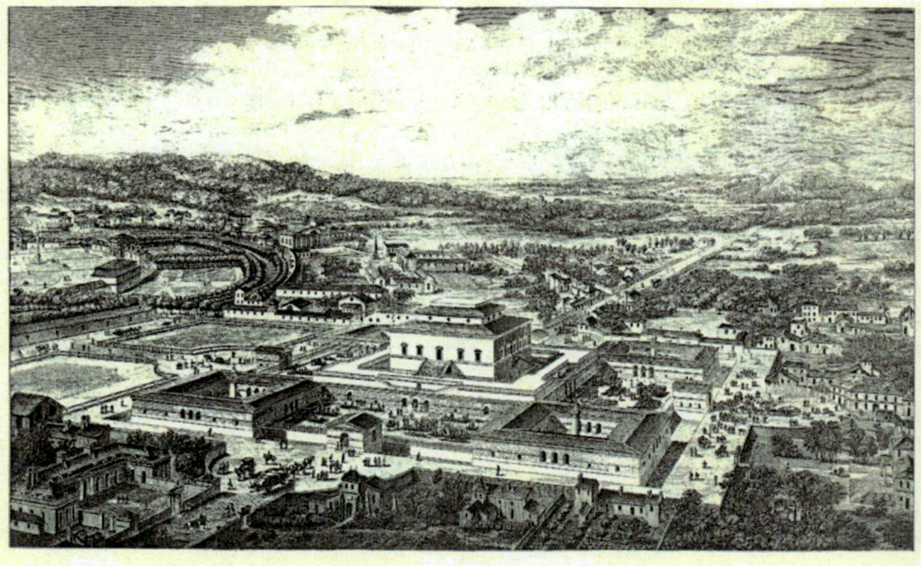

Fig. 2 Chaux, marketplace, overall view

lactarian column) that resembled the one located in ancient Rome's *Forum Holitorium*, where abandoned infants were left at its base with the hope that a "wet nurse" could be found for them. The moral was clear: Society had to take care of orphans, because no utopia could ignore the fate of its weakest members.

As for religious buildings, Ledoux included them only partially in the process of sacralization of civic values. Their function was to suggest to beholders an allegorical interpretation of architecture, in which the allegory was always motivated by moral reasons.

The project of the church at Chaux (Figs. 3 and 4), modeled after a Greek cross with four burial sites at the corners, did not contemplate traditional uses for the inner spaces. In fact, at least two of the four arms of the cross led to a central altar using a flight of steps. The significance of

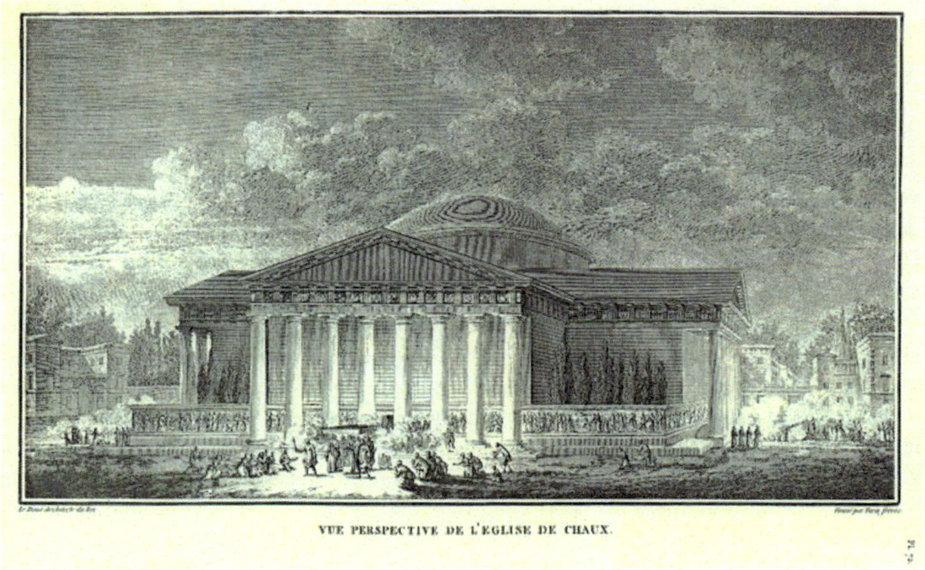

Fig. 3 Chaux, church, perspective view

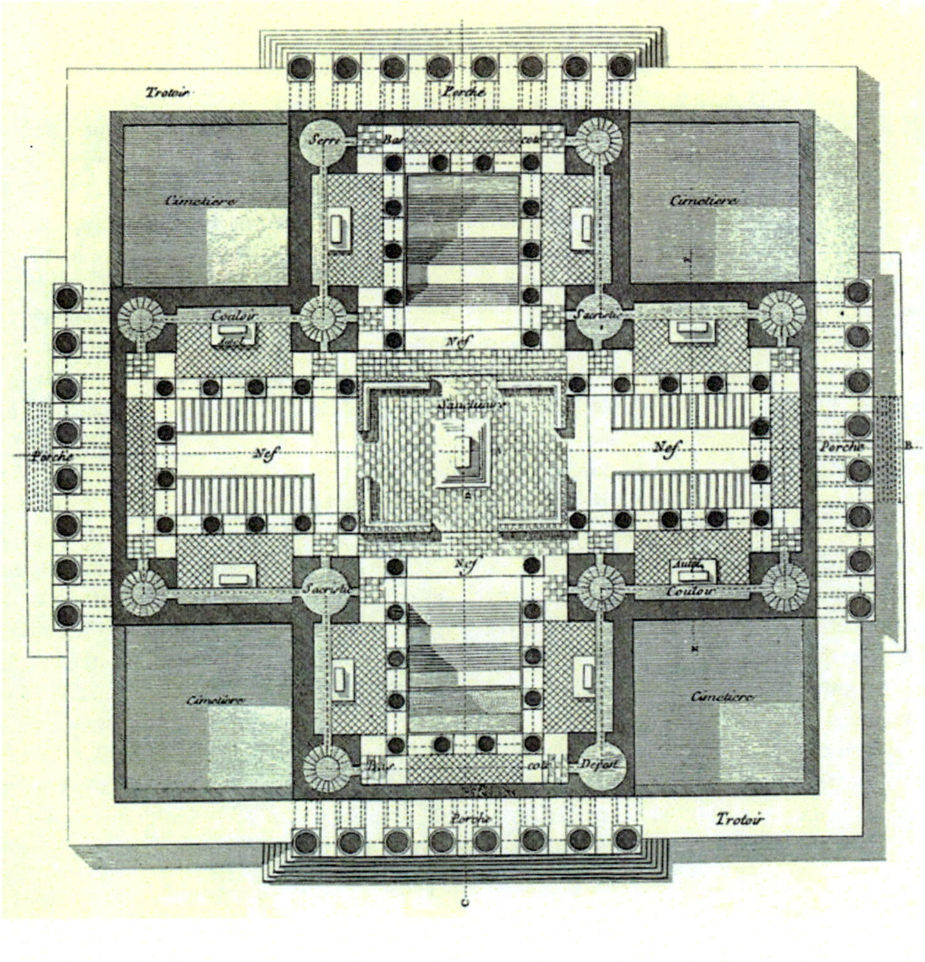

Fig. 4 Chaux, church, plan

the building was made clearer in the accompanying text, which begins with a poetic description of a sunset, a metaphor for the inevitable end of human life. A similar *memento mori* (a reminder of mortality) summoned men not to waste their time and to adopt a moral conduct beyond reproach. The only way men could distinguish themselves was through worthy acts or important works, which were promptly suggested by the bas-relief images that had been sculpted on the large *plutei* (high stone parapets) that walled off the burial sites.

The project for the cemetery at Chaux (Fig. 5) evoked ancient catacombs and reinterpreted them as a three-story, square-shaped structure below ground, which was reserved for community mausoleums. But the fundamental symbolic feature was a large sphere, half below ground and half above ground, which was located at the center of the building and had both a technical purpose and a symbolic one: On the one hand, it featured an *oeil-de-boeuf* at the top, an exhaust point for the pestilential odors that lay below; on the other, it correlated the spherical structure with the shape of the globe, suggesting that the souls of the deceased would be reabsorbed into the earth.

In any case, Ledoux's projects that more clearly exemplified the relationship between sacredness and morality were "les monuments intentionnels du civisme" (Rabreau 2000), i.e., civic buildings with an intentional monumental character, whose names implied the invention of a peculiar and new vocabulary.

The *Pacifère*[9] or *Conciliateur* (Fig. 6), which was also referred to as *temple au bonheur* (temple of happiness) and *monument à la conciliation* (monument to reconciliation), was a monumental building intended for the administration of justice, where a judge both wise and understanding would restore harmony among people who had been brought into conflict by either jealousy or selfishness, which were the only minor grievances that could be raised against the civil denizens of Chaux. This reasoning, which was apparently based on an excess of optimism, was actually a ruse to reinforce Ledoux's criticism of the big city: If Paris was fertile ground for the most heinous of crimes, rural areas could at worst be faulted for temporary interpersonal resentments that could be easily overcome. According to Ledoux, the monument to reconciliation should be as simple as the laws it was intended to represent. Consequently, he conceived its volumetric arrangement as a symbolic system, in which the base, with its geometric exactitude, evoked an ideal of perfect justice resting on a solid foundation.[10] The façades were decorated with great slabs of stone that very closely recalled the tablets of Mosaic law. Separated by an order of fasces, they reproduced quotes by great philosophers, both ancient and modern.

Another cube-shaped building, which was raised on a platform, was the *Maison d'union*, or House of Union (Fig. 7),[11] where unity was celebrated as a virtue capable of assigning moral order to force. Ledoux believed that unity was the best path toward both individual and collective happiness, which was the end game of all utopias. Convinced that happiness is derived from education, he came up with a functional plan for the building, which included study halls, libraries, and galleries devoted to the study of subjects such as agriculture, commerce, and literature.

The *Temple de Mémoire*, or Temple of Memory (Fig. 8), is yet again a cube-shaped structure that rests on a large podium and is adorned with four celebratory columns at the corners. On one hand, Ledoux dedicated the temple to the procreative and protective qualities of women, in contrast with the belligerent world of men. On the other hand, he chose Trajan's Column, which commemorates the exploits of a great battle commander and emperor, as an architectural model. His intent to provoke controversy is clear: henceforth, these *colonnes triomphales* (triumphal columns) should no longer be used to celebrate military conquests but peaceful actions instead, based on the ideas of love and knowledge or, more generally, on any act that is driven by the kind of creative will that women personify and that the Graces inspire. The relationship that links women with memory seems to make way for another symbolic meaning: If the building could be viewed as the temple of Mnemosyne, mother of the Muses, patron goddesses of the arts, then the principle of memory could be legitimately linked with the principle of creation (or procreation), with all that would ensue on an aesthetic level; for example, a free interpretation of the above-mentioned historical experience, which was regarded as the foundation of the theory of character.

The *Panarétéon* (Fig. 9)[12] was to be a school of morals, with the ultimate goal of teaching men their duties and, above all, civics. In this instance, Ledoux assigned to virtue a sacred role, capable of absorbing and of blending all of the most significant religious values of ancient cultures, from the most primitive to the culture of the Greeks, the Romans, and the Goths. He believed that virtue stood in opposition to vice, and that it would lead to collective happiness. Therefore, he honored it with the design of a cubic temple that had become a symbol of timelessness, just as Nature's laws governing morals were—in his opinion—timeless themselves. Sculptures would embody the character of the

[9] CLAUDE NICOLAS LEDOUX, *L'Architecture…*, pages 113–115 (*planche* 40).
[10] Ibid., page 115, note 1.
[11] Ibid., pages 117–118 (*planche* 43).
[12] Ibid., pages 184–186 (*planche* 92). Within the text, the building is referred to as the *Panarétéon*, whereas in the engraving it is called *Panarèthéeon*.

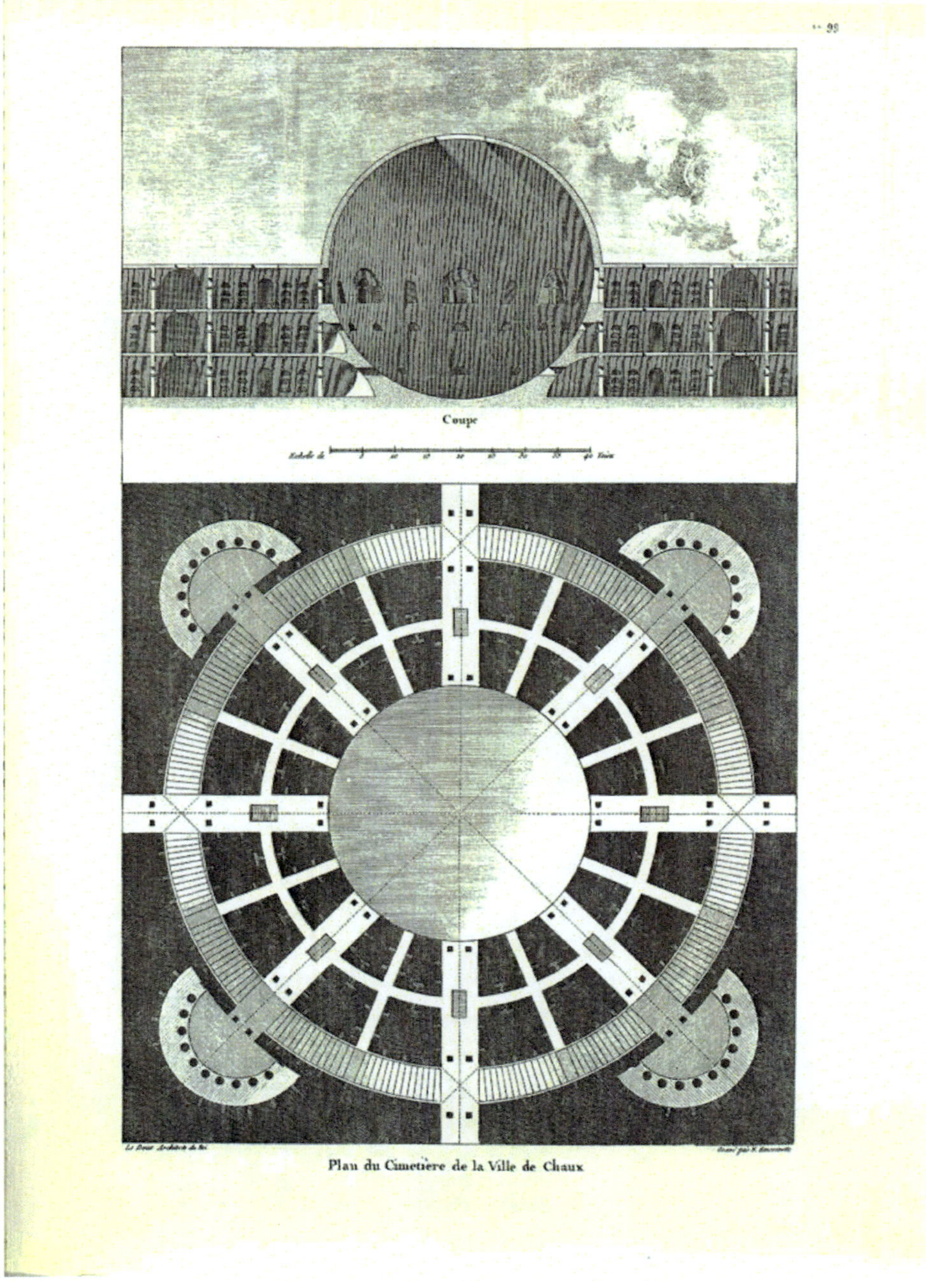

Fig. 5 Chaux, Cemetery, plan, and section

building and were meant to appeal both to philosophers who were capable of interpreting them and, above all, to the young, to whom the architect wanted to offer virtuous models to follow.[13]

In Ledoux's mind, the *Oikèma* (Fig. 10)[14] was meant as a reformed brothel where the display of vice should lead visitors to find their way back to virtue. How such a process should take place is not made clear at all. The building, which rested on a base that was used to connect two split-level parcels of land, did not have the customary cubic shape, but a phallic configuration instead, even if the configuration was only visible in the building plans. The architect presented his building as *Fragment d'un Monument Grec* (Fragment of a Greek Monument), an image that he reinforced through the use of a temple-like motif in the Ionic style for the main entrance and a peristyle of the same order at the opposite end of the structure.

[13]Ibid., page 185, note 1.
[14]Ibid., pages 199–204 (*plances* 103–104).

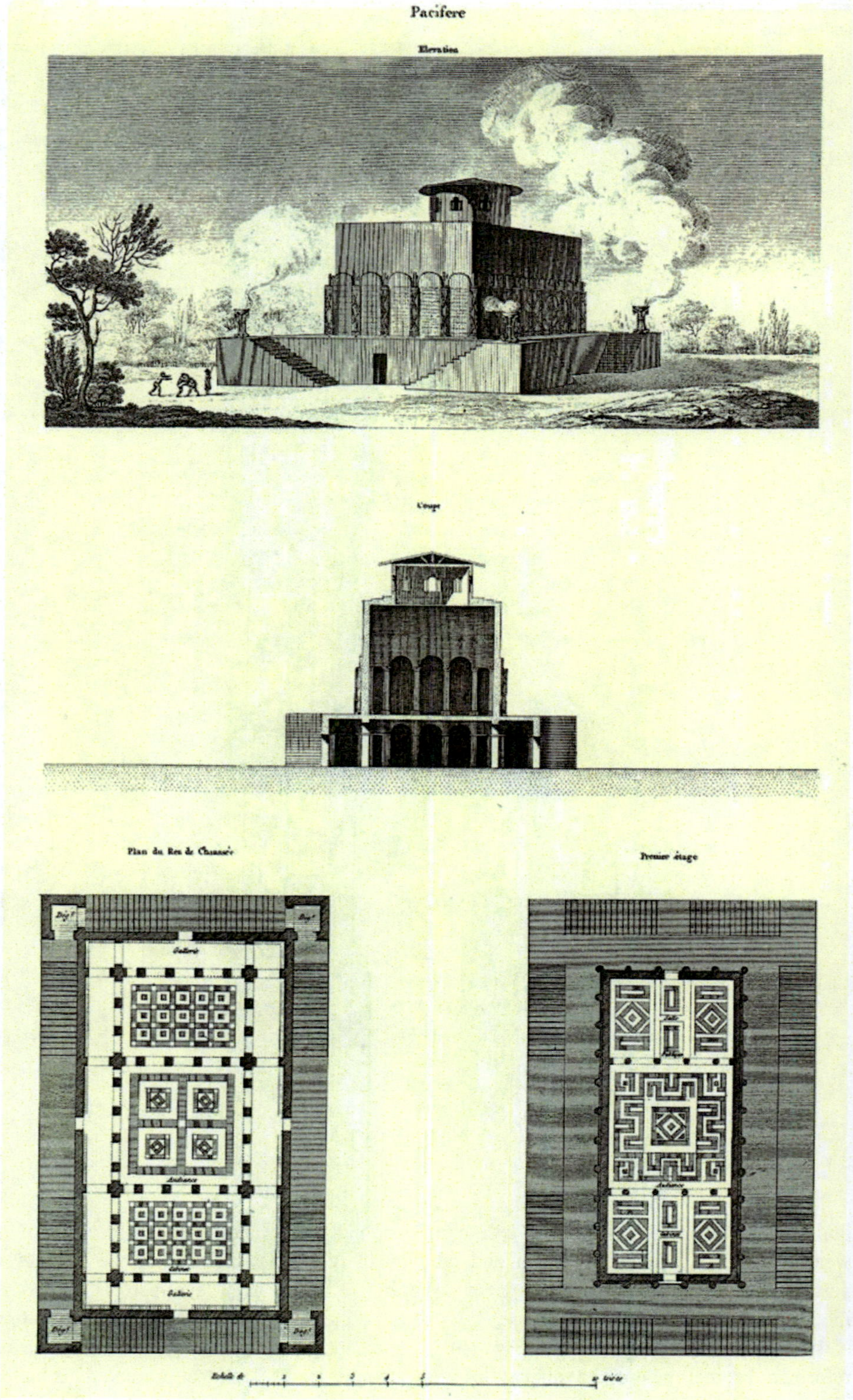

Fig. 6 Chaux, *Pacifère*, perspective view, section, and plans

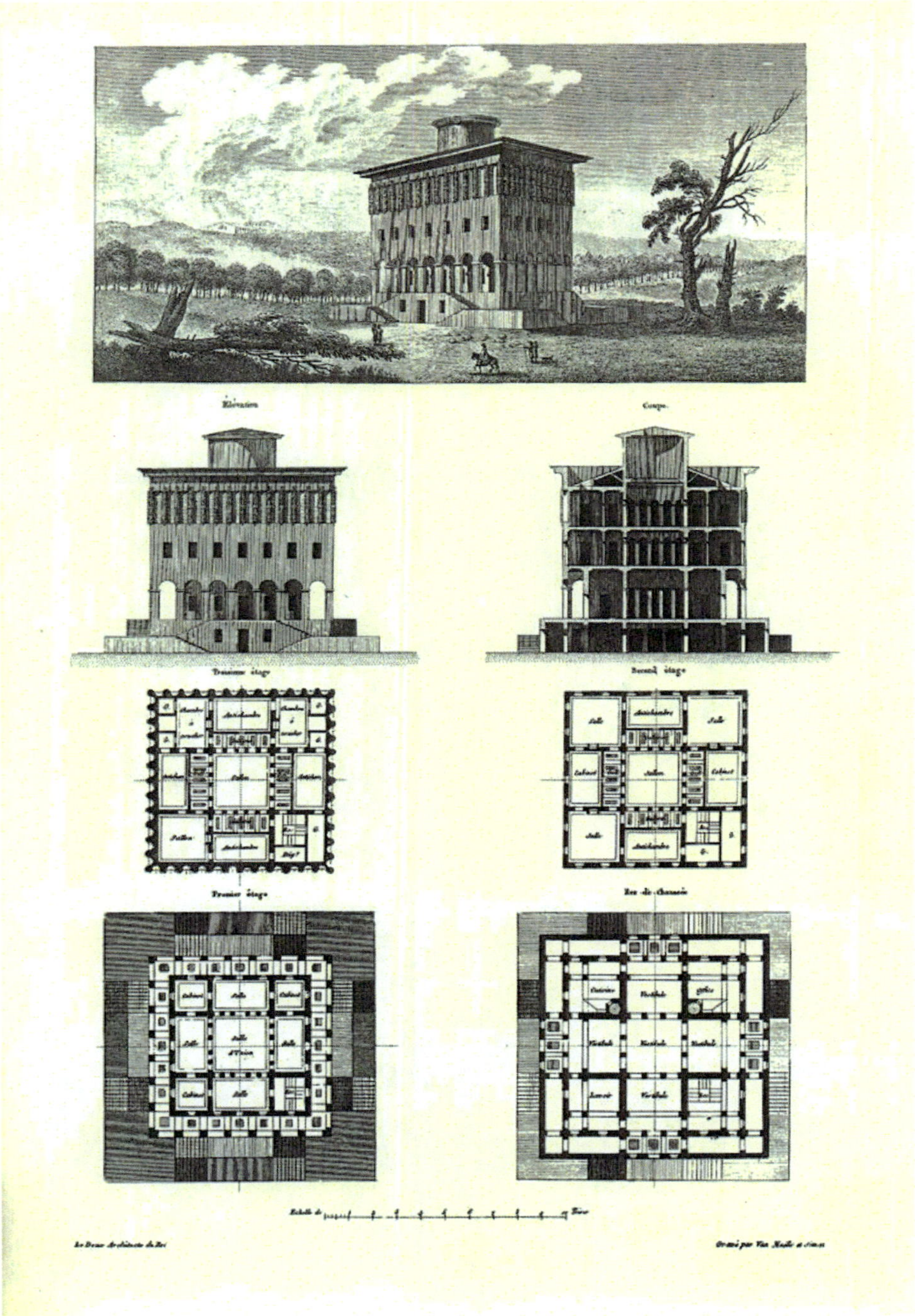

Fig. 7 Chaux, House of Union, perspective view, elevation, section, and plans

Finally, we should mention the buildings used for housing. In the heyday of his professional career, Ledoux had designed *châteaux* (castles) and *hôtels particuliers* (the urban houses of the aristocracy), which he had used to express the social *status* of his noble clients. At Chaux, however, he had focused on the *métier* (the profession) of its hypothetical denizens and had devoted special attention to dwellings used for communal living. Many of them were located at strategic points in the forest of Chaux and were reserved for crews of lumberjacks, artisans, forest rangers, and coalmen. Primarily, Ledoux envisioned them as wood structures with a large, centrally located stone chimney, and conceived them as meeting places for workers who had joined a gild or a corporation. Examples include the *Atelier*

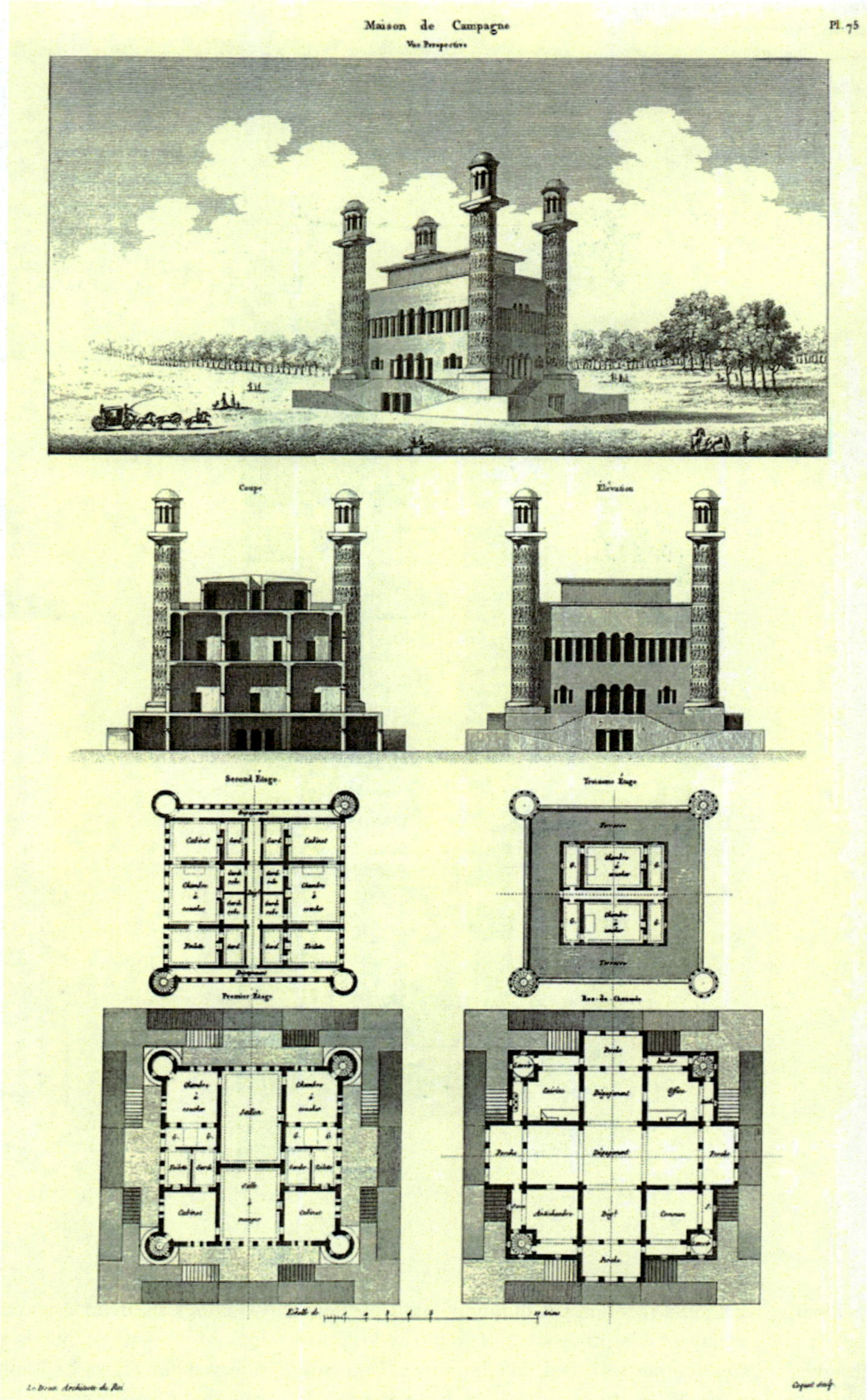

Fig. 8 Chaux, Temple of Memory, perspective view, section, elevation, and plans

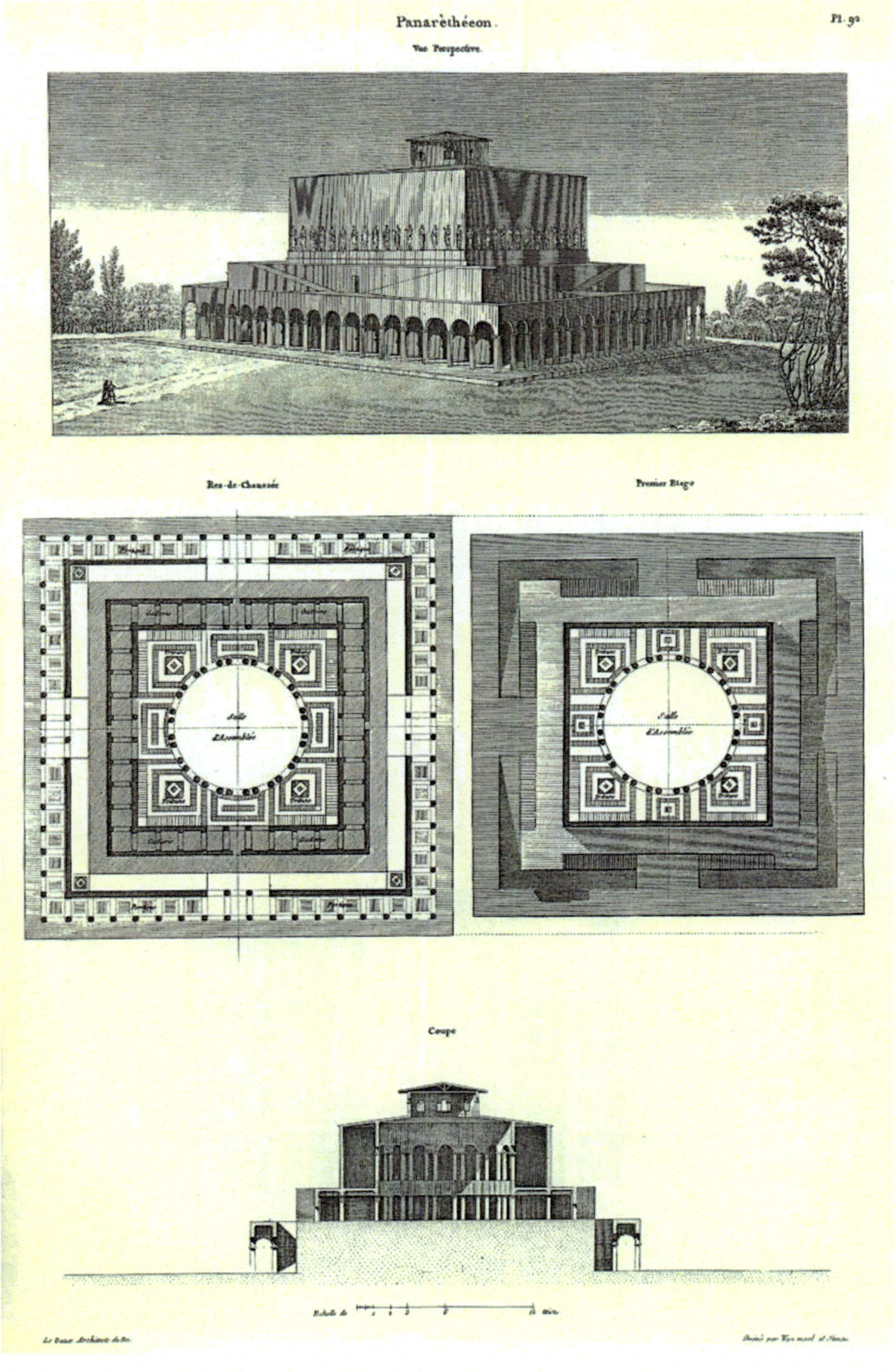

Fig. 9 Chaux, *Panarétéon*, perspective view, plans, and section

Fig. 10 Chaux, *Oikèma*, elevations, sections, and plan

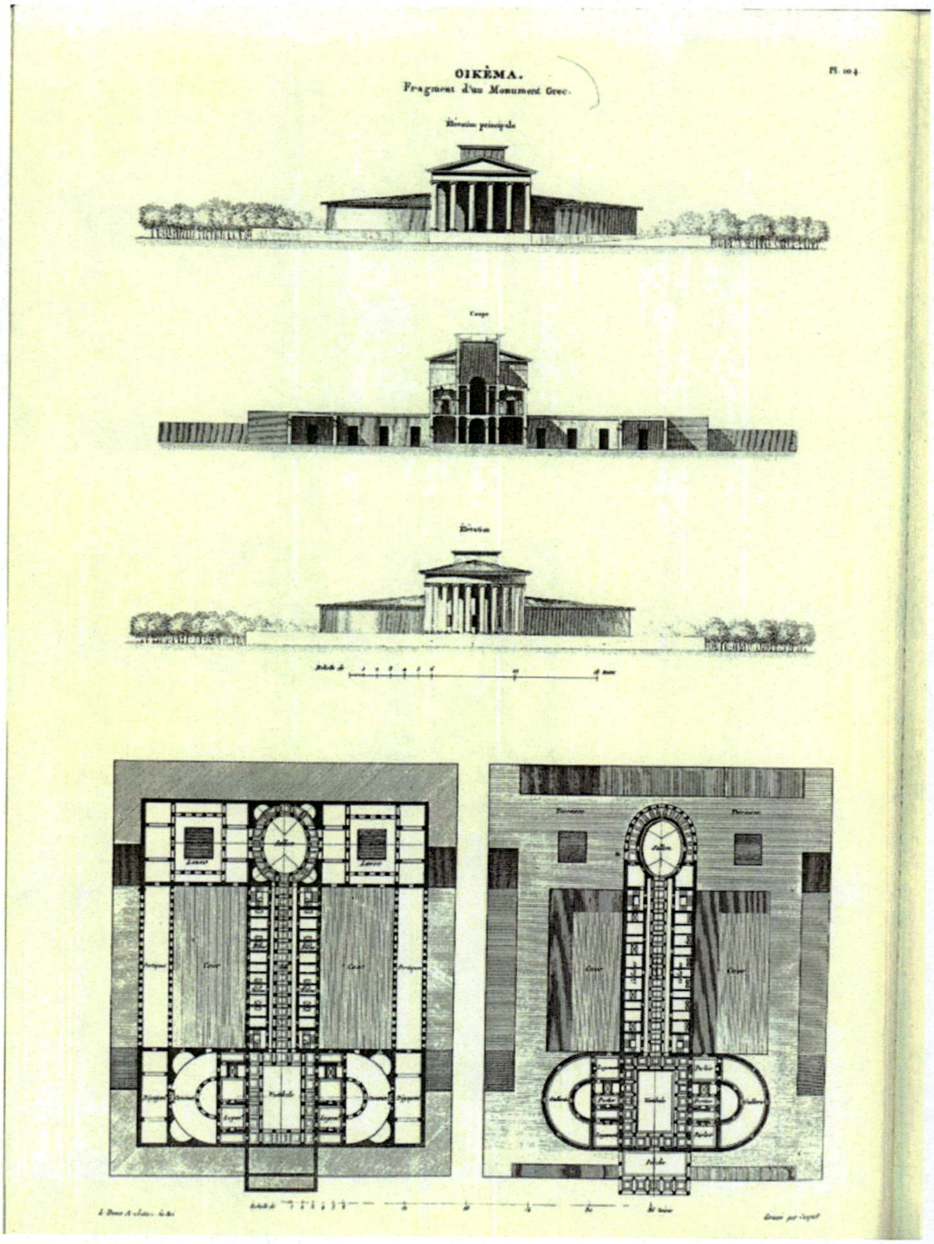

des scieurs de bois (Fig. 11),[15] the *Atelier des bucherons, gardes de la forêt* (Fig. 12),[16] and the *Logement destiné au gardes de la forêt de Chaux* (Fig. 13).[17] In all of these projects, the concept of "family" that lay at the foundation of residential architecture morphed into the concept of *compagnonnage*, i.e., trade guilds (Vidler 1990). Ledoux describes these forest workers as simple folk, content with their jobs, and yet untainted by civilization. Their personality is perfectly mirrored by the clear geometries of their dwellings, where no kind of decoration is needed.

[15]Ibid., pages 102–104 (*planche* 32).
[16]Ibid., pages 197–198 (*planches* 101–102).
[17]Ibid., pages 191–192 (*planche* 97).

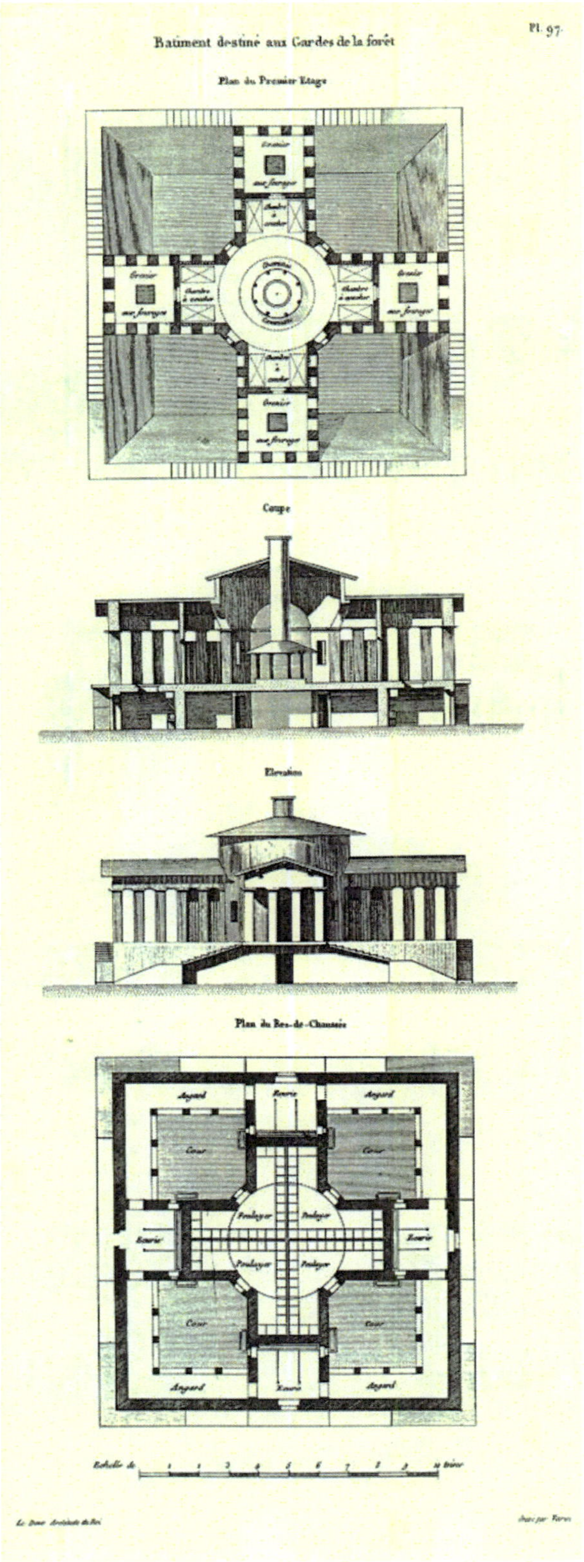

Fig. 11 Chaux, Woodcutter's lodge, plans, section, and elevation

5 Conclusions

Ledoux's utopia, viewed from a critical and historical distance, seems more interesting today precisely because of its lack of cohesion. Unlike classical texts on utopia, which presumed an exact organization of the city—like Plato's *Republic* (Plato, IV Century BCE), Thomas Moore's *Utopia* (Moore 1516), and Tommaso Campanella's *The City of the Sun* (Campanella 1623), as well as Aristotle's *Politics* (Aristotle, IV Century BCE) with his knowledge of Hippodamus, even though it was not a utopian book—Ledoux's *L'Architecture…* presented a series of fragments or a collection of architectural projects with accompanying text that were much like the pieces of a puzzle with an uncertain outcome. Rather than theorize a utopia, he described and illustrated some of its partial applications, which implied either the creation of new institutions or the reformation of existing ones. He focused on preeminent civic functions, such as justice and education, but he also dealt with more general notions, for example, freedom of assembly and communal life. In approaching these things from a moral standpoint, he conferred upon them a sacred character, i.e., the kind of character that would demand the utmost respect from society. In this sense, sacredness, and morality are analogous concepts.

Still, Ledoux had to face a powerful contradiction: on one hand, he aspired to achieve a renewed social order that he considered within reach and that could be accomplished through goodwill; on the other hand, such an order existed only in his imagination. In other words, the desire of the architect to grant architecture a moralizing role was met by an institutional vacuum that Ledoux attempted to fill using a series of narrative and project-related fragments, conceived as snippets of social life. Consequently, the utopia of Chaux was presented as an ideal but infinitely distant objective. Such an objective could perhaps be pursued, but it could certainly not be attained: not only because the physiocratic thought that inspired it was about to be swept away by the industrial revolution, but also because the contents of Chaux's utopia, far from being obvious, had to be invented by the architect on the basis of the above-mentioned philosophical reflections, as soon as he attempted to translate them into architectural expressions.

Accordingly, Ledoux's utopia was neither a "utopia of escape" nor a "utopia of reconstruction" (Mumford 1922); rather, it was an "exercise in personal poetics," albeit not one that sought to retreat onto itself. His objective was to inspire the reader of *L'Architecture…* to reflect upon the moral value of public institutions and of the kind of life that came with it.

Nowadays, we no longer need to demonstrate or to emphasize how utopian thinking has failed. And yet, the

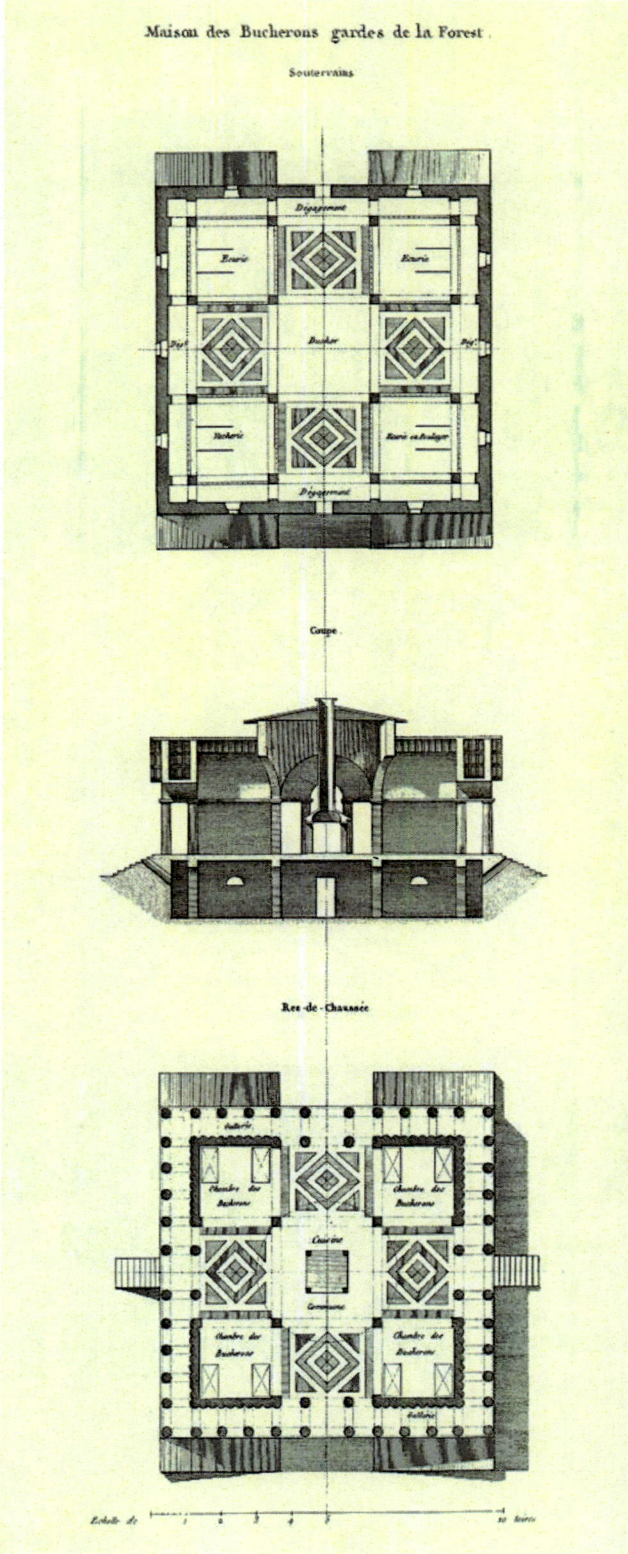

Fig. 12 Chaux, Lodge of lumberjacks and forest rangers, plans, and section

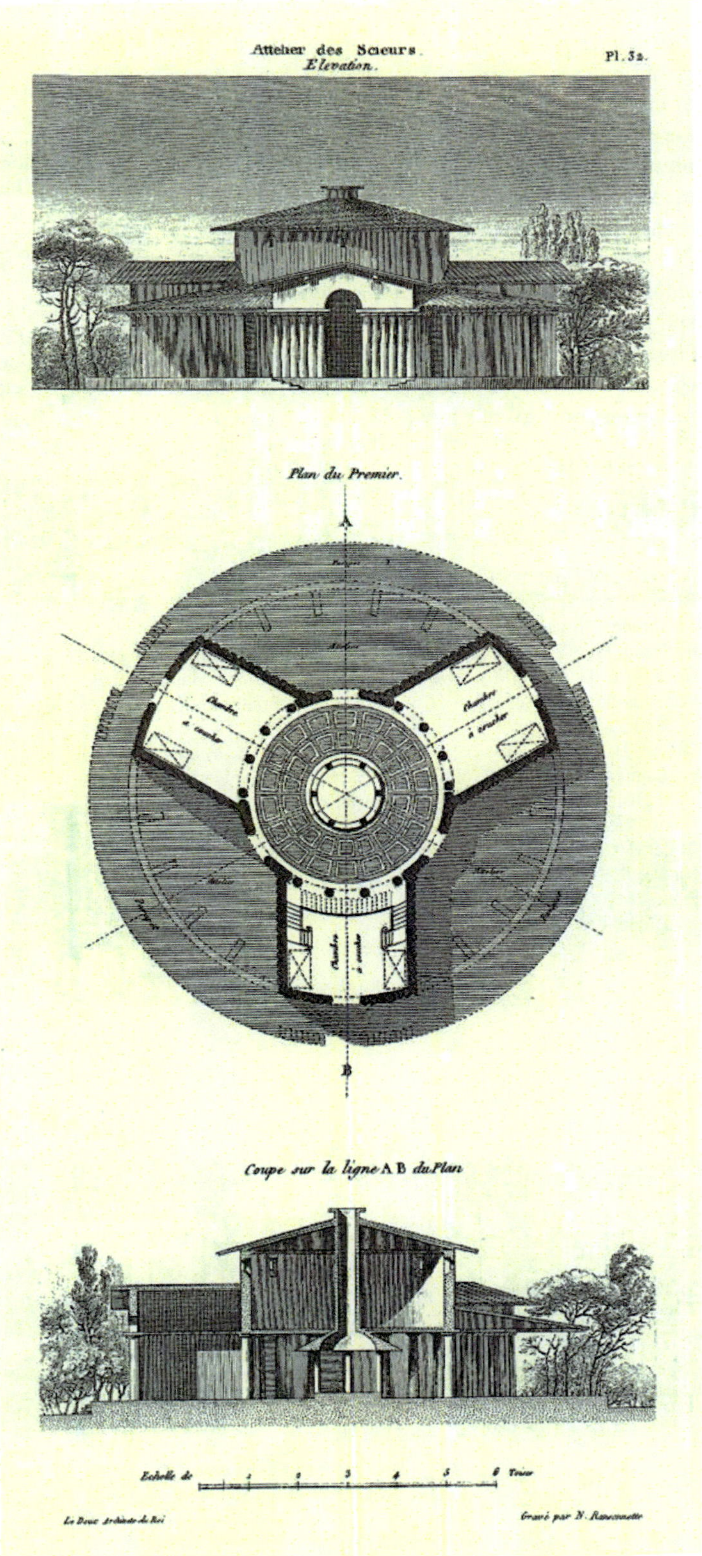

Fig. 13 Chaux, Lodge of Chaux forest rangers, elevation, plan, and section

standing issue of institutional reforms should be tackled with a certain utopian enthusiasm at times like these, when the great financial powers seek to diminish the authoritativeness of civil institutions and to belittle their meaning. And, with the same enthusiasm, we should interrogate ourselves on the meaning of a life of association, because *social media* have made us all less sociable, since they have made people accustomed to speaking via technological means rather than directly with each other. With these premises in mind, we would do well to reconsider the partial approach of Ledoux, and its lack of absolutism. Far from being naïve, as it may have seemed in the beginning, it raises the issue of a shared ethic in which we can recognize the sacred character, if only metaphorically so, of utopia.

References

Aristotle (IV century BCE). *Politics*.

Blondel, J.-F. *Course d'architecture, I.* (1771–1774). Paris, France: Jombert (6 volumes completed by Pierre Patte in 1777).

Boffrand, G. (1745). *Livre d'architecture*. Paris, France: Cavelier père.

Boullée, E.-L. (ca. 1790). *Architecture. Essai sur l'art*. Manuscript.

Campanella, T. (1623). *Civitas Solis*. Frankfurt, Germany: Emmel-Tambach.

Didier, B. (1986). *Ledoux écrivain*, in *Soufflot et l'architecture des Lumières* [Actes de Colloque du C.N.R.S. organisé par l'Institut d'Histoire de l'Art de l'Université de Lyon II, 18–22 juin 1980], pp. 252–259. Paris, France: ENSBA.

Gallet, M. (1980). *Ledoux 1736–1806*. Paris, France: Picard.

Girard, R. (1972). *La violence et le sacré*. Paris, France: Grasset.

Honour, H. (1968). *Neo-classicism*. Harmondsworth, England: Penguin Books Ltd.

Kaufmann, E. (1933). *Von Ledoux bis Le Corbusier. Ursprung und Entwicklung der autonomen Architektur*. Wien, Austria and Leipzig, Germany: Passer.

Kaufmann, E. (1952). *Three Revolutionary Architects. Boullée, Ledoux, Lequeu* (New Series—Vol. 42, Part 3). Philadelphia, USA: Transactions of the American Philosophical Society.

Kaufmann, E. (1955). *Architecture in the Age of Reason. Baroque and Post-Baroque in England, Italy, and France*. Cambridge-Massachussets, USA: Harvard University Press.

Le Camus de Mézières, N. (1780). *Le Génie de l'architecture, ou l'analogie de cet art avec nos sensations*. Paris, France: Chez l'auteur et Benoit Morin.

Ledoux, C.-N. (1804). *L'architecture considérée sous le rapport de l'art, des moeurs et de la législation*. Paris, France: Perroneau.

Moore, T. (1516). *Utopia*. Habsburg, Netherlands: Arte Theodorici Martini.

Mumford, L. (1922). *History of Utopias*. New York, USA: Boni and Liveright.

Ozouf, M. (1966). Architecture et Urbanisme: L'image de la ville chez Claude-Nicolas Ledoux. *Annales, 6*, 1273–1304.

Plato (IV Century BCE). *The Republic*.

Quatremère de Quincy, A. C. (1788). "Caractère", *Encyclopédie méthodique. Architecture, I*, pp. 477–521. Paris, France and Liege, Belgium: Panckoucke & Plomteaux.

Quatremère de Quincy, A. C. (1832). "Caractère", *Dictionnaire historique d'architecture*, pp. 302–308. Paris, France: Le Clère.

Rabreau, D. (2000) *Claude-Nicolas Ledoux (1736–1806). L'architecture et les fastes du temps*. Bordeaux: William Blake & Co.

Santangelo, G. (1998). La rêve pétrifié de C.-N. Ledoux. *Culture, 11* (publication en ligne).

Sedlmayr, H. (1948). *Verlust der Mitte: Die bildende Kunst des 19. und 20. Jahrhunderts als Symptom und Symbol der Zeit*. Salzburg, Germany: Mueller.

Sedlmayr, H. (1955). *Die Revolution der modernen Kunst*. Hamburg, Germany: Rowohlts Deutsche Enzyklopädie.

Vidler, A. (1990). *Claude-Nicolas Ledoux: Architecture and social reform at the end of the ancient régime*. Cambridge-Massachusetts, USA: The MIT Press.

Sustainable Approaches to a Utopian Future

The final part of the book provides researchers with an exploration of a utopian future, without neglecting the importance of following utopian themes back through the history of sustainable development thought, and allows for diversity of future scenarios rather than a single vision for the future utopia.

In the chapter "Yugoslav Utopia of Sustainable City—The Synthesis of Living Environment and Social Order", the author addresses the Yugoslav perception of sustainable structures of future cities taking into account the prerequisites and potentials of shaping the mega-structural architectural volumes that were based on the values of socialist self-management system. Through the works of three Yugoslav atchitects, Richter, Mutnjaković and Delalle, the chapter offers a comprehensive analysis of urban models seen as sustainable and based on an integral organic planning in the Socialist society.

In following chapters, such as "Architecture and Sacred Landscape in Prehispanic Peru: A Comparative Approach from a GIS Perspective" the authors present a comparative multi-temporal study of landscape connections with sacred elements from various pre-Hispanic cultures in Peru, using remote sensing techniques and Geographic Information Systems. In specific, the Author identifies and contrasts influences of visibility, geomorphology, trace and metaphor concerning distinctive landscape characteristics such as mountains, rivers, lakes and deserts. The author has studied the influences of urban and site design and their interaction with their environments. After detailed analysis, the author has suggested a connection between the morphology of architectural designs in the Andean region and the framework of beliefs that have come together to model each location, shaping sacred landscapes.

Religious orders conditioned the urban development outside the Naples walls. Especially in the last decades of the sixteenth century, the new regular religious Orders will often promote radical buildings renovations and the opening of roads and small squares in front of their estates. Through analysis of unpublished documents, the author of the chapter "Old and New Settlement Strategies in a Marginal Area of Viceregal Naples: Benedictines and Jesuits in the Vomero Uphill Road" examines deeply the troubled events that leaded—between 1622 and 1627—to the foundation of a small Benedictine monastery in a farmstead along the Vomero uphill road. This event, totally ignored by Neapolitan architectural historiography, represents instead an extraordinary example of smart adaptation to a marginal site of the western urban expansion.

In the chapter "The Paradigm of the Possible City", the author starts from the thinking of a popular Italian writer, Italo Calvino, who in a lecture given in New York in 1983 denounces the failure of the "futuristic" Utopia. Due to the exhausting race towards innovation, a disharmony has generated towards nature. The chapter highlights the ineffectiveness of utopian projects based on an up-down approach, focused on the only material form of the city built neglecting the intangible values underlying its meaning.

In the chapter "Beyond Functionalism: How the Everyday and the Utopian Meet in Reused Parish Churches", the wide issue of the reuse of ancient building is focused on considering the specific aspects of the parish churches of Flanders. An 'everyday' and a utopian potential is generally recognized in this kind of sacred buildings; this helps to go beyond a too narrow vision of functionalism, acquiring a more sustainable orientation.

Yugoslav Utopia of Sustainable City—the Synthesis of Living Environment and Social Order

Danica Stojiljković

Abstract

Following the breakup with the Soviet model of socialism in 1948, the search for an authentic Yugoslav path offered a novel politico-cultural agenda that was promoted, among other, via new humane architecture and urban forms. In the early 1960s, utopian concepts have been developed all around the world under the influence of brutalism, structuralism, English avant-garde, as well as the Japanese architecture that delivered a large number of utopian projects within the concept of metabolism. Instead of static architectural and urbanistic compositions, these projects were based on the kinematic idea of generative, uniquely structured urban forms that are flexible and adaptable to future development of the city. Yugoslav architects Vjenceslav Richter, Andrija Mutnjaković, and Radovan Delalle accepted the ideas of Western avant-garde and offered unique sustainable urban models that were based on an integral organic planning in the socialist society. This paper addressed the Yugoslav perception of sustainable structures of future cities taking into account the prerequisites and potentials of shaping the mega-structural architectural volumes that were based on the values of socialist self-management system. Aspiring to transform the self-management concept into a real and palpable political function of living space, Richter focused on the problem of synthesis. His central thesis of 'Sinturbanism' was based on systematic approach to urban planning. On the other hand, Mutnjaković's urbanistic model 'Biourbanism' promoted the ideas of naturality and open and spontaneous growth, which he extracted from biological principles. His ideal was for the society to reach dynamic homeostasis as seen in the intact nature. The idea of continual growth was also present in the model 'Urbarchitecture' of Radovan Delalle. He presented urban structure as an aggregate of relations between all elements of that structure within a specific social-economic milieu. This paper starts from the premise that the ideas of sustainable city in utopian projects reflected an original ideological agenda of Yugoslav socialism which was based on the values of international and Marxist humanism. The establishment of humane principles and the overcoming of alienation through the synthesis of living environment was the priority because of the need for harmonic relationships between an individual and the society which is focused on the human as an integral biological and social being.

Keywords

Self-management socialism • Sustainable utopian concept • Synthesis • Integral organic planning • Marxist humanism

1 Introduction

Radical sociopolitical changes that took place in Yugoslavia between 1949 and 1953 induced the development of a unique model of self-management socialism. Attempts to articulate the approach of Marxist humanism and to offer a novel platform for responsible critique of Soviet model of Marxism were important segments of politics and culture of self-management socialism. Yugoslav intellectuals considered Marxism as a renunciation of any form of alienation, exploitation, injustice, or repression against the new socialist human. One of the starting points of Marxist humanism was that the shaping of human environment should correspond with the dignity of human nature and revolutionary and humanistic values of the society. Yugoslav political system represented an avant-garde Marxist concept that tried to establish utopian society, giving the priority to the visions of future over present. This system nourished the development

D. Stojiljković (✉)
Institute for Multidisciplinary Research, University of Belgrade, Belgrade, Serbia
e-mail: danica@imsi.rs

C. Gambardella et al. (eds.), *Advances in Utopian Studies and Sacred Architecture*,
Advances in Science, Technology & Innovation,
https://doi.org/10.1007/978-3-030-50765-7_24

of avant-garde cultural scene, which directed research in architecture and urbanism toward utopian projects of sustainable city (Ristic-Trajkovic and Stojiljkovic 2016). The vision of new socialist city called for creation for the sake of future generations that will be on a higher technical, economical, and social level of development.

In the early 1950s, a tendency to oppose modernistic functionalism and to promote a more formal and semiotic approach in architectural theory and practice, emerged. Structuralistic concepts spread all over the world under the influence of Aldo van Eyck and Team X, which rejected functionalism doctrine in urbanism. Structuralism represents the most important avant-garde movement in the 1950s, and developed as a reaction to CIAM's functionalism. Starting from the early 1950s, Yugoslav politics initiated the implementation of ideas of the main movements of Western architecture, in relation to the aspirations after freedom of thought and expression, and democratization of scientific and cultural practice. The end of 1960s and the early 1970s in Yugoslavia was a period of strong influence of Japanese architecture, which underwent phases of brutalism, metabolism, and structuralism (Kiyonori Kikutake, Noriaki Kurokawa, Arata Isozaki, Sachio Otani, and others), American New Wave (Ieoh Ming Pei, Kevin Roche, Moche Safdie, and others), and English avant-garde (Archigram group, James Stirling, Denys Lasdun, Norman Foster, and others) (Štraus and Jojić 1977). The application of avant-garde attitude, which was an important apparatus of critical thinking in Western societies in the 1950s, had a strong sociopolitical impact in Yugoslavia. Architects criticized functionalist urbanism which did not meet the real needs of socialist society, and promoted creative improvements of the environment which are interrelated with the progress. Within the critique of functionalist modernism during the first half of 1960s, theoretical models of integral space in utopian projects of cities that were developed by Vjenceslav Richter, Andrija Mutnjaković, and Radovan Delalle, stood out. These utopian models shared the premise that the environment represents a system of intertwined functions, and that living space and humans are integrated in interactive processes, which show fully functional correlativeness in achieving sustainable urban living.

2 Yugoslavia's Unique Model of Socialism

2.1 Yugoslav Socialist Avant-Garde

Following the breakup with the Soviet model of socialism in 1948, Yugoslav government searched for the ways to articulate new politico-cultural agenda that required a defined intellectual and avant-garde platform. The critical perception of socialist heritage and the search for path toward the communist future was aimed to change old and develop new face of socialism in Yugoslavia (Perišić 2013). The aspiration for the freedom of thought and expression, free communication and action, academic freedom and autonomy of scientific and cultural institutions led to cultural collaboration with the West and to the introduction of socialist democracy and 'workers' self-management in 1952. An important segment of cultural imagination of socialist system of self-management in Yugoslavia was an attempt to articulate the alternative Marxist 'new left' approach, which was based on free human action and humanistic Marxism. In the early 1950s, radical interpretations of Marxist theory were starting to develop in the cultural atmosphere that was inspired by the official critique of Soviet Union. Yugoslav 'socialistic democracy' focused on the Marxist concept of state and people, according to which the state has to be weakened, whereas the direct democracy of people has to be strengthened. The central idea of the new concept was the Marxist idea of direct democracy that included a gradual replacement of the state with social self-management in the transition period between capitalism and communism.

Political and cultural avant-garde critically revised the reality of socialism: 'Nothing that was built can be sacred to us to a degree that prevents us to surpass and replace it with something that is more progressive, free and humane' (Horvat 1970). At the same time, it was emphasized that Communist Party has to be 'avant-garde,' and that Party members have to be morally and politically apt for the education of others and for directing the society toward communism (Jović 2000). The avant-garde role of Communist Party and the perpetual critique can be found in the program of The VII Congress of the League of Communists of Yugoslavia (Savez komunista Jugoslavije 1958):

> Taking into account that practice is the final judge of the correctness of theory and the main interpreter of ideological developments, Yugoslav communists will continue to learn from their own practice as well as the practice of other socialist movements, and they will correct, with no hesitation, anything that is labeled by practice as obsolete or unsustainable. Therefore, it is in the very spirit of this Program that it will also be surpassed and replaced by new concepts in practice and theory of socialism.

Yugoslav socialism adopted the concepts of rational edification and technological development and promoted the idea that humans can subdue the Nature and shape the society according to rational criteria and esthetic norms (Čalić 2013). Economic recovery, broader intellectual freedom, and international exchange enabled the flourishing of art, culture, and consumerism. As a result of economic development and democratization of art in Yugoslavia during 1950s, experiments in architecture started to develop in different directions. In accordance with the need to establish an authentic Yugoslav political system, Yugoslav architecture showed a tendency to become avant-garde.

2.2 Sustainability and Socialism

The concept of sustainable development, which emphasized complex relations between environment and economy, came into focus in 1970s in relation to the intensified global process of urbanization and industrialization. One of the main characteristics of urban life was the omnipresent need for social and material metabolization of nature that should enable sustainability. Socio-ecological changes promoted values that encompassed cultural and emotional relationship with the environment and reflected the interest in prosperity. The concept of sustainable development contemplated the relations between the urban environment and nature through interconnected aspects in the fields of ecology, economy, politics, and culture. Socialist socioeconomic system, which was founded on the public property of means of production and on the production that is not motivated by profit, approached the ecological issues in a different manner than capitalist societies. In the reference of Yugoslav regime to the original theses of Marxism, attitude toward the nature became one of the key strategic points (Ignjatović 2013). The Marxist theoretical perceptions of ecological issues start with the Marxist concept of the human as a natural and social being that is interconnected with the nature. According to this concept, humanity will be more estranged from nature as more brutal exploitation of nature is applied. In the ideology of growth and progress, the legitimacy of socialist system was provided by the idea that socialism delivered a new quality of life by abrogating the exploitation between humans. However, the need to stop the exploitation of nature by humans emerged (Radojković 1989). According to Erich Fromm, Yugoslav self-management system truly reflected ideas of individual actions in coherence with the Marxist concept of humanism as completed naturalism (Marković 1986). In Yugoslav socialism, Fromm recognized an attempt to perform industrialization in a 'humane manner' by 'combining in practice the process of industrialization with social system that prevented bureaucracy' (Petranović 1988).

The system of environmental protection that has been developed in Yugoslavia was original and unique. It was based on the concept of socialist society as a structure which has production that is not driven by profit and which recognizes humans as its greatest value. The attitude to ecological problems included the protection of nature from degradation as well as the improvement of the environment. Such approach ought to show that self-management society is capable to prevent the degradation of nature and to introduce 'more programming, rationality, and imagination into the management of society and human relations' (Đorđević 1981). The reconciliation of society and nature represented the goal of socialist society. Edvard Kardelj pointed out at the Statutory meeting of Yugoslav Council for Protection and Improvement of Human Environment:

> It is necessary to improve the awareness of the working human as self-manager that his attitude to nature represent an intrinsic component of self-managing function, which involves managing in the broadest sense, i.e. it does not only involve the management of means that are in his own hands, but also of the entire environment that is built by him and that is provided by nature. The whole environment will be, through inter-dependence, the source of continual usefulness and it will ensure a more fulfilling and cultural life (Čiček 1975).

The humane approach to socialism with the human as the main 'anchor point' established the goal of new quality of life. Starting from the point that the protection of environment represents an inherent element of human personality, and 'an existential human need and a component of culture, tradition, and individual ethical and social code' (Kambovski 1990), new constitutional and legal concepts were developed. The Constitution of the Socialist Federative Republic of Yugoslavia from 1974 represented the first constitution in the world that established the individual right of humans to healthy living environment. It pointed out that socialist society: 'warrants conditions and improvements of natural and other qualities of human environment that are of interest for healthy, secure and productive life and work of present and future generations' (Marković 1986).

3 Utopia of City as a Living Organism

An aspiration to comprehend local social and cultural phenomena within a universal context developed after the World War II with the acceptation of structuralist theory. In line with this, social and political relevance of structuralism can be viewed from the point of universal values and internationalism. The aspiration to build universal society emerged as the need to create the city of future that is responsive to social and individual requirements. Structuralism examined relationships between the elements and it offered a radical criticism of functionalism and formal approach. The criticism was initiated by Aldo van Eyck, Team X member, who pointed out at CIAM Congress in Dubrovnik in 1956: 'There is a tendency to change from small cellular units of cluster which are used additively, to the formation of a major structural element, increasing the scale in order to make it more comprehensible' (Lüchinger 1981). Team X members endorsed an idea that an urban structure ought to show a high level of originality and identity, which is achieved by an individual focus to city functions. Static architecture and urbanistic compositions were replaced by the concept of generative structure and by urban forms that are shaped in accordance with unique characteristics and specific associations of human relationships. The shapes are open to change and growth and respect local nature and climate, which allow a temporally flexible planning and constant changes over time (fourth dimension). In the early 1960s, Dutch journal *Forum* stressed out the importance of organizing city as a living organism:

> The modern city is anatomically split up into a host of lifeless specialties and can only become a living organism again—one in which every nuance makes its contribution to the whole—by a sensibility towards everything that shapes life as a totality. We have to find a new definition for the points of reference themselves in order to be able to take hold of life in pieces of manageable size; but we are thereby enabled at the same time to give form to the real structure of life in the community in all its nuances (Lüchinger 1981).

In the 1960s, a large number of utopian projects were developed in Japan that were based on the ideas of metabolism, which promoted the elements of growth, change, flexibility, and reciprocal displaceability of architectural elements, as well as the formation of group forms or clusters in mega-structural projects. As a rule, such mega-structures required specific constructive solutions, so many of them represented visionary projects. The most widely known metabolist in Japanese architecture—Kenzo Tange had a strong impact on Yugoslav architectural scene through his winning project for the reconstruction of the city of Skoplje in 1965. In this project, he presented a metabolic theoretical concept that introduced metaphorical and symbolic elements and that showed a return to traditional values of the city (Fig. 1). Metabolism theory and practice relied on the structure of the living organism with biological connotations and metaphors, creating a new dictionary of urban architecture that was opposed to functionalism and its mechanical metaphors. By applying the theory of metabolic cycles, the city becomes a dynamic system that grows and develops through the elimination of obsolete elements and regeneration of new components (Ioannidou 2011).

Fig. 1 Project for the reconstruction of the city of Skoplje presented by Kenzo Tange in 1965 (Photo was obtained from the Archive of Yugoslavia 112141kmb-22508_17)

Yugoslav structuralism developed an integral theory of space and the synthesis of organic construction of the environment, and revised the measure of shape in relation of the processes of living. This corresponded to the criticism of Team X, which promoted the re-establishment of organic living environment. At the same time, Yugoslav Marxists considered that the cooperation between architecture, urbanism, art, humanistic sciences and revolutionary–humanistic practice, is necessary for true synthesis. For example, Zvonimir Radić, architect and theoretician of design, pointed out the importance of synthesis as an essential aspect of social environment: 'The future of any generation is inevitably in synthesis, in the destiny to change the matter, in the power to integrate, decompose, and organize into one world' (Radić 1963). Pertinent to this, architect Božidar Janković noted that form is not frivolous and that requires an integral approach to elements (Brkić 1992). Janković pointed out that the creative process of construction of collective shapes that intertwines all relations of life brings in itself complexity, diversity, and specificity of particular space. Within the criticism of functionalist modernism of the first half of 1960s in Yugoslav architecture, the most important utopian models were delivered by Vjenceslav Richter, Andrija Mutnjaković, and Radovan Delalle, which proposed organic and integrative planning of the environment. Their mega-structural compositions allowed development, integration, evolution, and adaptation to typography and socioeconomic reality. These integrative theories were aimed at establishing humane principles in the framework of synthesis of living environment. In line with this, the formation of the city as a living organism was focused on needs and capacities of socialist humans to participate in the changes (interventions) of urban space. This approach required the formation of an adaptable environment according to behavior, experience, and tradition. The self-regulation of construction within such environment would be based on the needs of the development of self-managing society.

3.1 Sinturbanism by Vjenceslav Richter

Starting from the transformation of structures of society, Vjenselav Richter developed in 1964 the concept of organization and shaping of new poly-functional urban environment of socialist city through the project of synthesis urbanism—Sinturbanism. Richter criticized Le Corbusier's urbanism which was analytical and had distinctly separated functions of the city and tried to develop an urbanistic concept that could enable rational use of time via synthesis of functions of urban living. The basis of Sinturbanism city lays in the integration of the following interests: (a) economic interests—financial, technical, and productive; (b) organizational interests—economy of time, transport and connections, integration of living functions; and (c) ideological interests—collective conscience, the management of complex functions, positivistic philosophic approach to life, and artistic expression (Richter 1964).

In the utopian concept of ambient structure of cities of the near future, Richter contemplated the requirements and potentials of construction of mega-structural architectural volumes that should enable the fulfillment of objectives of socialist society. Richter's million city was composed of one hundred sinturbanistic units. Each unit—ziggurat ('cikurat') represented a separate complex that autonomously provided an integration of the main functions of living (production, housing, recreation) for 10,000 people (Fig. 2). Richter designed the unit of Sinturbanism—ziggurat, as a small town with all the basic functions, which should promote the sense of collectiveness and responsibility. The idea was to convert the vision of self-management into real and palpable political function of living space. All the objects in the city were designed according to the equal structural and functional qualities in order to follow the principles of socialist society. Ziggurat was designed as a megastructure of pyramidal shape with dimensions of 300 × 300 × 150 meters. It is important to note that visionary projects with pyramidal design and terraces, which required specific constructive solutions, were relatively frequent in the 1960s. In contrast to such projects, Richter proposed a utopian model of sustainable city which was in accord with self-management model of socialism. The life in ziggurat is not static. The core is elastic with multiple possibilities for the remodulation of space that are aimed to support 'pulsations of ever-changing life' (Horvat-Pintarić 1970). The hall ('asambleja') for 6000 people represents the instrument of administration, where all decisions are made by plebiscite/referendum. The production was also placed inside the ziggurat. The explanation was that zonal industry is not efficient in creating a healthy urban atmosphere, and that other measures are needed, such as 'waste filtration.' A working day in sinturbanistic city was named by Richter as 'sintak,' referring to the synthesis of thinking and kinetics. Sintak was a form of organization of life that was based on the synchronicity and complexity of actions which were possible only in socialism. Richter noted that capitalism was burdened by individualism and analytical thinking, which promoted the development of individual components that may be a good solution per se but do not deliver a coherent and interconnected whole (Richter 1964). Richter contemplated that the concept of sinturbanistic city would result in the synthesis of life and compactness, since the ziggurat system should impact all manifestations of life and spread integrative concept on production, system of thinking, behavior, and organization of time.

3.2 The Concept of Biourbanism by Andrija Mutnjaković

Andrija Mutnjaković found the theoretical starting point for sustainable socialist city in biological principles. Mutnjaković referred to visionary urbanistic research of Yona Friedman ('Ville spatiale'), Paul Maymont ('Ville flottante'), and Geoffrey Jellicoe ('Motopia') that delivered 'post-Corbusier' upgrades of design of contemporary ambient. For him, the city is a complex organism of human life, where specific functions, such as house, street, park, cannot be viewed separately; there is only one function—the city (Mutnjaković 1964). Mutnjaković tried to design a city that functions as an 'organism of inter-pervaded spaces' with free and organic composition that is incorporated into the landscape. He was inspired by biological homeostasis and harmony and compared the society with biological ecosystems (Kritovac 1975). An ideal was to establish a dynamic balance as in intact nature.

Promoting the ideas of naturalness and open and spontaneous growth, Mutnjaković defined the urbanistic model of 'Biourbanism' (Mutnjaković 1982). Mutnjaković applied the concept of Biourbanism in the urbanistic design project 'Biograd for Belgrade' for residential complex for 30,000 people in New Belgrade in 1965 (Fig. 3). The concept of Biourbanism was manifested in formally concave lines, symmetrical arrangements, and terrace organization along all three axes which resulted in opened panoramic view toward green areas of all residential units. Mutnjaković concluded that the quality could be achieved only by well-developed concept. Mutnjaković further proposed that microarea center ('mikrorajon') should be in the visual and organizational focus of the residential space and that all apartments should be oriented toward the center. Microarea centers are further organized into macroarea centers ('makrorajon'). The principle of addition and growth builds a community of microarea centers that retain their characteristics. The city has number of macroareas that are organized in a circular way and have 'rajon' center in the middle. The main theses of Biourbanism are growth, naturalness, openness, richness in individuality, spontaneity, freedom which is related to the concept of freely developing and spontaneously growing organic shape. Mutnjaković's views encompassed a tendency toward a 'total space' and integral approach in the design of living environment.

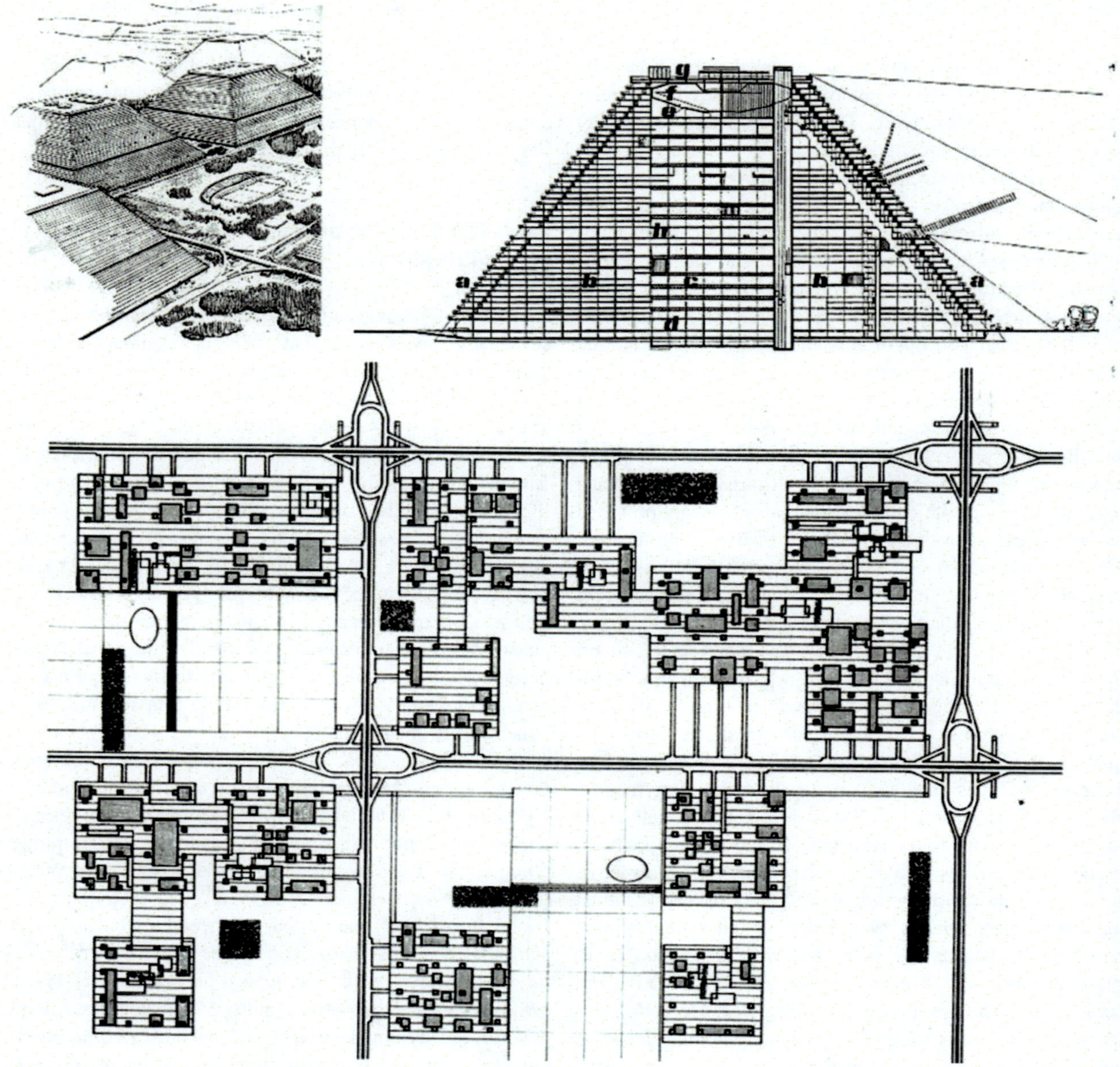

Fig. 2 Ziggurat—axonometry, cross section, and urban planning (Richter 1964)

The possibility of personal involvement in creating a personalized residence according to the wishes, hobbies, needs, and taste of the owner had an important place in the concepts of Andrija Mutnjaković. Mutnjaković supported the idea that the design should be left to the owner in order to make the community a living organism. The involvement of residents in the building of their own homes is aimed at enabling human biological instinct to have an impact. He noted that the charm of Mediterranean towns is based on different personal demands that were spontaneously controlled by collective concepts. Pertinent to this, Mutnjaković questioned the concept that a 'stranger' should design the space that will be used by a family that he/she does not know and which then has to live in that space without individualization and relation with the object. He concluded that it is necessary to have a collective concept in order to reach a sufficient quality of the urban environment. Mutnjaković built the idea of self-creation on progressive architectural, psychological, social, and esthetic theses, trying to contribute to solving the problems of socioeconomic situation in Yugoslavia. The commitment of socialist society to provide each citizen with a proper living space remained just an

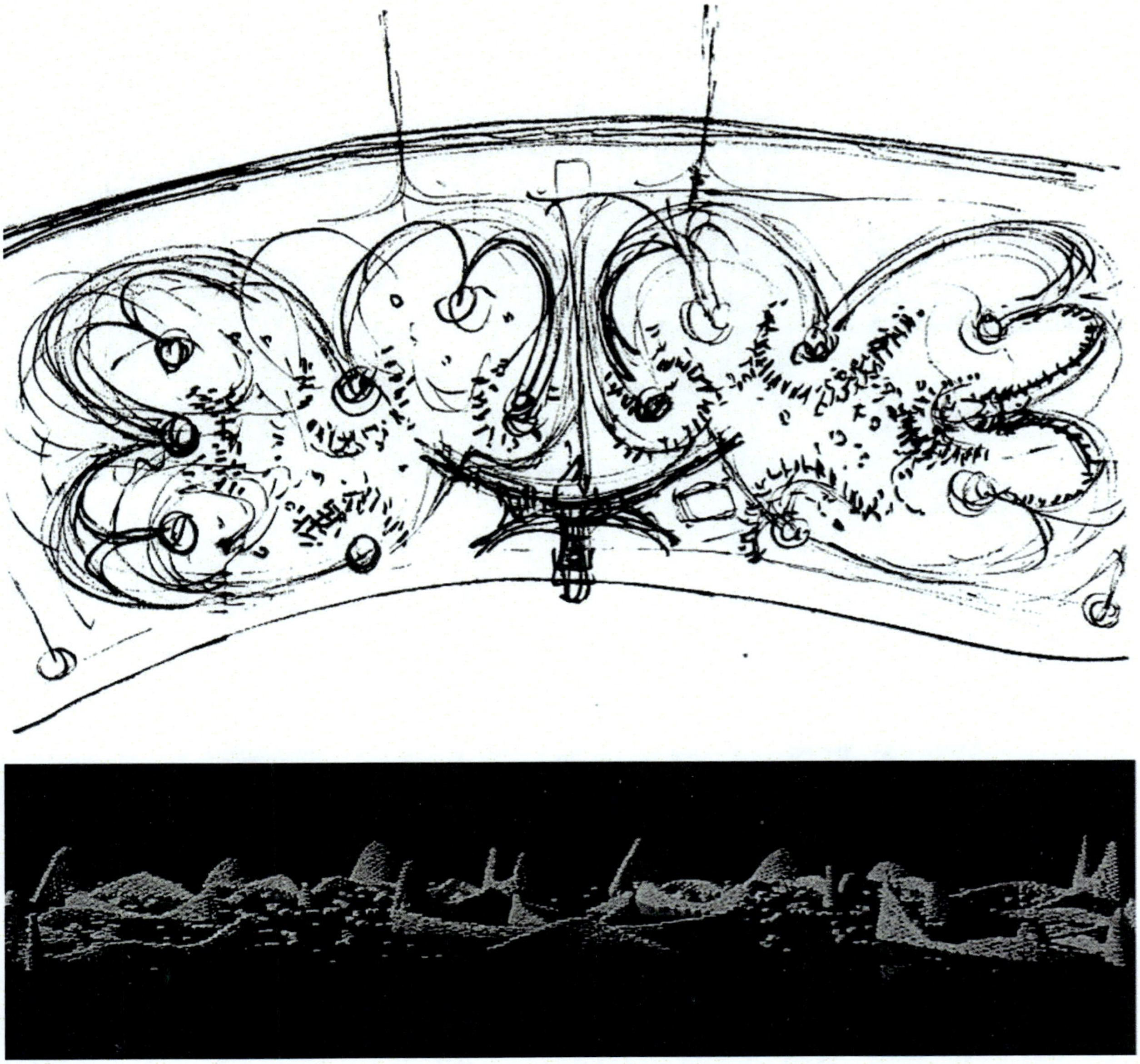

Fig. 3 Competition for urbanistic design of residential complex for 30,000 people on the bank of Sava River in New Belgrade—urban planning and axonometry (Mutnjaković 1982)

aspiration in relation to the limited finances. In relation to this, Mutnjaković proposed that the involvement of public funding should be decreased and that individual contribution in financing construction of residences should be increased.

3.3 Urbarchitecture Concept by Radovan Delalle

Radovan Delalle first proposed Urbarchitecture, a utopian concept of organic urban development, in 1964, during his work on experimental urbanistic project of Marindvor in Sarajevo. The concept was further developed in competition projects, studies, and exhibitions. The starting point of Urbarchitecture was the relationship between the human and the physical structure, which reflected in urbanism and encompassed specific socioeconomic relations (Fig. 4). Urbarchitecture analysis investigated interactions and dependences between: (a) the elements of physical structure (topography of terrain, architectural topography, and urban typography); (b) humans and physical structure (phenomenological approach, ecological psychology, cognitive

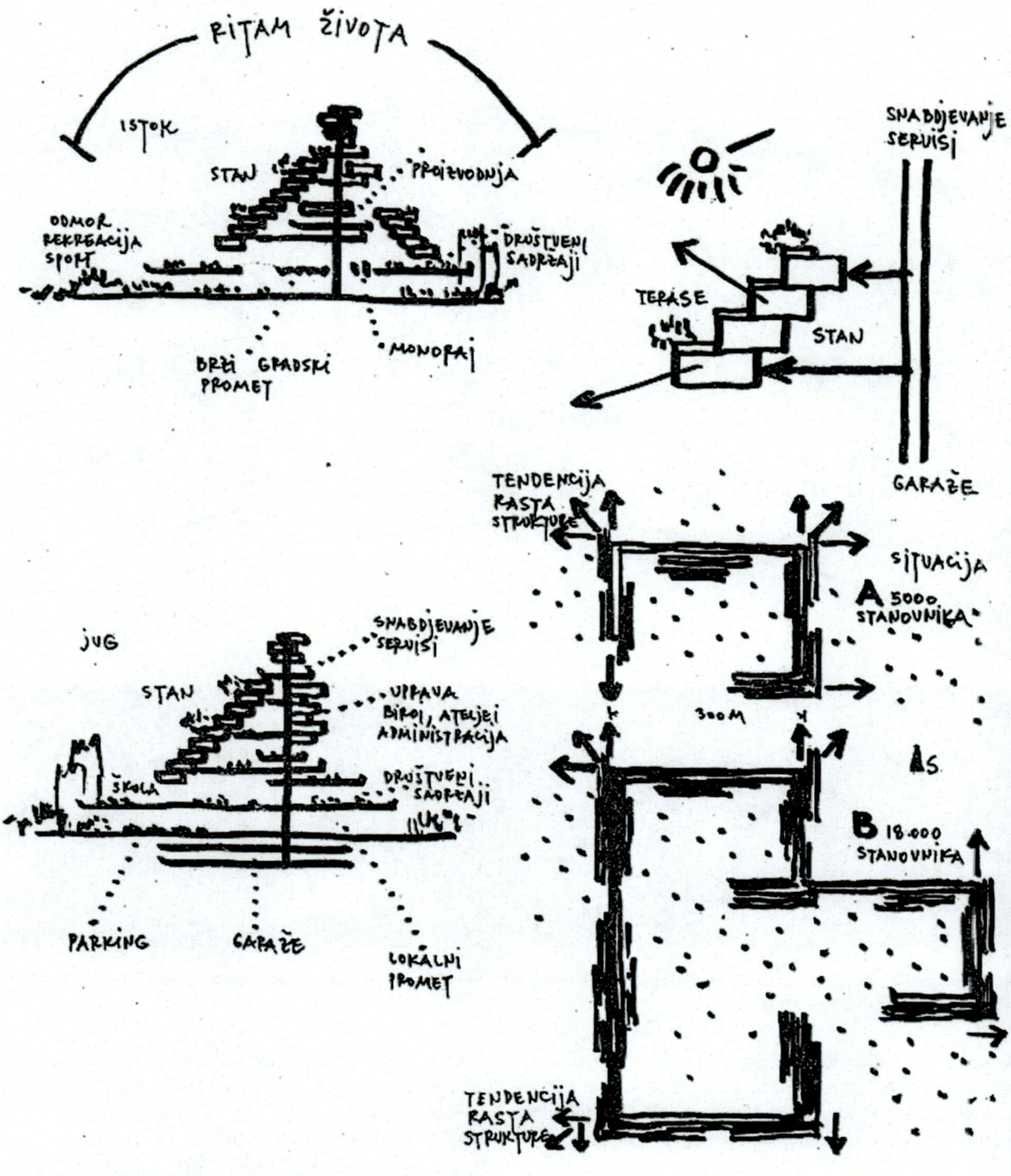

Fig. 4 Analysis of Urbarchitecture (Delalle 1970)

maps, mental images, ambient and symbolic values of urban space); and (c) social relations (production, interpersonal relationships, space use, urban sociology, urban practice, etc.) (Delalle 1988). An important starting point of his method was the possibility of continual and evolutionary growth and development of urban structure:

> If we take urban environment as an object (urban center), we have to reveal all elements that are structured in such spatial entity through the examination of their inter-relationships in perpetual change and development. It is important to point out that urban centers do not represent just a small part of future city. They represent urban structure which is an organic, complex, evolutionary and unique entity in all phases of development. Each element of urban center is defined by inter-relationships with elements of the surroundings (Delalle 1973).

The main goal of Delalle's Urbarchitecture concept was sociopolitical engagement within the self-managed socialist community in order to stop the lotting of urban space and to give back the human values to the city. Urbarchitecture was aimed at overcoming the alienation, lostness, and an incapability to act in the 'rhythm of society.' He pointed out that human experience urban space through active and dialectic relations, in continuous change, movement, and action. Urban space is not only defined by form. It contains hidden dimensions which were described by him as 'the possibility of action in space, current event, the possibility of adventure, ambiental experience, touch, orientation, light, temperature, color, communicability of architectural shapes, visual-artistic experience, smell, mental image of space, time, poetics of space' (Delalle 1973). He advocated perpetual alterations of dynamic rhythm that change the conditions of spatial relations in order to promote new discoveries and increased quality. He reflected on the claim of Claude Lévi-Strauss that the structure represents a content within logical setup that is comprehended as an attribute of reality. In relation to this, Delalle contemplated the concept of totality, including questions of connections between structure and content, content as continuous movement and change of singularities within the whole (Delalle 1970).

Delalle proposed that a microspace, an imaginary minimal space that is needed for each human to achieve harmony with the surroundings, has to be present within the spatial structure. The continual process of growth represented the main characteristic of the new urban structure, which was defined by the entirety of relationships between all elements (Delalle 1973). In this concept, he created a new urban life in which change and growth of urban space follows the changes in the society. Function, structure, and form are not analyzed separately; they are components of an integrative approach (Fig. 5). On the Yugoslav competition for the 'Urban design of Mišeluk area in Novi Sad' (1979–1980), Delalle further developed the concept of socio-spatial organization with maximal openness, flexibility, and polyvalence of space. His project was designed to enable the development of individual, family, social groups, and

Fig. 5 Urbarchitecture project of experimental residential community (Delalle 1973)

society needs. The principle of pluralism of territorial needs and practice of different social groups was based on specific form of self-management (Delalle 1988). Urbarchitecture was conceived to become an integral part of life of society. The flexibility and polyvalence of space are aimed at enabling a permanent orientation toward changes in intrinsic urban structure and new ways of life of new generations.

4 Conclusion

The realization of a new quality of life was an integral part of Yugoslav self-management socialism, avant-garde culture, and critical attitude. The politics of Yugoslav society used a specific cultural concept that embraced progressive ideas of international and Marxist humanism to promote the research in architecture and urbanism in the context of humanistic values and the synthesis of living space. The change in the functionalism practice was initiated by revalorization of previous experiences and by the 'infiltration' of humanistic approach and environmental sciences into Yugoslav urbanistic discourse. The avant-garde in architecture and urbanism that spread all over the world in the second part of the twentieth century was mainly related to the design that promoted technology development, and to utopian projects. Yugoslav architects accepted the ideas of Western avant-garde and offered unique sustainable urban models that were based on integral organic planning in the socialist society. Utopian concepts were asserted by the carriers of radical social and cultural thought with the aim to create a better society.

The presented case studies illustrate different theoretical models of integrated space and synthesis that were developed as critique of functionalist architecture in the first half of the 1960s. Aspiring to transform the self-management concept into a real and palpable political function of living space, Richter focused on the problem of synthesis. His central thesis of 'Sinturbanism' was based on systematic approach to urban planning. On the other hand, Mutnjaković's urbanistic model 'Biourbanism' promoted the ideas of neutrality and open and spontaneous growth, which he extracted from biological principles. His ideal was for the society to reach dynamic homeostasis as seen in the intact nature. The idea of continual growth was also present in the model 'Urbarchitecture' of Radovan Delalle. He presented urban structure as an aggregate of relations between all elements of that structure within a specific socioeconomic milieu. The presented utopian models showed common starting principles. Conceptual consciousness of space was built through realization of socio-urbanistic synthesis in the context of self-regulated development that followed the needs of self-management society. The establishment of humanistic principles in the context of synthesis of living environment was the priority of socialist system and politics of avant-garde Marxist humanism, which led to the concept of the human as an integrated biological and social being. Instead of static architectural and urbanistic compositions, utopian projects promoted the concept of generative structure of urban forms which showed unique characteristics, specific associations of human relationships, openness to change and growth, appreciation of local nature and climate, and flexibility in planning over time. Modular space facilitated different routes of further development of urban community. The critique of utopian projects, which emerged in the early 1970s, showed that the proposed structures suffered from one-sided formal and technical viewpoints. Nevertheless, utopian projects of sustainable city increased the awareness and interest of society for environmental improvements, which became a predominant theme in Yugoslav socialist society in the late 1960s.

Acknowledgements This work was supported by the Ministry of Education, Science and Technological Development of the Republic of Serbia under Grants: OI179048: 'Theory and practice of science in society: multidisciplinary, educational and intergenerational perspectives'.

References

Brkić, A. (1992). *Znakovi u kamenu: srpska moderna arhitektura: 1930–1980*. Beograd: Savez arhitekata Srbije.

Čalić, M. Ž. (2013). *Istorija Jugoslavije u 20. veku*. Beograd: Clio.

Čiček, J. (1975). Samoupravne interesne zajednice za zaštitu čovjekove okoline. *Čovjek i prostor, 269*, 21–22.

Delalle, R. (1970). Urbarhitektura —kontinuitet života. *Arhitektura, 106*, 8–12.

Delalle, R. (1973). Urbarhitektura put prema humanijem oblikovanju ljudske okoline. *Arhitektura, 146–147*, 29–34.

Delalle, R. (1988). *Traganje za identitetom grada*. Rijeka: Izdavački centar Rijeka.

Đorđević, J. (1981). Ekologija, čovekova sredina i uloga nauke u savremenom društvu. In *Čovek, društvo, prirodna sredina*, 52–60. Beograd: SANU.

Horvat, B. (1970). *Privredni sistem i ekonomska politika Jugoslavije: problemi, teorije, ostvarenja, propust*. Beograd: Institut ekonomskih nauka.

Horvat-Pintarić, V. (1970). *Vjenceslav Richter*. Zagreb: Grafički zavod Hrvatske.

Ignjatović, A. (2013). Usavršena priroda-Beogradski sajam (1953–1957). *Zbornik Matice srpske za likovne umetnosti, 41*, 181–203.

Ioannidou, E. (2011). Structuralism and metabolism. In Tomas Valena, Tom Avermaete, & Georg Vrachliotis (Eds.), *Structuralism reloaded—Rule-based design in architecture and urbanism*, 222–227. Stuttgart: Edition A. Menges.

Jovic, D. (2000). *The breakdown of elite ideological consensus: The prelude to the disintegration of Yugoslavia (1974–1990)*. PhD thesis, London School of Economics and Political Science (United Kingdom).

Savez Komunista Jugoslavije. (1958). *Program Saveza komunista Jugoslavije usvojen na Sedmom kongresu saveza komunista Jugoslavije*, 22–26. Beograd: Kultura.

Kambovski, V. (1990). *Ustavno-pravni aspekti i pravci razvoja pravnog sistema u zaštiti životne sredine*. Beograd: Bezbednost i društvena samozaštita.

Kritovac, F. (1975). Uz'angažiranu arhitekturu'Andrije Mutnjakovića. *Čovjek i prostor, 267*, 10–13.

Lüchinger, A. (1981). *Structuralism in architecture and urban planning*. Stuttgart: Karl Kramer.

Marković, D. (1986). *Socijalna ekologija*. Beograd: Zavod za udžbenike i nastavna sredstva.

Mutnjakovic, A. (1964). Grad našeg radosnog sutra. *Arhitektura urbanizam, 25*, 22–23, 58.

Mutnjaković, A. (1982). *Biourbanizam*. Rijeka: Izdavački Centar Rijeka.

Perišić, M. (2013). *Diplomatija i kultura: Jugoslavija: prelomna 1950. Jedno istorijsko iskustvo*. Beograd: Cicero Print.

Petranović, B. (1988). *Istorija Jugoslavije: 1918–1988. Book 3, Socijalistička Jugoslavija: 1945–1988*. Beograd: Nolit.

Radić, Z. (1963). Industrijsko oblikovanje i kadrovi. In Vjenceslav Richter (Ed.), *Industrijsko oblikovanje*, 13–30. Zagreb: Centar za oblikovanje rukovodni kadrova u industriji.

Radojković, M. (1989). Ekologija i socijalizam. *In Ekologija i kriza—Zbornik sa jugoslovenskog savetovanja o ekologiji "Ekologija—ekonomija—politika—društvo"*, 85–96. Novi Sad: Polja.

Richter, V. (1964). *Sinturbanizam*. Zagreb: Mladost.

Ristić-Trajković, J., & Stojiljković, D. (2016). Sustainability and socialism: Socio-ecological ideas in the urbanization of New Belgrade. *Facta Universitatis-Series: Architecture and Civil Engineering, 14*(3), 343–353.

Štraus, I. (1977). *Nova bosanskohercegovačka arhitektura: 1945–1975*. Sarajevo: Svjetlost.

Architecture and Sacred Landscape in Pre-Hispanic Peru: A Comparative Approach from a GIS Perspective

Carlos Zeballos-Velarde and Gonzalo Ríos Vizcarra

Abstract

One of the most relevant characteristics of the "landscape-friendly societies" is their ability to recognize aesthetic attributes to portions of the territory that exceed their functional purpose. In ancient Peru, there were many ways of relating architecture (religious, civil and production) with a broader "borrowed landscape".This article presents a comparative multi-temporal analysis of the landscape relationships with sacred elements from different pre-Hispanic cultures in Peru, using remote sensing tools and geographic information systems. This approach allows to distinguish and compare factors of visibility, insertion in geomorphology, trace and metaphor, in relation to distinctive elements of the landscape such as mountains, rivers, lakes and deserts. Five representative cases have been selected for territorial studies: Moche, Wari, Pisac, Ollantaytambo and Sacsayhuaman, analyzing factors of urban and site composition and their relationship with their landscapes. Also, for an architectural level, a cloud-point model of Sacsayhuaman has been used for the study of solar alignment. Finally, a correlation is proposed between the morphology of architectural developments in the Andean territory with the system of beliefs that came together to model each place, forming sacred landscapes.

Keywords

Sacred landscape • Ancient Peru • GIS

C. Zeballos-Velarde (✉)
Universidad Católica San Pablo, Arequipa, Peru
e-mail: crzeballos@ucsp.edu.pe

G. R. Vizcarra
Universidad Católica Santa María de Arequipa, Arequipa, Peru

1 The Importance of the Sacred Landscape

1.1 Introduction

The relationship of ancient Peruvian civilizations with their surrounding landscape has been largely debated from the historical, anthropological and archeological perspectives. Some studies on the Peruvian history of architecture have also pointed out the importance of the early settlements as well as the alignment of buildings with stars on particular dates.

However, there is still a knowledge gap in terms of the understanding of these constructions as part of sacred territories. As in many cases in general, architecture is studied as an isolated object, deprived of any connection to surrounding elements.

As Nobel laureate Mario Vargas-Llosa points out "Very few human beings are capable of accepting the idea of the "existentialist absurdity", that we are "thrown" here in the world by an incomprehensible chance, by a stellar accident, that our lives are mere casualties devoid of order or harmony" (2012). It is for that reason that mankind from all times and places sought to articulate those signs that he perceived as divine indications of a supreme order in which he could sustain the meaning of an uncertain life. From that need, humans began to read the territory and to inhabit it in expectation of meeting the superlative.

The primordial dwelling, as recognized by the German philosopher Heidegger, is "waiting for the signs of the advent of the divine without ignoring those of its absence" (1951). Humans sought to underline and articulate these signs of the presence of the sacred to an entirely inclusive narrative, which happened in a tangible physical environment, which he delimited and granted the category of sacred landscape.

Therefore, this paper aims to understand ways in which ancient Peruvians used to structure their lives, buildings and cities involving their surrounding cosmovision (both

C. Gambardella et al. (eds.), *Advances in Utopian Studies and Sacred Architecture*,
Advances in Science, Technology & Innovation,
https://doi.org/10.1007/978-3-030-50765-7_25

territorial and stellar) in order to create a sacred landscape. This study puts in evidence these relationships by means of different uses of GIS software, analyzing five historic monuments from ancient Peru, which belong to different spatiotemporal scenarios, different cultures and served different purposes, but share a common interest for embedding themselves in a sacred landscape.

1.2 The Construction of the Notion of Sacred Landscapes in Peru

While the "holy" is associated with religious elements and has to be designated by law (either divine or human), the "sacred" it is such by nature. The sacred is separated from daily life, and it is a factor of social cohesion (Spinetto 2018).

A landscape is a human construction that emerges especially in environments where the relationships between the territorial elements are evident and induce a perception that exceeds the merely functional. A landscape cannot be reduced to its physical characteristics, but rather to the aesthetic relationships that a human group recognizes in it.

For the sacred to take place in the landscape, the observer must go beyond the perceptual instance (which implies distancing himself from the perceived object) in order to access a contemplative status that is to say to vibrate in an intimate and internal conjunction.

In the case of pre-Hispanic Peru, as in the vast majority of ancestral cultures, the aesthetic perception is inextricably linked to the sacred dimension. As mentioned by Jean Pierre Crousse (2016: 22), "The wide landscape diversity is the product of a multiplicity of very diverse geographical events that occur in a small extension of the territory, inserted in a tropical latitude and with the influence of the South Pacific anticyclone".

1.3 The Recurring Elements in the Peruvian Sacred Landscape

The Andean space, characterized at the same time by its variety and inhospitableness, was inserted within a "sacred geometry, for it had been ordained by the gods in the remote time of the Origin and its maintenance depended on rituals" (Regalado 2000). Landscape features such as mountains, big rocks, rivers, lakes or springs were thought to "possess supernatural powers or constitute the materialization of supernatural beings" (Reinhard 2002). As Makowski points out, "unlike in the Greek Hellenistic, Roman, and Christian traditions, it is the landscape, transformed or not, rather than buildings or iconography, that comprises the sacred and ritual spaces of the Andes" (2015).

The following elements stand out as remarkable and recurring elements in several pre-Incan and Incan landscapes.

1.3.1 The Apus, the Waka and the Llacta

The very presence of the Andes that spread from north to south along the Peruvian territory is a constant and an almost indispensable element in the configuration of the landscape system of the Andean cultures. Even from the most archaic societies, the ancient Peruvians have attributed magic-religious powers to the predominant features of the landscape, calling them "*apus*". The *apu* or Wamani "is the mountain or tutelary deity that harmonizes life in the Andean meta-visual landscape, both on a territorial scale and on a micro-regional scale" (Crousse 2016, 84). Prominent mountains or *apus* are understood as natural landmarks of sacredness where a certain deity materializes. Apus "are revered as earthly spirits that protect, but may also punish, their human constituents, and were often linked to distant ancestors" (Williams and Nash 2006), and they also acquire the value of distinctive symbols of an ethnic group, further strengthening its territorial roots considered the most important local deities.

Despite living amidst a rough topography, the inhabitant of the Andes did not renounce his desire to model the territory and invested enormous amounts of energy and resources to adapt natural promontories or build their own artificial hills which he called "wakas". Unlike the tutelary presence of the *apu*, the waka coexists in a much closer way with the community and is often the site of ritual ceremonies. It was used to elevate social status, to separate itself from the earth, to stand out visually (Williams 1981).

The llactas were human settlements, concentrations of buildings arranged around roads, terraces and steep terrains. While there was a basic pattern to organize the buildings called "cancha" or court, this was modified to adapt to different types of geomorphology, and the location of the llactas made the courts and platforms adapt their layout, which was based on a sacred geometry (Guzmán 2013).

1.3.2 The Traces

One of the most distinctive elements and of greater validity in the ordering of the territory is the trace that originates the accommodation of the artificial over the natural. Pallasmaa says that the echo of a city "depends on the layout and scale of its streets and its preponderant architectural materials" (2014). Concordantly, we can assert that in these primordial axes remain the intentions of subjection of the imagined with the material.

It is important to note, however, that the traces in the Andean world differ from the Euclidean visions of the West, as ancient Peruvians considered the natural rock fractures to obtain the different geometries of the blocks used in the buildings. The use of blocks with fractal geometry carefully located to resemble the shape of the prominent mountains suggests the use of a "petrometry" (geometry of the stone) that was used in different scales (Menegat and Porto 2007).

1.4 The Recurrent Relations for the Conformation of the Peruvian Sacred Landscape System

While the near presence of natural elements in conjunction with those created by men can form unitary perceptual fields, this does not guarantee the constitution of a landscape. In order to do this, it is necessary to highlight relations through territorial interventions or by the architecture itself, which fulfills the role of articulating both worlds. Pallasmaa reminds us that "architecture clearly articulates primary human experiences of "being-in-the-world", such as gravity and mass, horizontality and verticality, sky and earth, center and periphery, nature and culture, landscape and artifice, individuality and collectivity, the past and the present" (2014).

Furthermore, for these perceptual portions of territory to be considered as sacred, it is necessary for the system to acquire symbolic connotations with the possibility of being transmitted and commemorated through mythical and ritual narratives. Seldom does the sacred reveal itself in a conclusive way, in most cases, it insinuates itself, necessitating the complicity of the believers who, clothed with their faith, assimilate and accommodate it as something concrete in their life. The landscape in pre-Hispanic Peru was the symbolic vision of the universe through which the gods communicated.

Among these relationships that constitute the Andean sacred landscape, the authors have determined that the most recurrent are

1.4.1 Orientations

The sunrise from the east is the first reference that early civilizations take to isolate themselves from chaos and avoid their own dispersion. The detection of the cardinal points will be a concrete way to ally with the cosmic laws, understood as divine designs. From time immemorial, humans arranged their creations from a marked intention of conciliation, and their sacred architecture is based on this guideline.

In the case of the Peruvian landscape, we propose to demonstrate the importance of this relationship, which is seen as essential in the genesis of the layout of sacred buildings, especially considering the veneration of the sun as a direct manifestation of the principal deities that chose to manifest themselves in different seasonal periods. "In the landscape, the common orientation of several elements can create a unitary character and the repetition of directions that the landscape suggests can link the buildings to their surroundings" (Norberg-Schulz 1998).

1.4.2 Alignments

Among these intentions of territorial order, the Andean constructors found useful to have different elements within a same line to favor visual connections.

In the case of the sacred landscape, Peruvian cultures created a network of lines with the intention of encompassing the entire territory and gradually making it sacralized. These lines were called "ceques", being able to verify their territorial validity. But there are also shorter alignments that seek to link natural elements such as the peak of an "*apu*" with the top of a "waka" and thus connect the natural with the artificial, the divine with the human creation.

1.4.3 Directions

The abundance of elements in the territory makes it appear as of an omnidirectional appearance. Men will be compelled to emphasize the directions that they believe most convenient to fulfill their goals and to highlight the sacred, and the appropriate use of the horizontal and the vertical was essential.

Gaston Bachelard reminds us of the ability of images to transmit values. As for verticality he mentions: "Being in communion with the verticality of a straight object through imagination implies being benefited by the ascendant forces" (2015, 51). The Peruvian sacred landscape understood is full of tangible references that indicate this ascending path.

On the other hand, horizontality is what keeps men positioned amidst the sky and the earth inviting them to journey.

1.4.4 Centers

The detection of a center in the territory is not only the recognition of a generative point from which a whole complexity can be irradiated, but it must contain the necessary information from which the rules of growth, expansion and regeneration are issued. In the pre-Hispanic Andean territory, ritual centers fulfill a regenerative function to which the faithful must reach on pilgrimage routes in order to reaffirm their convictions (Ríos and Zeballos 2018).

1.4.5 Polarities

Just as lines insinuate routes, a landscape structure needs elements that indicate partialities, a beginning and an end that concretes relationships. In the case of sacred landscapes, polarities are necessary to determine equilibria or to show contrasts as the only way to explain something through its opposite.

1.4.6 Limits

One could not speak of landscape if not able to recognize the limits within which its properties maintain constant. When detecting limits, men discriminate their environment and make decisions based on reliable information. In the case of a landscape, the limits are not always precise nor do they have a physical corporeity, they depend in many cases on the senses with which they are recognized or on the degree of influence that a centrality is capable of radiating.

2 Methodology

2.1 The Use of GIS in Historical Landscape Analysis

Although not yet widely spread, the use of geographical information systems is very suitable for the mapping, analysis, display and communication of the characteristics of sacred landscapes in general and particularly from a historical/archeological approach.

GIS has been used mainly, but not exclusively for geomorphologic, topographic, hydrologic and slope analysis (Parcero et al. 1998), land use, vegetation and population change (Zeballos 2020), predictive modeling (Diggs and Brunswig 2013), visibility and viewshed analysis. Most recent applications of remote sensing, point clouds and LIDAR allow a more refine study of the architectural features in the landscape (Štular et al. 2012). In addition other software used in combination with GIS, such as AutoDesk Revit, can help in the understanding of solar analysis in particular dates of the year.

However, in the case of the study of Peru's sacred landscape, very little work has been carried out using this technology. One of the few examples was carried out by Williams and Nash (2006) who analyzed the visibility of Wari settlements at Cerro Baúl and the surrounding mountains at a regional level (100 km) in order to underline the relationship of the sacred landscape. Their study was coupled with an anthropological approach of rituals of worship, still carried out in our days.

2.2 Materials and Methods

The methodology used in this article is mixed, since it uses quantitative data from the GIS analysis and then interprets the results based on qualitative data, such as historical documents or archeological findings carried out by other researchers.

To this end, three levels of analysis using different GIS techniques are carried out in order to understand the structure of sacred landscapes in ancient Peru (Table 1).

2.3 Criteria of Selection

While several other sites from the 6000-year history of Peruvian civilization would have been suitable for this analysis, the selected five sites have a clear relation with the landscape. They also belong to the early, intermediate and late periods of pre-Hispanic history, and they are located in a coastal valley, in an inter-Andean valley or above a mountain.

Also, while Machu Picchu contains all the elements discussed in this article and it is considered a remarkable example of sacred landscape, it has not included in this study because in some cases it has been already studied.

3 Landscape Analysis of the Sites

3.1 Waka of the Sun and Waka of the Moon, Moche

The Moche or Mochica culture flourished in the Northern Peru from about 100–700 AD, following several experiences in the coastal Andes (Gavazzi 2010). The Moche society carried out major hydraulic engineering works such as irrigation canals in order to expand their agricultural frontier on a large scale. They were remarkable architects, building monumental religious complexes made of adobe or mud bricks. The most impressive Moche structures were the so-called Waka of the Sun and Waka of the Moon. Those

Table 1 Summary of the materials and methods

Level	Software	Source	Action	Outcome	Site
Macro	Global mapper 19 ArcGIS 10.5 Google Earth	ASTER GDEM v2 worldwide (1 arc-second resolution)	Topography visibility (range 5–50 km)	Major connectivity with *apus* (mountains)	Moche Wari Tipon
Meso	Global mapper 19 ArcGIS 10.5 Google Earth	ASTER GDEM v2 worldwide (1 arc-second resolution)	Topography slope hydrology visibility (range of 5 km) viewshed	Visual and geometric relations of the wakas and Llactas	Ollantaytambo Sacsaywaman

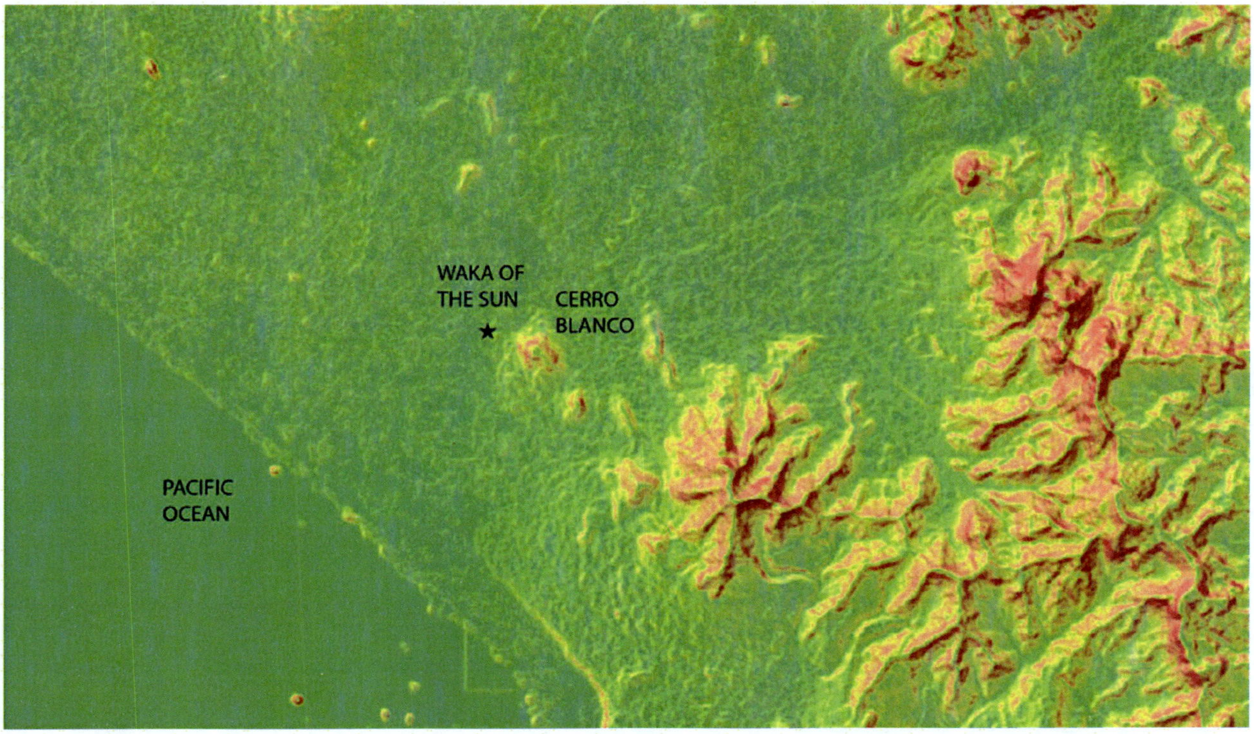

Fig. 1 Slope analysis of the Moche complex. Image by C. Zeballos

names were given to these structures in reference to the pyramids of the Sun and Moon in Teotihuacan, Mexico, although there is no evidence that the Moche wakas were dedicated to those luminaries (Williams 1981).

Both wakas are located in the flat Moche valley, 500 m from the coast (Fig. 1). It is located next to the Cerro Blanco, a hill (elevation 500 m) of a natural pyramidal shape whose profile is very observable in the landscape (Fig. 2).

A visibility analysis of a 25 km range from Cerro Blanco was carried out that shows how distinguishable this *apu* was from its surroundings (Fig. 3). Because of its distinguishable shape, this mountain enjoyed good visibility, as it could be seen from a long distance along the coast, where most settlements used to be located. The waka is not clearly distinguishable from the highlands as other mountains block the visibility, however in terms of symbolic landscape the views from the coast were the most important, as the Moche were mostly farmers and fishermen and they settled nearby the sea.

The Waka of the Sun used to be a truncated pyramid made of adobe which originally measured 345 × 160 m, with a height of 35 m (Canziani 2009, 197–202). The monument was partially destroyed by Spanish conquistadores, who deviated the Moche River in the search of treasures inside the structure. The complex of the Waka of the Moon comprised of several buildings and a ceremonial court and extended in an area of 6 Ha (300 × 200 m). It was 25 m high. As mentioned before, both wakas were located in a flat valley, and therefore, they could have been placed anywhere. However, they were built next to Cerro Blanco, and both the orientation and alignment are directed toward Cerro Blanco (Fig. 4). Moreover, a sacred rock was located in a small court between the Waka of the Moon and the Cerro Blanco (a sector called Platform II), which was the place for human sacrifices, probably offered to this *apu* in times of extreme events such as El Niño (Canziani 2009, 211).

3.2 The Capital of the Wari Civilization, Ayacucho

Wari was a civilization that flourished in the central Andes of Peru between the seventh to the thirteenth century AD. Their capital, the city of Wari, was located about 25 km NE of the city of Ayacucho on a plateau of irregular topography delimited by the ravines of Pacaycasa and Tarawayqo, at an elevation of approximately 2700 m (Lumbreras 1974). 5 km to the east of the city, a prominent mountain can be observed, which was considered a protective deity for the capital. (Fig. 5).

The Wari capital did not follow a geometric layout such as other cities of this civilization, such as Pikillacta. Its

Fig. 2 Cerro Blanco and the Waka of the Moon, Moche valley, Peru. *Photo* C. Zeballos, 2009

location was very strategic due to the rough topography. Even though, the city was protected by a series of walls due to the permanent military activity in the area (Fig. 6).

In the sector of Cheqo Wasi, southwest of Wari, there are notable lithic chambers, which apparently would have served as burial grounds. These chambers are made of large stone monoliths which are finely carved. Their alignment with the tutelary mountain is notorious, suggesting eternal protection beyond death, which is also present in other cultures in the world (Jansen and Raffa 2015) (Fig. 7).

The mountain has good visibility from Wari and especially from Cheqo Wasi (Fig. 8); however, there is no evidence of any special influence in the organization or the urban trace of the city, whose layout is very organic.

3.3 The Tipon Complex

Tipon is remarkable for its insertion in the territory, its significance in the relationship between man and water, its role as an agricultural laboratory and its impressive hydraulic technology. Kenneth Wright (2006) highlights the importance of water management in an agricultural society such as the Inca, due to climatic variations resulting from phenomena such as El Niño and La Niña.

Tipon is located at 3600 m high, 27 km southeast of Cusco, near the town of Oropesa, set in a small valley overlooking the Huatanay River. In this valley, several microclimates are presented according to the height at which the observer is located.

Guzmán (2013)suggests that there is a sacred system of *apus*, going from the smallest and nearest, called Sinkunakancha, a direct *apu* about 200 m to the northeast of Tipon, containing an *intihuatana* or astronomical observatory. Subsequently, the intermediate *apu* called Cruzmoqo, where a spring is located and according to Bauer (2011) used to hold a waka or center of worship as part of the Antisuyo ceque. Finally, as mentioned by Dammert (2007), Tipon was associated with the superior mountain Patachusan, one of the fundamental *apus* of Cusco (Figs. 9 and 10).

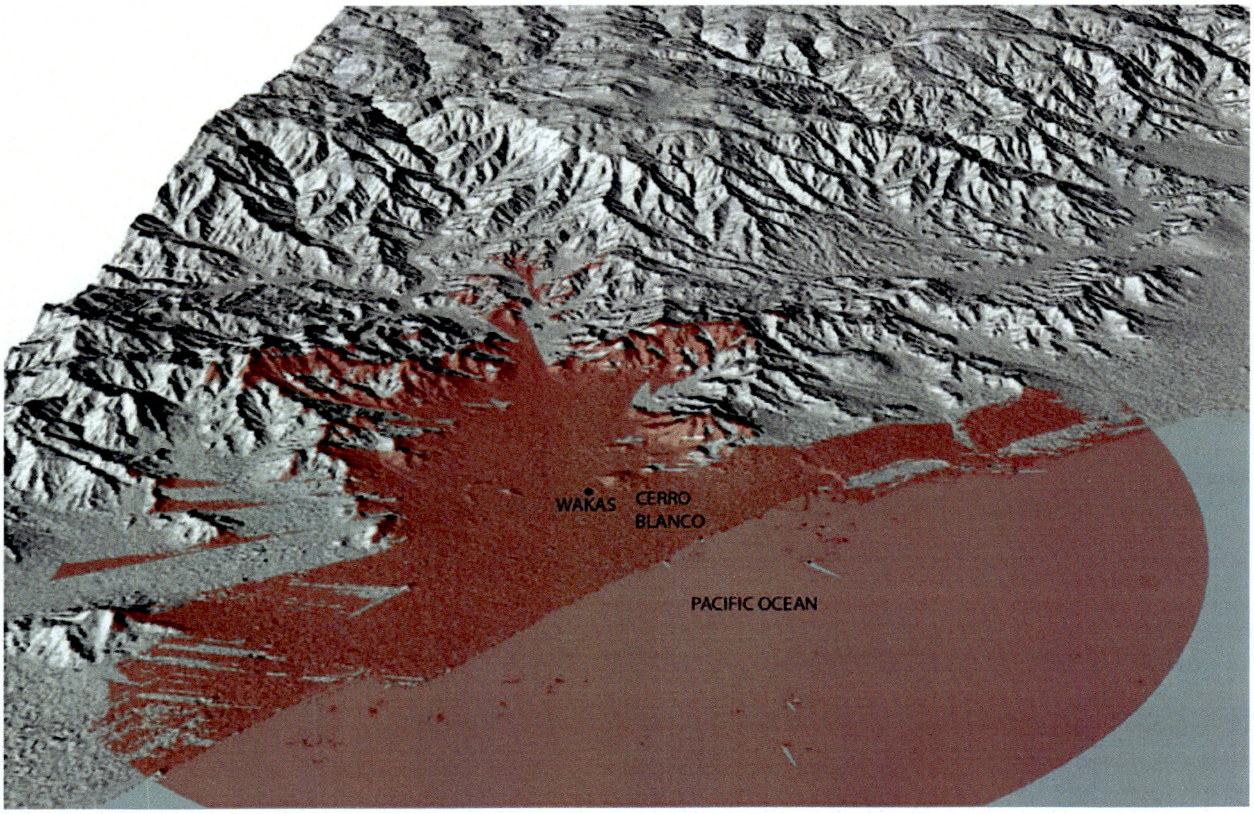

Fig. 3 Visibility from Cerro Blanco and the Moche valley. Image by C. Zeballos

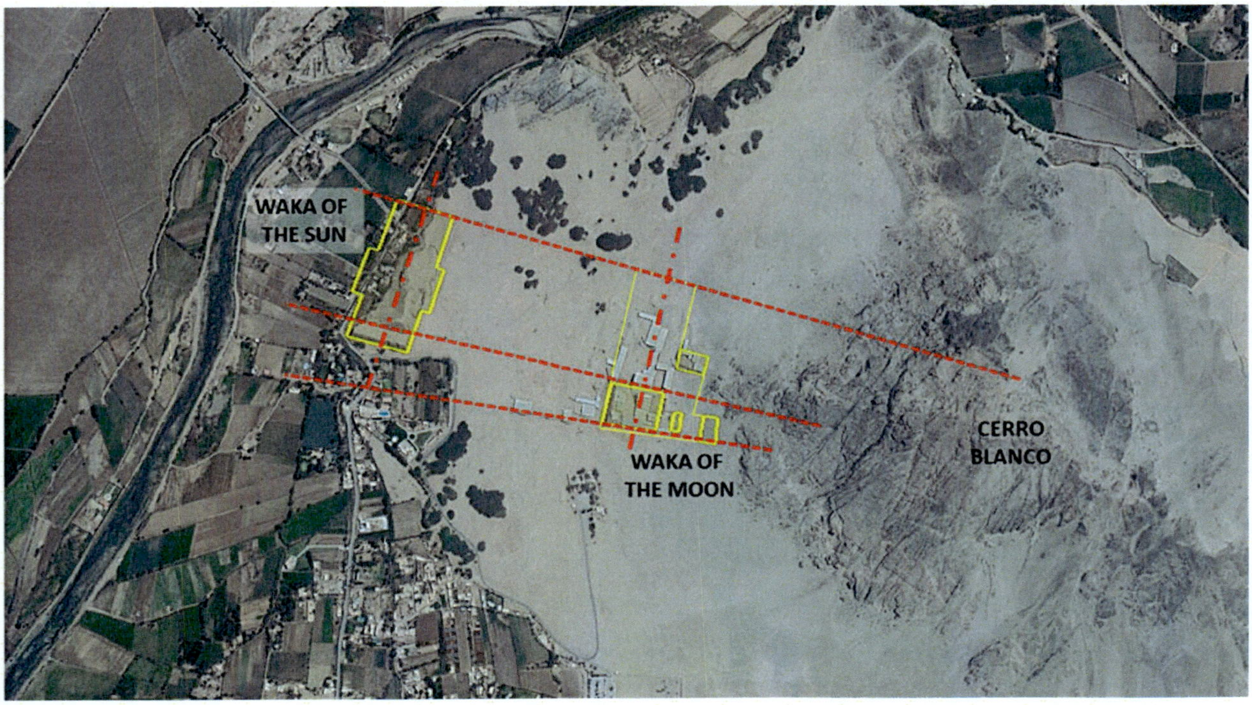

Fig. 4 Traces over the Moche structures aligned with the Cerro Blanco. Redrawn by Zeballos over map by Canziani, 2019

Fig. 5 Location of Wari and surrounding landscape features, Peru. Model by C. Zeballos, 2019

The Tipon complex is born at the top of the Cerro Cruzmoco and forms a series of 13 terraces or platforms built in imperial style masonry and hydraulic channels that descend toward the ravine (Fig. 11). At the other end of the axis is a natural promontory which is the Apu Sinkunakan. The most important sector of the complex that is the spring is aligned between both *apus* (Fig. 12). The geometry of the terraces is a rational abstraction of the surrounding topography of the ravine (Zeballos 2009).

3.4 The Ollantaytambo Site

Ollantaytambo is located 97 km northwest of the city of Cusco at an elevation of about 2800 meters, settled at the confluence of the Urubamba valley and the Patacancha River, between the very steep mountain Pinkuylluna and the Bandolista hill (Fig. 13).

The Patakancha River defines two clear zones in the settlement: to the east, a trapezoidal grid with a ceremonial plaza and more or less regular dwellings, known as Qosco. To the west, a sacred area called Arakama, formed by the unfinished Temple of the Sun and an urban pattern, is organized around a large space called Manyaraki Square, as well as a set of ceremonial fountains and channels. Fine terraces rise up the hillside and others, carved on the plain, make up an unusual fan arrangement (Fig. 14). The quality of these terraces, as well as their extension (the cultivated area would exceed the population settled in Qosqo), suggests that they were used for some type of cultivation of a sacred nature (Protzen 1993).

It is even more surprising to observe them from the Pinkuylluna hill, located in front of Arakama, since the platforms have a configuration in the shape of a llama, on whose head the Temple of the Sun is Zeballos (2007). In addition to the visual connection with mountains, the creation of a sacred landscape also involved the urban trace and modeling resembling sacred animals or deities located (Villavicencio 2012) (Fig. 15).

One of the most conspicuous symbolic elements in the landscape is a rock carved in the shape of a human face located in the Pinkuylluna hill, which would represent Wiracocha or Tunupa, a civilizing deity, who managed the system of production, treatment and storage of products (Eliorreta and Eliorreta 2002). Hence, it was associated with the storages or "colcas" located on the slope of the Pinkuylluna (Fig. 16).

Fig. 6 Protective wall, Wari city, Peru. *Photo* C. Zeballos, 2018

3.5 Sacsayhuaman

The Inca Pachacutec transformed a small village into Cusco or Qosqo, the capital of his empire. To this end, he used carved stone models and designed an elongated city in the form of a crouched puma (Williams 1981). The head of the sacred animal was the Sacsayhuaman *apu* and its body stretched along the Shapi and Tullumayo rivers, now channeled (Agurto 1987) (Fig. 17).

The city was divided into two areas: the Hanan Qosqo or upper part, to the north and the Urin Qosqo, or lower part to the south. The ceremonial and military complex of Sacsayhuaman is located on top of a hill presiding the Hanan Qosqo. Its northern facade consists of a sequence of three zigzag-shaped parallel walls resembling a lightning, built with monumental stone blocks and assembled with impeccable precision. In the center of the complex, there would have been a circular tower (Mar and Beltrán-Caballero 2014). The Koricancha or Temple of the Sun was located in the Urin Qosqo, forming a duality with Sacsayhuaman.

According to Guzmán (2013), the system of *apus* extended much further, including Sacsayhuaman, Koricancha and the Apu Muyuorco and Apu Wanakauri (Fig. 18). An analysis of visibility from Sacsayhuaman also suggests this visual relationship. From the Urin Qosqo, this connection is less visible due to urbanization; however, it can be seen from some areas such as Avenida El Sol, which runs on the channeled river Shapi (Fig. 19).

4 Results and Findings

Not every society can be classified as landscape-friendly that is, not every culture is capable of attaching annotations to the territory that exceed the utilitarian, entering in the realm of aesthetics. In the case of pre-Hispanic Peru, the artistic does not exist if it is not linked to the mystical; therefore, all this network of visual relationships without a functional purpose but that of forming landscape units has its origin in the deepest intentions of the Andean men to be part of a place in which the sacred happens.

In western societies, the perceiving subject and the perceived landscape are two isolated entities that establish their relationships from an appropriate distance. In the examples that we have analyzed for the Peruvian case, this distance (which sometimes includes territories of several kilometers)

Fig. 7 Lithic chambers in Cheqo Wasi, near Wari. *Photo* C. Zeballos, 2018

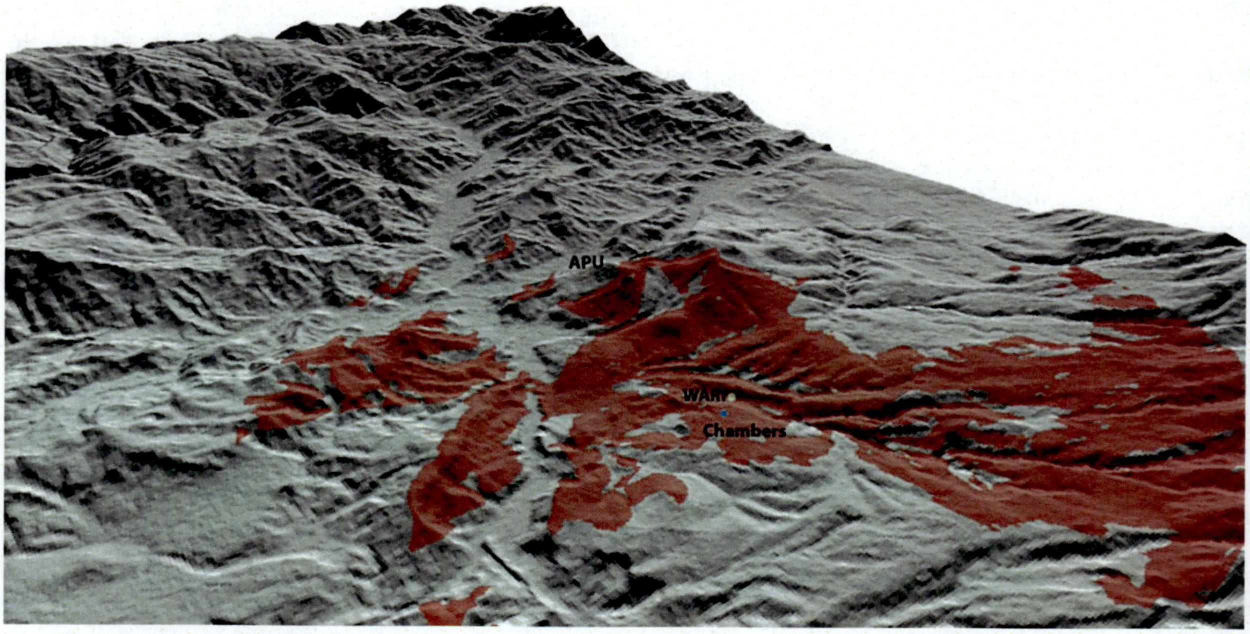

Fig. 8 Visibility of Wari and Cheqo Wasi. Image by C. Zeballos

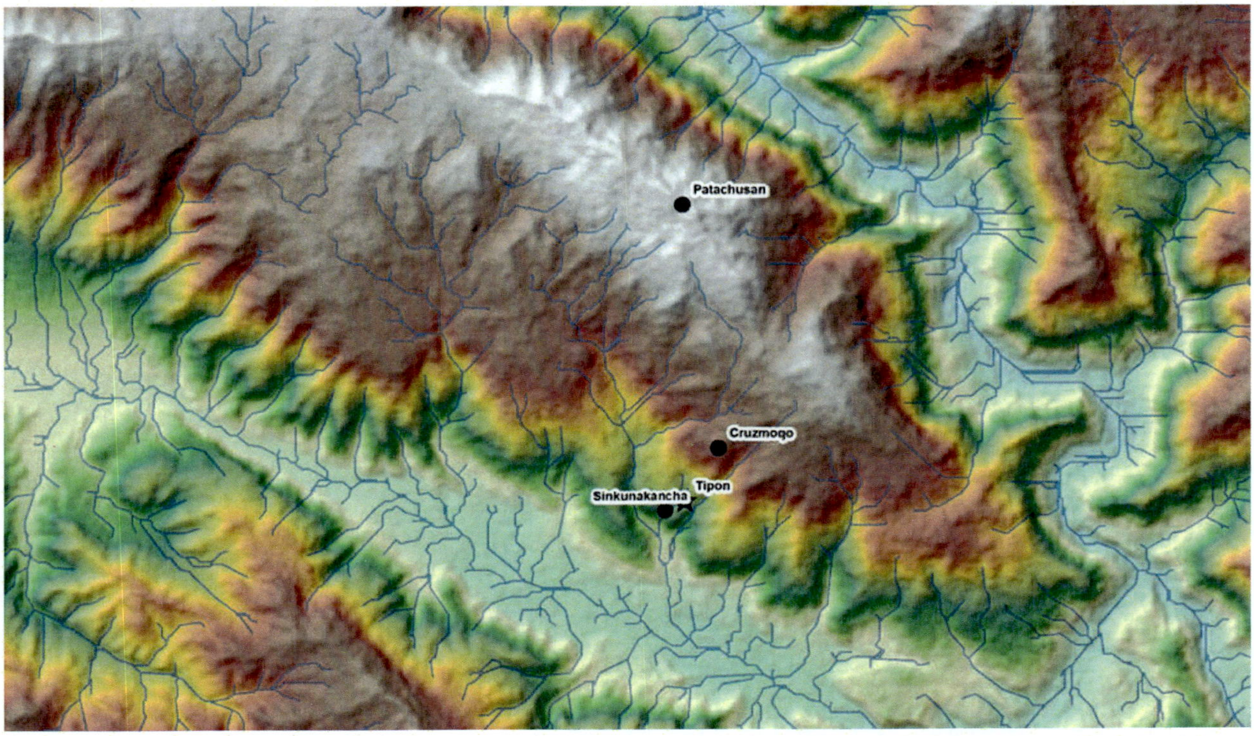

Fig. 9 *apu* system of Tipon. Image by C. Zeballos

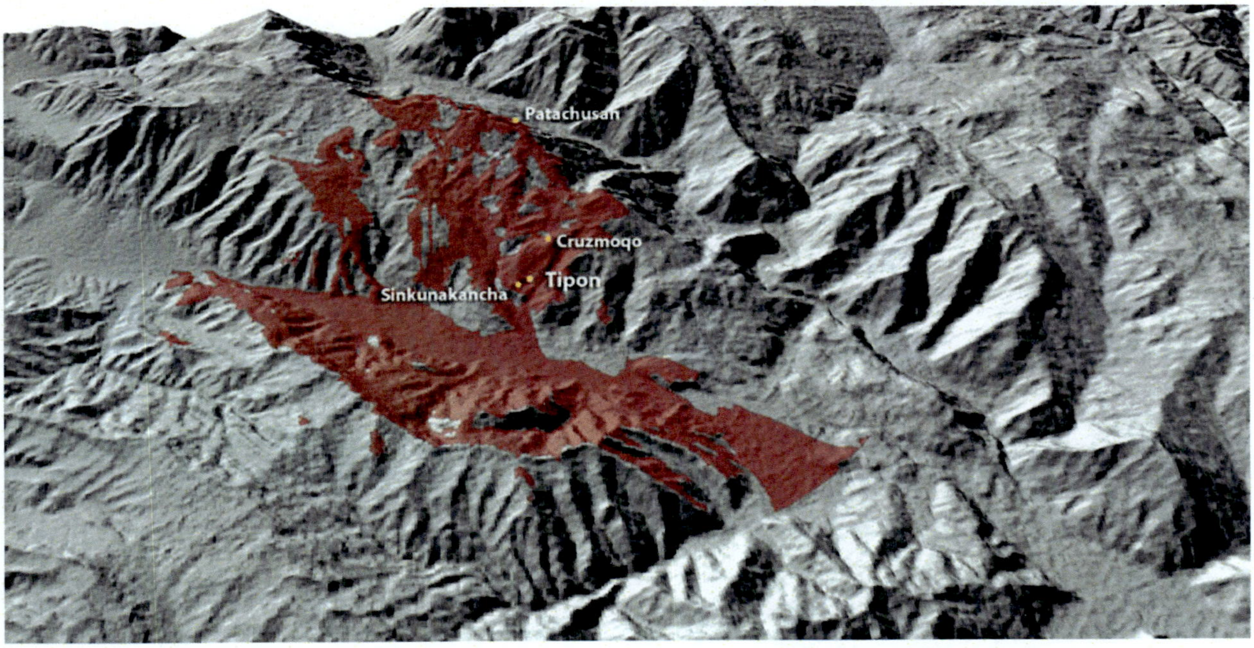

Fig. 10 Visibility analysis from Cruzmoqo hill. Image by C. Zeballos

Fig. 11 Tipon showing Cruzmoqo hill. *Photo* C. Zeballos, 2018

Fig. 12 *apu* system in Tipon. Image by Google Earth

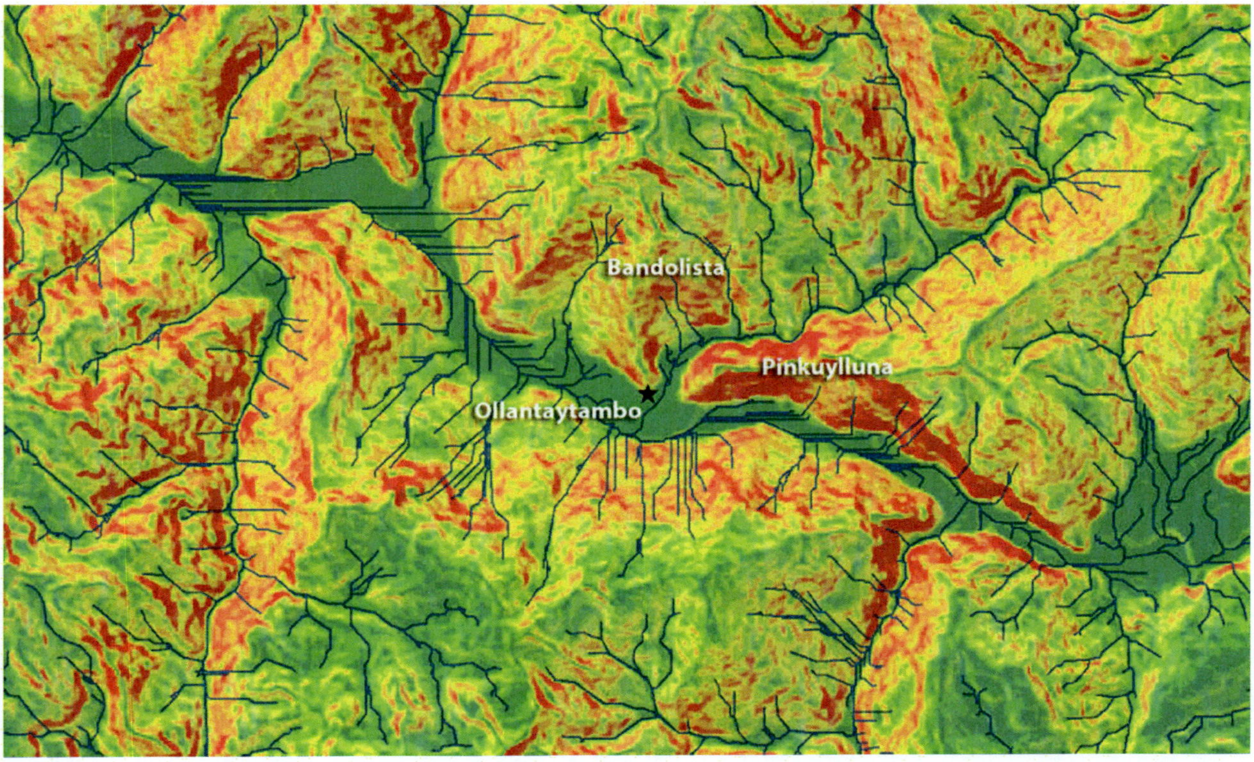

Fig. 13 Slope analysis of Ollantaytambo site. Image by C. Zeballos

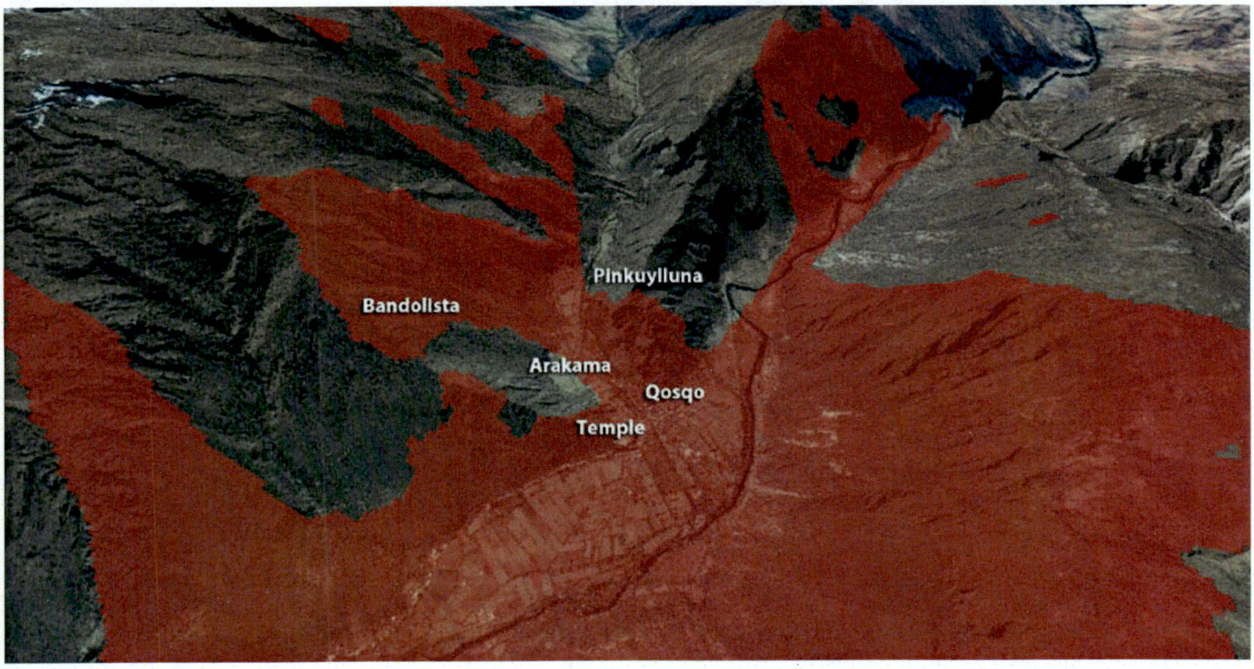

Fig. 14 Visibility analysis from the Temple of the Sun, Ollantaytambo. Image by C. Zeballos

Fig. 15 Terraces at Arakama, modeled as a sacred llama, as seen from Pinkuylluna hill. *Photo* C. Zeballos, 2009

is shortened until man and territory are merged in a single organism. This is achieved through rituals that propose the displacement of these networks that cease to be only visual and involve other types of senses.

The Andean cosmovision that manages conceptions such as duality, equilibrium, centers is reproduced in the attempts to underline these relations in the territory. This can be corroborated in the studies carried out in monumental complexes belonging to different cultures, settled in dissimilar territories and that, nevertheless, have the same cosmovision, developed throughout the Andean territory.

Given the multiplicity of sites, separated in space and time, that have been selected to carry out this research, different approaches can be determined regarding the ways in which the forms modeled by humans to interact with the natural features in a given territory. However, there is an evident constant regarding the relationships that are intended to be established between them, which reaffirms the existence of a very ancient and reiterated narrative.

The efficiency of the Andean territorial planning is due in large part to an exhaustive knowledge of the morphology of the territory and to the mastery of a series of principles that are reiterated and transmitted from generation to generation. For this transmission to be efficient, transcendental myths, legends, foundational narratives, etc., emerge and are considered divine designs.

5 Conclusions

The idea that everything created continues to be administered by a guardian deity that is present everywhere, is a common idea of any religion. In the Andean world, the presence of the "*apus*", or sacred mountains, is inherent to the daily life of the inhabitants and defines the omnipresence of the sacred. By means of GIS analysis, it was possible to verify the visibility of these geographical elements and their different degrees of visual interaction with the most relevant

Fig. 16 Huge carving of Wiracocha on the left, see the scale of the terraces in the left at Pinkuylluna hill, as seen from the Arakama sector. *Photo* C. Zeballos, 2009

places in the territory, as is the case of Cerro Blanco in the Moche culture.

The geometry embodied in the territorial organization and the architecture of the Andean world is far from having an exclusively functional connotation. The urban generatrix lines start from a sacred conception, which is corroborated only if they are perceived from wide territorial scales that allow visualizing their relations with places considered sacred. In the case of the Andean cultures, aligning their urban and architectural layouts to sacred features in the landscape was a way of aligning themselves to a divine dimension.

The growth and overlap of contemporary cities with respect to pre-Hispanic cities make it possible to simply see fragments of what it used to be entire territorial units. Thus, we perceive pre-Hispanic platforms, channels, roads, etc., from a pedestrian view that show extreme beauty but are unintelligible in their global conception. The aerial views provided by the new technologies make it possible to recompose these pieces in order to decipher the macro-conception of the territory as is the case analyzed by Tipón, whose skillful water and agricultural design is associated with a larger geomorphology that is imperceptible at short distances.

Accustomed today to value things based on their practical utility or their market value, it is difficult to understand the construction of urban organisms that concatenate in one piece the profane and daily use of spaces with the sacred and ritual. From GIS technology, symbolic configurations can be appreciated, those which are not designed to be appreciated by the ordinary citizen, but rather are conceived as a kind of encrypted language for the gods.

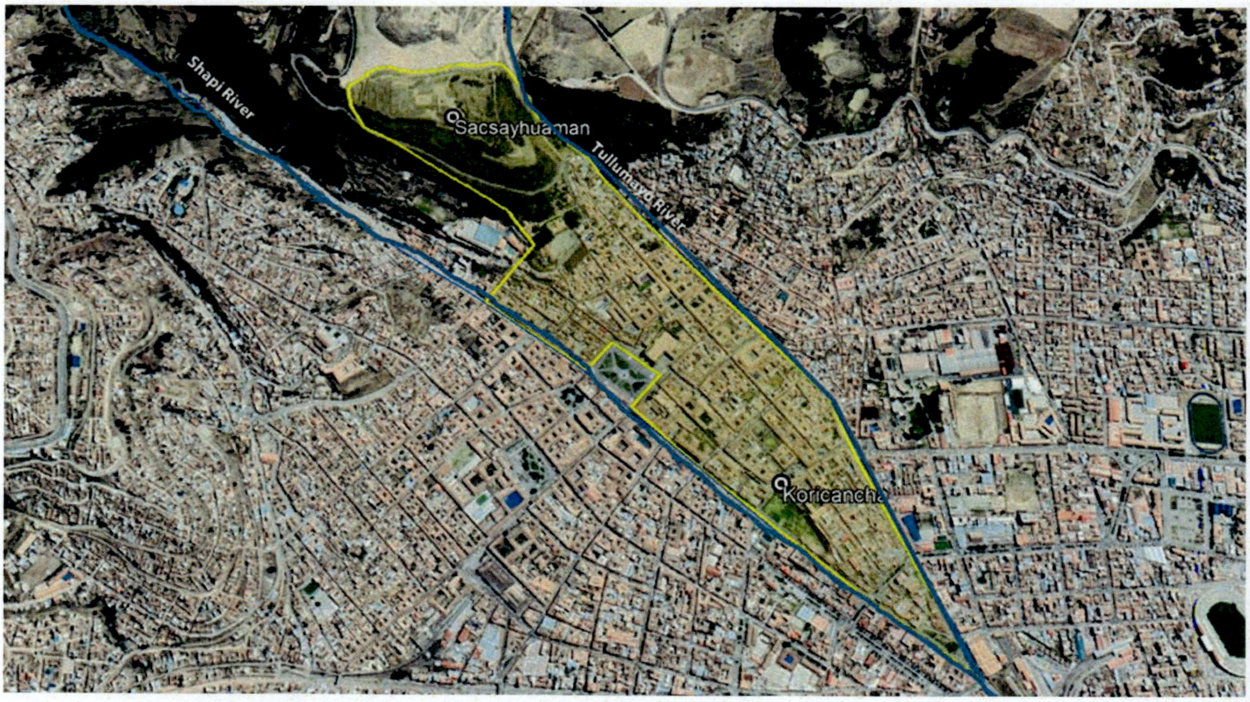

Fig. 17 Layout of Cusco in the shape of a puma. Redrawn from C. Williams by C. Zeballos

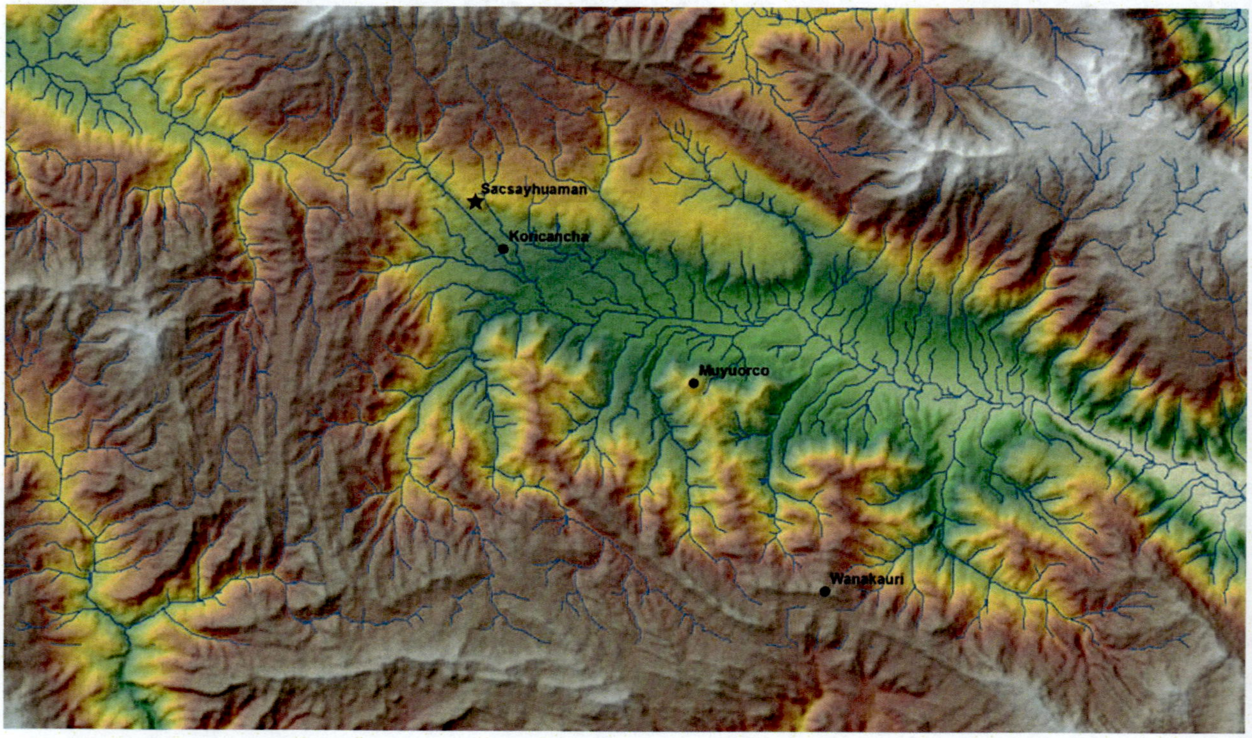

Fig. 18 *apu* system at Cusco. Image by C. Zeballos

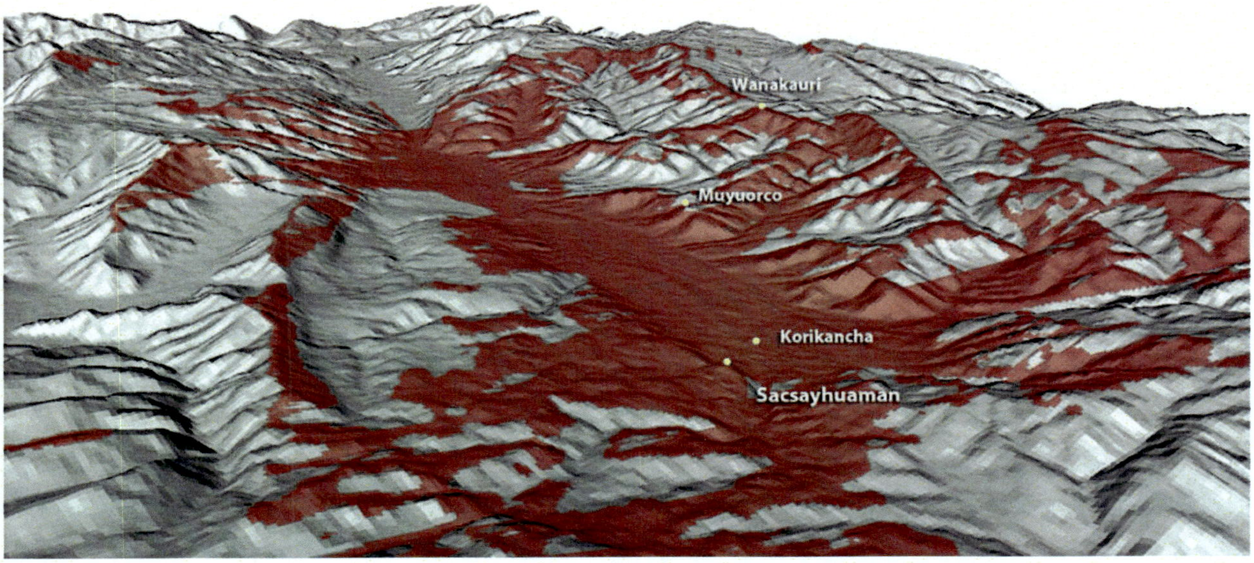

Fig. 19 Visibility analysis from Sacsayhuaman, showing the *apu* system. Image by C. Zeballos

Acknowledgements We would like to thank Architect Ruben Jeri for his guidance during the visit to Wari city.Also, many thanks to Edwin Ríos for carrying out a drone flight in Sacsayhuaman, which was very useful for visual analysis.

References

Agurto, S. (1987). *Estudios acerca de la Construcción, Arquitectura y Planeamientto Incas*. Lima, Peru: Cámara Peruana de la Construcción.

Bachelard, G. (2015). *La llama de una vela*. Buenos Aires, Argentina: Cuenco de Plata.

Bauer, B. S. (2011). *The sacred landscape of the Inca: The Cusco Ceque System*. Austin, USA: University of Texas Press.

Canziani, J. (2009). *Ciudad y Territorio en los Andes. Contribuciones a la historia del urbanismo prehispánico*. Lima, Peru: Fondo Editorial Pontificia Universidad Católica del Perú.

Crousse, J. (2016). *El paisaje peruano*. Lima: Pontificia Universidad Católica del Perú.

Dammert, M. (2007). *La red de parques arqueológicos*. Cusco, Peru: I. N.C. Dirección Regional de Cultura.

Diggs, D., & Brunswig, R. (2013). The Use of GIS and Weights-of-Evidence in the Reconstruction of a Native American Sacred Landscape in Rocky Mountain National Park, Colorado. In En L. Lozny (Ed.), *Continuity and Change in Cultural Adaptation to Mountain Environments*, (pp. 207–228). Springer.

Eliorreta, F., & Eliorreta, E. (2002). *Cusco y el Valle Sagrado de los Incas*. Cusco, Peru: Tanpu S.R.L.

Gavazzi, A. (2010). *Arquitectura Andina. Forma e Historia de los espacios sagrados*. Milano, Italy: Jacka Book SpA.

Guzmán, C. (2013). *Llactas Incas. Concepción del Planeamiento e interacción con ek medio natural*. Lima, Peru: Universidad Nacional de Ingeniería.

Heidegger, M. (1951). Construir, Habitar, Pensar. Darmstadt.

Jansen, M., & Raffa, V. (2015). *Tiempo y comunidad*. Leiden University Press.

Lumbreras, L. G. (1974). *Las Fundaciones de Huamanga. Hacia una prehistoria de Ayacucho*. Lima: Nueva Educación.

Makowski, K. (2015). Makowski, Krzysztof. Pachacamac—Old Wak'a or Inka Syncretic Deity?: Imperial Transformation of the Sacred Landscape in the Lower Ychsma (Lurín) Valley. In B. Tamara (Ed.), *The Archaeology of Wak'as: Explorations of the Sacred in the Pre-Columbian Andes*, (pp. 127–166).

Mar, R., & Beltrán-Caballero, A. (2014). El conjunto arqueológico de Saqsaywaman (Cusco): Una aproximación a su arquitectura. *Revista Espanola de Antropologia Americana, 44*(1), 9–38.

Menegat, R., & Porto, M. (2007). Relações entre a paisagem e a cidade inca de Machu Picchu: elementos para decifrar sua construção. In *I Encontro IALE-BR*, (pp. 33–35). Brasil: Rio de Janeiro.

Norberg-Schulz, C. (1998). *Intenciones en arquitectura*. Barcelona: Gustavo Gili.

Pallasmaa, J. (2014). *La imaginación corpórea. Imaginación e imaginario en la arquitectura*. Barcelona, España: Gustavo Gili.

Parcero, C., Criado, F., & Santos, M. (1998). Rewriting landscape: Incorporating sacred landscapes into cultural traditions. *World Archaeology, 30*(1), 159–176.

Protzen, J. (1993). *Inca architecture and construction at Ollantaytambo*. Oxford: Oxford University Press.

Regalado, L. (2000). La cosmovisión Andina. En *Mundo Peruano Antiguo. Una visión interdisciplinaria*. Lima: PUCP/PROMPERU.

Reinhard, J. (2002). *Machu Picchu. El Centro Sagrado*. Cusco, Peru: Instituto Machu Picchu.

Ríos, G., & Zeballos, C. (2018). *Poética de un Mundo Habitado* (Vol. I). Arequipa, Perú: Universidad Católica de Santa María.

Spinetto, N. (2018). *Homo religiosus? L'esperienza del sacro nella storia delle religioni (video)*. Obtenido de Fondazione Collegio San Carlo.: https://youtu.be/JxOD_-qz35Y.

Štular, B., Kokalj, Ž., Oštir, K., & Nuninger, L. (2012). Visualization of lidar-derived relief models for detection of archaeological features. *Journal of Archaeological Science, 39*(11), 3354–3360.

Vargas Llosa, M. (2012). *La civilización del espectáculo*. Madrid: Alfaguara.

Villavicencio, M. (2012). *Perú Mágico II. Homenaje a Macchu Picchu*. Lima, Peru: Centro Cultural de España.

Williams, C. (1981). *Arquitectura y Urbanismo en el Antiguo Perú*. Lima, Peru: Editorial Juan Mejía Baca.

Williams, P., & Nash, D. J. (2006). Sighting the Apu: A GIS analysis of Wari imperialism and the worship of mountain peaks. *World Archaeology, 38*(3), 455–468.

Wright, K. (2006). *Tipon: Water engineering masterpiece of the Inca Empire.* American Society of Civil Engineers.

Zeballos, C. (2007). *Planeamiento Inca y Paisaje: Ollantaytambo.* Obtenido de Mi Moleskine Arquitectónico: http://moleskinearquitectonico.blogspot.com/2007/03/planeamiento-inca-y-paisaje.html.

Zeballos, C. (2009). *Tipón, el Agua y los Incas*. Obtenido de Mi Moleskine Arquitectónico: http://moleskinearquitectonico.blogspot.com/2009/07/tipon-el-agua-y-los-incas.html.

Zeballos, C. (2020). Atlas Ambiental de Arequipa. Arequipa, Peru: Universidad Católica Santa María. ISBN: 978-612-00-5115-3.

Old and New Settlement Strategies in a Marginal Area of Viceregal Naples: Benedictines and Jesuits in the *Vomero* Uphill Road

Giuseppe Pignatelli

Abstract

Visibility and accessibility have constituted the indispensable premises for the settlement of religious communities in the modern city. In particular, especially in the sixteenth century, the Neapolitan religious orders conditioned the urban development outside the city walls: in the last decades of the century, exactly in the extra-moenia areas, the new regular religious orders will be joined to the Mendicants in the laborious search for unbuilt estates or edifices to be re-adapted to monasteries, convalescent homes, and novitiates with the contribution of influential and wealthy subsidizers. Replacing theirselves to Viceroy authorities in the utilization of the hill slopes, the religious often will promote radical buildings renovations and the opening, even if with privatistic purposes, of roads and small squares in front of their estates. Through analysis of unpublished documents, this essay examines in depth the troubled events that leaded—between 1622 and 1627—to the foundation of a small Benedictine monastery in a farmstead along the Vomero uphill road. This event, totally ignored by Neapolitan architectural historiography, represents instead an extraordinary example of smart adaptation to a marginal site of the western urban expansion.

Keywords

Religious orders architecture • Settlement strategies of religious orders • Sixteenth–seventeenth centuries • Viceregal naples

G. Pignatelli (✉)
Department of Humanities and Cultural Heritage, University of Campania 'Luigi Vanvitelli', S.M. Capua Vetere (Ce), Italy
e-mail: giuseppe.pignatellispinazzola@unicampania.it

1 Introduction

The analysis of the different settlement strategies of religious orders is nowadays an essential instrument for the study of the modern city; in particular, visibility and accessibility—two necessary conditions for the territorial dominance—constituted the presuppositions for the foundation and development of *intra* and *extra-moenia* religious complexes on the basis of precise rules inevitably conditioned by the historical periods, the places of settlement and, above all, the different functions that the monastic communities had to take on in the many urban environments in which they were based.

Since the seventeenth century, the closer and closer relationships between State government and ecclesiastical authorities, and the consequent autonomy gained by the religious communities in the administration of their own assets, especially outside the cities, supported their control—first and foremost social and economic—on some still scarcely urbanized areas, through the foundation of coenobies, small monasteries and chapels, real outposts placed near the urban gates or along the main penetration routes (Guidoni 1990, 306; Pellegrini 1990, 27–29).

Only from the second half of the thirteenth century, the primitive methods of settlement (characterized by the lack of any strategy and the absolute precariousness of the seats) were replaced with well-established practices that allowed the Mendicans to settle permanently inside the city, strengthening their importance in the urban society, which they were therefore able to forge mutual relations of support and services (Pellegrini 2003, 12).

2 The Religious Orders Role in the Neapolitan *Extra-Moenia* Areas

This phenomenon appears in all its complexity in the city of Naples, already characterized in the Ducal era by the birth of *extra-moenia* Basilian settlements that were flanked, then, by

C. Gambardella et al. (eds.), *Advances in Utopian Studies and Sacred Architecture*,
Advances in Science, Technology & Innovation,
https://doi.org/10.1007/978-3-030-50765-7_26

the much more influential Benedictine presence. This slow process reached its climax between the thirteenth and fourteenth centuries with the decisive contribution of the Angevin sovereigns, willing to promote the foundations of the new Mendicant orders (in particular Augustinians, Dominicans, Franciscans, and Carmelites) not only inside the city walls—more and more inadequate to hold an unstoppable population growth and a chaotic building development—but also far beyond the traditional urban limits (De Seta 1981, 28; Colletta 1985, 14).

In the Capital city, the fast building development around the more recent religious poles represented the result of notable land donations, of royal subsidies for the construction and maintenance of new headquarters and, above all, of countless fiscal privileges guaranteed to religious orders.

Without considering in this paper the *intra-moenia* settlement of the Mendicants and its inevitable repercussions on the transformation of the ancient center into a modern city, the religious orders must be recognized to have cleverly adapted to their own needs—not always in observance of their monastic rules—a morphologically complex suburban territory, proposing themselves as the real fuel of the unstoppable development outside the city walls, always being seen as a symbolic filter between the city and its suburb area, grown in a totally disordered and parasitic way (Fig. 1).

In this perspective must be so read the establishment of the large conventual complexes at the margins of the city (St. Eligio Maggiore, the Carmine Maggiore and St. Pietro Martire along the southern walls between the market area and the commercial port, St. Maria la Nova and St. Pietro a Majella close to the western walls, outside the most ancient nucleus of the city), and above all the Mendicant communities settled in *extra-moenia* areas still not very urbanized, as St. Maria dei Vergini on the northern hills, St. Antonio Abate, the Maddalena and St. Maria Egiziaca a Forcella (or all'Olmo) outside the eastern walls, St. Maria Incoronata, St. Maria della Croce, the St. Trinità, and St. Martino between the slopes and the peak of the western hills.

3 The Religious Orders Role in the Chiaia Hamlet

The settlement dynamics of the religious communities behind the Chiaia seaside, west of the consolidated urban area, confirmed the extraordinary contribution offered by the religious orders to its passage from a modest coastal village to the status of a modern city neighborhood (Gentile 1953, 27–28).

The hamlet of Chiaia (from the Greek πλάγα, playa, beach) began to develop around the fifth century BC, when small groups of houses placed independently along the coastline, in a peripheral area between the western slopes of Pizzofalcone and Mortelle hills and the first offshoots of the Posillipo promontory. Only in the second century BC, after the definitive Roman conquest of the city and the establishment of heavy customs duties, Naples assumed a marginal role for the progressive affirmation of the Phlegraean area. This condition represented for the western zone of the city an extraordinary opportunity for development due to the coexistence of the main streets to the military and commercial ports of Pozzuoli, Baia and Miseno.

In particular, the so-called "via *per colles*" went up from the Lake of Agnano and crossed the Vomero hill to descend into the city through the "cavone" of the Infrascata; a second

Fig. 1 Bastiaen Stopendaal, Napoli, 1663. The viceregal city and his suburban territory

road, even more uncomfortable, climbed from the Bagnoli plain along the tuff cliff of Coroglio, crossed the Posillipo hill and licked the beach up to the valley between the hill of Mortelle and Pizzofalcone, to reach finally the port area (De Stefano 1963, 227; Pignatelli 2014a, 19). Along this path, in the Augustan period was built the so-called "Grotta di Pozzuoli" (or "*Crypta Neapolitana*"), a narrow and long tunnel dug in a straight line across the Posillipo hill to facilitate the connections with the Phlegraean area in a commercial and, above all, defensive perspective. The new "via *per cryptam*", became a public road at the end of the fourth century AD, followed the usual coastal route to the end of the Chiaia beach branching off at the end of the tunnel into two distinct sections, one toward Bagnoli, the other linked straight to the road to Pozzuoli (Johannowsky 1952, 139). Albeit in a fragmentary way, from the end of the first century AD a compact settlement began to consolidate along the winding path behind the coastline, keeping the original agricultural functions as highlighted by the discovery of several ancient rural buildings. This phenomenon modified the hilly landscape of the hamlet, a limited but important area extra-urban autonomous from the rest of the city which, instead, was struggling to reaffirm itself politically and economically (Napoli 1959, 209).

In spite of its still marginal role, the western coastline soon established itself as a residential area coveted by the Roman ruling class, very far—not only physically—from the chaotic city life: leaving aside the well-known maritime villas along the Posillipo coast, to the late Imperial Roman age are datable, for example, the residences of the general Lucio Licinio Lucullo between the hill top of Pizzofalcone and the underlying islet of Megaride, the nearby villa of Pollio Felice or that of Silio Italico, placed near the entrance to the Pozzuoli cave.

Substantially immune to the involution phenomenon that characterized the city at the end of the Roman Empire, the Chiaia area retained its own self-sufficiency also in the late Ducal era, thanks to the unprecedented forms of territorial management introduced by the new land regime based on the contracts between the settlers and the religious institutions. So, the built area in Chiaia started to grow slowly around small coastal coenobies, real attractive poles for agricultural, trade, and fishing activities, only later expanding toward the hill behind the seaside thanks to the Mendicants presence.

Replacing the pre-existing religious communities, between the thirteenth and fifteenth centuries the Olivetan Benedictines of St. Maria a Cappella, the Celestinians of the Ascension, the Dominicans of St. Leonardo, the Augustinians of St. Maria di Piedigrotta and the Servites of St. Maria del Parto identified specific areas of influence on a territory economically less and less separated from the rest of the city. The traditional functions related to the fish and agricultural products trade, which remained unchanged only in the most remote areas (in particular, in the extremity of the Posillipo promontory), were replaced by the management of a considerable real estate consisting not only of coastal houses and warehouses but also of vast hill properties of difficult exploitation, but later made particularly attractive by a precise town planning, aimed at the necessary dilatation of urban boundaries now insufficient to meet the defensive needs and unstoppable population growth of the city (Pignatelli 2014a, 21–22) (Fig. 2).

Despite a still uneven development, the Chiaia hamlet took on the characteristics of a pleasures place, heart of the social life of the Angevin and Aragonese city. In particular, during the fifteenth century, the new representative functions also contributed to the progressive abandonment of the saturated and chaotic coastline toward the quieter hill slopes: as in the case of the residences of the Caiafa and Terracina families, a large number of farmhouses joined the existing rural buildings, redeveloping the hilly landscape after centuries of neglect thanks to complex operations to contain the land and to recover the ancient "cavoni" (a series deep grooves naturally carved by rainwater) in new settlement routes along which it began slowly to grow a still fragmented building fabric (De Seta 1981, 59; Giannetti 1994, 11).

This particular settlement phenomenon underwent a further and decisive acceleration during the sixteenth century, when Chiaia—defined as the "parte mejor de la ciudad" ("the best part of the city") by the Spanish authorities—became the focus of an ambitious urban plan which, in about thirty years, led to the definitive inclusion within the city walls of the hilly areas of San Martino, Pizzofalcone, Mortelle and, partially, Chiaia (Pessolano 1998, 60–65).

Of course, the regular religious orders born after the Council of Trento, could not avoid to take part in this unstoppable process, determined as they were to work alongside private citizens and pre-existing monks communities in the laborious acquisition of soils or real estates to be transformed into monasteries, convalescent houses, novitiates and boarding schools, initially destined only to a limited number of brethren, and later on expanded in order to escape the decrees of suppression of the less numerous religious communities (Fig. 3).

In this view, it should be noted that if for some communities (for example, Jesuits, Theatines, Oratorians and Mercedaries) coming into the hamlet was only one further step of their settlement inside the city, for other ones (for example, the Barnabites of St. Carlo alle Mortelle or the Clerics Regular of the Mother of God of St. Maria in Portico), the settlement in Chiaia coincided with their first Neapolitan seat, making the choice of the site and the construction of their complexes particularly delicate.

Being the public interventions almost totally absent, the new religious orders were therefore decisive in the future

Fig. 2 Joris Hoefnagel, Elegantissimus ad mare Tyrrhenum ex Monte Pausilipo Neapolis montique Vesuvij prospectus, 1578. The Chiaia hamlet at the end of the sixteenth century

Fig. 3 Settlements of the religious orders in Chiaia hamlet, sixteenth to eighteenth centuries

urban planning of the area, replacing the viceregal authorities in promoting radical renovations of buildings, the construction of rental house complexes and the opening, even if for own private purposes, of roads and small open spaces in front of their estates in the precise optic of territorial dominance.

4 Benedictine and Jesuit Settlement Strategies in the Vomero Uphill Road

Exemplary are, in this perspective, the events related to the foundation of a small Benedictine settlement built in the early twenties of the seventeenth century along the Vomero uphill road. Guiltily ignored by the Neapolitan architectural historiography—only in the seventeenth-century guide of the city by Carlo Celano, is just mentioned "un piccolo monastero benedettino lungo la bella strada che noi chiamiamo Imbrecciata, e che arriva fino al Vomero" ("a little Benedictine monastery along the beautiful road we call Imbrecciata, which goes to the Vomero village" (Celano 1692, 41), these events really clarify the complex, and in many ways intertwined, plot between public needs and private and religious interests in the configuration and management of the *extra-moenia* urban space between the 16th and seventeenth centuries. As mentioned in a document stored in the Neapolitan State Archive (Processi Antichi f. 3036, fs. 69,826), the Vomero road uphill, an ancient and very steep "cavone", became a real street only in 1579 (Fig. 4), when a group of inhabitants asked the viceroy, Marquis of Mondéjar, to pave and rectify the uncomfortable route now become "una bella via che cala dalla montagna […], nella qual si van edificando nobili et bellissimi edificii" ("a beautiful road that comes down from the mountain […], where noble and wonderful edifices are being built") (Strazzullo 1975, 232–235). Among these, the previous year Giovan Battista Terracina had sold to Andrea Terracciano a small farm, an original portion of the vast rural properties of the female Benedictine monastery of St. Marcellino sold in 1480 to Eliseo Terracina, and since then part the family's assets (Croce 1901, 129).

In October 1621, the farm was bought back at a fiscal auction by Giulio Cesare Guadagno, husband of Delia Terracina and influential royal notary who in the same year had been appointed by the viceroy, the Cardinal Zapata, to end the long dispute over the division of the Neapolitan properties of Bernardino Sanseverino, last prince of Bisignano, passed away in 1606 without direct heirs (Galasso 1992, 43).

Perhaps, Guadagno's interest for the Chiaia hill estates—a place strongly coveted by the city elite—had been raised by the definitive assignment to Michele Antonio Orsini, Duke of Gravina, of one of the largest estates in the hamlet, the same "palazzo grande con giardino" ("big mansion with garden") inherited a few years later by his sister Felicia Maria Orsini and then donated in 1627 to the Clerics Regular of the Mother of God for the foundation of the new church of St. Maria in Portico (Pignatelli 2017, 74–75).

In the viceregal Naples, the co-participation of powerful and wealthy citizens in the foundation of new ecclesiastical institutions was not unusual, but this custom appears to be absolutely decisive in the settlement of the post-Tridentine religious communities above all in the Chiaia hamlet, that was progressively becoming an area of high residential value, privileged seat of the most recent and rich Neapolitan and Spanish ruling classes. I refer, for example, to the conspicuous donations of Michele Vaaz (a very rich Portuguese merchant who moved to Naples at the end of the century) for the reconstruction of the Celestinian complex of the Ascensione (1622), to the financings of the noblewoman Isabella Mastrogiudice and of the Viceroy, Count of Pigneranda "molto affezionato all'ordine degli Scalzi" ("who had a great devotion to the Order of Discalced Carmelites") for the foundation of the Carmelite monastery of St. Teresa (1622–1650), or to the crucial support of Giovanna d'Austria Branciforte (Emperor Charles V's nephew) for the transformation into a Theatine convent of the ancient Carmelite church of St. Maria della Victoria, later used as a Jesuit congregation but abandoned by the followers of St. Ignatius around 1620 (Celano 1692, 40–41)

According to similar customs, in the spring of 1622 Giulio Cesare Guadagno strongly wanted to contribute to the establishment of the new regular Orders in the Chiaia hamlet, donating his farm along the Vomero uphill road to the influential Society of Jesus, that had been looking for quite a long time for a building in that hamlet to be used as a boarding school for the education of young people (Speme 1957, 238). This offer was initially accepted by Father General Vincenzo Maggio but then refused, after a careful consideration, because the property resulted to be unsuitable for school functions, difficult to access and, above all, too far from the usual main roads (Fig. 5).

To demonstrate a *modus operandi* well consolidated and calibrated in a precise view of dominance—not only territorial, but also social and cultural—on the hamlet population, the following year the Jesuits persuaded the royal notary to buy for them a small chapel and a little group of houses along the coastal Riviera road "in excambium dicti alterius loci eis donate" ("in exchange of the previous other place they had been donated"), and securing themselves at the same time as many as 36 thousand ducats for the construction and the maintenance of a boarding school "per coltivar con sommo profitto il rozzo e numeroso ceto de' pescatori" ("to nurture with the greatest result the uncouth and numerous class of fishermen"), as asserted by Father Saverio Santagata in 1757 (Santagata 1757, 235–236); according to the same author, also Felicia Maria Orsini

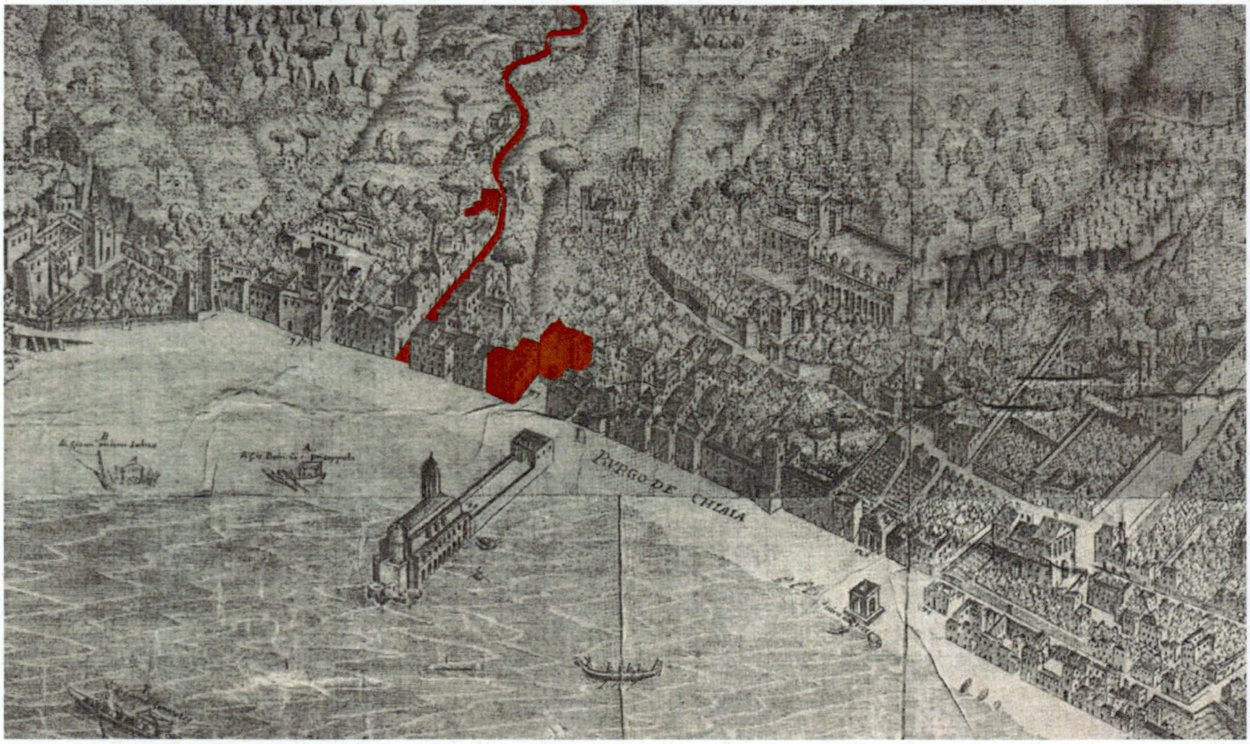

Fig. 4 Alessandro Baratta, Fidelissimae Urbis Neapolitanae…, 1629. Detail of the Chiaia hamlet. In the picture are highlighted the Vomero uphill road (1), and the Benedictine (2) and Jesuit (3) settlements

Fig. 5 Gaspar Butler, Naples from Chiaja, 1725. On the left are visible the religious complexes along the Vomero uphill road

would have promised her estate to the Jesuits, but this hypothesis is not supported by any documents.

Determined to tie his name to the likewise powerful Benedictine monastery of Montecassino, the following year Giulio Cesare Guadagno offered the use of the hill farm to a modest and less demanding monks community coming from the Neapolitan eastern hamlet of Barra, willing to move in a peripheral but strategic area for the commercial traffics with the hill farms of Vomero and Antignano, and to which the newly renovated climb was the main connection. In April 1625, the contract with which the royal notary—indicated as Founder of the new monastery—donated to the Benedictines its own farm was signed. He also donated them a thousand ducats, to be paid in five years "in fabricam ecclesiae et monasterii in dicta nova abbatia construendae et aedificandi, iuxta modellum conficiendum per architectos ad ambarum partium soddisfactionem" ("for the building of the church and the monastery in the new abbey, according to the project that pleased bothof them"), as mentioned in a further document stored at the Neapolitan State Archive (Corporazioni religiose soppresse, v. 1331).

The works for the monastery and the new church, designed at the end of 1624 as mentioned in an unpublished document recently kept in a private archive, never started because in 1630, at Gaudagno's death, every donation was suspended by his sons, determined to return in possession of the paternal hilly property (Pignatelli 2014b, 127).

Deprived of any possible annuity except for the rents on the back garden, the small Benedictine complex underwent modest expansions of a practical nature: in 1628, for example, a cistern and a little stable were built inside of the courtyard, two rooms were added for kitchen and refectory use, as well as two private rooms to be reserved for the Prior. In the middle of the century, as mentioned in a manuscript kept in the Neapolitan State Archive (Corporazioni religiose soppresse, v. 1322), the complex consisted of a small chapel with a sacristy (temporarily set up in a pre-existing room on the ground floor), a large common room, just ten private rooms, a cellar and a stable, as well as a small garden planted with citrus fruits, figs and vineyard, still recognizable in its original borders in the eighteenth-century plans.

Other modest extension works were promoted between 1666 (when a large room and other five small rooms with a connecting corridor were built on the first floor) and 1693, when the infirmary and a small guesthouse were added. In 1682, a new water cistern was dug for the three small rooms rented along the public road, proving that it had been necessary to rent a part of the monastery as well as a substantial portion of the garden, divided and rented for ninety ducats annually (Pignatelli 2014b, 129).

Thanks to the new and constant annuities guaranteed by the Benedictine Monasteries of Montecassino, Montescaglioso, St. Lorenzo in Aversa and St. Severino e Sossio in Naples, in the middle of the eighteenth century, the interior spaces and the courtyard were reorganized, the entrance door was enlarged to facilitate carriages entry and new six rooms were built on the first floor. Reading the monastery very detailed economic reports, it emerges that only at the end of the eighteenth century, the small community occupied an important role in the westernmost part of the Chiaia hamlet, despite the presence, along the same Vomero uphill road, of the new monasteries of St. Francesco degli Scarioni (founded in 1704 by the Tuscany merchant Francesco Scarioni for a large women Franciscan community) and, above all, of the Saints Giovanni and Teresa. This prestigious institution was founded in 1746 by the Neapolitan Discalced Teresian nuns of St. Giuseppe in Pontecorvo in place of the pre-existing farm of the judge Ottavio Gaeta, transformed into a hospice and dormitory with the decisive contribution of the sovereigns, Carlo and Amalia di Borbone (Amirante 2004, 147).

The construction of the new church of S. Benedetto (promoted by the religious since 1693 to replace the primitive chapel, now insufficient to meet the needs of the increased population of the hamlet as prescribed in the deed of donation), were started in 1706 and ended three years later (Fig. 6). In 1709 was indeed finished the stucco interior decoration, and the choir, the confessionals and the wooden organ were placed; in the same year were also delivered two Nicola Malinconico's paintings for the lateral chapels (Rizzo 1987, 158). Only in 1717 was completed the main altar, and in 1730 the lateral altars and the polycrome "riggiole" flooring.

The project was entrusted to the Royal Engeneer Arcangelo Guglielmelli, a leading figure in the XVIII century renewed Neapolitan architectural panorama (Pignatelli 2020 in press): since the same 1693, on the other hand, he was the Montecassino Ordinary architect on the initiative of the Abbot Erasmo Gattola (not by chance the Prior of the Chiaia monastery from 1700 to 1702), also providing various projects for some Order's new churches (Amirante 1990, 217).

The original stucco decoration and the singular elliptical plan with a deep choir and two little lateral chapels dedicated to the Holy Crucifix and to the Madonna (Fig. 7), were unfortunately altered by the many alterations that the church underwent as early as the middle of the eighteenth century, preventing from a correct reading of the overall Guglielmelli's project. In 1743, due to serious structural problems, the small dome—still visible in a coeval Gaspar Butler's painting (Fig. 5)—was demolished and substituted by a vaulted ceiling; also the simple porticoed façade along the Vomero uphill road (still visible in eighteenth-century plans, and similar to the solution adopted in 1688 by the same Guglielmelli in the nearby church of S. Maria in Portico), disappeared at the beginning of the nineteenth century after the suppression of the Benedictine community, when the church was progressively incorporated inside new buildings.

In September 1806, the monastery was indeed one of the first Neapolitan religious complexes to be abolished by the French government, and confiscated among the State assets together with the church and the garden. In December of the same year, the entire building was purchased at the fiscal auction by the lawyer Giovanni Tarantini, and gradually transformed into a residential complex.

Until the beginning of the last century, anyway, the church, the sacristy, and the crypt remained open to worship at the expense of the Tarantini family. Then abandoned for over twenty years, the building was purchased by the Reverend Luigi Spitaleri, and raised to a parish in 1927 by

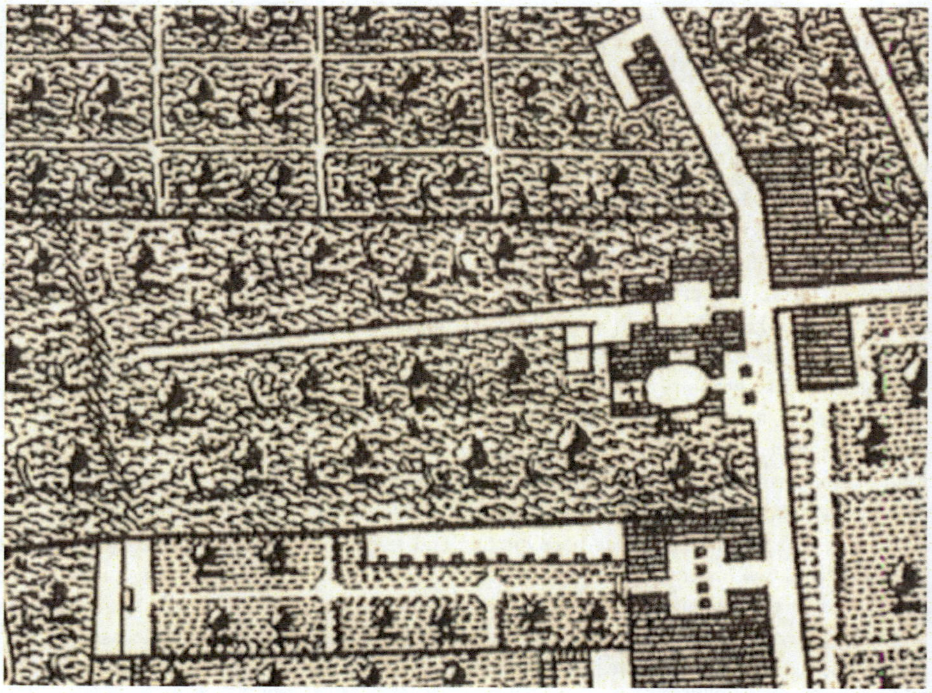

Fig. 6 Giovanni Carafa, Mappa Topografica della Città di Napoli, 1775. Detail with the monastery and the new St. Benedetto church

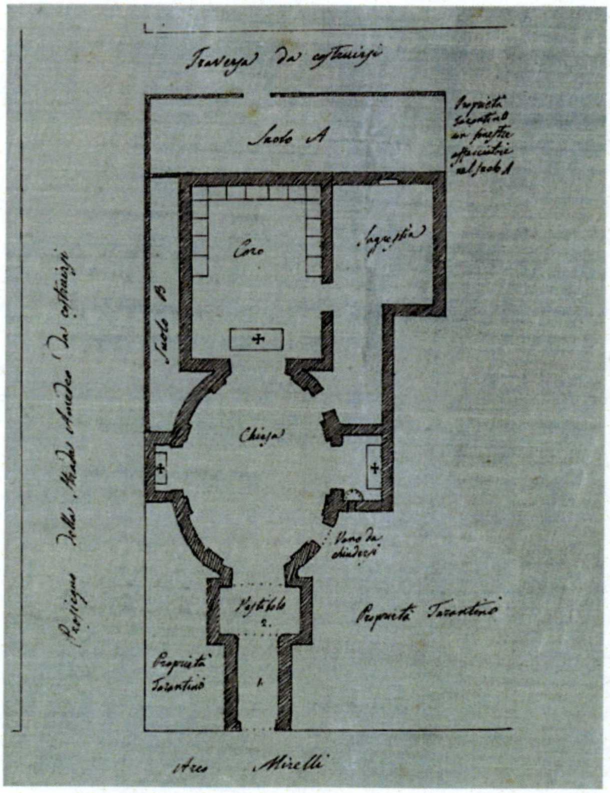

Fig. 7 Plant of the church of St. Benedetto, 1918 (St. Benedetto parish Archive)

Neapolitan Archishop Alessio Ascalesi. Between 1953 and 1965 the simple altars of white marble, the polychrome "riggiole" flooring, the wooden choir, the eighteenth-century organ and all the original stucco decoration disappeared forever (Speme 1957, 274).

The constructive events of the St. Giuseppe a Chiaja Jesuitic complex (also founded, as mentioned above, with the decisive contribution of Giulio Cesare Guadagno) are widely documented, and well known to the Neapolitan architectural and artistic historiography since the second half of the seventeenth century.

In 1623, the Jesuits took possession of a little group of houses at the end of the Riviera road, progressively adapted to be used as a boarding school, and later as a seminary (Divenuto 1990, 141). In 1666 works began on the construction of the new church, traditionally attributed to the Jesuit architect Tommaso Vanneschi, completed in 1673. Here too, the general organization of the building—extremely functional—responded to the Company custom. The boarding school symmetrically grew around the wide two-story cloister, whereas the church (a unique square space separated from the two large side chapels by a double serliana) (Fig. 8), as well as the double façade (completely redesigned in 1836), were clearly inspired by the Theresian church of St. Giuseppe delle Scalze a Pontecorvo, designed by Cosimo Fanzago in 1643 (Divenuto 1992, 131; Cantone 1992, 44) (Fig. 9).

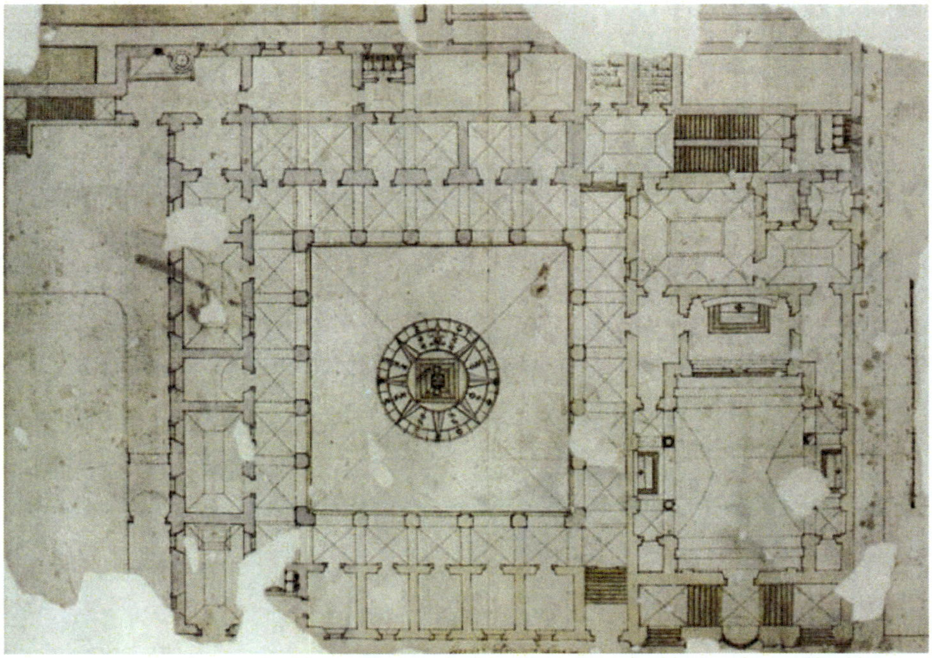

Fig. 8 Plan of the monastery of St. Giuseppe a Chiaja, 1690 ca. (State Archive of Naples)

Fig. 9 Façade of the church of St. Giuseppe a Chiaja (in Petrini 1718)

5 Conclusions

Different settlement strategies leaded therefore to different outcomes despite the two sites being very close to each other. During almost two centuries of life, the Benedictine complex will never get to grow and to establish roots in the hamlet because of its condition of marginality (the Vomero uphill road continued to represent the western limit of the Chiaia build development until the middle of last century), the absence of annuities and new backers (in addition to the few donations of the faithful, and to the incomes guaranteed by the monasteries of Montecassino, Cava, Montescaglioso and St. Severino and Sossio, the Chiaia monks could only count on the rent of the garden and of the four "bassi" along the public road) and, above all, because of the settlement of the influential monasteries of St. Francesco degli Scarioni and Saints Giovanni and Teresa, both subjected to the Royal Patronage. Since its foundation, the settlement has been so forced to the ambiguous role of temporary house and convalescent home more than a real monastery, in a continuous succession of Abbots, Priors and monks coming from the different Cassinese communities of the Kingdom of Naples until the final suppression at the beginning of the nineteenth century.

The Jesuit complex had, instead, a very different destiny: strategically located—as required by the strict and forward-looking Jesuit rules—along the Riviera di Chiaja, the most

Fig. 10 The residential complex built around 1930 in place of the Seminary of St. Giuseppe a Chiaia (Municipal Historical Archive of Naples)

important and busy street of the hamlet, during the seventeenth century, it became in fact the principal point of reference for the whole western Neapolitan coastline. Thanks to the financial support of numerous private donors and, above all, of the same Society of Jesus (in 1649, the Jesuit father Scipione Cosso donated almost 100thousand ducats to resume expansion work), after the death of Giulio Cesare Guadagno the new boarding school was enlarged, and certainly completed around 1670.

Abolished on the occasion of the expulsion of the Jesuits from the Kingdom of Naples in 1767 (well before the Benedictine monastery), the St. Giuseppe complex will nevertheless continue to guarantee educational and assistance functions, partially adapted two years later in a Marine boarding School and in 1818, after the French occupation, in a blind-hospice supervised by the Royal Albergo dei Poveri under the naming of St. Giuseppe and Lucia (De Luzenberger 2001, 29); the remaining portion of the complex was, instead, turned into a cloth factory (Romanelli 1815, 168).

Definitively closed after the Unification of Italy, the complex was sold, and replaced in early twentieth century by a modern residential building (Fig. 10), whereas the church had become a parish already at the end of the eighteenth century.

References

Amirante, G. (1990). *Architettura napoletana tra Seicento e Settecento: l'opera di Arcangelo Guglielmelli* (Vol. 6). Naples: Edizioni scientifiche italiane.

Amirante, G. (2004). Il tempio di Carlo: la chiesa dei SS. Giovanni e Teresa all'Arco Mirelli. In Pessolano MR & Buccaro A, (Eds). *Architetture e territorio nell'Italia meridionale tra XVI e XX secolo. Scritti in onore di Giancarlo Alisio* (pp. 147–163). Napoli: Electa.

Cantone, G. (1992). *Napoli barocca*. Bari-Roma: Laterza.

Celano, C. (1692). *Delle notitie del bello, dell'antico e del curioso della città di Napoli, per i signori forastieri*. IX. Napoli: Giacomo Raillard.

Colletta, T. (1985). Napoli la cartografia pre-catastale. *Storia della citta, 10*(34–35), 5–178.

Croce, B. (1901). La casa di una poetessa (Laura Terracina). *Napoli nobilissima* (pp. 129–138). Arte tipografica: X. Napoli.

De Luzenberger, M. (2001). *S. Giuseppe a Chiaia e Carminello al Mercato: storia di due collegi popolari napoletani.* Edizioni scientifiche italiane.

De Seta, C. (1981). *Napoli.* Roma-Bari: Laterza.

De Stefano, R. (1963). Storia urbanistica di Chiaia. *Napoli Nobilissima. II. Napoli: Arte Tipografica, 2*(6), 227–239.

Divenuto, F. (1990). Napoli sacra del XVI secolo. *Repertorio delle fabbriche religiose napoletane nella cronaca del gesuita Giovan Francesco Araldo.* Napoli: Edizioni Scientifiche Italiane.

Divenuto, F. (1992). Una pianta centrale per la Compagnia di Gesù a Napoli: le chiese di San Giuseppe a Chiaia e Sant'Ignazio al mercato. In Della Torre S. & Patetta L, (Eds). *L'architettura della Compagnia di Gesù in Italia. XVI-XVIII secolo* (pp. 131–138). Genova: Marietti.

Galasso, G. (1992). *Economia e società nella Calabria del Cinquecento.* Guida editori.

Gentile, E. (1953). I benedettini a Napoli. In *Benedictina* (Vol. 7, pp. 25–60). Roma: Istituto Grafico Tiberino.

Giannetti, A. (1994). *Il giardino napoletano dal Quattrocento al Settecento* (Vol. 1). Electa Napoli.

Guidoni, E. (1990). Storia dell'urbanistica, il Duecento, ed. *Laterza, Roma-Bari.*

Johannowsky, W. (1952). Contributi alla topografia della Campania antica. *Rendiconti dell'Accademia di Archeologia di Napoli, 27,* 139–185.

Napoli, M. (1959). *Napoli greco-romana* (Vol. 5). F. Fiorentino.

Pellegrini, L. (1990). Territorio e città nella dinamica insediativa degli Ordini mendicanti in Campania. *Gli Ordini mendicanti e la città. Aspetti architettonici, sociali e politici, Milano, Guerini studio* (pp. 27–59).

Pellegrini, L. (2003). Presentazione. In Villetti G. *Studi sull'edilizia degli ordini mendicanti* (pp 11–15), Roma, Gangemi.

Pessolano, M. R. (1998). Napoli nel Cinquecento: le fortificazioni "alla moderna" e la città degli Spagnoli. In *Restauro* (Vol. 146, pp. 59–118). Napoli. Edizioni Scientifiche Italiane.

Petrini, P. (1718). *Facciate delle chiese più cospicue della città di Napoli.* Napoli, Paolo Petrini.

Pignatelli, G. (2014a). Come una città separata. Chiaia da borgo extramoenia a quartiere borghese. Napoli, Edizioni Scientifiche Italiane.

Pignatelli, G. (2014b). Un insediamento religioso ai margini della città: il monastero napoletano di S. Benedetto a Chiaia. In Cioffi R. & Pignatelli G. (Eds). *Intra et extra moenia. Sguardi sulla città fra antico e modern* (pp. 125–132). Napoli, Francesco Giannini.

Pignatelli, G. (2017). Palazzo Nobile a Chiaia. Un poco noto episodio di edilizia residenziale tra Cinque e Ottocento. *Archivio Storico per le Province Napoletane., 135,* 69–87.

Pignatelli, G. (2020). *Il monastero ritrovato. S. Benedetto all'Arco Mirelli.* In Press.

Rizzo, V. (1987). Altre notizie su pittori, scultori ed architetti napoletani del Seicento (dai documenti dell'Archivio Storico del Banco di Napoli). *Ricerche sul '600 napoletano, Saggi e documenti per la storia dell'arte* (pp. 153–175), Milano, L&T.

Romanelli, D. (1815). *Napoli antica e moderna* (Vol. 1). nella Tipografia di A. Trani.

Santagata, S. (1757). *Istoria della Compagnia di Gesù appartenente al Regno di Napoli.* Nella Stamperia di Vincenzo Mazzola.

Speme, S. (1957). Il monastero di S. Benedetto a Chiaia. In *Benedictina* (Vol. 11, pp. 235–274). Roma: Istituto Grafico Tiberino.

Strazzullo, F. (1975). Documenti del '500 per la storia dell'edilizia e dell'urbanistica del regno di Napoli. *Napoli Nobilissima Napoli, 14* (2), 229–238.

The Paradigm of the Possible City

Mariacarla Panariello

Abstract

"What is the city for us today? I think I wrote something like a last love poem to the cities, when it became increasingly difficult to live in them as a city. Perhaps, we are approaching a moment of crisis in urban life and invisible cities are a dream born of unliveable cities". With these words, Calvino, in a lecture held in New York in 1983 denounces the failure of the "futuristic" Utopia that through the exhausting race towards innovation, a disharmony has generated towards nature (entropy) making the city calibrated within the human dimension and in some cases uninhabitable. The soul of a city is generated by those who pass through it, while walking through the "city of stone" it is alive because it illustrates to those who stop to read it, the past, the events and the space. The memory of the city emerges from the relationship between cultural and religious phenomena (symbols, rituals and institutions) and the analysis of spatial contexts where they took place. The heritage to be transmitted to future generations is constituted by the legacy of the past which reaches us through tangible works (monuments, sculptures, landscapes of uncommon beauty) "joint creations of man and nature" and intangible; perhaps, we could say "invisible" which allows us to interpret the city into mental images. In the anthropocentric idea, social capital is opposed to the individualist agent because it is aware of the role connected to active citizenship through communication and information. A participatory technological neo-humanism is desirable. We can observe through a functional and aesthetic reorganization study of the city, a perspective of sustainability which is able to enhance the harmony between the parts. This draws inspiration from the balance of the classical renaissance city to improve the relationship with its fruits of nature and good food, giving an enhancement of the memory and identity of the places with the aim of a better quality of life.

Keywords

Ethics • Urban planning • Cultural heritage

1 The Failure of the "Futuristic" Utopia

"*Perhaps, we are approaching a moment of crisis in urban life. Today, we speak with the same insistence of the destruction of the natural environment as the fragility of the great technologic systems, which can produce chain failures, paralyzing entire cities. The crisis of the city is too great compared to the other face of the crisis of nature. (…) What is at heart is to discover the secret reasons that led men to live in cities, which can be valid beyond all the crisis. Cities are a combination of many things: memories, desires, signs of a language, cities are places of exchange, but these exchanges are not just exchanges of goods, but exchanges of words, desires, memories*" (Calvino 2016).

With these words Italo Calvino held a conference on March 29, 1983 to the students of the Graduate Writing Division of the Columbia University of New York. The reflections that Italo Calvino subjects to the audience pose a series of interrogatives that lead to investigate the transformations of the city by questioning human nature. This gruelling race to the new, the lights, the crowd, the movement, the rapid ageing of the neighbourhoods, the great empty spaces of the demolitions, the new buildings, the continuous making and undoing, capture the city's existence in constant transformation, a place of perennial building sites, confusion, spatial and temporal disorientation. The extreme rationalism that gives life to a world where everything is coordinated and organized rationally in order to remove every element of the unforeseen and disorder seems to contain in the Cartesian observation of the "*Discourse on the*

M. Panariello (✉)
University of Campania "Luigi Vanvitelli", Caserta, Italy
e-mail: mariacarlapanariello@gmail.com

C. Gambardella et al. (eds.), *Advances in Utopian Studies and Sacred Architecture*,
Advances in Science, Technology & Innovation,
https://doi.org/10.1007/978-3-030-50765-7_27

method" its essence: the zeroing and refoundation of knowledge, a manifesto of modernity and its radically innovative spirit. The philosopher sad it is better to build from nothing, having first destroyed every pre-existing reality and starting from scratch, rather than trying to improve what is there with corrections or additions. The city stands as the material exemplification of Utopia demonstrating the possibility of its realization but at the same time demonstrates its limitation: The consciousness that through its material determination stops the evolutionary possibility that comes from man, from his evolution from the change of his yearnings and aspirations. The meaning of this model city is reflected in those architectures painted to grasp the sense of the Utopias which are represented there. Imagine the manifesto of futurist architecture signed by Antonio Sant'Elia (Figs. 1 and 2), can be seen as "*creating a futuristic home, an architecture that has its "raison d'être" within special conditions of modern life*", highlights how "*each generation must build their city*", just think of his "*housing block with external elevators, gallery, covered walkway, on three street levels (railway line, road for cars, metal walkway) lights and wireless telegraph*" (Adorno 1992) or to the silent and mysterious squares of De Chirico, until you arrive at functionalist experiences.

At a time, when the city tries to give a material answer to the Utopia which it implies to, instead of affirming a possible balance between social conflicts, which translate into the illustration of a stasis and become a denial of social Utopias.

The city which is brought back to us by these experiences is a dismembered and dismantled city whose essence is lost

Fig. 1 "La Città nuova", Antonio Sant'Elia-web render

Fig. 2 "La Città nuova", Antonio Sant'Elia-drawing (Adorno 1992)

and disappears. It is a city that does not value interpersonal relationships and is far from the human dimension as a city that has generated a discrepancy with nature by producing excessive systemic entropies. In some cases, a conflict emerges between the functionalism of a part of the city and its semantic power, a conflict between meaning and function. The city is like a factory of marked and unmarked elements which "*speaks to its inhabitants, while we live inside it by walking through its streets and appreciating its history*" (Barthes 1967).

Today, international journeys are not just journeys through a series of different places but routes from one point to another with a gap in the middle, above the clouds for international travel, underground in the city.

For Italo Calvino, the ideal city is very far from the futuristic metropolis of today that have lost harmony with landscape and nature to the point of being considered unliveable; so a sequence of images, which dialogue with analogical, metaphorical and symbolic thought, flow in a suspended and indefinite time where the future is mirrored, and at the same time is anticipated by the past. Calvino discovers the relationships of the city, whose factories are made up of invisible threads that link the poles of matter and images. A memory of the past emerges that the city "*which remains like the lines of a hand*" is made up of "*links between the measurements of its space and the events of its past*". "*Cities are an expression of desire and memory, of diversity and of routine, of what is temporary and of what is permanent, of life and death, of what has yet to be born, of images, of symbols and of maps, of identity, of ambiguity, of reflections of the liveable and the invisible, of harmony and disharmony, of justice and injustice, of labyrinths and traps, of beauty and brutality, of metamorphosis, destruction, rebirth, continuity, possibility and change (Rapetti 1974).*

Since the Greek period, we can see in the Utopian tales, the metaphor of the shipwreck which was used as a symbolic element, a purification, while today we are aware of being in a reality of a perennial condition of drifting. *Dialogue is a broad panorama of Utopias and Cacotopias that explore a marvellous aerial fantasy world, where the relationships between people belong to places"* (Moneti 2012).

In the book "The Invisible Cities" there is a transformation of the cities into a unique and continuous city, similar to the busy life we have today, moving from one city to another pursuing your daily routine in the metropolis where you originally started your journey. A location must become an interior landscape, so that the imagination begins to inhabit that place and make it its own theatre; the question of anonymity, being able to be in the crowd, watching everyone disappear feeling almost unnoticed. A city absorbs a series of information, similar to an encyclopaedic work or department stores which have their own language, resembling newspaper pages or encyclopaedia entries.

The journey through Paris in the company of Calvino takes place on two levels; the first part is that of the writer who reads the city as if it were a book. The second is that of the man who sees his past reflected in the city's past. Italo Calvino describes "his" Paris (Figs. 3 and 4), the city where he lived, as an iconic city of the twentieth century, with a keen eye for history and changes in progress. In Paris, he is fascinated by cheese shops which are exposed to hundreds of different labelled descriptions, "*the triumph of the spirit of classification*" in some drugstores "*the mercantile esoterism from the first colonialism of the last century, a spirit of universal exposure, a type of shop in which one feels that Paris is the city that has moulded museums in a particular way. Civilization and the museum have shaped the various activities of daily life so that there is no solution of continuity between the rooms of the Louvre museum and the shop windows, passing from the street to the museum, keeping the old one next to the new one allowing the inhabitants to make history travel. One feels that Paris is a city that has influenced museums in a particular way towards civilization. In Paris, even the historical places seem to have an order, observing the perspective that starts from the Louvre, Place de la Concorde, Champs-Élysées, Arc de Triomphe, as if to say that Monarchy, the Revolution, the first and second Empire, are opening towards a forest of skyscrapers in the new district of La Défense*" (Rapetti 1974).

Fig. 3 Metro-Paris France (Rapetti 1974)

Fig. 4 Shop-Paris France (Rapetti 1974)

To review the various stages, where this has evolved, we need to give a new image of a "Futuristic" Paris, which is now difficult to imagine as a living neighbourhood, a set of geometric shapes put there by chance. It is not the randomness of a living city grown as a biological organism, it is a programmed randomness. It gives the impression of a desire for voluntary modernity; it is not possible to gather that aesthetic value that sometimes even ugliness transmits, as a load of contents belonging to an era of astonishment, turmoil, mute facades and distant rumbles.

The enormous excavations that Calvino encountered in the heart of Paris gave him the impression of a terrible violence, a destruction of something that we have lost forever. Since the Middle Ages, in the historic district of the merchants, Zola describes open-air markets which were pavilions built under Napoleon III's reign then later demolished. A city where Merovingian tombs and ruins of Churches of the tenth century have been found, interrupting bulldozers beating against the vestiges of dead ages.

The past remains close to the present, the ages add up and mix, the city becomes a collective memory, in which the architectural detail is a sign of an era, we think of the Gothic Cathedrals where every architectural, and ornamental detail was a sign of something, we can read as a collective unconsciousness, like a large collection, a great piece of clothing. This is a key that we can use to interpret Paris, as an album of our thoughts, like a catalogue of exhibitions. Monsters and ghosts of the unconscious visible outside of our minds.

Calvino said that Paris was the capital of Surrealism and Cinematography, a museum and an encyclopaedia to consult, both for the quantity of films and for the whole network in the study of the Latin Quarter in which it is possible to reconstruct piece by piece the history of Cinematography. Paris is a city that invites collectors, even those who have impalpable memories to a city where nothing is lost and everything is accumulated, as ideological layers, which are deposited, collecting the knowledge as an existential search in which through objects one searches for oneself. This is an existential adventure linked to the sense of discovery, while the attitude of consultation is similar to the catalogue of a museum (Rapetti 1974).

1.1 The City of Stone

The city built along this historical process is the material effort of diverse Utopias put together. It is determined through a conflict of ideas to which the material act of building has given shape.

The urban phenomenon is declined from a historical, cultural and spatial point of view according to a differentiated branching, connected to a plurality of ways of building the city and the urban social structure.

The "*primacy of the cognitive*", which has characterized the tradition of western thought for many centuries, since Plato's time, has been expressed through the rational elaboration of ideal models of life and society. (Mumford et al. 2017).

In modern times, contradictory and confused reality were perceived as a disturbing dimension to be subjected to the control of rules based on rigorous principles of theoretical knowledge. In the age of technology, efforts have been made to reduce the complexity of life to the dimension of productivity and material well-being and to lock reality into the formal structures of technical bureaucratic rationality. The specialization processes linked to the production activity have led to a crushing of knowledge and an absence of absolute foundations of knowledge and moral life. The radical design of Utopia is expressed in the enterprise of zeroing knowledge and in the geometric regularity of urban planning. The extreme rationality leads to a carelessness of the concrete conditions of social life, deprives it of a recognizable public space, leading to a crisis of individual and collective identities and to the emergence of the values of the private and individualistic interest, "*favouring the return to the conception of a primordial/primitive ego predetermined by non-cultural factors and immune to culture and the substantial abandonment, both in social sciences and in widespread opinions, of the notion that the civil order possesses some essential characteristics that are presumed to be non-negotiable and sine equa non*" (Bauman and Cupellaro 2017). The city is in fact concretized in a design that leads to continuous mobility for work, study, free time, etc. The metropolis imagined by the futurists is the "City of machines" corresponding to a new vision of the world. This tendency to impose on the social reality the patterns of a rationality, understood in a preponderant way according to the models of technical manipulation and control derived from scientific knowledge, enter into an irreversible crisis; all the strong paradigms of modern rational thought, from the geometric-mathematical one of the first Utopias to the philosophical-historical one of 'scientific' socialism, no longer find acceptance, because the idea of man has entered into crisis.

According to Heidegger, the modern world has separated man from his space, and it is precisely in this lack of relationship that the real crisis originates, in that the relationship between man and space "*is nothing other than living thought in its essence*". The crisis does not exist in the lack of housing linked to the economic problem but in the eradication of the contemporary man who inhabits space, from the house to the city, without having more roots: "*However, hard and painful, however, serious and dangerous the scarcity of housing*—writes Heidegger—*the real housing crisis does not consist in the lack of housing. The real housing crisis is older than the world wars and their destruction, also older than the increase in the earth's population and the condition of the industrial worker. The real crisis of living consists in the fact that mortals are always in search of the essence of living, that they must first of all learn to live*" (Heidegger 2018).

Heidegger's suggestion is to return to live in the proper sense of *Oikos* and *Heimat*, or the *Polis* when the city was at one with thought. The Greek *Polis* starts from rational bases, the idea of the City of the ancient world that transpires from Platonic Utopias, from urban planning regulations, from descriptions of historians and travellers and from archaeological evidence, is static and spatially delimited. It is characterized, both in the sacred city and in the agricultural area, from the aspiration to dimensional, economic and social stability. The ancient city is a spatial manifestation of its contents, an expression of a community that inhabits those territories, a city of men for men. The Roman territorial organization itself structures the territories through the installation of a typical city, according to a universal model which manages to guarantee an economic and social balance since the territorial units were integrated into a central political system. The relationship of the Romans with the host land is complex and articulated; the hardness and simplicity of rural life remained an ideal always present in Roman consciousness and considered the noblest standard of perfection to be proposed to man in order to form energetic and virtuous natures.

This is accompanied by a desire for a home where you can go once you are free from political activity to spend rest periods, direct agricultural work and monitor vilicus. The centre of these large properties was the villa. "*Following the increase in manpower, the complication of the processes of oil and wine production, ended up creating a type of rustic villae, widespread in Campania and in the richest regions of Italy. Soon, this kind of rustic villa, conceived specifically for the economic exploitation of an agricultural estate, appears too modest to the wealthy Romans who give life to a new architectural model, the suburban villa, very sumptuous, where they spent their free time away from their city occupations*" (Foresta 2012). The garden where beauty and wisdom mixes and responds to an intimate sensitivity to remain faithful to the ancient virtues takes on a particular value. "*The layout of the rooms of the suburban villa had been designed so that Nature was always present: On one side the sea, on the other the countryside*" (Foresta 2012), the owner was never deprived of the view of the surrounding landscape and the spectacle of nature.

In the historically consolidated pattern, urban reality represents a semantic reading of the specific characteristics of the identity of that place, underlining the immaterial and material cultural heritage, from architecture to landscape environments. One of the environmental realities typically developed by the established relationship with the territory it belongs to is the Amalfi Coast, an anthropized territory that has been structured starting from the construction of individual units in spontaneous aggregation. In the process of growth, there is the need to overcome the difficulties caused by the territorial slopes that have been realized by the fusion of the town with the landscape, the sea and the mountains. The construction of urban landscapes takes place through a design for the modification of an area in complex environmental conditions respecting the identity of the places of belonging (Giordano 2012). Next to the "*spontaneous germination*"(Giordano 2012) of the growth process; culture has developed through primary needs, social economy, skills, spirituality and religiosity. These elements structure one's identity, through the elaboration of behavioural and religious codes, knowledge connected to transformation and conservation of places, through actions implemented in respect of the fragile man–nature balance. The impervious nature, characteristic of geographical identity, has conditioned the changes to the territory implemented by man over the centuries. The territory has imposed its orography through changes that it has absorbed over time. The landscape that presents itself is a sequence of steep vertical slopes interrupted by terraces planted with citrus groves and vineyards in a relationship of continuity with the coastal centres and the hilly urban areas.

In these places, the inhabitants have continued to work on the terracing, traced paths, furrows and take care of the up keep of the woods, water courses and possible ways to prevent the outbreak of fires. In the past, they used building materials found on site and implemented ancient Roman construction techniques (lapilli and cocciopesto) to build anything from a rural house to an architectural monument and macerines. This was combined with the Arabic knowledge of blade rotation in the mills and bright decorative ceramic domes. The presence of different types of citizens and environments such as monks, pastors, traders, philosophers and artists who taught the use of spices, medicinal herbs and constituted those mental images that have brought these places closer to the dream-like dimension (Buonincontri et al. 2013). The convention for the safeguarding of intangible cultural heritage, concluded in Paris on October 17, (2003), "*considering the need to create greater awareness, especially among the younger generations, of the relevance of the intangible cultural heritage and its preservation,*" in art 2 defines the intangible cultural heritage:

1. "*intangible cultural heritage" means the practices, representations, expressions, knowledge, know-how-as well as the tools, objects, artefacts and cultural spaces associated with themselves—that communities, groups and in some cases individuals recognize as part of their cultural heritage. This intangible cultural heritage, transmitted from generation to generation, is constantly recreated by communities and groups in response to their environment, to their interaction with nature and their history and gives them a sense of identity and continuity, thus promoting respect for cultural diversity and human creativity. For the purposes of this convention, this intangible cultural heritage will be used solely for the compatibility with existing instruments on human rights and with the requirements of mutual respect of community, groups and individuals as well as sustainable development*".
2. *The "intangible cultural heritage" manifests itself among other things in the following sectors: (a) oral traditions and expressions, including language, as a vehicle of intangible cultural heritage; (b) the performing arts; (c) social customs, ritual and festive events; (d) knowledge and practices relating to nature and the universe; (e) traditional craftsmanship (UNESCO 2003).*

In the light of what has been highlighted up to now, in conjunction with material heritage, as an expression of a community that inhabits the territory, "*traditions become important as an intangible heritage as a representation of an expression of adaptation with the environment through its use*" (Buonincontri et al. 2013). The territorial area of the Amalfi Coast is characterized by the symbiotic relationship between the specific identities of the landscape places and the spiritual, cultural, architectural and artistic heritage. This area was chosen as "*locus amoenus*" by Roman aristocrats and tribunes to build leisure villas along the whole coastline, in fact there are four, specifically in Positano, Minori, Tramonti and Vietri sul Mare, not all in excellent condition. In the Territory of the Municipality of Tramonti, a Villa Rustica was found whose economy was based on viticulture and still continues today, centuries later. Wall structures, a cistern and ceramic fragments have been found. The maritime villa of Vietri sul Mare is located in Bagnara di Marina di Vietri; a "*riggiole*" factory was built on a part of the villa in 1700 while various rooms of the baths are kept inside a pizza restaurant, such as a plant circular and one with rectangular plants and an abyss whose floor was suspended on small pillars under which hot air passed. The circular room had two basins, a circular one of white marble and another rectangular one which was placed right under a rocky wall from which a spring gushed. "*The natural cave which was open at the top was an integral part of the thermal structure datable between the first century B.C. and the first century A. D. The villa of Positano (Fig. 5) was discovered under the oratory of the Church of SS. Maria Assunta but only a part of it has been excavated. Thanks to the collaboration between institutions (Superintendency, Municipality, Curia Vescovile) the project culminated with the opening to the public of the MAR—Museo Archeologico Romano of Positano, making it possible to visit places of extraordinary scientific importance, with its triclinium, mosaic flooring and wall paintings in the IV Pompeian style*" (MAR, n.d.). "*The official date of the start of the Roman Villa excavations in Minori (Fig. 6) was in 1932, when the foreman Luigi D'Amato, while carrying out some works on the property of Mr. Sammarco, opened a big hole in the floor where he noticed a void of considerable size below, which was almost immediately checked. Furthermore, during the river's regimentation works, structures related to environments that still retained the pictorial decoration, perhaps belonging to another villa, or to an oecus, came to light in analogy to other noble buildings of the time. Following the flood of October 26, 1954, the structure was buried again and only later was brought to light while cleaning up the area above, which was occupied by houses and gardens that were gradually demolished and that had used some rooms such as cellars. At the end of the work, the villa is the aspect we see today*" (Villa Romana–Minori, n.d.). The Otium for the Romans was, therefore, a comfort and liberation from daily worries (relaxation animi); the most important feature remains the assimilation of Greek culture, the passion for the literature, philosophy and art. In the Villa, the Roman aristocrats could take on the patterns of Hellenic life and culture much more easily than in the city; the whole course of the day was accompanied by readings or performance, at the spas, during meals or on the hunt. "*Do not agitate me with neither hopes nor fears, do not bother me with any noise. I speak only to myself and my books. Oh innocent and straightforward life! O rare and honourable idleness, more beautiful, almost, of any shop! Oh sea, bathing, real-time secret of the Muses*" (Pliny the Younger, Ep. 1-9).

The territory has preserved its vocation as a pleasant and spiritual place. It is known as the Divina Costiera and was declared a "*World Heritage Site*" in (1997) by UNESCO where it still retains a visceral connection with the *Genius Loci*. In this context, the meaning intended by the Romans was the patron divinity of a place. *The word "Genius" comes from the Latin verb "gignere" which means "to generate, create" and was used to identify the name that constituted the creative force, the "vis generandi" of man. The genius was a central figure in Roman religion; he was consecrated above all "Dies Natalis", the birth of man (but we can think of it as consecrating to our genius the day of the "birth" of our house); and at the same time, he accompanied man in everyday life until his death*". (Viacava n.d) "*Nullus locus sine genio*" "*no place is without genius*" (Battarra n.d.) wrote Servius Mario Onorato, a Latin scholar who lived between the fourth and fifth centuries D. C. every place has its soul, its essence, its tradition.

The tradition of the "*Battenti*" (Fig. 7) has been handed down from the encounters between the Christian religion and pagan rites, a living custom continuing since the fourteenth century by the Archiconfraternita del S.S. Sacramento di Minori, a town on the Amalfi Coast in the province of

Fig. 5 Roman maritime villa—Positano (SA) Amalfi Coast Italy. https://marpositano.it/en

Fig. 6 Roman maritime villa—Minori (SA) Amalfi Coast Italy. http://www.villaromanaminori.com/. https://storico.beniculturali.it/mibac/opencms/MiBAC/sito-MiBAC/Luogo/MibacUnif/Luoghidella-Cultura/visualizza_asset.html?id=152155%26pagename=157031

Salerno. The "Battenti" represents the Passion of Christ in a dramatic key, representing sinners who by covering their faces and bodies in a white dress, encircled with a hemp rope, sing songs of pain. The lights of the city go out and torches and candles are lit. From the sea to the hilly area of the city, in a unique atmosphere, shrouded in mystery, the religious followers advance in the darkness of the night illuminated by torches, step by step they walk along the streets of the city intoning century-old texts and melodies handed down orally from generation to generation. The procession takes place at dawn on Good Friday and they sing "*e vascie*" (below). On Good Friday night, they sing "*e ncoppe*" (above), this distinction arises from the fact that previously there were two brotherhoods in that area, one located in the level ground and the other in the hilly area of the country, carrying on their shoulders the dead body of Christ throughout the procession, which leaves and arrives at the Basilica of Santa Trofimena. The faithful unite in prayer and participate in a conspicuous manner, condoling themselves, "*souls mourn not for the punishment itself but for the sin of which the just punishment is a consequence*" (Grabher), more than compassion for suffering, one sees in the condoling an assimilation to the state of the penitents. The songs were studied by the musicologist Roberto De Simone (n.d.). "*From the observation of the manifestations of Minori's doors, one is struck by a multiplicity of elements whose analysis involves different scientific fields. These range from anthropology to ethnomusicology; from history, to sociology, to collective psychology. And they also touch the direct analysis of the local territory and the field of archive research*".

The inter-religious secular event, one of the richest and most complex in the South, is a penitential rite of Holy Week, handed down, with tenacity, strength, without compromise within the ecclesial authority, by the confreres of the Archconfraternity of the Most Holy Sacrament. Sacrament through the songs of pain. "*These are liturgical chants, of a heterodox liturgy and historically there is little to say because the present speaks for itself. We only need to support this tradition. The world of these men is certainly not that of those who go to discos because the consumer society has destroyed religiosity. These traditions express the richness that is inside each person and that expresses itself when we are together. Emotionality is something that leads to interiority. Interiority starts from a human band similar to the vibrations of the soul. We can say that these songs comfort us because they recognize that the soul is there. It is the soul that vibrates. Believing in the soul or interiority that is manifested through these religious elements is art. Even an artist when he listens or creates is emotional and the inner vibrations are fed by a light of truth. In these situations, it is the soul that vibrates. On Good Friday, these men sing death: a scandal for life. But can the soul die? This is the crulest question. These manifestations tell us that the soul does not die but manifests itself in these things*". (amalfinotizie. it) For about nine years, the song of the "Battenti" was declared a

Fig. 7 The hooded "Battenti" of Minori (SA) Amalfi Coast Italy

Historical Cultural Heritage of the City, as well as "Immaterial Demo-ethno-anthropological Good" by the Mibac, the Ministry for Cultural Heritage and Activities.

1.2 The Legacy of the Future

In a suspended and indefinite time, the past anticipates the future and this is reflected in the past, in an urban "*conflict between meaning and function*" (Barthes 1967) where the city lives at the same time with tradition in the present and future. The social and economic transformations of the territory challenge the inherited models. Pursuing a Utopia of a city without errors, without imbalances, without waste means referring to a timeless, mindful city that wants to become a model of a possible city. If we think of the Renaissance painting "La Città Ideale" (Fig. 8), the painting depicts a city using the scientific principles of the central perspective, made evident by the geometry of the pavement which represents the ideals of perfection and harmony of the Italian Renaissance. In the orderly and symmetrical shape of a city, the absence of people is the dominating element which gives life and makes it Utopian and imaginative. It represents a piece of the city, like an enormous prototype of how we would like the city to be, equilibrium, geometric rigour, it is a historical moment in which a relationship of perfect symbiosis is created between the will of those who govern the city and the image they want to represent; the meaning of the ideal city is reflected in the perfect scan of the empty spaces and is fully painted in perspective to represent the sense of Utopia. The teaching of Renaissance idealism that we can learn and translate into good practices is to study about the roots of our territory, to be aware both at a local, national and international level of our surroundings, of our material and immaterial cultural heritage, safeguard it, enhance it and promote it.

"*If we consider three benchmarks for the systemic study of urban spaces dimension, organization and functioning, then it emerges from the studies that the larger a city the more entropies it presents and the more difficult it is to approach in a way that can intervene according to best practices obtaining satisfactory and lasting results. The city defined as intermediate by several studies (between 50 and 200 thousand inhabitants) has according to different experiences become the one in which it is easier to implement a series of interventions, monitoring the trend often managing to achieve the expected results*" (Fistola 2013). Finally, small cities in which it is possible to implement a series of interventions and actions in order to orient the urban system towards sustainable and efficient states.

"*In the cities, regardless of their size, the religious presence is manifested not only by street celebrations, but also by the services it manages to offer to the territory such as hospitals, educational facilities, meeting places in order to radicalize and territorialize the presence on site. Through these actions religion asks for recognition from public authorities and society while urban space represents the place, the framework in which social bonds are established" (Gonzàlez and Gusman 2016).*

In a perspective of lower land consumption and exploitation of abandoned urban spaces, art is often denied to denounce a situation of urban discrepancy and as a proposal for an improvement project. The territory regains a decisive centrality, because thanks to its particular characteristics (suggestive landscapes, historical buildings, tourism, etc.); it is an ideal mediation element between the manifestations of art and the people who have an affinity level more or less elevated with the unique characteristics of that territory. It becomes the element capable of favouring the opportunity to enjoy art, which in turn generates effective communication methods, aimed at creating a social face-to-face relationship.

A fundamental role is played by art, which integrating with the urban context gives voice and form to the spiritual part of the intangible heritage, often with a social value. Since the 60 s, the authors of artistic practices have felt the

Fig. 8 Città Ideale 1480–1490, Urbino Italy

need to act outside the exhibition of art, reflecting on the idea of public and private space. The doubts that the artists posed were in line with the historical context and assumed an ideological political attitude through a strong self-criticism. The places of contemporary art are born from a questioning of the problems of the current city to which the artist tries to answer by dealing with a different cultural area.

Jorit Agoch, who lives in the reserves of allegory and metaphors, is active mainly in Naples and supports the social recovery of the suburbs through street art. He designs his murals in the suburbs of difficult neighbourhoods such as Ponticelli, Scampia, Quarto, San Giovanni, Forcella (Fig. 9), using actors, footballers, singers, revolutionaries as subjects, to give a face to San Gennaro. He uses common people representing, as Caravaggio did, a sort of sanctification together with the social messages indicated in his works where the face is the fulcrum of all his works, and through the faces, he tells the story of humanity that are bent by the abuses of life and do not stop in demonstrating the nobility of their souls.

On the basis of these reflections, we can affirm that in the future city Renaissance anthropocentrism reappropriates its value, we cannot think of the city except as a function of its inhabitants.

The basis of these reflections is the recovery of social capital, understood as awareness of the citizen to be an integral part of the urban fabric and its evolution. Henri Lefebvre in his book "*The right to the city*" states: "*The right to the city presents itself as a superior form of rights, as a right to freedom, to individualization in socialization, to habitat and to living. The right to work (to the participating activity) and the right to use (very different from the right to property) are implicit in the right to the city*" (Lefebvre 1972).

Through the sense of belonging to the city, social capital acquires value, which is concretized in active participation in urban regeneration.

If we consider the city as a physical unit expressed in a cultural entity combined with the complexity of social capital, this finds its balance in a constructive connection between past and future, in the awareness and memory of places and the sense of reality.

"*Building the living in its various forms, principles and values, taking care of the territory while preserving its quality and ensuring its safety in order to achieve the best conditions in the dialectic between development of society and protection of the environment. These are the results of a complex relationship that involves different subjects, single or associated, driven by different purposes and orientations in the public sphere. It is here that public action, exercised with regulatory devices by entities and institutions, is expressed in some cases through a hegemonic and generative will of other types of behaviour. In other forms it is combined with that of private operators, groups of citizens, social movements,* etc. "…" "*Paths of affirmation of a multiplicity of needs and questions that are not always easy to satisfy on the one hand, and an institutional public action that is too often late and ineffective on the other, have revealed the plurality of ways in which the various communities act by transforming cities and territories*" (*Atti della XX Conferenza Nazionale SIU—Società Italiana degli Urbanisti* 2017).

Fig. 9 Jorit Agoch–San Gennaro, Napoli Forcella Italy

In the face of the degradation and abandonment of public spaces, by using participatory planning, an attempt is made to rediscover the possible relationships between urbanized and empty spaces of nature by creating new opportunities for interaction and social cohesion and inclusion at local level.

Since the 70's, various cultural experiments have materialized in spatial interventions based on values of the memory and historical identity of the places, just think of the creative workshops with the children of the Rione Traiano in Naples by Riccardo Dalisi, which fits into a disadvantaged social context with the aim of realizing a new experimentation of the theory and method of the project through participatory planning aimed at urban and social regeneration. Dalisi establishes a new relationship with the urban context by joining students and PhD students of architecture to the urban lower classes in a process of participation in the planning process not without difficulty. First and foremost the involvement of human capital, specifically children who Dalisi tries to entice initially with the expedient of a small economic incentive and then encourages participation by promoting the practical element of the projects and the experiential playful aspect, to these factors add the use of waste materials such as cardboard and wood by reusing them, based on the creative expression of the drawings, in a creative recycling.

"*Riccardo Dalisi is in the intermediate space between creative involvement and Utopian radicalism, in a sort of ideology of participation through form, especially clear in its deeper and more mature act, which still remains Traiano*" (Gambardella 2005).

In the first decade of the twenty-first century, Tactical Urbanism began to spread, first in America and then in Italy, a methodology of planning and redevelopment which provides for short-term and low-cost actions aimed at urban modification. The movement stems from the awareness that the human capital of the city can contribute in various ways, in a process of active citizenship, to the transformation of the territory aimed at reusing the environment, historical heritage and contemporary architecture. This can happen through different forms, such as using spontaneous actions from private citizens who regain the urban space from local administrations and redevelop them into areas for leisure activities such as parks. The proposals of the municipal administration, such as the requests to adopt a road, a flowerbed, a square, (project Open Squares of the Municipality of Milan) or a monument (adopt a Monument) which takes the form of a concerted action by individuals, groups of citizens, associations and local government for the development of an effective project, with low costs to remove forgotten places from degradation and oblivion. The inhabitants take care of these projects because they feel part of a community, which promotes their valorization and protect them also from a spiritual point of view.

The lesson we can learn from ancient society and the latest experiences of the contemporary city is that "*civitas*", the cooperating community of citizens, is fundamental for the survival of the "*urbs*".

1.3 Conclusions

The human capital within the social and anthropological system becomes a catalyst element from which all possible sustainable processes of evolution arise. The starting point of the possible evolutions is based on the awareness of the part of man where the territory is a resource. However, this should not be exploited until it is definitively exhausted. Any action takes place in conditions of scarce resources and in a scenario of continuous global changes, it is inevitable that every growth, every development process produces entropies, and on the basis of this, we should perhaps talk about possible sustainable developments. Consider the differences between the cities regarding architecture together with the quality and lifestyle over the centuries where history cannot be levelled in favour of a univocal planning forecast. A correct policy for managing a territory is based on the assumption of diversity and a balance deriving from the mutual relationship of these. Particular attention must be paid to the intrinsic characteristic values of the identity of a place, the "*intangible heritage, social behaviour and all the rites, customs and traditions to be valued as an instrument of conscious management of a territory and its culture*" (Buonincontri et al. 2013). However, such differences constitute a wealth of the territory, human relationships with mankind and man's relationship with nature makes him an inhabitant of the city and a conscious being of the Earth. As Aristotle argued, "*the universality that we find in poetic creation is basically very similar to what we can find in science and technology*" and precisely through technological innovation the stone city can be innovated. The value of anthropocentrism manifests itself in the bottom-up approach, as C. Ratti states in his publication "*We are facing a substantial change: there is no longer the idea of the architect, the designer or the engineer who proceeds*" from above "*and how Le Corbusier can afford to impose his vision, but rather a vision that can be born from below*" (Ratti and Mattei 2014) and grow from all of us together. In a new system of networks, man is reflected in a new way of living, conscious, both in social terms, achieving inclusiveness as a factor of social cohesion, enhancing his dimension of belonging and collective responsibility in a vision of cultural enrichment and international projects remaining faithful to the identity of the places combining technological innovation and sustainability which can guarantee developments aimed at a better quality of lifestyle assessed in terms of "*health, education, work, material wealth, and the environment*". There are above all, interpersonal relationships and the ability to live in a participating society. "*These seven categories measure well-being, two horizontal measures adding: equity and sustainability. As has been demonstrated, the addition of these two elements should have an effect on macro variables in the future we build today*" (Ferraro 2010).

References

Adorno, P. (1992). *L' arte italiana: le sue radici greco-romane e il suo sviluppo nella cultura europea. Il Novecento: dalle avanguardie storiche ai giorni nostri.* Messina: G. DAnna.

(2017). *Atti della XX Conferenza Nazionale SIU-Società Italiana degli Urbanisti*, Roma, 12–14 giugno 2017. Retrieved from http://www.planum.net/xx-conferenza-siu-2017-pubblicazione-atti.

Barthes, R. (1967, September). Semiologia e urbanistica. *Op. Cit. Selezione Della Critica D'arte Contemporanea, 10.* http://www.opcit.it.

Battarra, E. (n.d) *Genius gloci: l'arte del territorio.* Retrived from https://www.larcaelarco.it/genius-gloci-larte-del-territorio/.

Bauman, Z., & Cupellaro, M. (2017). *Retrotopia.* Bari: Laterza.

Buonincontri, P., Caneva, G., Maurano, C., & Simeon, M. I. (2013). Patrimonio Culturale materiale e immateriale. *Il Futuro Dei Territori Antichi. Problemi, Prospettive e Questioni Di Governance Dei Paesaggi Evolutivi Viventi. CUEBC—Territorio Storico Ed Ambiente, 4.* ISBN 978-88-7228-716-3.

Calvino, I. (2016). Le citta invisibili. *Mondadori.* ISBN 8804668024.

Ferraro, I. (2010). L' economia della felicità e l'arte di lavorare sul nostro futuro. *Testimonianze Bancaria, 9.*

Fistola, R. (2013, April 8). Smart city: Thinking about urban intelligence. *Tema Journal of Land Use, Mobility and Environment.* Retrieved from http://www.tema.unina.it.

Foresta, S., & Pacetti, P. (2012) *Le Ville Romane della Costa d'Amalfi.* Zelig srl.

Gambardella, C. (2005). *Radicalmente Napoli: Architettura e design.* Napoli: Clean edizioni.

Giordano, P. (2012, June). Il disegno dell'architettura costiera. La costiera amalfitana _ il territorio, le città e le architetture. *La Fabbrica Della Conoscenza, 17.* ISBN 978-88-6542-124-6.

Gonzàlez Dìez, J., & Gusman, A. (2016, June). Religioni e città. Approcci emergenti in antropologia urbana. Introduzione. *ANUAC, 5*(1), 91–106.

Heidegger, M. (2018). *Costruire abitare pensare.* Milano: Lotus.

Lefebvre, H. (1972). *Spazio e politica: il diritto alla citta.* Milano: Moizzi.

MAR. (n.d.). Roman Villa Positano. Retrieved from https://marpositano.it/en/.

Moneti, M. (2012). *Città e Utopia* (Vol. XVII). Firenze, Italia: Annali del Dipartimento di filosofia. ISSN 1824-3770.

Mumford, L., Crespi, F., & DAgostino, R. (2017). *Storia dell'Utopia.* Roma: Feltrinelli.

Ratti, C., & Mattei, M. G. (2014). *Smart city, smart citizen.* Milano: EGEA.

Rapetti, N. (1974, February). Italo Calvino: un uomo invisibile. Retrieved from https://www.youtube.com/watch?v=6jdiCztTLQw.

Roberto de Simone. (n.d.). Retrived from https://amalfinotizie.it/settimana-santa-minori-canti-battenti-roberto-de-simone/.

Unesco Italia. (1997). Operational guidelines for the implementation of the World Heritage Convention. Retrieved from http://www.unesco.it/.

Unesco Italia. (2003, October 17). Convenzione per la salvaguardia del patrimonio culturale immateriale. Retrieved from http://www.unesco.it/.

Viacava, C. (n.d.). La casa è uno spazio "sacro": Il Genius loci. Retrieved from http://www.viacavaclaudio.it/la-casa-e-uno-spazio-sacro-il-genius-loci/.

Villa Romana—Minori. (n.d.). Retrieved from http://www.villaromanaminori.com/

Beyond Functionalism. How the Everyday and the Utopian Meet in Reused Parish Churches

Roel De Ridder

Abstract

The parish churches of Flanders are under permanent scrutiny. Most of the actors involved in the future of the church buildings, however, happen to reason in very 'functional' terms. The main issue—or so it appears—is to find the correct new 'function' for each parish church in order for this 'problem' to be solved once and for all. Building on field research and literature, this paper argues that such a line of reasoning is at odds with the very nature of church buildings. The parish church, this paper holds, is a unique building because of its public 'potential'. Not only does a parish church offer a possible glimpse of utopia, it is also strongly connected with the sphere of the everyday. The main argument of the paper consists of theoretical and practical explorations that connect the everyday with the utopian. Aspects of public space, local engagement and the 'autonomy' of the church architecture are the threads that are being stretched between the everyday and the utopian. From a continuous mediation between literature and practice, this paper seeks to build a theory that, in its dealings with these (old) buildings, is explicitly 'church building related', rather than merely generic and simply 'functional'. Thus, the present situation—architecturally and socially—informs possible futures for the church buildings.

Keywords

Parish churches • Everyday life • Public space • Local engagement

R. De Ridder (✉)
Hasselt University, Hasselt, Belgium
e-mail: roel.deridder@uhasselt.be

1 Introduction

Contrary to so many readings of the Western world today, often characterized by dystopian images of doom that, on top of that, are not even *that* far-fetched—one is reminded of the climate challenge here, for example—the parish church might appear as a quasi-utopian 'potential', a reminder of a simpler time that might carry in it a sometimes even compelling promise of community, meaning, and so on. In Belgium, parish churches that are not decommissioned (yet) are always *public* buildings, property of municipalities or so-called fabric committees,[1] needles to add here that real public spaces are becoming rare, in Belgium and in the region of Flanders as well. What authors like Anna Minton (2017) say about the UK is happening in Flanders on a smaller scale: land, especially when it is zoned as residential area, is becoming a speculation good and therefore scarce. There is a constant pressure on open space and consequently on public spaces too, especially in the more or less densely populated areas. This persistent force is being intensified by municipalities that in turn experience the pressure of the European Union to balance the budgets. Parish churches are not only precious public spaces, but also (and some of them *were* until their recent reuse) being managed by *public* authorities, that is, by the same fabric committees. These committees are public legal bodies consisting of volunteers. And because parish churches are so 'inefficient' in economic terms—since they are voluminous, cold, sometimes dark, and not rarely listed as immovable property heritage and therefore demanding in terms of maintenance, management

[1]Decommissioning, or the act of extracting a church building from the public domain is called 'desaffectation' in French ('desaffectatie' in Dutch), or 'profanation' in colloquial English. The French 'desaffectation' and the Dutch counterpart are legal procedures, whereas 'profanation' and 'ontwijding' in Dutch (literally: 'desecrating') point towards a Catholic service that may or may not take place when a church building loses its public statute, and during which a relic may or may not be taken away.

C. Gambardella et al. (eds.), *Advances in Utopian Studies and Sacred Architecture*,
Advances in Science, Technology & Innovation,
https://doi.org/10.1007/978-3-030-50765-7_28

and use—they can flaunt this almost natural resistance against commodification (as a matter of fact, this resistance becomes operative and legally binding when the building is listed, hence: protected). Following the Italian critic Manfredo Tafuri, one can read in Pier Vittorio Aureli's (2011, 2016) written work a way to understand the church building as a possible 'obstacle' in an urban environment that is most of the time dominated by a generic and flattening 'urbanization'—which is an economically driven process in the first place. Thus, parish churches can boast both a public 'potency' *and* an obstacle-like character. Hence, it is not even strange to view them as 'utopian' places. But, in order to actualize this potency and to hold on to their obstacle character—those are two ambitions, moreover, that are firmly intertwined, certainly when it comes down to church buildings as public spaces—this paper argues that practical and, above all, *human* agency will be needed, as well as a vison of what a church building is (still) capable of. As Franco Berardi (2017, p. 8) says: 'Potency is the energy that links a possibility inscribed in the present with its subject'. As will be shown later on, this 'subject', in its most active and engaged form, *is* present in (some places in) Flanders. And moreover, the present-day Flanders turns out the be a place where the circumstances are good for developing a vision on the many parish churches.

In Flanders at the moment, not all church buildings are still being used as churches. For the most part, this is a quite recent phenomenon, propelled by a 2011 policy note (Bourgeois 2011; De Ridder 2013), which in turn was an answer to a fairly new reality. The futures of many parish churches in Flanders are currently being rethought because of a lack of priests, the consequent redesign of the 'pastoral map' and the closing down of churches by (some of) the bishops. Municipalities and cities contribute to plans regarding the reuse—partial or full—of parish churches because they still are the main financial backers of the buildings, while at the same time they know that the churches have to deal with a decreasing attendance (for traditional mass *and* for 'special occasions' such as funeral ceremonies and wedding masses). Lots of feasibility studies—for more than 100 of the approximately 1800 parish churches in Flanders—are currently being conducted or were recently finished, most of them under the auspices of the 'Projectbureau Herbestemming Kerken' (the project agency for the adaptive reuse of church buildings, launched in 2016 as a temporary collaboration between the Centre for Religious Art and Culture, the Flemish government architect, The Flemish Agency for Immovable Property Heritage and some intermediary actors). This paper draws from several studies on church buildings, commissioned by municipalities and/or fabric committees and conducted by the author; most of these projects started before the 'Projectbureau', although the author is involved in that as well.[2]

To somewhat delineate the issue at hand—the future of the Flemish parish churches—this paper focuses on the more fundamental aspects of that issue. The emphasis is on 'the everyday' and, more precisely, on possible ways in which 'the everyday' can be both inspirational and effective to allow for a revaluation of the quasi-utopian qualities of parish churches mentioned above. This focus on everyday life is fundamentally linked to the Belgian/Flemish situation, characterized by fabric committees and other local volunteers who take care of the church buildings. Underlying, throughout the text, there is the hypothesis that 'another' way of conceptualizing the future of parish churches is possible, a way that is not about the kind of 'functionalization' (freely after Henri Lefebvre, as cited by Gardiner 1995, p. 100) that is present in the adaptive reuse of parish churches discourse (propelled, again, by the 2011 policy note, albeit to a lesser degree now than in 2011 and directly after). Whether it is the direct result of local governments pushing towards functional, i.e. problem-solving solutions for the emptying church buildings in a quest for economically sound schemes, or whether architects and other process managers suffer from the complexity of the issue and hunger for similar functional answers, it is striking that only in particular cases, the full social and cultural potential of church buildings is being taken into account—rather than to look for a quick-win that is. So, next to focusing on the everyday, this paper tries to show—in an introductory way—how the full potential of parish churches can be taken from there. For that purpose, the text looks into social, legal, urban, architectural and—to a lesser degree—pastoral aspects of the matter.

In this rather explorative paper, roughly, three threads are being stretched between the everyday and the utopian; they are public space, engagement or 'praxis', and to a lesser extent the autonomy of the church architecture. Although theoretical insights are often the mainspring in texts like these, in practice, most of the theoretical input will be nuanced along the way through the experiences of real cases. This is characteristic for the current explorative phase in Flanders: preconceptions are constantly being challenged through working in practice, with real buildings and real people. Before the paper will zoom in on the three

[2]This paper draws from (services) research projects conducted by the author, from within the research group ArcK (Hasselt University), and commissioned by the central fabric committee of the municipality of Kortessem (Flanders), the city of Hasselt (Flanders), the municipality and the central fabric committee of Maasmechelen (Flanders), the city of Genk (Flanders), the central fabric committee of the city of Bree (Flanders), the municipality of Heers (Flanders, research project together with 'a2o-architecten'), and the city of Tongeren (Flanders). Some insights have grown from working for the 'Projectbureau' (in a team with the private firms of RE-ST, TRANS, Petillon Cueppens and Endeavour).

consecutive threads, however, some challenges with regard to dealing with church buildings in present-day Flanders will be introduced.

2 Between Autonomy and Pragmatism: The Complex Nature of Parish Churches

For the Flemish parish churches in this day and age to operationalize even the slightest glimpse of the already mentioned utopian potential, many challenges must be overcome. To begin with, parish churches are rather complex matters. Therefore, the slightest intention to intervene in a church building one way or another will confront the building with many and very diverse actors. Policy-wise, one can think of the involvement of immovable property heritage, spatial planning, culture and local governments. And apart from that, there is also Canon law (ecclesiastical law) and the sometimes practical involvement of the different levels of the Catholic Church—since Flanders is (or was) mainly Catholic. It is also not uncommon for citizens—religious or not—to feel involved with regard to 'their' church building. Furthermore, the contemporary study of church buildings is not limited to one field of expertise only. Several experts—acting interdisciplinary or apart from each other—are concerned with the future of church buildings. Scholars from several fields of expertise—theology, anthropology, pedagogy, sociology, history, heritage studies, philosophy, (Canon) law, architecture and so on—made valuable contributions to the issues, directly or indirectly (De Ridder 2013). So, when the current situation is being examined in theory and in practice, a clear-cut separation between the many discourses and areas of expertise is rather hard to maintain. In practice, traditional church-related matters meet with the more substantive or fundamental aspects and with practical questions of adaptive reuse. What is more, church buildings seem to challenge the norm—if there is one, that is. They do not respond to one voice only. With Bruno Latour's (1991/1993) schemes about the fact that there never really was a clear-cut separation between the nature, culture and other spheres, and also his schemes about the proliferation of bridging networks—even in the modern era,—in mind, the parish church presents itself as an illustrative case of 'hybridization' as the French sociologist of science calls it.

Apart from 'complexity', parish churches suffer from other challenges. For one thing, it is problematic to simply equate all parish churches with public spaces. Undeniably, a fairly general 'will to public space' is present in the operations of the Projectbureau and therefore in more or less the entire discourse on parish churches in Flanders. Both the organization itself and the architects involved strive for parish churches that remain to be meaningful and public. But the 'clients' of the Projectbureau, soliciting feasibility studies for 'their' churches, are the Flemish municipalities. And as already mentioned, these municipalities feel the pressure of the generally underused parish churches on the municipal budgets. Accordingly, the goals of the municipalities that contact the Projectbureau are not of a fundamental nature, but instead are very practical and pragmatic. The cities and municipalities are mainly investigating, via the Projectbureau, whether the parish churches on their territory (and depending on their financial interventions—more on that below) can become more 'relevant'. The 'functionalization' of parish churches is not far away then. Most of the time this does not happen out of bad intentions, but it is clear that under these circumstances parish churches will not easily become 'universal' public spaces. A tension remains between fundamental intentions and the pragmatics of the field. Much is relying on the goodwill (and the right practical preconditions) of the municipalities *and* the dioceses, because the latter—through the bishop—can decide whether or not the church will be reused.

Even when there is the local will to invest in a truly public church building, other challenges are there to be conquered. As many (feasibility) studies conducted by the author have shown, working on the future of parish churches is not solely about good ideas and intentions, but even more about the right process and, finally, about solid management. The two latter traits happen to be firmly entangled, as the right process is a stepping stone to decent management. When a process is designed in such a way that possible managers are involved or motivated to stand up during the process, bridges between ideas on the one hand, and the actual future performance of parish churches on the other, can be built. Still, feasibility, the moderation of the conversation between all the actors involved and future management of the church building should not be taken lightly. If properly monitored, a difference in favour of a real *public* building can be made, thereby contributing to operationalizing the public potential that is present in churches.

Concerning the Flemish parish churches, questions arise: who will pick up the gauntlet and start tackling these practical issues (management and daily maintenance being the most poignant ones), and who will come up with feasible financial plans? How will the more substantive, often fundamental differences between local governments, dioceses and the local public be dealt with? And last but not least, there is the methodological challenge: who will draw the path forward to a renewed future for the parish churches, and how will this path look like? A new practice that is flourishing in Flanders—the supervising and moderating of processes that aim at developing new ideas for one or more church buildings—is sometimes architecturally oriented, sometimes oriented towards community work, sometimes it is or tries to be both at the same time.

Arguably, the best results stem from processes in which architectural or spatial expertise is involved (but maybe not *solely* spatial expertise). Because the future of the parish churches has become an issue at the grassroots level as well, there are intermediary organizations that try to cover just that: ideas citizens have for their church building are mapped by these organizations and passed on to the municipalities, but only rarely are they tested against the existing architectural and urban boundary conditions.[3] It happens that architects, at a later stage, are asked to investigate the feasibility of one or more options for the reuse a church building and that they soon have to conclude that the brief itself needs to be questioned. Some options just would not fit the local parish church, or could never work because, for example, the church building is listed as a monument and not all interventions will be allowed by the immovable property agency. Sometimes, the context of the church building was not taken into account during the previous stages, and when architects or urban designers become involved, it turns out that there are better locations or existing buildings to fit that particular function asked for in the brief. Architectural and urban skills can help to prevent such bumps in the road—especially when architectural and urban skills are actively combined. That is why, spatial practitioners such as Aldo Rossi remain so inspiring for this specific issue.

Aldo Rossi, as his book *L'architettura della città* (1966/2009) proves, never failed to connect the urban and the architectural. For Rossi (1966/2009), 'vital' old buildings, or 'permanences', contribute to the city. And the city in turn, with all its 'permanences' or 'monuments', acts as a backdrop for human life and human flourishing. The word 'permanence' already says it all: for the famous Milanese architect, buildings—and therefore parish churches—cannot simply be reduced to a style period. Their meaning transcends a plain chronological classification. Monuments—he uses a broad definition of this term—*make* the city; they have a public purpose. Spatial skills and understandings are able to operationalize these insights with regard to parish churches.

When the future of parish churches is under scrutiny, the intrinsic link between architecture, the city and human life is far from the only lesson to be learnt from Aldo Rossi's teachings. His emphasis on 'permanences' and the 'longue durée' of the city as a whole is deeply connected to a strong rejection of architectural functionalism. A building can never be limited to just a style period, nor to one 'function' (Rossi 1966/2009). As Diane Ghirardo (2019) also mentions in her recent book on Rossi, the Italian architect was one of the first and certainly the most prominent architect of the post-war period to dismiss the dominance of 'function' in architecture. For a building to keep on *actualizing* the city, it cannot be confined to just one function. That is one of the reasons why *The Architecture of the City* remains inspiring for everything that has to do with adaptive reuse. It keeps on accentuating the larger scale (the city), the relativity of what exactly happens inside compared to the permanence of the building itself—its relative autonomy,—and therefore the *potential* of buildings. With Rossi (1966/2009), it can be argued that a functionalist approach does not fit parish churches, because most—if not all—actors involved want them to be buildings of the 'longue durée'. Certain smaller interventions probably will not hurt; they might even prolong the existence of the building since they are able to ensure its ongoing relevance in the urban fabric. But in this line of reasoning, it is of the utmost importance not to let an all too functional approach jeopardize the (potential of the) building. Architectural functionalism, as in designing for one specific purpose only, might be a trap, but the total rejection of it, as in the mere celebration of the building's autonomous character, might be as well. Hence, for parish churches in Flanders, within this situation determined by so many actors and boundary conditions, it seems to be key to find the right balance between the autonomy of the building and a form of pragmatism. Rossi's theory, however, can be interpreted freely enough to allow a certain dose of pragmatism.

When dealing with parish churches, the more fundamental questions are never far away. The sheer size of the total challenge is so vast that the effects of ill-considered interventions will be just as enormous. According to insiders, seven out of 10 parish churches are being questioned today—that is: more than 1.200 churches in total.[4] What is more, well-considered interventions and strategies will be able not only to revitalize the buildings themselves: they will have an effect on the quality of life in the Flemish villages and neighbourhoods as well. For this (enormous) potential to be realized, it is definitively worth investigating the fundamentals. A mere functional approach will not suffice, even when that is the (silent) standard, especially for the municipalities—the main financial backers of the still operative parish churches—as the functioning of the Projectbureau proves. Consequently, for practitioners engaged in church buildings, it might not be a bad idea to sometimes challenge the norm. Aldo Rossi—among many others of course—paved the way.

In case not (only) the short term, but also the long term will be taken into account, the 'public' will become part of

[3]This was a topic during a seminar organized by the cultural organization Ar-Tur, 22 February 2019, in Beerse, Belgium.

[4]This information comes from informal talks of the author with the CRKC (Centre for Religious Art and Culture, named *Parcum* since 2020); several Projectbureau meetings and the review of several policy plans for parish churches ('kerkenbeleidsplannen' in Dutch).

the equation almost naturally. That is one of the reasons why it is not strange to consider 'utopia' when dealing with church buildings and not start 'solving' the present day 'problems' right away, without considering the longer term. The long term and the public, two categories that are explicitly church building related, challenge the current standard, since it is rather hard to reconcile these categories with quick solutions that try to 'functionalize' parish churches. Thus, the unique characteristics of parish churches might inform the future of these building. It is therefore possible to keep at least part of the meaning of the church buildings open for new interpretations that are in line with the old. The next section will further investigate the deep connection with 'the public' that parish churches maintain. The public character of the parish church will be introduced as a first thread between the everyday and utopia.

3 The Public Character of the Parish Church and Its Current Challenges

One of the important reasons for the conceptual haziness of parish churches in Flanders *and* a connection between the everyday and the utopian is to be found in the genealogy of the parish church as we know it today. Since Napoleon—more specifically the 1801 Concordat between Napoleon and Pope Pius VII—that is since before the founding of Belgium (let alone the investment of the Flemish region), parish churches are sponsored by the state (i.e. the municipalities) and parish priests are paid by the federal state (the department of Justice). There never was a clear church-state separation in Belgium and/or Flanders, not in the strict sense. Scholars of Canon law even talk about a 'benevolent' or 'positive neutrality' (De Pooter 2003). This mutual dependence exists on the level of the state (and/or the regions) and the dioceses as well as on the scale level of every operational parish church: liturgy is 'made possible' by the financial backing of the municipalities and also by the volunteers that take care of the flowers and the cleaning of the building, that make sure the sacramental bread and wine are available, etcetera. Precisely, because Christianity is so 'radical' compared with other religions—only God is sacred, as the Gospels explain through Jesus' teachings—it is not hard to understand that the difference between the content of the Christian faith and the daily operations that support this content is so enormous. In practice, however, many 'mediators'—humans and non-humans (such as parish churches)—stand between the everyday and the sacred, as Gauchet (1985/1997) has shown in *The Disenchantment of the World* (*Le désenchantemend du monde*).

The daily operation and the statute of the Flemish church building—both rather mundane—allow us to imagine the building apart from its traditional stewards. 'A building is a building' as the ex-Jesuit Geert Bekaert (1990) used to say. Exactly because a church building *is* a 'functional' building as well, and not only a sacred place, it is perfectly possible (now) to think of another future for a church building. But that does not mean that every Flemish church building is just an ordinary structure and nothing more. For one thing: the official statute of the parish church, as a place of public worship, makes it a *public* building. This is very important: a parish church is a public building, *de jure*, precisely because of its statute of parish church. Since the Concordat, the parish church in the region of Flanders (that did not exist as such then) is owned by a public legal body: either the municipality (because the French Revolution confiscated the church buildings) or the fabric committee. That of course does not mean that it is de facto being used as a public building. Important to add here is that, in order for a church building to accommodate a new type of use, the withdrawal from the public domain of the parish church at hand is not always obliged. The final decision is the bishop's. It is up to him whether secondary use—the 'alternative' use of a church building that is still used for public worship as well, with this 'alternative' use usually taking place in another part of the building or at other points in time—is acceptable or not. In secondary use, the everyday and the sacred can come extremely close together, and because in that case, the building is still officially a parish church and thus a public building; the public potential mentioned in the introduction is safeguarded.

Imagining the potential of church buildings however, in a more or less secularized society such as the Flanders Region, is not a matter of course. Not so long ago—and this phenomenon is still firmly rooted, not in the least in public debates—so-called progressive politicians were outspokenly anticlerical. Still, politicians that want to abolish as many as possible religious symbols from public life are plenty. This renders some options for abandoned parish churches which are being discussed in the villages and cities of Flanders, such as a general quiet space, a columbarium or even a community centre, not as obvious as they seem at first sight. Cemeteries and community centres are municipal affairs in Belgium and some—especially liberal or socialist—municipalities prefer to keep their cemeteries and community centres 'neutral'. Sociologists might call attention to the decline of the traditional compartmentalization (the Christian pillar, the socialist pillar, …), the sociologist Mark Elchardus (2007, p. 232) nevertheless shows that in practice these compartments still exist. This de facto compartmentalization may hinder the public potential of reused church buildings.

Some dioceses as well 'fail' to separate (decommissioned) church buildings from their former use. They are not welcoming 'alternative religious use' (user groups from outside of the World Council of Churches) or types of use that have to do with non-religious (existential) meaning, whether mixed with commercial activities (as in for example funeral homes) or not. In the Netherlands, when church buildings are closed down and being sold, perpetual clauses are becoming fashionable to prevent church buildings from becoming 'spaces with inappropriate use'.[5] What is appropriate and what is not, is not always that easy to ascertain though. There are no clear rules, not in Canon law and not in the Bible. What *is* certain, however, is that—as mentioned before—the bishop is the only one able to detach a church building from its institutional links. Architecturally speaking, an undertaker's business, for example, which these days has its own auditorium, might respect the interior and the exterior of a church building perfectly—imagine, for example, a modern church which was designed as a lecture space—but fundamentally, there is a problem in several dioceses. The most common explanation local priests and church board members that were reprimanded for genuinely considering such options give, is that all secular services are seen as 'competition' by the dioceses (this is not an official standpoint, but the interpretations of local priests and committee members involved in such discussions). Options that are in fact similar to church use, and that therefore would be able to gain local support, are often stopped from above—from civil or ecclesiastical authorities. This way, the more daily aspects of both religious life and local social life (De Bleeckere and De Ridder 2014) are often being neglected when it comes to the future of parish churches.

The public character of church buildings, however, has already proven itself as an asset. It is because of this historic 'publicness'—which implies financial backing by (local) governments—that there never really was the urge to question the future of the parish churches. As explained earlier, in Flanders, it all started as late as 2011. Secularization kicked in a bit later and slower in Belgium (De Ridder 2013) than in The Netherlands, but still, the way church buildings in Belgium and The Netherlands are being treated differs greatly. In Belgium/Flanders, there seems to be a certain inertia with regard to dealing with the parish churches issue. On the one hand, (governmental) institutions such as the Projectbureau are well aware of this history and are themselves responsible for the apparently thoughtful Flemish method; on the other hand, (some) municipalities are more or less forced to think in a more economic and therefore more functionalistic and less 'public' ways. When, by decommissioning the building, the typical legal framework of the Belgian parish church falls away, the parish church becomes literally outlawed and possibly the subject of uncontrollable events. Many meetings related to the operations of the Projectbureau prove that this does rarely really occupy the actors involved—that is: the municipalities and the religious leaders (in a lot of cases the members of the fabric committees *do* feel involved in the public future of their church building). Unfortunately, it is precisely at that level that a fundamental difference can be made. Some very important decisions are being made or are about to be made by the municipalities and the dioceses. Somewhat dramatically expressed, it is there that the utopian potential of the parish churches is either recognized or ignored. The next section will look into how to understand the role of the fabric committees and other local volunteers in bridging the gap between the everyday and utopia. However bleak the public future of the parish churches in Flanders may look, the role of locals is an integral part of the operation of parish churches, and regarding the future, there might be some bright spots right there.

4 A 'Praxis' Between the Everyday and Utopia

In order to obtain a sharper image of the potential and the dangers related to the future of parish churches, it is inspiring to revisit Michael E. Gardiner's essays that were bundled in the book *Weak Messianism. Essays in Everyday Utopia* (Gardiner 2013). More than once, Gardiner involves Walter Benjamin, as in this citation: 'In the dream in which every epoch sees in images the epoch which is to succeed it, the latter appears coupled with elements of pre-history—that is to say of a classless society' (Walter Benjamin, as cited by Gardiner 2013, p. 164). Benjamin is called into bridge the everyday (the fact that people dream, and the fragments of the past people are confronted with) with the utopian (the *content* of the dream). Fragments of the past seem to carry a promise, but, as Gardiner goes on in conversation with Agnes Heller's work: 'There is no *living* connection between the ruin and the here and now; the former is not part of the lifeworld as experienced by real individuals. [As] a form of anamnesis or recollection, the discontinuous folk memory of a lost 'Golden Age' is prone to mythologization and co-option by regressive and irrationalist political forces [...]' (ibid., p. 170). Gardiner (2013, p. 165) makes clear that Benjamin was well aware of this danger of co-option. What is more, Benjamin would make sure that in order for 'dreams, wish-images and artefacts' (in this case, of 'mass culture') to really feed the 'utopian imagination', they needed a 'connection to praxis'. He adds that '[the] task at hand was one of retrieving or rescuing the 'truth-content' of such images from the phantasmagoria of the commodity and the mythification of the social world' (ibid.).

[5]The perpetual clause was also a topic during the seminar organized by the cultural organization Ar-Tur, 22 February 2019, in Beerse, Belgium. Several Dutch architects were present, see https://centrum.ar-tur.be/2019/01/11/de-kerk-buiten-de-stad-werk-en-studiedag/. The Belgian diocese of Ghent is investigating a similar clause.

Related to the existing church buildings in Flanders, that also remind of another era—even the modern buildings—(De Bleeckere and De Ridder 2014), Benjamin's emphasis on 'praxis' to avoid 'mythologization' is telling. It is precisely what can be felt in today's parishes. On the one hand, local volunteers (see also De Ridder and De Bleeckere 2015) fear that there will be no follow-up for their activities. In some Flemish dioceses—the diocese of Ghent for example—parish churches are being closed down. When a church is closed down, the local fabric committee, which is responsible for making liturgy possible, is no longer needed. Its possessions go to the committee of the parish with which the former parish will be merged. In that case, chances are real that the daily maintenance and management of the decommissioned church will not be a local affair anymore. Above that, the possibility exists that the closed down church will be sold to a new owner. Daily life threatens to flow away from the church. Not only fabric committees partake in the 'daily life of a church building', other volunteers do as well, mostly parishioners (De Ridder 2013). The social glue that kept the church building from fragmenting is disappearing fast, and with that the Benjaminian 'truth-content' of the utopian promise a church building might express. On the other hand, Benjamin hints at the constant 'danger' that other forces will recuperate and commodify the fragments of the past between which the praxis had disappeared. Strictly speaking, this is not (yet) the case in Flanders, or only on a small scale. Not too many commercial developers are involved in former parish churches and only rarely are religious symbols being exploited for commercial gain (as was the case in the Netherlands before Catholics started to erase all symbols from decommissioned church buildings) (Beekers 2017a, b). In itself, the recuperation or commodification of (parts of) parish churches may not be that harmful, but when a parish church is being dealt with in a 'real estate way' however—which is perfectly possible—it is probable that not only public space will be lost, but that local social life too will lose one of its strongholds, while the church building itself, as a kind of token, will still be present.

Benjamin's double warning—the need for praxis and the danger of mythologization—is not only insightful in itself, it is also a possible step towards the proper management of reused church buildings. Precisely, because parish churches are so inefficient (as mentioned in the introduction), and also because there is little practical demand for spaces such as church buildings, it will not be easy to find good management forms for these buildings. The more that is true, the greater the importance of local support. In Flanders, the awareness is growing that a realistic way to look at the management is to start from local support, and to build on that. The Open Churches Network indirectly stimulates those insights.[6] By motivating local people—committee members and others—to open the doors of the buildings (and thus valuing them as public spaces), the network 'forces' the locals to think beyond liturgy and beyond the traditional system of fabric committees. Locals are required to become creative in keeping the doors open and in looking after the building. That way, the church building and the local community are actively being prepared for a next 'epoch', in which the current institution (i.e. the fabric committee) will not be present anymore, but the volunteers themselves could still be there nonetheless. A possible threat as to local support for church buildings (and options regarding management) is the merging of parishes, and therefore fabric committees, that is in the pipeline. There is already intense collaboration between parishes in so-called pastoral units or 'pastoral zones' which consist of several parishes (up to all parishes in one municipality). The real danger is that these changes are generally being imposed from above. Daily life is evaporating quickly from between the existing local stones, and chances are that the prospect of a meaningful, local future is doing the same.

Not only the Open Churches Network offers hope with regard to imaginable quasi-utopian futures of parish churches, but also the many participatory processes related to churches conducted by, or in collaboration with, at least one spatial practitioner. It must be said that in many cases, either municipalities or senior clergymen (in collaboration with a central church board) were the ones asking for such studies in the recent past.[7] A major advantage of participatory processes is that it is possible to connect the local support base with (other) potential future 'managers'. The process can be designed in such a way that both the existing support base and 'new managers' are motivated to stand up and express their engagement. To conceptualize this way forward from support base through participatory process to the actual management of a building, one can think of Bruno Latour's lack of a clear difference between product and process: a product—in this case an otherwise oriented parish church—is always the result of a network or a process.[8]

[6]For more information about the Open Churches Network, see http://opencharches.be/.

[7]See footnote 2 for the 'cliënts' of the studies the author conducted.

[8]This understanding, of a process and its product being the same ontologically, can be substantiated by references to Bruno Latour's books Pandora's Hope (1999) and Politics of Nature (1999/2004) and to Michael Guggenheim's 2009 text on 'mutable immobiles' (which in turn refers to Latour's work). In *Pandora's Hope*, Latour makes clear that all things are constructed and *therefore* they are able to act autonomously. The construction is part of the thing. In *Politics of Nature* he positions 'matters of concern' against matters of fact; matters of concern are entities to which the mode of production remains explicitly tied. Guggenheim (2009) explains that buildings actually are networks.

Within this movement from everyday volunteering to the continuation of the utopian potential of the church building, neither the attitude of the locals, nor their 'praxis' must change. Locals can continue to appreciate their church building as this special place that is set apart from the daily routines, yet made possible in a very 'worldly' way. As the next section seeks to exemplify, local volunteers are assisted in this by the building itself. In a very material way, the church building takes part in the production of a very distinct place. That might sound tautological, but there is more to it.

5 Architectural Autonomy and the Everyday

The Italian architectural theoretician, historian and critic Manfredo Tafuri (1973/1978; 1968/1980) was clearly influenced by Walter Benjamin. Tafuri too investigated the connections between material artefacts, their (social) production and their meanings. The conceptualization of architecture as an obstacle that resists urbanization (and thus the adverse effects of capitalism), already mentioned in the introduction, that Pier Vittorio Aureli (2011) works out in *The Possibility of an Absolute Architecture* is a direct interpretation of Tafuri's work on modern architecture. Tafuri famously declared modern architecture (almost always) incapable of being 'engaged', for the economic forces are too strong to resist. For Tafuri, modern architecture is nothing more than a 'façade' to make the ubiquitous industrialization acceptable. For Aureli however, 'absolute' architecture, as in architecture that is 'being resolutely itself after being 'separated' from its other' (Aureli 2011, p. ix), *is* a possibility. Throughout his 2011 book, Aureli positions 'absolute' architecture as actually able to be engaged with the city. Therefore, Aureli is more optimistic than Tafuri ever was, no matter how much the former is influenced by the latter. In the more recent book *Rituals and Walls. The Architecture of Sacred Space* (Aureli and Shéhérazade Giuduci 2016) Aureli co-edited, the authors bring a similar logic specifically to sacred spaces. In church buildings and temples, Aureli (2016) finds obstacle-like qualities intensified by their 'walls' and thus by a strict separation of what is 'inside' (the ritual dimension) and what is not. This—as processions and pilgrimages do too—'organizes' the city. What is inside is always public—not really belonging to anyone—and this 'public' in turn 'invites' the very economical, as in church buildings and market places that are located next to each other—thus organizing the city.

Aureli links the sacred aspect and the public aspect of church buildings quite strongly. In post-Catholic Flanders, such a move is perfectly justifiable, certainly on a legal level. Belgian—and therefore Flemish—parish churches truly *are* public buildings, as indicated before. That, combined with a still prevailing post-Vaticanum 2-atmosphere (in the ecclesiastical world at least)—the world is inhabited by the people of God, meaning that for the Catholic church there is no necessary and clear separation between *the* public and *their* public—and with the fact that parish churches are subsidized by the state, makes for a powerful correlation sacred-public in Flanders: everyone is welcome and no one needs to pay. Aureli's insights, though, seem to stress that architecture itself must not be forgotten in order to actualize the promises that are part of the unique Flemish situation.

Aureli's appreciation of architecture *as* architecture—with the wall as the most basic and maybe most 'everyday' element—is very timely, for not only church buildings but architecture in general becomes a favorite topic of several other professions and disciplines (Awan et al. 2011, p. 28). Architectural expertise, though, is able to propel 'processes', as mentioned before. It is also able to bring in sometimes very basic insights about buildings, their performance and their (possible) meanings. These basic comprehensions seem to be missing in the current churches discourse. A new functionalism is developing in the Belgian region of Flanders. Fundamental aspects of architecture, such as the insights Tafuri, Aureli and also Rossi share, are being repressed by the functional demands of builders—i.e. mostly municipalities—who are suddenly confronted with this new challenge. If architecture threatens to become a service provider of essentially economic forces—a 'façade' as Tafuri would say,—a lot can be lost, especially regarding the utopian potential of parish churches. It is not generally understood that church architecture is able to be the dam against commodification Aureli holds it to be, and therefore a step up to a (modest and/or local) utopia. In many plans that are being developed, for example, church walls are literally being opened (most of the time partly, via glass doors that open to an adjacent square), reducing the obstacle-like qualities of the buildings. It will be hard to convince municipalities of the idea that precisely that is a move *against* the publicness of church buildings, as, according to Aureli, a clearly separated *sacred* space is more *public*. The new pragmatism in architecture (and adaptive reuse) is not without dangers to the intrinsic qualities of church buildings. As said before, the balance between the relative autonomy of church buildings and a more pragmatic view will be key.

Aureli's work is inspiring, and in that, it not only values the relative autonomy, or 'absoluteness', of architecture, but is also able to link it (in other texts of which the content wanders through his entire oeuvre) to the 'architecture of autonomy', that is: the architecture of self-organization (Aureli 2008). As in Rossi's *L'architettura della città* (1966/2009), behind the seemingly strong emphasis on architecture proper lies a deep appreciation for the social (and the political). When combined, Aureli's texts on absolute and sacred architecture, and his texts on 'everyday autonomy' as in self-organization, form a steady body of work. It can be that the Flemish discourse on the future of church dings is precisely in need of such an integral vision for long term. Because right now, neither the buildings nselves—for their public potential and their obstacle-like lities—nor the local communities—for their possible axis' that is currently one of the only ways out of the ure management question—are valued as they could, ile the utopian potential is actually right there.

For a parish church to be a 'dam against commodifica- n' may sound rather pompous, but actually, this notion is led by something as mundane as a wall. And that wall is rinsically part of every church building in Flanders. tholic churches have always been those mediators tween the everyday and something higher. According to anon law, parish churches are 'temporal' goods, but of urse, they were—or are still—used to remember, to pray, preach, to serve and to celebrate. Parish churches thus nbody a way of life that is related to utopian redemption, nd they do that right here, in real villages and real neigh- ourhoods, and with equally real materials such as stone and ood.

Conclusion

Every church building, it can be said, maintains a relationship between the everyday and utopia—at least potentially. However, when it comes to the future of parish churches, it is not self-evident to substitute a general 'utopian' for the former 'sacred'. Much more than the shrinking number of visitors for mass, the real challenge is the lack of a clear vision on the future management of churches, while not devaluating the unique architectural qualities of the buildings. This situation is being intensified by the strong link between the buildings and priests that the dioceses make (the traditional management of parish churches by the fabric committees only exists when liturgy still takes place). When the dioceses close down churches, local management and volunteering threatens to disappear. But not only that: there is the loss of legal protection of that genuine public space the parish church is, and there is the potential loss of a unique dam against commodification. Under the current circumstances, however, local support and thus local capacity building are the most convincing bases to build a local future upon, but there is a gap that needs to be bridged—between the promises of the past and those of the future—to make the most of church buildings.

Nowadays, the architects are in charge, at least this is what the Flemish authorities signal. But space, let alone the kind of space church buildings promise to be, is not made by architects alone. Local practices are equally important, if not *more* important, to capitalize on the societal and the cultural potential of parish churches. A first step towards this is possibly the radical opening of the church doors, in order to invite diversity, to think beyond the dioceses and the priests, to rationalize management, and to make the potential of parish churches visible and tangible. This is not to say that architects will have no function. An 'Aurelian' appreciation of the obstacle character churches traditionally possess and a feeling for the value of local support might help to build sustainable futures—sustainable as in protecting in the first place the potency of church buildings for the long term. Municipalities and many architects alike, however, seem to like their church buildings somewhat more conformist, preferring them to move with the (economic) stream and not against it. Instead, even the very basic architectural features of a parish church could be understood as assets in themselves, gateways to more meaning maybe, and need not be seen as hindrances that must be overcome to achieve a generic or strictly functional solution. The legal status of the parish church, its role in the urban fabric and the *affect* of local residents can be assets too and, moreover, building blocks of an approach that is specifically church-related.

The existing system of fabric committees gains from its strong legal base and its financial support from the municipalities, which will be gone once the church building is decommissioned. If a future (local) 'praxis'—maybe comparable to the commons, cooperatively managed or not—can be found in which the parish church will remain public, the church building will appear as a kind of (public) easement, a (public) right of way, to something 'higher' or 'other' (religious or not). The more this kind of potential is being emphasized, the less appropriate the present approach of 'functionalizing daily life' will become. Accordingly, it is the traditional church building —architecturally, socially and legally—that informs possible futures for decommissioned church buildings.

References

Aureli, P. (2008). *The Project of Autonomy. Politics and Architecture within and against Capitalism.* New York/Princeton: The Temple Hoyne Buell Center for the Study of American Architecture & Princeton Architectural Press.

Aureli, P. (2011). *The Possibility on an Absolute Architecture.* Cambridge: The MIT Press.

Aureli, P. (2016). Rituals and Walls. On the Architecture of Sacred Space. In P. Aureli, & M. S. Giudici (Eds.), *Rituals and Walls. The Architecture of Sacred Space* (pp. 10–25). London: AA Publications.

Aureli, P., & Giuduci, M. S. (Eds.). (2016). *Rituals and Walls. The Architecture of Sacred Space.* London: AA Publications.

Awan, N., Schneider, T., & Till, J. (2011). *Spatial Agency. Other Ways of Doing Architecture.* London/New York: Routledge.

Beekers, D. (2017a). De Waarde van verlaten kerken. In O. Verkaaik (Ed.), *Gods huis in de steigers. Religieuze gebouwen in ontwikkeling* (pp. 161–92). Amsterdam: AUP.

Beekers, D. (2017b, November 17). De omstreden kerk: Herbestemming, erfgoed en religieuze transformatie in Nederland. *lecture.* Leuven, Belgium: Kadoc.

Bekaert, G. (1990). Een gebouw is een gebouw. *Beschouwingen over secularisatie van kerkgebouwen. Kultuurleven, 57*(2), 32–43.
Berardi, F. (2017). *Futurability. The Age of Impotence and the Horizon of Possibility*. London: Verso.
Bourgeois, G. (2011). *Conceptnota "Een toekomst voor de Vlaamse parochiekerk"*. Brussel: Vlaamse Regering.
De Bleeckere, S., & De Ridder, R. (2014). *Het open kerkgebouw. Heden, verleden en toekomst van het kerkgebouw in Vlaanderen.* Kalmthout: Pelckmans.
De Pooter, P. (2003). *De rechtspositie van de erkende erediensten en levensbeschouwingen in staat en maatschappij.* Brussel: Larcier.
De Ridder, R. (2013). Kerkgebouw genereert publiek: Een actuele strategie voor de architectuurtheoretische herdefiniëring en de toegepaste herwaardering van de Vlaamse parochiekerk. *unpublished doctoral thesis*. Hasselt: Universiteit Hasselt.
De Ridder, R., & De Bleeckere, S. (2015). Making Church Happen. Architectural Methods for Transforming the Parish Churches of Flanders into Civic Collectives. In L. Hustinx, J. von Essen, J. Haers, & S. Mels (Eds.), *Religion and Volunteering. Complex, contested and ambiguous relationships* (pp. 169–90). Cham: Springer.
Elchardus, M. (2007). *Sociologie. Een inleiding*. Amsterdam: Pearson Education Benelux.
Gardiner, M. (1995). Utopia and Everyday Life in French Social Thought. *Utopian Studies, 6*(2), 90–123.
Gardiner, M. (2013). *Weak Messianism. Essays in Everyday Utopianism.* Bern: Peter Lang.
Gauchet, M. (1997). *The Disenchantment of the World. A Political History of Religion [Le désenchantement du monde].* (O. Burge, Trans.) Princeton: Princeton University Press. (Original work published 1985).
Ghirardo, D. (2019). *Aldo Rossi and the Spirit of Architecture*. N Haven/London: Yale University Press.
Guggenheim, M. (2009). Mutable Immobiles. Change of Use Buildings as a Problem of Quasi-Technologies. In T. Bender, Farías (Eds.), *Urban Assemblages: How Actor-Network The Changes Urban Studies* (pp. 161–78). Londen/New Y Routledge.
Latour, B. (1993). *We have never been modern [Nous n'avons jam été modernes: Essais d'anthropologie symmétrique].* (C Por Trans.) Cambridge: Harvard University Press. (Original w published 1991).
Latour, B. (1999). *Pandora's Hope. Essays on the Reality of Scier Studies*. Cambridge: Harvard University Press.
Latour, B. (2004). *Politics of Nature. How to Bring the Sciences ir Democracy [Politiques de la nature].* (C. Porter, Trans.) Ca bridge: Harvard University Press. (Original work published 199
Minton, A. (2017). *Big Capital. Who is London For?*. London: Pengu Books.
Rossi, A. (2009). *De architectuur van de stad [L'architettura del città].* (E. Kurpershoek, & H. Hoeks, Trans.) Amsterdam: Su Architecture. (Original work published 1966).
Tafuri, M. (1978). *Ontwerp en utopie. Architektuur en ontwikkeling va het kapitalisme [Progetto e Utopia. Architettura e svilupp capitalistico].* (U. S. Barbieri, C. Boekraad, F. Denissen, & K Vollemans, Trans.) Nijmegen: Sun. (Original work publishe 1973).
Tafuri, M. (1980). *Theories and History of Architecture [Teorie e storic dell'architettura].* (G. Verrecchia, Trans.) London: Granada. (Original work published 1968).